# A HUMANIST IN THE BIBLE BELT
## Collected Papers 1974-2002

Wholly Babble

WILLIAM HARWOOD

ISBN: 1-4107-0984-1 (e-book)
ISBN: 1-4107-0985-X (Paperback)

This book is printed on acid free paper.

1stBooks - rev. 02/25/03

They came for the homosexuals, and I wasn't a homosexual so I didn't try to stop them. They came for the abortion providers, and I wasn't an abortion provider so I didn't try to stop them. They came for the Muslims, and I wasn't a Muslim so I didn't try to stop them. They came for the atheists, and I wasn't an atheist, so I didn't try to stop them. Finally they came for me, and there was no one left to stop them.

*(Graffiti painted on an Ottawa sidewalk ten years after the election of the Western Rednecks Alliance of Religious Nutcases as the Government of Canada.)*

When Dr William Harwood wrote a series of letters to a Canadian newspaper, they were responses to a present and local situation. They were not designed to elicit responses that would prove he was living in the redneck anus of the universe, among the same kind of wonderful folks who gave the world the Crusades, the Inquisition, the Thirty Years War, the Salem witch trials, the Moral Majority, Ayatollah Khomeini, ethnic cleansing, and the Taliban. That was just the way it turned out.

To a victim of god addiction, right and wrong are whatever his god's speechwriter says they are. Are such believers insane, intestinally challenged—or simply untaught?

Persons handicapped by unintentional ignorance despite being capable of rational human thought will find the antidote in this collection of letters, journal articles, book reviews and satire. Unteachables will not.

## WARNING TO THE INTESTINALLY CHALLENGED

The facts of history can be hazardous to a moral coward's mental health. Persons who need the mind-deadening opiate of an afterlife belief to overcome their terror of death and get them through the day without having to be institutionalized and diapered are warned to read no further.

THIS IS RELIGION              THIS IS YOUR BRAIN ON RELIGION

Cartoon by Don Addis for FREE INQUIRY
used by permission of the Council for Secular Humanism

# TABLE OF CONTENTS

ix

# STOP THE WORLD. I WANT TO GET OFF

When the blind lead the blind....

Courtesy of *Discover*

The doctrine of the double motion of the earth about its axis and about the sun is false, and entirely contrary to Holy Scripture.

Pope Paul V

The earth is flat, and anyone who disputes this is an atheist who deserves to be punished.

Sheik Abdel-Aziz ibn Baaz
supreme religious authority, Saudi Arabia

If the Bible said that Jonah swallowed the whale, I would believe it.

William Jennings Bryan

# 1. Education: A Faculty of Incompetence

**I have passed Educational Trivia 261, Educational Nonsense 450, Educational Garbage 313, and Educational Horse Manure 491. I am a Bachelor of Education.**

Teachers today are, or have the potential to become, reasonably well paid. Only a few years ago that was not the case. Teachers' salaries were so inadequate that, with few exceptions, no person with the capacity to enter any other profession would consider teaching.

As a result, teachers were in such short supply that teacher-training institutions were strongly discouraged from failing any but the most implausible candidates. Thus "education" courses came into existence with the sole virtue that they could be passed by students who, lacking the ability to obtain any other qualifications, had decided to become teachers.

Times have changed.

Increased salaries and status have made teaching an attractive alternative to students who, as recently as ten years ago, would have sought degrees in arts or science.

Unfortunately, "education" courses have made no attempt to cater to this new clientele. Courses designed for intellectual mediocrities are now being offered to students whose intelligence equals that of students in other faculties. It is hardly surprising that such students are insulted and offended at the puerile balderdash to which they are subjected.

The Faculty of Education has long been recognized as a licensing bureau for Arts and Science's rejects. Naturally the Faculty has striven to upgrade its image. However, the obvious solution of instituting courses actually designed to teach something has not occurred to the distinguished "professors of education" who stagnate therein.

Instead they have concocted a peculiar vocabulary expressly designed to disguise the fact that the Faculty of Education serves no useful function. So effective is this unintelligible gibberish in concealing the vacuum of ideas from which it springs, that lectures, articles and even theses written entirely in Faculty of Education doubletalk have been minutely studied by other "professors of education," and even praised by outsiders who have assumed that anything that cannot be understood must be profound and significant.

Teachers have long claimed to be highly trained professionals. That claim is not and never will be valid as long as teacher training is entrusted to an organization as devoted to the propagation and perpetuation of mediocrity as the Faculty of Education.

Education is a faculty of incompetents, run by incompetents for incompetents. Its most popular "professors" are those who make no pretence of teaching. They entertain, and there is no denying that some are excellent entertainers. But with few exceptions they do not teach. And even the exceptions teach that which is not worth learning.

For example, a course laughingly called "Educational Psychology" involves primary statistics. Future teachers are instructed to mark their pupils by a system known as the "curve." Under this system, if a class of twenty contains three pupils of I.Q. 140, two of them would receive Bs!

Similarly, a test score of 95%, if it happened to be the lowest in the class, would be an F! No teacher with even an elementary sense of justice would ever use such a heinous and pernicious monstrosity.

But the Faculty of Education not only teaches this outrageous piece of fraud. It uses it on its own students. Of 34 courses examined (the writer's own), 12 in "education" and 22 elsewhere, this system was not encountered outside of the Faculty of Education.. Only a professor of education, it seems, is capable of believing in such iniquity.

There are, of course, injustices and incompetents in other faculties. Education does not have a monopoly on persons who refuse to acknowledge the validity of any viewpoint but their own. But whereas elsewhere the fault lies in the individual, in Education it is an integral part of the system.

Professors of education do not teach because they are themselves graduates of a system that teaches them not to teach. And what would they teach? If they knew anything they would be teaching it—in another faculty.

One example of how professors of education operate should suffice. Admittedly, even in a faculty dedicated to incompetence, this incompetent is a prize; but other courses are not startlingly different.

The course was "Reading in the Secondary Schools." The professor was an expert in Faculty of Education doubletalk, and had actually written a thesis in that peculiar dialect. The course was devoted to the testing of children who cannot read. Of the tests themselves, the best that can be said is that only a professor of education could take such bunkum seriously.

After several weeks during which the teaching of reading was not mentioned, the professor was asked: "Having administered a test to a child that I already know can't read, and having learned from the test that he can't read, what do I do about it?" That question was never answered.

The Faculty of Education is a gigantic hoax. It does not do the job for which it was created, and should be painlessly put out of its misery. An alternative method of licensing teachers would need to be found, but this presents no problem. A teacher trainee would intern in a school for one year, under the supervision of a professor of the subject he would be teaching. At the end of that internship he would be granted a teaching certificate by the Department of Education.

The minimum qualification for admission to an internship would be a university degree—a real degree, in a legitimate subject, since the Bachelor of Education would no longer exist.

## AFTERWORD

The same issue of the *Gauntlet* in which the foregoing appeared, also carried a petition drawn up by the Students' Union President, reading, "The undersigned call upon the University of Calgary's Senate to form a Task Force to investigate the quality of instruction in the Faculty of Education." So many signatures were obtained, that the requested Task Force was formed. Several weeks later, its report upheld all of the complaints in my article, and described education classes as "a mutual exchange of ignorance."

## 2. Alice in Faculty of Education Land

### Alice

Dean, I want to complain about Doctor Bunker's "Reading in the Secondary Schools" course. It's been going on for six weeks and so far she hasn't taught anything. She just spouts inanities. I think she's incompetent.

### Queen of Hearts

Of course, Mister Dickens. I've read your *Hard Times*, and I believe I can help you. Tell me, how long have you had these feelings of hostility?

### Alice

She's supposed to be teaching us how to teach reading, and so far all she's talked about is a whole bunch of competing tests that we can give a kid we already know can't read, in order to prove that he can't read.

### Queen of Hearts

I see. And you feel that eight-year-olds working sixteen hours a day in the mines for a shilling a day is a bad thing? It makes you angry?

### Alice

Of course it makes me angry. I took the course because I want to learn how to teach reading, and I'm convinced now that she can't teach me because she doesn't know herself.

### Queen of Hearts

You're convinced that the factory owners are getting rich at the expense of their employees' health and well-being?

### Alice

I've even asked her direct questions about what to do in specific circumstances, and she shrugged them off with Faculty-of-Education doubletalk designed to intimidate me into believing she knows what she's talking about.

### Queen of Hearts

And the factory owners told you that any child who thinks the job is unhealthy is free to quit? I see.

### Alice

We had an exam last week. She hasn't taught anything, but that didn't stop her from giving us an exam. On one of the questions I dared give a reasoned opinion on why I could not agree with her, and she gave me a zero. She's trying to frighten me into going along with the system, by reminding me that she has the power to give me an F if I make any waves.

### Queen of Hearts

Now let me get this straight. Your book criticizes the people who own the textile mills, and when you tried to read excerpts from it in a textile town the mill owners hired a mob to run you out of town? And you don't think they were justified?

### Alice

Justified? In trying to shut me up when I drew attention to her incompetence?

### Queen of Hearts

Are you aware that the person whom you are criticizing runs one of the most successful textile mills in Birmingham? Or that he has written an extremely sound dissertation on the economics of the prudent use of child labor in the textile industry? It just happens that I've read his dissertation, and I must say I was very impressed by it.

### Alice

You read her Faculty of Education doubletalk and you think it actually said something? Ma'am, the reason Faculty-of-Education doubletalk was invented was to enable the criminally incompetent to hide their ignorance behind jargon that nobody could understand because it didn't mean anything in the first place. Frankly, if you read her thesis and didn't recognize it as a close relative of *The Jabberwocky*, you oblige me to question *your* competence.

### Queen of Hearts

So you believe that I am also part of the conspiracy to oppress the poor? You think that just because I own shares in a coal mine that employs children, that I must have a vested interest in maintaining the status quo?

### Alice

Look, I came in here to complain about a teacher who isn't teaching anything because she's too stupid even to recognize the extent of her ignorance. Not only can she not teach anything because she doesn't know anything herself. She doesn't even know that she doesn't know. She seriously believes that the ability to juggle Faculty-of-Education doubletalk into the semblance of sentences is an adequate antidote for the brain amputation I am becoming convinced all professors of education have undergone.

### Queen of Hearts

Would you mind at this point if I try to summarize what you are saying? Please correct me if I'm wrong, but this is what I hear you saying. Factory owners are getting rich by not paying their workers enough to live on. Factories are unsafe because the owners won't spend any money to rectify obvious health hazards. Children are dying from coal dust in their lungs, and the owners have no incentive to change anything because there are plenty of other children to replace those that get consumption. You think the government should intervene to protect the workers. Is that what you're saying, Mister Dickens?

### Alice

I've been saying for twenty minutes that I want to be taught something and she's not teaching anything, yes.

## Queen of Hearts

Apart from the factory owners, the government, the mine owners, the poorhouse managers, and me, is there anyone else toward whom you feel this hostility?

## Alice

Look, whether one student is hostile or complacent about a situation that needs rectifying is irrelevant. As it happens, I strongly resent having to spend three hours a week in a class that has as much educational value as an astrology column. That's why I'm here to see you, Dean.

## Queen of Hearts

Well I'm certainly glad you decided to come to me. I'm sure talking out your problem has helped more than you realize. If it hasn't been enough, and you find your feelings of hostility returning, don't hesitate to come back and talk to me again.

## Alice

*#+=^<">/

# 3.  No Sex Please.  We're Masochists

I once asked a believer in what conservatives call "morality," liberals call "conventional morality," and rationalists call "taboo," if his god created sex because it has a dirty mind.  While the question was admittedly facetious, it also raised a legitimate problem that no supporter of biblical sex taboos has ever faced, let alone tried to solve.

The strongest believers in the evil of sex and the virtue of celibacy tend to be religious fundamentalists who go to extraordinary lengths to prevent their teenaged children from exercising the sexuality they were born with, while placing no such restrictions upon themselves.  They make it all but impossible for their children to obtain the condoms that constitute the difference between safe, commendable, joyful recreation, and dangerous, morally reprehensible, exploitative self-gratification.  As a result, unwanted pregnancies and AIDS in "sex is dirty" families are significantly higher than among families with the rationality to recognize, "If it doesn't hurt somebody, then it's not a sin."  There has not been a single prosecution of a parent whose enforcement of a "no condoms" policy led to a child's AIDS death.  Perhaps it is time such a law was introduced.  Jehovah's Witnesses and Christian Scientists (at least in some states) are not permitted to deprive their dependent children of lifesaving medical procedures.  Fundamentalist rednecks should not be permitted to deprive their dependent children of lifesaving condoms.

The notion that self-inflicted celibacy is a virtue can be traced back to perhaps the most chronic masochist who ever lived, a renegade Brahman named Siddartha Gautama who defected from Hinduism 2,500 years ago.  But specific sex taboos originated much earlier, starting with the invention of "adultery" around 3500 BCE.

Prior to 3500 BCE, humans had no more awareness of the relationship between recreation and procreation than lions, horses, chickens or insects.  They mated to satisfy a genetically programmed orgasm need, and in the process reproduced as an unknown and unintended side effect.  If any lawgiver had instructed them to "be fruitful and multiply," they would have had no idea how to comply with that demand.

Obviously there were no sex taboos, since the reasons for the taboos did not yet exist.  There were, however, monogamous matings, although without the exclusivity that would have been absurd among people who viewed mate lending as having no more long-term consequences than lending one's bow and arrow.  A woman who was willing to gratify a particular man's sexual needs on demand, gave him a tremendous incentive to stay with her and find food for more children than she could have supported alone.  The man might even have developed paternal feelings for the woman's children, despite not knowing that they were also biologically related to himself.

Then came the Big Discovery that children have fathers as well as mothers.  With the knowledge that babies grew from seed planted in women by men, it became desirable to ensure that only the right seed got planted.  Just as women had always been able to identify their daughters, now men wanted to be able to identify their sons.  For that reason men imposed on women the breeding-slave status that came to be known as marriage, and the extramarital impregnation taboo called adultery.

Adultery was a crime against property.  It differed qualitatively from illegally using a man's chariot and returning it undamaged, because an adulterated wife could not be returned undamaged.  Her womb was carrying usurper's seed that made it impossible for her to bear her legal owner's legitimate heir.  For the discovery that semen was baby-seed was not accompanied by any awareness of the short lifespan of sperm or the duration of pregnancy.  Nor did anyone know that every pregnancy was triggered by a single sperm so tiny that more than two million were ejaculated in a single mating.  The myth of Zeus needing thirty-six hours to pump enough sperm into Alkmene to generate Herakles reveals the ancient belief that sperm was cumulative.  Not until much sperm had been intromitted into a woman over a period of many weeks did it reach "critical mass" and start growing into a baby. And just as wild seed could lie dormant in a cornfield for years before sprouting and producing a weed, so could a usurper's seed lie dormant in an adulteress for years before combining with her husband's to produce a "bastard," defined as a child of many fathers.

Men had no wish to raise bastards. So they sacrificed them to whichever god was likely to be most appreciative. And since women came to the marriage couch carrying the dormant seed of several years of casual copulations, it followed that every firstborn was by definition a bastard. In every society with an adultery concept, the sacrifice of firstborns as bastards was automatic. The laws that the Egyptian prince Moses imposed on the Israelites included a decree from his god that "the firstborn belongs to me" and must be sacrificed (Exodus 13:1-2. see also Leviticus 27:29). Following the sacrifice, the woman thereafter observed an adultery taboo, and henceforth each additional birth could be attributed to her legal owner with reasonable certainty. To encourage women to observe the adultery taboo, the phallusocracy invariably imposed a death penalty for its violation.

It cannot be overemphasized that adultery was an impregnation taboo, not a recreation taboo. Banning recreation that gave pleasure to the participants and hurt nobody would have been as absurd as banning handshaking. For that reason new wives were not subjected to the taboo, since their first offspring were going to be sacrificed anyway, and the sooner their wombs could be pumped up to critical mass and the resultant bastard sacrificed, the sooner they could start bearing legitimate heirs. Had condoms existed in 3500 BCE, adultery would have been defined as the extramarital mating of a married woman without using a condom. By any definition that conforms to the taboo's original purpose, a mating cannot constitute adultery unless the *woman* is involved in an exclusive breeding agreement, the man is not her contracted partner, and the mating has the potential to cause a pregnancy that the woman's husband did not authorize.

The real meaning of adultery continued to be recognized well into historical times. The Priestly author of Leviticus, around 615 BCE, spelled it out in unambiguous terms (Leviticus 18:20): "You're not to engage your compatriot's [*fellow Jew's*] woman in carnal copulation in which sperm is intromitted, for that would pollute her [*i.e., destroy her ability to produce a legitimate heir*]." And when Sir Lancelot, in Malory's *Le Morte d'Arthur*, swore to his god that Queen Guinevere was not guilty of adultery, even though he was himself tupping her, he was neither a liar, a hypocrite, nor a madman (as he must seem to Malory's modern-day readers who see adultery as a recreation taboo). Rather, he knew that he always practised *coitus interruptus*, and without sperm intromission there was no adultery. To Malory, who died in 1471, adultery still meant fraudulent impregnation. Courtly love, copulation without internal ejaculation, did not qualify.

The second sex taboo, premarital mating by women, was a corollary of the first. With the realization that live births were too rare to be wasted, infant sacrifice fell into disfavor. But that meant that some other method had to be found to ensure that a new wife's first child was fathered by her husband alone. The solution was to decree that women must spend their premarital years celibate. And to enforce that decree, a death penalty was imposed on any new bride whose husband failed to detect an intact hymen the first time he inserted his investigatory finger between her legs (Deuteronomy 20-21). The Italian custom of publicly displaying a bloodstained honeymoon sheet is a throwback to a time when no other means existed to prove that a woman's first child was her husband's legitimate heir. A "virgin" was a woman capable of producing that legitimate heir. The survival of such a word in a culture in which it is common knowledge that menstruation is clear proof that no future pregnancy can be attributed to any mating that occurred prior to menstruation is absurd. Even more absurd is the concept of a male virgin, a *man* capable of producing a legitimate heir because he has never been implanted with bastard seed.

The third sex taboo, like the first two, was motivated by population control. But unlike the first two, it was designed to promote breeding, not to restrict it. It was first invented as far back as four thousand years ago, although it did not reach isolated cultures such as much of Micronesia until

William Harwood

Christian missionaries introduced it in colonial times. The taboo was "incest," from a Latin word meaning impious or unchaste.

For generations after the Big Discovery, women were snapped up as breeding slaves, "wives," by the first man who learned that they had reached puberty. Rarely was that anyone but a brother. So traditional did sibling marriage become, that many fathers kept their daughters unmated in the hope of wedding them to a sibling who either was never born or did not live to maturity. The consequence was that men without sisters and women without brothers often died childless, and the family that produced them ceased to exist.

The solution was to encourage juveniles to marry the first stranger who made an offer their father could not refuse. But encouragement was not enough. The belief that only a sibling was an acceptable mate caused exogenous marriage to be viewed as, while legal, nonetheless repugnant. The way to make it tolerable was to outlaw the alternative.

Prohibiting sibling marriage was not difficult. Since most of the gods were married to siblings, sibling marriage was pronounced taboo, meaning sacred, too sacred to be permitted to anyone but a god. And since few people were willing to risk the wrath of the gods by defying their taboos, the new taboo worked. Exogenous marriage became as universal as endogenous marriage had been before, while the older practice became a "sin" called incest. And like adultery, incest was also a breeding taboo. Just as copulation using a condom (had such an item existed) would not have been deemed adultery, neither would it have been deemed incest.

The next sex taboos were not invented until the late seventh and early sixth centuries BCE when, within a century of each other, two lawgivers came up with moral codes that were identical in some respects and diametrically opposite in others. The earlier of the two was the inventor of the still-existing Parsee religion, Zoroaster.

When Zoroaster founded his new religion, he recognized that there were two ways of increasing the number of tithe-paying believers. Proselytizing worked, but breeding followers worked even better. He accordingly invented the first set of sex taboos designed solely to maximize the birth rate. He declared that prayers uttered while copulating were ten thousand times more effective than those chanted in any other context. He categorized celibacy as a monstrous sin, a rejection of his god Ahura Mazda's greatest gift. He prohibited masturbation, in the belief that depriving masturbators of an alternative form of orgasm-release would force them to contribute to his breeding program. And he prohibited male homosexual recreation for the same reason.

Zoroaster did not ban lesbianism. In a culture in which it was not economically feasible for a woman to live without a man, relationships between women did not prevent such women from breeding. Banning them would therefore have served no useful purpose. And all sex taboos down to and including Zoroaster's had a practical purpose. Besides, he probably saw lesbianism as a safe, sinless alternative to adultery for unsatisfied women. The Jewish priest who copied Zoroaster's taboos thought likewise. No ban on lesbianism is to be found anywhere in the Jewish bible.

Homosexuality and masturbation were not banned anywhere in the world prior to Zoroaster. Biblical fables written two centuries earlier and often cited in support of those taboos in fact meant nothing of the sort. The crime of Onan was not sperm wasting, but refusing to father children who would be obligated to offer ancestor-worship to his dead brother rather than himself. The crime of the men of Sodom was not homosexuality (although the rapes they planned would have been homosexual), but violation of the universal hospitality code, in that they planned to violate Levit's protected houseguests. Calling masturbation "onanism" is anachronistic, and calling homosexual coupling "sodomy" is anachronistic.

Zoroaster was barely cold in his grave when an innovator in India promulgated a moral philosophy so antagonistic to Zoroaster's copulation-glorification that it seems incredible that both could have contributed equally to Essene Judaism and its gentile offshoot, Christianity. But Christianity's antigod, Satan, along with homophobia and the flamethrowers of Hell, can all be

10

traced to Zoroaster, while its equation of celibacy with virtue dates back to the inventor of Buddhism, Siddartha Gautama, whose followers called him *Buddha*, "The Enlightened." If Gautama was enlightened, then so was Leopold von Sacher-Masoch.

Gautama decreed that all sexual activity was a sin, including the mating of husband and wife, in bed with the lights out, in the missionary position, while thinking of the motherland. The Jewish Essenes, three hundred years later, adopted the philosophy preached to them by Buddhist missionaries and became celibate masochists (tautology). So universally is joy now equated with sin, that even the names of ancient sex goddesses are deemed vulgar: India's Cunti, Germany's Frig, and Egypt's Nuki.

Zoroaster's strategy to maximize breeding was borrowed by the Priestly author of Leviticus within a generation. Besides banning male homosexuality, the Priestly author had his god Yahweh instruct Abraham, "Be prolific and increase in number." That teaching made sense in a world that was underpopulated and a culture that figured that the best way to enslave the surrounding gentiles was to outbreed them. Today, twelve million children a year are dying as a direct consequence of overpopulation (source: World Vision). Yet the Catholic pope and his co-criminals continue to enforce birth control prohibitions that, if continued, will exterminate the human race by the year 2300 CE. The kindest thing that can be said for such anthropocidal morons is that they expect their obedient attempts to exterminate the human race to be rewarded by a *deus ex machina* intervening to save them in the last act. Even a superficial glance at more than two thousand years of history would show them that no god has ever done that.

The most recent sex taboo, dating from no earlier than the eighteenth century CE, was the raising of the age of mating from the day after the onset of puberty to the sixteenth, seventeenth and even eighteenth birthday. With most women, prior to the invention of surgical forceps, dying in childbirth, it is not surprising that fathers who loved their daughters eventually decided not to sell them into marriage and probable death until they reached an age at which preventing them from taking lovers would have been all but impossible. The average age of marriage did not rise much beyond eighteen until the invention of an effective birth control pill enabled women to satisfy their sexual needs without having to become some man's privately owned breeding slave.

But even in a society which recognized that a woman was not ready for the responsibilities of marriage until her late teens (if then), men continued to view younger women as sexually desirable. To protect such women from sexual predators, laws were passed imposing an "age of consent" below which no woman could be mated without committing a criminal offense. The age differed from country to country. In England it was originally fourteen, but it was raised to sixteen at a time when, due to malnutrition and other inescapable evils, many fourteen-year-olds observably were still children.

In America and nowhere else, the current age of consent is eighteen, although criminal charges are seldom laid when the woman is sixteen and uncoerced, and her chosen lover is neither a person in authority over her nor someone a local despot wants to prosecute on any pretext he can find.

Age of consent laws, especially in America, deny basic human rights to the very persons they purport to protect. Nonetheless, young persons do need protection from men who would use them for self-gratification. A compromise is needed, and the obvious solution is to abolish the concept of a single age of consent, and instead restrict juveniles to sex partners insufficiently older to be deemed predators.

Children up to the age of twelve should be permitted by the state (but not necessarily by their parents, who could still enforce their taboos) to engage in sex play only with partners of precisely the same age, age being determined by the most recent birthday. Thirteen-year-olds could take lovers not more than one year older, fourteen-year-olds two years older, fifteen-year-olds three years older, and sixteen-year-olds four years older. The delusion that seventeen-year-olds are infants incapable of giving an informed consent should be abandoned altogether. And to avoid the absurdity of lovemaking being legal one day and illegal the next day, as would happen when a sixteen-year-old's twenty-year-old lover turned twenty-one, a mating that was legal when first consummated would never again be illegal.

All taboos restricting female sexual freedom are holdovers from the time when male absolutism was unchallenged and women's status as a slave caste was not in dispute. In the most tyrannical phallusocracies, prepubescent females are subjected to clitoridectomy, surgical removal of the pleasure organs. The rationale is that a woman who has been permanently deprived of the ability to feel sexual pleasure will have no reason to engage in premarital or extramarital intercourse. Clitoridectomy is often describes as "female circumcision," a terminology that trivializes a monstrous atrocity by downgrading it to a mere perversion. A better analogy would be "female castration." Clitoridectomy remains a cultural norm in much of the Muslim world, despite attempts by some governments to ban it as non-Koranic child abuse.

The ultimate sexual hoax is the pretence that persons who have been observably married for many years and have produced and raised children together are not really married if they have not formally registered their cohabitation with the designated government statistician. The term "living in sin" has all but disappeared for unlicensed relationships, but such terms as "de facto," "common law," and "living together" are still widely understood to mean "not really married." Yet for most of human history, being married was the observable fact of living together, and a wedding ritual was merely a public celebration of a new reality. It did not bring a marriage into existence.

For more than one thousand years the Christian church, with its Essene-buddhistic glorification of celibacy, viewed marriage as at best a tolerable sin. Priests were under strict instructions to stay away from weddings in case, by their presence, they appeared to give church approval to a contract to engage in voluptuous immorality. Then in the twelfth century the church finally recognized that, by dissociating itself from a cultural reality, it was forfeiting a tremendous potential for wealth and power. For if the church were to claim the right to bring marriages into existence, and to be the only power capable of dissolving a marriage, its ability to extort bribes from the rich could be increased enormously. So not only did the twelfth-century church recognize marriage as sinless. It went so far as to declare matrimony a sacrament.

Two elements of wedding rituals are holdovers from the time when a wife's status as her husband's privately owned sex slave was beyond dispute. One is the ceremonial question, "Who gives this woman to this man?" which leaves no doubt that a piece of property is being transferred to a new owner. The other is the practice of carrying a new bride across her husband's threshold.

So long as official Christianity viewed all carnal copulation as diabolic, and women as evil incarnate ("a tool of Satan and a pathway to Hell," as the sainted Jerome described them), it followed that when a man took a new bride into his residence, it was for an immoral purpose. But the powers of darkness were believed to be incapable of entering a Christian home uninvited. A tool of Satan could not cross a man's threshold under her own power. Her husband was therefore obliged to carry her across, since otherwise her ungodly capacity to engage in sexcrime (all copulation was sexcrime) could not be unleashed.

Since marriage is an observable cohabitation entered into by private agreement between the participating parties, it is clear that any law equating marriage with the uttering of magic words or placing more restrictions on the making or terminating of a marriage than are imposed on any other private contract, is a law respecting an establishment of religion. It follows that a secular state can no more authorize or refuse to authorize a marriage than it can authorize or refuse to authorize a birth or a death. All that the state can do is record that a birth, death, marriage or termination of marriage is an observable fact, and perhaps for administrative purposes require the registration of that observable fact. The pretence that a marriage does not exist when it observably does, or that a marriage continues to exist when it observably does not, constitutes state enforcement of a religion.

The registering of a marriage or divorce would be analogous to the registering of a birth or death. The instituting or terminating of a legally registered marriage would be accomplished by filing the appropriate form with the registrar of births, deaths, marriages and divorces. The state would recognize no distinction between marriage and cohabitation, and would be obliged to record a marriage or divorce without delay and without question as soon as the appropriate notification form was filed. Questions of child custody and financial obligations would remain the province of the courts. As with wills, a new notification would annul all previous relevant documents.

There would be no restriction on who could marry, with one minor exception. Provided that one person signed the notification as the male partner and the other as the female partner, the actual sex of the contracting parties would be irrelevant, as would any blood relationship between them beyond the first degree (since permitting first-degree matings would be an invitation to abuses of authority). Where a notification of divorce was signed by only one of the former partners, a notary would have to attest that either he had served notice on the non-signing partner, or he had been unable to locate the non-signing partner.

# 4. Do American Sex Laws Violate the First Amendment?

In the beginning, God said, "Thou shalt not screw."

Actually, the Bible contains no such commandment, although the Catholic Church imposes such a restriction on its professional pushers. The Bible in fact encourages matings between a male husband and a female wife, in the missionary position, in bed with the lights out, while chanting sectarian prayers and conscientiously attempting to further the already chronic overpopulation. (The further stipulation that copulators must try to minimize their enjoyment is of papal origin, not from any bible.)

The First Amendment states that, "Congress shall make no law respecting an establishment of religion." Yet America enforces a multitude of laws restricting consensual sexual behavior, because the upholders of such laws believe that any activity disapproved by their god must be prohibited, and to hell with the First Amendment. For the god's obedient little mind-slaves live in constant terror that their omnibenevolent Sky Führer will torture them with flamethrowers if some day, somewhere, somebody is allowed to experience a pleasure that the enforcers are denied.

The strongest believers in the evil of sex and the virtue of self-inflicted joy-deprivation tend to be religious fundamentalists, who go to extraordinary lengths to prevent their teenaged children from exercising the sexuality they were born with, while placing no such restrictions on themselves. The First Amendment does not prohibit parents from taking measures to impose their religious taboos on their children. It is the imposition of their taboos on other people's children (such as forcing parochial schools to disallow condom machines in bathrooms) that are in clear violation of the Bill of Rights. Parents and priests can legally ban condoms. Secular lawmakers cannot.

In many states, "adultery" is a legally recognized concept, either as a criminal offense or as grounds for divorce. But without exception, those states define adultery as a purely recreational activity, not necessarily involving the fraudulent impregnation that was the taboo's original purpose. When the taboo made sense, adultery was a breach of an exclusive breeding contract. It robbed a man of the right to pass on his inheritance to his legitimate heirs, by saddling him with a cuckoo's chick (hence: cuckold). Nor could an attempted adultery turn out to be nonconsequential, since it was believed that a baby was jointly fathered by every man with whom a woman had previously coupled, whether the coupling occurred five months before birth or ten years. Adultery as a form of fraud or larceny could be restricted without violating the First Amendment. Adultery as a nonconsequential recreational act cannot.

Laws protecting children from sexual predators, by imposing an "age of consent" below which a child can be deemed incapable of giving an informed consent, would not be unconstitutional if the recognized age of consent was not significantly higher than anywhere else in the world. In Canada, women of fourteen are recognized as having the sole right to determine whether or with whom to copulate. In England and Australia, the age is sixteen. In America and nowhere else, women are legally deprived of a basic human right up to the age of eighteen. Since no reasonable person really believes that a woman of seventeen is a child, the religion enforcers rationalize that they are protecting taboo violators from the wrath of a dice-tossing god. And the First Amendment makes that a no-no. As for laws defining consensual behavior involving a young adult as "statutory rape," anyone who cannot recognize "rape with the consent of the victim" as an oxymoron is probably also incapable of recognizing "holy war" as an oxymoron. And when laws are made that criminalize a mature woman who provides sexual gratification to a post-pubertal schoolboy, since even the most chronic religion addict cannot be so insane as to imagine that such behavior is any more injurious than providing him with a hot fudge sundae, it should be self-evident that enforcement of such laws constitutes an establishment of religion.

A law restricting consensual homosexuality is a law respecting an establishment of religion, since it stems solely from the pretense that a god authored the prohibition of male-male sexuality found in Leviticus 18:22. That the author of Deuteronomy (13:6) endorsed the legitimacy of "the male lover who means as much to you as your own breath" does not change the reality that anti-gay laws are based on Leviticus and no other consideration, and are therefore incompatible with the First Amendment. And laws denying lesbians equal treatment with heterosexual couples are similarly religion-based, even though the Leviticus author saw lesbianism as a sinless alternative to adultery for unsatisfied wives and did not ban it.

Prostitution is a commercial activity not involving the hurting of a nonconsenting victim, and consists of acts that would not be illegal if no money changed hands. So why is it restricted? The answer is that the first practitioners of copulation for remuneration were nuns in the service of a sex goddess, and male gods' speechwriters saw such matings as false-god worship. A law permitting a woman to provide a man with secretarial services for money, but denying her the right to provide sexual services for money, is a law respecting an establishment of religion.

Marriage is a private contract between consenting parties. Divorce is the observable fact that cohabitation has ceased. The pretense that two persons who observably live in cohabitation are not married, or that two persons who have separated are still married, constitutes enforcement of a law respecting an establishment of religion. The State may legitimately require the filing of a marriage certificate or a divorce certificate before it grants official recognition to any change in an individual's marital status. And financial and custody issues arising out of a divorce ought to be the province of the courts. But laws treating the making or terminating of a contract of cohabitation any differently from any other private contract are unconstitutional.

"Obscenity" and "pornography" are religious concepts. Banning *Hustler* is as much a violation of the First Amendment as banning *TV Guide*. And the failure to invoke obscenity statutes against the Judeo-Christian Bible constitutes selective enforcement, not unconstitutional but certainly illegal under equal rights laws.

Inbreeding was first banned to put an end to the situation of fathers keeping their daughters unmated in the hope of marrying them to brothers who either were never born or failed to reach maturity. The present-day ban on sibling marriage can be justified on the ground that intensive inbreeding is now known to harm a species' chances of survival. But a law labeling nonprocreative recreation between siblings as incest and subjecting it to criminal sanctions is a law respecting an establishment of religion.

Religious fundamentalists continue to pass laws depriving women of sovereignty over their reproductive function, even to the point of banning abortion in the first two trimesters when a fetus has no brainwave activity consistent with self-awareness. The Supreme Court, in "Roe versus Wade," placed restrictions on third trimester abortions, in recognition that a fetus with human brain activity might reasonably be seen as a human being, particularly if it is capable of surviving outside of the womb. But instead of accepting "Roe versus Wade" as a reasonable compromise, fundamentalists show no ability to distinguish between a self-aware sentient being and a mindless pre-human tadpole. And there is a real risk that a future Supreme Court, stacked by ultra right wing presidents, could some day reimpose a total ban on abortion based solely on the tenets of a religion.

America has become the theocracy that the framers of the Bill of Rights specifically tried to prevent. And it will stay that way as long as persons who recognize the problem do nothing. To quote graffiti written on a Berlin wall in 1945:

"Finally they came for me, and there was no one left to stop them."

## AFTERWORD

Since writing this, I have learned that the supposedly anonymous graffiti
with which I ended the essay originated with German Lutheran theologian

Dietrich Bonhoeffer, hanged by the Nazis a few days before the end of World War II.

# 5.  Do Laws Permitting Designated Religions to Kill Their Children Violate the First Amendment?

There are American states in which Jehovah's Witnesses and Christian Scientists who kill their children by denying them lifesaving blood transfusions or other medical procedures can escape the consequences of their crime by pleading "freedom of religion." Currently thirty-nine states have religious exemptions from child abuse or neglect charges in the civil code, while thirty-one have a religious defense to a criminal charge.[1] In a study of 172 child deaths where medical treatment was withheld on religious grounds, it was found that 140 would have had at least a 90 % likelihood of survival with medical care.[2]

A law passed in 1996 and still in effect states that the Federal Child Abuse Prevention and Treatment Act (CAPTA) did not include "a Federal requirement that a parent or guardian provide a child any medical service or treatment against the religious beliefs of the parent or guardian."[3] A Senator from Indiana and a Congressman from Pennsylvania, both Republicans (so what else is new?), actually argued that parents have a First Amendment right to withhold medical care from their children.[4] Even in those states where homicidal child neglect is prosecuted, defendants are allowed to offer the jury a defence based on sectarian beliefs not held by other religions. Why?

The First Amendment states, "Congress shall make no law respecting an establishment of religion, or prohibiting the free exercise thereof." Under that Amendment, any law, court ruling or jury verdict that denied adherents of a minority religion or belief system the rights granted to practitioners of all other belief systems would be unconstitutional. Christians are permitted to swear oaths on the sacred book of their choice; therefore Jews and Muslims must be permitted to swear oaths on the sacred book of their choice, and nontheists permitted to "affirm" rather than "swear," under the same penalty of perjury.

But just as the First Amendment prohibits denying adherents of designated religions rights granted to all others, so does it prohibit granting designated religions rights not enjoyed by all others. Laws criminalizing homicide by neglect are applicable to the whole population. They are not laws respecting an establishment of religion. Since the killing of children by substituting prayer for necessary medical procedures is a criminal offense for Catholics, that makes it a criminal offense for Christian Scientists and Jehovah's Witnesses. Any law giving those religions the right to withhold lifesaving procedures from children is therefore unconstitutional. Congress cannot make such a law, and a jury cannot make such a law.

Nobody ever suggested that Thuggs and Assassins, whose religions demanded that they ritually sacrifice outsiders, should be able to violate laws prohibiting homicide by pleading "freedom of religion." Where is the difference? Freedom of religion does not include the freedom to kill. And what of the religions that are trying to force whole populations that disagree with them to grant a "right to life" to pre-human tadpoles with zero brainwave activity indicative of human thought? Are they screaming with outrage and demanding the same "right to life" for children whose parents choose to let them die? Their silence is deafening.

### NOTES

1   Iowa and Ohio allow a religious defense for manslaughter. Delaware and West Virginia allow a religious defense for the murder of a child. Arkansas allows a religious defense for capital murder. Oregon allows a religious defense for homicide by abuse. States with a religious defense for child endangerment, criminal abuse or neglect, and cruelty to children, include Alabama, Delaware, Georgia, Idaho, Indiana, Iowa, Kansas, Louisiana, Maine, Minnesota, Missouri, Nevada, New Hampshire, New Jersey, New York, Ohio, Oklahoma, South Carolina, Tennessee, Texas, Utah, Virginia, West Virginia, and Wisconsin. Some

similar laws have been repealed since 1990 in Arizona, Colorado, Hawaii, Maryland, Massachusetts, Minnesota, North Carolina, Oregon, and South Dakota.

2   *Pediatrics*, 101 (April 1998) 625-629.

3   Sec. 113, Rule of Construction.

4   For additional information see www.childrenshealthcare.org

## 6. Can a Theist be Moral?

Even the moderately religious will often concede that religious fundamentalism is a form of mental dysfunction. Not all fundamentalists go as far as the Flat Earth Society, asserting in all sincerity that, since their various bibles endorse a flat earth, therefore the earth *is* flat. But they do all accept the dictum that right and wrong are whatever their respective gods say they are, and attempt to give taboo-enforcement the status of secular law. At the extreme, fanatics such as Karol Wojtyla relentlessly pursue policies that can only lead to the extermination of the human race by forcing it to overpopulate until it cannot feed itself, in the obstinate belief that they are obligated to give blind, unquestioning obedience to what can only be an anthropocidal sadist in the sky. While the fundamentalists' misanthropy must be evaluated in the light of their belief that, if they obediently attempt species suicide, a *deus ex machina* will intervene to save them in the last act, their willingness to place an alleged god's alleged orders above the welfare of the human race is nonetheless evidence of moral bankruptcy.

But what of the moderate theist, the person who, for example, would never participate in an abortion, or eat pork, or drink alcohol, or read *Playboy,* but who does not deny others the right to do so? Such a person may not commit the overt immorality of persecuting dissent, but can he/she ever be truly moral?

Godworshippers regularly raise the question: Can a nontheist be moral, having no belief that certain acts are "sinful," *i.e.,* that they contravene a divinely revealed lawcode? A more pertinent question, however, would be: Can a godworshipper be moral, having renounced the capacity to judge right and wrong on a rational basis (*e.g.,* does an action unnecessarily hurt a nonconsenting victim?), and instead accepting the right of a lawgiver to base its whims on: Heads it's a sin and tails it's a virtue?

In principle, *only* a nontheist can be moral, since only a person not bound by laws written centuries ago is free to recognize that what was virtuous under one set of circumstances may be evil under another set of circumstances, and vice versa. A theist, on the other hand, is bound to see any law attributed to a god as permanently valid. "Be prolific and increase in number" was sound advice when Genesis 1:28 was composed 2,600 years ago, but today is a prescription for disaster. Laws restricting sexual activity, which depended for their validity on the belief that every such act contributed to eventual pregnancy, and that only a virgin bride could produce a legitimate first offspring, continue to be endorsed by believers in an age of birth control and better scientific knowledge. Since the theists' objection to (responsible, safe) sexual activity among the young is based on the delusion that "God said so," are we not justified in concluding that they would be incapable of recognizing murder and robbery as evil if their gods (or their gods' speechwriters) had omitted to say so? Before answering that question, let us compare theists' observable behavior with the behavior prescribed by the most prominent Western religions' sacred books.

The doublethink that enables some Christians to view parts of the Jewish Testament's law codes as valid (*e.g.,* Exodus 20:1-17), and other parts as no longer valid (*e.g.* most of Leviticus), is beyond the scope of this article. But even the doublethinkers maintain that all biblical proscriptions were valid some of the time, and some of them all of the time. They therefore acknowledge that their paramount god ordered them to execute all persons guilty of: heresy (Exodus 20:20); working on Saturday (Exodus 31:14-15); disobeying a parent (Deuteronomy 21:18-21); gay recreation (Leviticus 20:13); zoophilia (Exodus 22:19); adultery (Leviticus 20:10); failure to possess an intact hymen at the time of marriage (Deuteronomy 22:20-21); fortunetelling (Leviticus 20:27); being a priestess of the wrong god (Exodus 22:18); and insulting the god's name (Leviticus 24:16). While few believers, certainly no one but a fundamentalist, would vote to restore all or any of those death penalties, I have yet to meet a believer capable of comprehending that at least some of those capital punishments could never have been justifiable or virtuous.

So can a theist ever be moral? The answer depends on whether one is judging the morality of *beliefs* or *behavior*. In an absolute sense, it may be immoral even to *believe* that, for example, a woman does not have absolute sovereignty over her own body. But as long as the person with such a belief makes no attempt to prevent her from exercising freedom of choice, justice demands that the person be evaluated, not by beliefs, but by behavior stemming from those beliefs. A theist can be moral, but only if, while paying lip service to every teaching of his religion's sacred books, his observable behavior in fact repudiates those teachings.

## 7. Is Religiosity a Form of Unsanity?

In the concluding novel of Arthur C. Clarke's *Odyssey* series, he postulates a distinction between "insane" and "unsane." Essentially, the difference is that "insane" means neurologically programmed to be incapable of rational thinking and behavior (whether such a condition actually exists is not immediately relevant), while "unsane" means able to put one's mind in neutral in order to engage in irrational thinking for the purpose of nullifying evidence that contradicts a security belief.

It is possible that not a single godworshipper is insane, although Ruholla Khomeini and Pope Pius IX make such a postulation very tentative. But all may be unsane, since even those who could have been freed of god addiction if they had ever encountered the falsifying evidence, are able to rationalize that "when God does it, it's not evil."

A person who believes that the execution of every human being who will ever live, in reprisal for the crimes of his primeval ancestors, would have been evil if Hitler did it but is not evil when his god does it, may be unsane.

A person who believes that an omnipotent Master of the Universe sentences his imagined enemies to trillions of years of sadistic torture in a hell that even the current pope has repudiated, but is nonetheless a nice guy, may be unsane.

A person who believes that, "Not a sparrow falls without his consent," (Matthew 10:29) but when a loved one is killed in a plane crash goes to a church to thank the imagined executioner for his omnibenevolence, may be unsane.

A person who can read Luke 16:1-19, in which Jesus preaches a sermon whose message can be summarized, "Cheat those who are no longer useful to you, and use the stolen money to bribe those who are in a position to do you good," and believe that such a teaching, coming from the Bible, must epitomize true morality, even though he would have recognized it as morally repugnant if it had come from Machiavelli, may be unsane.

A person who can believe that victimless behavior, such as drinking coffee, engaging in consensual sexual recreation with another adult while avoiding undesirable consequences, eating pork, leaving home without a turban, or reading *Playboy* while wearing red socks on a Thursday in August, can be a "sin" simply because a lawgiver, living or dead, so decreed, may be unsane.

A person who can reason that a Bible author should be disbelieved when he says that the earth is flat, but believed when he says that a god revealed its existence, may be unsane.

A person who can argue that the life-supporting universe is too complex to have come into existence undesigned, but an even more complex entity capable of designing the universe could have come into existence undesigned, may be unsane.

A person who can read Exodus 18:11, "Yahweh is the greatest of all the gods, for they tried their hardest and he defeated them," and blot that passage out of his memory in order to continue believing that the Bible was written by monotheists, may be unsane.

A person who can recognize that creating a number that is more than ten but less than nine is definitively impossible, but nonetheless believes that his god could do so, may be unsane.

A person who believes that "argument from consensus" must be valid when more than half of the human race believe in a creator god, even though it was not valid when the entire human race believed the earth is flat, may be unsane.

A person who believes that his god can pre-know the future to the minutest detail, but that does not mean that the future is predetermined, may be unsane.

A person who believes in a god that is omniscient and committed to a plan that it already knows is "the right thing," but will change its plan at the request of a sycophant, may be unsane.

A person who can rationalize that Daniel's (12:2) statement that there is life after death nullifies Ecclesiastes' (9:5) statement that death is the end of existence, and shut out the reality that the presence of even one pair (there are many more) of diametrically opposite statements refutes the pretence that the Bible is "revealed truth," may be unsane.

A person who can believe that the first fifty savior gods to rise from the dead existed only in mythology, but the final copy of a copy of a copy really did it, may be unsane.

A person who believes that the masochism of self-inflicted celibacy can gratify a voyeur in the sky, while simultaneously denying that such a voyeur must be a sadist, may be unsane.

A person who believes that an imaginary playmate in the sky has the omnipotence to prevent such evils as disease, famine, war, natural disasters and transportation accidents, and chooses not to do so, but is nonetheless omnibenevolent, may be unsane.

A person who can believe that one plus one plus one equals one may be unsane.

A person who can rationalize that, even though continuing overpopulation is certain to cause massive starvation and food wars within a few decades, there is really no problem because a *deus ex machina* will intervene to save us in the last act, may be dangerously unsane.

A person who can rationalize that welfare and social security should be abolished, because the unemployed cannot starve to death unless Ol' Massa in the sky wants them to die, may be criminally unsane.

A person who can murder his god's imagined enemies in the belief that the god wants them dead but is impotent to kill them itself, perhaps goes beyond being merely unsane.

And finally, a person who believes in an omnipotent god that wants the entire human race to be Christian, but in 2,000 years has failed to win the allegiance of even one fifth of the human population, may be unsane, as may persons whose gods want all humans to be Jews, or Muslims, or Hindus, and have been equally unsuccessful.

All of the foregoing situations require the mind-neutralization that George Orwell called doublethink and Arthur Clarke categorized as unsane.

So does "unsanity" have any more objective reality than "insanity"? Or is it merely an eye-of-the-beholder label that does not make fuzzy thinking any more comprehensible after labeling than before? A better question would be: Is unsanity a useful concept, as curved space and imaginary numbers are useful concepts, even if it has no real existence? Since the behavior so categorized is real, even if the label is not, I suggest that it is.

## 8. Is Agnosticism Defensible?

Many prominent scientists who dared acknowledge that they did not subscribe to the god hypothesis have been roundly condemned by dogmatic theists, even though the scientists described themselves as agnostic, meaning "without knowledge" to answer the question of whether a god exists. Among the persons attacked for such a cause were Carl Sagan, Stephen Jay Gould and Michael Shermer, all compulsive compromisers who leaned over backward to avoid telling theists that they are plain *wrong*.

Sagan defended his agnosticism in the words, "An atheist is someone who is certain that God does not exist, who has compelling evidence against the existence of God. I know of no such compelling evidence." But Sagan was neither a historian specializing in religious origins nor a philosopher specializing in logic. He was therefore unaware of the "compelling evidence" known to practitioners of both of those disciplines.

Gould, an uncompromising opponent of what he called "the politically potent, fundamentalist doctrine known by its self-proclaimed oxymoron as 'scientific creationism,'" nonetheless went far beyond Sagan in defending religion per se as a worldview different from but not inferior to his own. He described science and religion as Non-Overlapping Magisteria (NOMA), methodologies that could not be harmonized but were somehow not incompatible. How he managed to reach such a conclusion is incomprehensible. But the fact that he did makes his claim to be an agnostic justifiable.

Shermer takes basically the same position as Sagan, writing, "I would have thought God's existence, from a scientific and rational perspective, remains an open question—it cannot be 'proved' one way or the other." Belief that the nonexistence of the protagonist of a fantasy novel cannot be proven is agnosticism by definition. But is that a reasonable belief for any person, particularly a scientist, who has grown up in a culture in which contact with the falsifying evidence is all but unavoidable?

The gods of all western religions are credited with omnipotence, omniscience and omnibenevolence. Agnostics surely know that. But those qualities, besides being incompatible with one another, are also impossible by definition. Omnipotence means the ability to do absolutely anything, including create a number that is more than ten but less than nine, or a triangle with four sides. Outside of an insane asylum, not even the most rabid fundamentalist would claim that his god could do that. Omnipotence cannot exist, and therefore a god that is omnipotent cannot exist.

Omniscience means perfect knowledge of the past, present and future, even though the future depends on decisions not yet made and effects not yet caused. The only way even a god could know the future is for information to travel backward in time. But even that would only be possible if the future cannot be changed, in other words if every future thought and action is predetermined. But Western religion does not preach an inflexible god who presides over a puppet world in which the Hitlers and bin Ladens innocently fulfill the destiny programmed into them before they were conceived. Rather, it postulates a god who already knows every move an individual will make, while insisting that such individuals simultaneously have the free will to trigger an infinite number of possible futures. Stipulating that predestination does not exist, it follows that omniscience cannot exist, and an omniscient god cannot exist.

Omnibenevolence, combined with omnipotence, means that a god with both of those qualities could and would prevent such evils as disease, famine, religious wars, natural disasters and transportation accidents, and would never have created urine, excrement, menstruation, AIDS or trisomy 21. Since those evils observably exist, therefore a god that is simultaneously omnipotent and omnibenevolent cannot exist. But while the existence of "a god" lacking the qualities that are intrinsic to the god of organized religion cannot be disproven (there may be aliens that would appear to us to be gods), the existence of "God" is clearly falsifiable, and has been falsified.

But philosophical arguments only prove that "God" *cannot* exist. Historical evidence shows that he/she/it *does not* exist.

Would any reasonable person claim to be agnostic on the question of whether Lilliputians and Brobdingnagians exist, after learning that they are characters in a fantasy novel that is incompatible with the known facts of geography, and that no evidence of their existence is to be found outside of that novel? Similarly, all claims of a god revealing its existence have been traced to allegedly sacred writings that also state as a fact that the earth is flat, the sky is a solid dome to which the sun, moon and stars are attached, stars are small enough to become detached from the skydome and fall to earth as meteors, humans and lower lifeforms were created separately with no common ancestors, and the universe is less than ten thousand years old. Can any agnostic legitimately claim that those biblical assertions are impossible to falsify, and therefore might be true?

Religion has been disproven. The existence of a "God" with the specific qualities attributed to it by religion has been disproven. So how can anyone not crippled by political correctness maintain that he is "without knowledge" that gods, including "God," are a product of the human imagination? A proclamation that one is an agnostic is a confession that he is either intentionally ignorant of facts that cannot reasonably have failed to attract his attention, or he is so politically correct that he would not tell flat-earthers they are wrong if there were enough of them to make waves.

An agnostic is an ignoramus by definition. The word means "one who does not know." But he is not simply ignorant of facts that cannot be known. By that definition, we would all be ignoramuses. An agnostic confesses to being ignorant of facts that *are* known, and of which he could only be ignorant by shutting out information and logic that would force him to admit that he *does* know. Religious dogmatists have labeled agnostics "chicken atheists." They just might be right.

## 9. Why I Am Not a Terrorist

Whenever I hear that religious fanatics in Algeria or Palestine/Israel or Florida have murdered persons who disagreed with them or who refused to be bullied into obeying the fanatics' taboos, I make a serious effort to put myself in the terrorists' place. That is not as impossible as it might seem. Consider the following analogy.

The earth is round. Its roundness has been proven, starting with theoretical arguments by Pythagoras and Aristotle, followed by physical exploration by Magellan, and culminating in NASA photos taken from outer space. I therefore do not merely believe that the earth is round. I know that the earth is round.

So how do I respond to the information that a religion called the Flat Earth Society in Missouri asserts that the earth is as flat as their bible says it is? Do I become outraged, and organize hit squads to execute, not merely the Flat Earthers, but also secular government officials who allow such heretics to live? The question is obviously rhetorical. Rational persons do not try to punish ignorance. They laugh at it.

Putting myself in the position of such fanatics as Hazballah, Hamaz, Likud, the Christian Coalition, Alberta rednecks, Serbian Orthodoxers, and the I.R.A., who view Jews, Arabs, abortionists, secular humanists, Catholics and Protestants as unfavorably as I view the Flat Earth Society, I ask myself what could motivate me to homicidal rage. It could not be the conviction that I am right and they are wrong, because in that case I would want to kill Flat Earthers. The only hypothesis that makes sense is that, despite the most intensive auto-reinforced mind-warping, there must still be a deeply hidden segment of a dogmatist's brain that religious programming cannot reach. And in that unused recess there is a subliminal awareness that their "God hates heretics" dogma is indefensible, and the only way to ensure its survival is to exterminate everyone whose brainwashing is equal but different.

As easy as it is to conclude that fanatics become addicted to dogma concocted in a pre-scientific age because they lack the capacity to recognize that parts of their sacred writings cannot possibly be true, putting myself in the fanatics' position leads to the opposite conclusion. Fundamentalists do recognize, better than anyone else, that if their source of Revealed Truth contains even the slightest error, then that error invalidates the whole Book. And if their Book is invalid, then the concept of eternal life must also be invalid. Given the alternative of accepting that they must some day cease to exist, totally, permanently and irreversibly, they instead opt for the only belief that gives them the delusion of immortality, namely, that they alone possess Absolute Truth, and all who disagree are hated by their hypothetical Master of the Universe and must be wiped out.

If that means believing that the earth is flat, as the Judeo-Christian Bible states in seven places, or that humans did not evolve from lower lifeforms, or that gods are allowed to play favorites and help their pets steal other people's lands, or that species suicide by overpopulation is virtuous, or that victimless behavior can be evil, or that eating different kinds of food from the same plate can be evil, or that right and wrong are whatever a dice-tossing god's scriptwriter says they are, "Heads it's a sin and tails it's a virtue," or that savoring the masochism of self-inflicted celibacy enables a sadist in the sky to get its orgasm-substitute, then for the mind-deadening opiate of an afterlife belief, accepting such absurdities is a price the intestinally challenged are willing to pay.

While my conceit that "I am right and they are wrong" is based on competent evaluation of the evidence, as the terrorists' is not, the end result is the same. I am satisfied that I am right, and they are satisfied that they are right. So why am I not emulating them, and massacring believers in religion, parapsychology, astrology, alien invaders, quack medicine, masochism as a virtue, and victimless sin? The answer is that, far from seeing such believers as worthy of my hatred, I can only view them as pathetic ignoramuses. And as Frank Herbert wrote, "There's nothing wrong with being

*William Harwood*

ignorant.  Guilt should attach only to anyone who remains ignorant in the presence of an opportunity to learn."

## 10. Social Incest:  Taboo in the Kibbutz

All sexual taboos were invented for reasons related to birth control.  The adultery taboo, dating from *c* 3500 BCE, banned the conception of babies of uncertain paternity.  The homosexual taboo for males, invented *c* 800 BCE, encouraged the breeding of more tithe-paying believers by taking away the right to recreate nonprocreatively.  The masturbation taboo was invented at the same time for the same purpose.  The incest taboo was likewise designed to maximize breeding.

A father with six daughters and no sons may have sold some of them to outsiders.  But where there was a surplus of only two or three, spare daughters would have been reserved for sons who were either never born or failed to reach mating age.  Consequently, an unacceptable proportion of the population would not have married at all, or not married until they were no longer breedable.  That in turn would have caused a serious drop in population, raising the fear that under the existing system one's family would become extinct.  The solution was to mate all of one's children as soon as they reached puberty, and that could only be done by marrying them to outsiders.

It was one thing to formulate a solution, and another to implement it.  If a society accustomed to seeking mates within the immediate family was to be persuaded to look elsewhere, the alternative of following the old way would have to be removed.  Thus the traditional mating practice of four million years was declared sacred to the gods, *taboo*, and a new word was coined for a breach of that taboo: *unchastity/incest.*  That happened *c* 2000 BCE.

So widespread has the incest taboo become in the 4,000 years since its invention, that today the reluctance of persons raised together to intermarry is often cited as a kind of artificial incest concept.  Persons raised in an Israeli kibbutz, for example, almost never marry members of the same kibbutz.  Not surprisingly, believers in biblical taboos see that as evidence of the "naturalness" of the incest taboo, rationalizing that persons growing up in the same household reject their familiars as prospective mates through some kind of instinct.  In fact the explanation is much simpler, and has nothing to do with any instinct to mate with outsiders.

The nuclear family forms the basis of most modern societies.  Almost without exception, those societies are conditioned to the metaphysical belief that sexual recreation is intrinsically immoral and can only be justified by a marriage certificate or a "meaningful relationship."  Sex play by children is totally forbidden.  Consequently, when children reach the age of sexual curiosity and wish to satisfy that curiosity with the nearest available playmate, who is bound to be a sibling, they are informed that such behavior is "wrong."  The warden means, of course, that all child sex is wrong.  But the child assimilates the teaching that sex with the most convenient partner is wrong.

Fifteen years of being kept away from that partner hardens into an unbreakable taboo.  When the age of sexual freedom is reached, a generation of regarding a particular individual as off limits causes the new adult to turn to potential partners who have not been so categorized.

The identical situation occurs in a kibbutz.  A three-year-old boy, showering with a girl, notices that her anatomy differs from his own and, curious, reaches between her legs.  The adults present intervene, just as happens in a nuclear family.  The children are immediately conditioned to the belief that one does not touch the sexual apparatus of *the other children in the kibbutz.*  Consequently, kibbutz graduates continue to regard each other as taboo long after the restrictions of childhood have been withdrawn. They could hardly do otherwise.

Incest is an artificial concept, no more a part of "natural law" than the now discredited meatless Friday.  Laws that criminalize even the nonprocreational violation of a biblical taboo are unconstitutional, since the First Amendment prohibits any law respecting an establishment of religion.

## 11. Sin Is Whatever My Führer Says It Is

Religiosity has long been recognized as a form of insanity. Only a person with the inability to tell right from wrong on a rational basis (the legal definition of insanity) could believe that an action can be simultaneously victimless and immoral, or that the Inquisition's fifteen million heretic-burnings must have been virtuous, since they were carried out at the Christian god's orders, and "When God does it, it's not evil."

Once it is recognized that, to a victim of god psychosis, right and wrong are whatever his god's speechwriter says they are, it becomes easier to understand why, for 1,400 years, almost no brainwashed believer has ever questioned the validity of a capricious list of "seven deadly sins" given to the world by Pope Gregory I (590-604). Even Eastern Orthodox Christians, whose hatred of everything Catholic causes them to cling to the Julian calendar with its eighteen-hours-per-century inaccuracy rather than adopt the more precise Gregorian calendar invented by a pope, nonetheless continue to accept an earlier Gregory's catalogue of sins without ever questioning whether it makes any sense.

Number one on Gregory's Big Seven list was anger. At first glance that might seem to be a reasonable inclusion. But an examination of what anger meant to a medieval pope and what it means today leads to the opposite conclusion.

Anger is a valuable human quality. Anger at robber barons' inhumane treatment of their workers led to fair-labor-practice laws. Anger at Adolf Hitler's contempt for international law led to the realization that sooner or later he had to be stopped. Anger that American citizens could starve to death once they had outlived their ability to support themselves, led Franklin Roosevelt to institute social security laws that a new generation of robber barons is conspiring to repeal. But Pope Gregory could not be expected to see that anger is intrinsically positive. To Gregory, anger meant the ability to recognize the injustice of a tyrannical papacy and respond with hostility.

Gregory also anathematized envy, meaning the ambition that led humans to emulate the flight of birds and aspire to the comfort of the ruling classes; gluttony, meaning the desire to eat as well as the clergy; greed, meaning the desire to own private property at a time when the Catholic Church wanted to be the world's only capitalist; and pride, meaning self-respect and the healthy recognition of one's worth as a sentient being, rather than the domesticated livestock of a petmaster in the sky.

Rounding out Gregory's Top Seven were lust and sloth. Even in Gregory's day, lust meant the tender love of sexually compatible partners. But in Gregory's day, and for 600 years afterward, celibacy was deemed the ultimate virtue, and marriage at best a tolerable sin. Without lust, the human race would become extinct in a generation. To Gregory, that was not a problem, because the imminent end of the world was a central tenet of Gregory's theology.

Gregory's concept of sloth was the willingness to live and let live instead of spying on friends and relatives and reporting suspected heretics to the vigilantes who foreshadowed the Inquisition.

All of Pope Gregory's deadly sins are virtues when practised rationally. Only when carried to excess do they degenerate into vices. For example, anger can become rage. Envy can mutate into the godly (Exodus 34:14) psychosis of jealousy. Gluttony and greed can become what those words denote today. Lust can become sexual predation, self-gratification at the expense of another. And sloth can become the depraved indifference that allowed thirty people to witness the murder of Kitty Genovese and not lift a finger to intervene.

The "seven deadly sins" are nothing of the sort, and the ability to believe that ambition, love, self-respect, and the capacity to recognize injustices are reprehensible, is one more proof that blind acceptance of the dogma, "Sin is whatever my Führer says it is," is a form of insanity. Equally insane is the inability to recognize the absurdity of the misnamed "golden rule."

The maxim that one should treat others as he would have them treat him is, theoretically, a laudable ideal. In practice it is impossible in some situations and morally reprehensible in others.

The admonition that one should not do to another whatever is hateful to himself was taught by Zoroaster, Confucius, Gautama, the Hindu *Mahabharata,* the Pharisee-authored *Tobit,* and the rabbi Hillel. It was and is the most perfect moral absolute ever propounded, and can be practised by anyone with the wish to do so.

Unfortunately, that epitome of morality and justice has in modern times been damned with faint praise as the "silver rule," and all but replaced in the popular imagination by its unachievable converse, the "golden rule."

Suppose, for example, that a person earning minimum wage wants one of this planet's billionaires to give him a million dollars. The impossible rule would have him give the billionaire a million dollars, as clearly he is incapable of doing. Similarly, he might want the United Nations to appoint him President of Earth. Again, the inverse is beyond his capacity. And what of the man who fantasizes about a supermodel ripping off his clothing and mating with him with callous indifference to his own desires or feelings? The golden rule says that he should do the same to her. In most jurisdictions, that would be called rape.

The golden rule, far from improving on its predecessor, took an ideal that was already perfect and destroyed it.

So why are believers able to accept such concepts as "seven deadly sins" and "golden rule," without ever recognizing that the alleged sins are in most circumstances positive qualities and the alleged golden rule is at best an absurdity? The answer is that dogmatic religiosity is a psychosis that deprives its victims of the rationality to question their self-appointed lawgiver's decrees, whether that lawgiver is a pope who says that the same act can be sinless six days a week but sinful on Fridays, or a televangelist who says that mass murder was evil when the Nazis did it with gas chambers but is virtuous when his sky-Führer does it with AIDS, cancer and natural disasters.

In the light of their brainwashing, it was inevitable that persons afflicted with god psychosis would continue the anthropocidal practices of air, water and population pollution, in the conviction that a *deus ex machina* will intervene to save them in the last act. And because of that insane behavior, the probability that the human race will still exist in three hundred years is already down to thirty percent, and will decrease a further one percent for every decade that religiosity continues to flourish. While that is a very rough estimate, time is running out too quickly for us to assume that it is too pessimistic. Of all the evils perpetrated by persons insane enough to believe that "Sin is whatever my Führer says it is," the attempt to exterminate the human race in the belief that they cannot succeed may well turn out to be the last.

## 12.  The Uzziah Syndrome

American President Ronald Reagan recently appointed a one-man committee to make a further "study" of the causes and effects of acid rain.  The discernible purpose of the appointment appears to be to dispel criticism of Reagan's do-nothing policy on this issue, so that he can continue to do nothing until he is safely out of office.  On previous occasions, Reagan has appointed and reappointed Environmental Protection Agency chiefs on whom he could rely to do absolutely nothing to protect the environment.  Since the ultimate effect of the poisoning of the water we drink and the air we breathe is the extermination of the human race, along with most other lifeforms above the arthropod level, the obvious question is:  Why?  Is the President of the United States callously indifferent to the fate of the human race?  Is he an evil madman who *wants* to commit anthropocide?  The answer is more insidious than either:

Ronald Reagan is a religious fundamentalist.

There is a story in Genesis (22:1-8) in which Yahweh ordered Abraham to commit genocide by sacrificing the as-yet-unmated ancestor of the Jewish nation, Isaac.  Abraham prepared to grant the mindless, unquestioning obedience that all tyrants, mortal and immortal, demand, and was rewarded for his submission when Yahweh intervened to prevent the disastrous consequences of his own order.  In a theocratic tyranny, it seems, obedience to the god's order to commit racial (or species) suicide is the route not only to survival, but also to prosperity.

There is also a tale, in 2 Samuel (6:6-7), in which the Chest of Yahweh's Treaty appeared to be about to tip over.  David's aide-de-camp, Uzziah, hurriedly reached for the Chest to steady it, and was promptly struck dead by one of Yahweh's thunderbolts for his presumption in thinking that Yahweh could possibly need his assistance.

It might be going beyond the evidence to suggest that President Reagan has the "Uzziah syndrome," blind terror that, if he makes any move to usurp Yahweh's prerogative to prevent the poisoning of planet earth, the tyrant in the sky will do to him what it did to Uzziah.  What he does have, however, is the "fundamentalist syndrome," the utter conviction that nothing he does or fails to do can in any way jeopardize humankind's continued existence.

Human extinction, according to Reagan, is definitively impossible, for the good reason that his imaginary playmate would never permit it to happen.  By the same logic, despite the carefully calculated inevitable effects of a nuclear winter, Mr. Reagan imagines that an all-out nuclear war can be survived.  And by the logic of the Isaac myth, he believes that, in a dangerously overpopulated world, rigid enforcement of his god's order to "be prolific and increase in population" will in fact prevent the starvation-extinction against which Malthus so long ago warned.

The most widespread, and therefore most necessary, form of population control in the world is abortion.  Mr Reagan is attempting to prevent women who would do so from obtaining abortions, first because it contravenes his god's order to "be prolific...," but also because he equates abortion with "murder."  In fact, when the author of Deuteronomy, in 621 BCE, put into the mouth of Yahweh a commandment, "You may not murder," his concept of murder not only excluded abortion, infanticide and the summary execution of even one's adult offspring for filial impiety, real or imagined; it also excluded the killing of gentiles. (Talmud, Sanhedrin 78b)

President Reagan believes that doing nothing to prevent the poisoning of this planet, and actively contributing to the advancing overpopulation, is pleasing to his god and therefore a positive contribution to human welfare.  There is nothing wrong with his reasoning—provided his initial assumptions are valid.  If his god exists, and if it is as capricious as he think it is, it will indeed do everything the President expects of it.  Unfortunately, self-inflicted anthropocide *can* succeed, and with a succession of Reagans in the White House and Wojtylas in the Vatican *will* succeed, for the

30

obvious reason that the *deus ex machina* that Reagan is counting on to save us in the last act may not exist.

So what is the solution? Do we execute the polluters? storm the White House? raze the Vatican? The answer is: none of the above. Reagan is merely a symptom; religious fundamentalism is the disease. If the human race is to survive, not the destroyers but the philosophy that motivates them must be eliminated.

The choice is simple: Either humankind must exterminate religious fundamentalism, or fundamentalism will exterminate humankind.

## AFTERWORD

Sometime after the publication of this article, Reagan was diagnosed with Alzheimer's disease, raising the question: How could they tell? And in 2000 the threat of "a succession of Reagans in the White House" became concrete with the overthrow of democracy and the treasonous appointment as President of a defeated candidate even more subhuman stupid than Reagan.

## 13. Is This 1984—Or What?

The past was unalterable. The past never had been altered. Oceania was at war with Eastasia. Oceania had always been at war with Eastasia. (p. 223)

Oceania was at war with Eurasia. Oceania had always been at war with Eurasia. (p. 231)

The Party said that Oceania had never been in alliance with Eurasia. He, Winston Smith, knew that Oceania had been in alliance with Eurasia as short a time as four years ago. But where did that knowledge exist? Only in his own consciousness, which in any case must soon be annihilated. And if all others accepted the lie that the Party imposed—if all records told the same tale—then the lie passed into history and became truth. (p. 31)

Does that sound familiar? It should. It is called the Big Lie. It was standard operating procedure when George Orwell's fictionalized Anglican religion (disguised as Russian communism) did it in 1949 in a book that his publisher forced him to rename *1984* to disguise a description of a nightmare present as a prophecy of a nightmare future. It has been standard operating procedure in all religions for as long as they have existed. And it is standard operating procedure to this day.

In 1992, I included in *Mythology's Last Gods* a differential equation that can be used to calculate the world population of Christians and Jews at any given time in the future. It showed a figure of zero for the year 2192, 200 years from the date of publication. Since 200 years was the time I calculated that the book would take to wipe religion from the face of the earth, the equation was based on the number of Christians living in 1992. I set the figure at one billion, because that was the figure given by a majority of reports in the mass media and even official religious publications.

Within the past few months, several alleged "news" broadcasts have stated as a fact that there are two billion Christians on earth. At a time when church attendance has dropped fifty percent in a generation, we are being asked to believe that the number of god addicts has doubled in that same generation.

Where did they get such a figure? The answer is that they made it up. In a desperate attempt to deny that religion is on its last legs, and only fanatics like Osama bin Laden are proving incurable, organized religion is resorting to the Big Lie that religion, specifically Christianity, is on the increase.

There are not two billion Christians on earth. Only by counting the entire population of countries with a professed Christian majority does the figure even reach one billion.

There are six billion human beings on this dangerously overpopulated planet. That figure is made up of one billion Christians, mainly in Europe and the Americas; one billion Muslims, mainly in southwest Asia and Africa; one billion Hindus, mainly in India, a half billion folk religionists, half of them in China, two billion atheists, agnostics and other nontheists, half of them in China, and a half billion minority religions, traditional and fringe, including Buddhism, Shinto and Judaism. (Figures for China are taken from *The Statesman Yearbook 2000*.)

So has this planet's population jumped from six billion to seven billion in the past few months, or are the extra billion Christians simply the latest spin on the Big Lie? Does anyone with a functioning human brain even need to ask?

## 14.  George W. Bush's Ultimate Hero

When a statistically insignificant but numerically large number of Afghans and Pakistanis express admiration for Osama bin Laden for murdering over three thousand Americans in a single day, it seems reasonable to conclude that their choice of Hero is a measure of their own moral evolution.  To admire evil, one must *be* evil.  The same conclusion can be drawn about persons who to this day express admiration for Adolf Hitler or Josef Stalin.

America's unelected president is on record as expressing the view that his favorite philosopher, from whom he tries to take his values, is one "Jesus Christ," actually Jesus the Nazirite ("Christ" is a title that implies that he was the mythical character he claimed to be).

So how heroic or admirable was the aforesaid Jesus?  By way of answer, let us consider his teachings.

In Luke 16:1-9, Jesus is shown preaching a sermon that can be summarized, "Cheat those who are no longer useful to you, and use the stolen money to bribe those who are in a position to do you good."  And when Jesus' followers asked him why he preached in riddles, he told them, "To you, permission has been given to understand the mystery of Allah's theocracy.  But to those outside, all things are to be told in fables, so that in seeing they may see but not perceive, and in hearing they may hear but not understand, in case they convert and obtain forgiveness for themselves." (Mark 4:11-12)  In other words, Jesus preached in riddles so that he would not accidentally cause the salvation of someone his Sky Führer was planning to damn.

Jesus was a liar, or perhaps a sincerely deluded madman, as his own family believed (Mark 3:21), who promised his adjutants, "There's no one who has abandoned home or brothers or sisters or mother or father or children or lands for my sake and the sake of the evangelion, who will not receive one hundred times as many homes and brothers and sisters and mothers and children and lands in this time." (Mark 10:29-30)  Even the most braindead biblical literalist (tautology) acknowledges that that promise was never fulfilled—or ever can be, since the persons to whom it was made are all dead.  Nor has his promise ever been fulfilled that, "Whoever says to this mountain, 'Be torn up and tossed into the sea,' and is uncritical in his heart and credulous that what he asks will happen, it will be done for him." (Mark 11:23)  No person who has attempted to exercise the power of telekinesis Jesus bestowed on him has succeeded in moving a single grain of sand, let alone a mountain.  And unless there really is a Wandering Jew (as even Bush probably does not believe), Jesus' promise has long been unfulfillable that, "There are some standing here who are not going to experience death until they have seen Allah's theocracy established by force." (Mark 9:1)

As a celibate communist, Jesus condemned all property owners to the eternal flamethrowers: "It's easier for a camel to pass through the eye of a needle than for a capitalist to enter Allah's theocracy," (Mark 10:25) and further urged all men who felt normal sexual desire to castrate themselves. (Matthew 19:12)  One is reminded of Aesop's fable of the dog in the manger.  Since it could not eat the hay itself, it saw to it that the cattle also starved.  And when Jesus felt hungry and wanted some figs, only to discover that figs were out of season, he actually cursed a fig tree. (Mark 11:13)  Sour grapes, anyone?

While the Christian gospels do not show Jesus specifically validating the concept of predestination, they do portray him as assigning eternal torture to persons who happened to be the venue of inescapable destiny:  "Even though scandals are inevitable, oy vay to the human through whom a scandal comes." (Matthew 18:7)  They show him validating slavery. (Luke 12:47; Matthew 18:23 ff.)  And he was an advocate of state sanctioned ritualistic revenge murder, even for actions that are not illegal under the laws of every civilized nation on this planet. (Mark 7:10)

Of the hometown and probable birthplace that rejected him as an upstart local boy, Jesus said, "As for you, Kafar Nahoum, weren't you exalted to the sky?  You're going down to Hades."

(Matthew 1:23) Nor was he contented to sentence his real and imagined enemies to extermination. So that he could savor their screams of agony for billions and billions of years, he consigned every person who refused to endorse his candidacy for King of the Jews to a prototype Auschwitz, "where their maggot never dies and the fire is never extinguished. For everyone is going to be pickled with fire." (Mark 9:48-49)

As the foregoing makes clear, Jesus was a sadist. He was simultaneously a masochist: "If someone punches you on the right cheek, allow him to punch the left as well." (Matthew 5:39) "If your hand scandalizes you, amputate it. If your foot scandalizes you, amputate it. If your eye scandalizes you, gouge it out." (Mark 9:43-47)

Jesus hated heretics and infidels, meaning everybody outside of his own little sect: "A woman of Syrian-Phoenician birth asked him if he would exorcise the demon from her daughter. And he answered her, 'It's not right to take the children's bread and throw it to the dogs.'" (Mark 7:26-27) "Don't go anywhere among the infidels, and don't enter any Samaritan town." (Matthew 10:5) And like religious fanatics of modern lunatic fringe sects, he demanded that his followers sever all connection with members of their families who did not join the cult: "Anyone who comes to me and does not despise his father and mother and woman and children and brothers and sisters cannot be my student." (Luke 14:26)

And that is the man—sadist, masochist, thief, unelected tyrant, psychopath—the President of the United States calls his favorite philosopher. With a role model like Jesus, it is no wonder that, as Governor of Texas, Bush became the most prolific serial killer in American history, with over 120 homicides on his resume. And since nobody ever elected Jesus dictator of the world, it is only logical that Bush should feel entitled to claim a similar position even though nobody elected *him*. Fortunately for Gee Dubya Shrub, while the five criminals who appointed him may one day face treason charges for their coup d'état, Bush has the legitimate defence of diminished responsibility.

## 15. The Most Dangerous Man in Canada

Robert Boston's unauthorized biography of Pat Robertson called him *The Most Dangerous Man In America.* So far no one has written a biography of the most dangerous man in Canada, a Robertson clone named Stockwell Day.

**[There is now a biography of Day, *Requiem For a Lightweight*, by Trevor Harrison, Black Rose Books, Montreal, 2002. See review # 100.]**

In 1987, Preston Manning, son of the theocrat who had ruled the province of Alberta as an absolute ayatollah for two decades, and had even banned Sunday football in the hope that he could force Albertans to listen to his "Back to the Bible" radio program, founded an ostensibly new political party called *Reform* that was in reality nothing more than a revival of his father's long-discredited pseudo-political religion of Social Credit. In the absence of American-type primaries, that have the virtue of reducing elections to two viable candidates, Manning was able to take advantage of a three-way split (liberal, conservative, socialist) in the anti-theocrat vote, and make his extremist party the Official Opposition in Canada's Parliament, winning three times as many seats as the Conservative Party despite winning the same number of votes.

For the next decade it seemed a reasonable expectation that, if Manning could be eliminated from the Canadian political arena, the hate cult that he founded would be eliminated with him. What that expectation did not anticipate was that Manning would be superseded as Party Leader, not by a less fanatic hatemonger, but by a more fanatic one, the aforementioned Stockwell Day.

At the beginning of his political career, Manning went on record as supporting a constitutional amendment that would deny Canadian women sovereignty over their own bodies by denying them the right to choose abortion. In recent months when Day has been questioned about his belief that abortion should be criminalized, he has denounced the news media for making an issue of his social agenda, even though he persistently refuses to declare that, given the power to do so, he will not attempt to make his personal religious beliefs the law of the land. When he was finally pushed to the point where "No comment" would have been tantamount to a confession, he weaseled his way out of giving a straight answer by insisting that only "the people," not an individual, can make laws for the whole of Canada. In other words, he will not turn Canada into a theocracy unless he can win a parliamentary majority willing to do so. That tactic of making a statement calculated to deceive people into thinking it means the opposite of what it really implies is called LYING. Day knows that Canadians will never elect a party that plans to enslave them to a god whose concept of right and wrong is whatever Day says it is, so he uses doubletalk to conceal his intent while allowing himself to say later, "I never said I would *not* impose such laws."

As letters to the editor in every Alberta city have made clear, western rednecks believe "democracy" means that a majority can do anything it wishes. Should homosexuals be deprived of equal rights under the law? Hold a referendum. Should abortion be criminalized? Hold a referendum. Should capital punishment be restored? Hold a referendum. If a majority votes "yes" on questions that should be none of the law's business, or which the fully-human have already rejected as subhuman, then the taboos of the majority religion should be imposed on all citizens, and minority rights be damned. And Stockwell Day is the archetype western redneck.

As Alberta's Social Services Minister, Day reduced welfare payments to the point where the unemployed had to choose between paying rent and buying food. Is that the kind of social agenda he thinks he should not have to reveal to voters in advance? As a religious fundamentalist, does he believe that no one can freeze or starve to death unless that is what his sectarian god wills? Does he think that persons who could not otherwise survive should be forced to seek charity from a church, along with the mandatory sermon that such encounters involve? Or does he believe that his god hates the eighty percent of the human race who, even though a majority believe in a god, do not believe in the god created out of the hatred and intolerance Day sees in the mirror?

Day was successfully sued (at a cost to Alberta taxpayers of $792,064.40) for libeling a lawyer whose offence, in Day's eyes, was that he agreed to defend a man charged under a pornography law that would make it a criminal offence for a parent to take a photograph of his/her naked baby, or to own a magazine containing a Coppertone™ advertisement. Whether the law would ever be invoked against those specific acts is irrelevant. The British Columbia appeals court struck down the law precisely because it *could* be so used. And Day wants such laws upheld, even if that can only be done by overriding the Canadian Charter of Rights and Freedoms. Does he really see nakedness as an unmitigated evil? Or does he simply want laws that will enable him to prosecute everybody outside of his hate cult?

It is uncertain to what degree the redneck faction led by Day can be identified as the puppetmaster at whose string-pulling Alberta Premier Ralph Klein promised to override the Charter of Rights if the Supreme Court upholds the right of gays to total equality under the law, including the right of same-sex couples to declare themselves married. It may be that Klein is himself a homophobic bigot. (He has made anti-equal-rights statements since Day's departure to federal politics.) But on past performances, it seems more likely that Klein's overriding compulsion is to be the biggest fish in the pond, and if that can only be achieved by pandering to the theofascist rednecks who constitute a controlling faction in his caucus, that is what he is prepared to do.

Stockwell Day is a religious fundamentalist. That is not in dispute. He is on record as expressing the belief that dinosaurs roamed the earth as recently as six thousand years ago, since his reading of his bible tells him that the earth did not exist seven thousand years ago. He was a teacher at an ultra-right fundamentalist church-school before entering politics, in defiance of laws declaring the unregistered school illegal, and continues to attend the same church on a weekly basis. While voicing objections to having his religious beliefs treated as a political issue, he simultaneously refuses to state that the government has no place in the nation's bedrooms, as Pierre Trudeau did back in the days when the possibility of Canada becoming a theocracy would have seemed absurd. It is a logical conclusion that he avoids every opportunity to endorse freedom of conscience, because that is the first thing his hate cult plans to wipe out.

As a fundamentalist, Day believes that the fantasy novel known as *The Bible* is literal truth whose decrees must be imposed on the entire population, whether they agree with it or not. And that Bible is the most obscene paean to evil ever written, with Adolf Hitler's *Mein Kampf* and the Marquis de Sade's *Juliette* fighting it out for second and third place.

For example, Matthew 27:25 has the Jews who allegedly condemned Jesus say, "Let his bloodguilt be on us and on our descendants." I have yet to hear a fundamentalist denounce that passage as anti-Semitic, or suggest that he is not bound to treat it as an order from his Führer in the sky. Day has specifically disputed a forecast that, once his hate cult has criminalized gays and believers in abortion, its next target would be Jews. But he has never dissociated himself from his Bible's anti-Semitism, and continues to maintain that his Bible's concept of righteousness is to be accepted without question.

The Judeo-Christian Bible is riddled with death penalties, for such crimes as homicide, adultery, working on Saturday, honoring an opposition god, failing to possess an intact hymen at the time of marriage, fortunetelling, and gay recreation. Day has never advocated executing gays. But he has enthusiastically endorsed the legitimacy of the death penalty, a barbarism that has been abolished by every civilized government on earth. And he has vehemently opposed equal rights for gays, even to the extent of running TV commercials during the Canadian federal election denouncing the Liberal government for "wasting money" on HIV research. It is uncertain whether that commercial was run in all provinces, or only in Alberta and B.C. where homophobic bigots constitute a voting majority.

Put those realities together, and anyone who tells himself, "He wouldn't..." needs to be reminded that Hitler was voted into power by persons who had read his proposals for the future in *Mein Kampf* and rationalized, "He wouldn't really do that." There is obviously a qualitative difference between a mass murderer and a nonviolent hatemonger. But it is an observable fact that politicians who rant against freedom of conscience and equal rights invariably turn out to practice what they preach. As an Alberta cabinet minister, Day argued against publicly funded abortions for rape victims, except where the victim's life was endangered. And he has made clear that he is willing to deprive women of sovereignty over their own bodies if a referendum permits him to do so.

I have never met anyone who has had the disease and recovered who is unaware that fundamentalist religiosity is a form of insanity. A person who was not insane before he started believing that mass murder was evil when Hitler did it with gas chambers, but is not evil when his god does it with disease, famine, religious wars, natural disasters and transportation accidents, is certainly insane once he does acquire such a belief. And the most notable symptom of religiosity is the conviction that the Bible is inerrant truth and that one is bound by its orders. In case anyone is unaware, the Bible's orders concerning heretics and infidels, usually interpreted as everyone, including moderate believers, outside of one's cult, are as follows:

"When Yahweh your gods [generic plural] has settled you in the land you're about to occupy, and driven out many infidels before you...you're to cut them down and exterminate them. You're to make no compromise with them or show them any mercy." (Deuteronomy. 7:1) "You're going to exterminate them in a massive genocide until they're eliminated." (Deuteronomy 7:23)

Does Stockwell Day take the position that those are "Old Testament" orders that are no longer binding? If so, he has not said so. Is it mere coincidence that his proposal for a flat tax, charging persons living below the poverty line the same tax rate as millionaires, so closely resembles Leviticus 30:13-15: "Every registered taxpayer is to contribute a half shekel...The rich are not to be charged more, and the poor are not to contribute less, than a half shekel"? And if he does reject the "Old Testament," he is still bound by his New Testament's hatred of the human race. To cite some random examples: "If any man come to me, and hate not his father, and mother, and wife, and children, and brethren, and sisters...he cannot be my disciple." (Luke 14:26, *King James Version*) "Think not that I am come to send peace on earth. I come not to send peace but a sword." (Matthew 10:34, *KJV*)

Moderate Germans did not believe that, if they gave political power to that good Catholic, Adolf Hitler, he would make hatred of an opposition religion the law of the land. Moderate Iranians did not believe that, if they gave political power to Ayatollah Khomeini, he would launch a bloodbath from which no one would be safe. Moderate Israelis did not believe that, if Ariel Sharon became prime minister, he would attempt to turn Israel into a theocracy and further the annexation of territory never recognized by the United Nations as part of Israel. Mainstream conservatives in Canada are being assured that, if they give political power to an extremist cult that deliberately refuses to make its social agenda clear, and when *Reform* became identified as Manningism even incorporated the word "conservative" into its new name, **C**onservative **R**eform **A**lliance **P**arty, in order to deceive voters into believing it is allied with the party of former prime ministers, John A. McDonald and John Diefenbaker, it will not establish a theocracy indistinguishable from Iran or Afghanistan. The Germans were wrong. The Iranians were wrong. The Israelis were wrong. Canada avoided making the same mistake in the November 2000 election. But Day increased his representation in Parliament, even to the extent of winning two seats in Ontario—and there will be another election in four years. By then the whole country (at least east of Winnipeg) may recognize the threat the theofascist rednecks pose. But there is a real danger that they may not.

Just as America's Nelson Rockefeller refused to place winning ahead of conscience, and continued to repudiate extremist Barry Goldwater even after Goldwater became his party's presidential nominee, so Canada's Conservative Party leader Joe Clark refuses to put his conscience in neutral in order to compromise with an extremism that is as much a threat to moderate Christians such as himself as it is to nontheists, homosexuals and family planners. He recognized Manningism as homegrown Khomeiniism, and he is clearly aware that Dayism is even more extreme. If Ontarians, whom Manning was unable to convert to his religion-posing-as-a-political-party, can be deluded that Day, of whom they currently know nothing, is less bigoted, less theofascist, and less dangerous than Manning, Canada can prepare for a tyranny such as even Alberta and British Columbia under a half-century of the Social Credit theocracy has never seen.

In recent days, Day has demonstrated such chronic, inept, foot-in-mouth disease, that even his own hate cult is muzzling his public speeches and openly urging him to resign. But he is desperately hanging on, and actually won a vote of confidence from party members in his home constituency. On the positive side, if Day is successful in remaining leader until the next election, he is likely to drag his whole pseudo-political hate religion into the toilet of history with him. But there is also the serious risk that his popularity could jump right back up in the two years or more until the election— and if that happens, Canada is in *BIG* trouble.

In the words of Judith Hayes (*The Happy Heretic*, p. 235), "All it takes for [this country] to become a theocracy is for nonbelievers to do nothing."

## AFTERWORD

On March 20, 2002, the Canadian Alliance Party voted overwhelmingly to remove Stockwell Day from the party leadership, and replace him with a man best known for advocating a "firewall" around Alberta to keep out such dangerous un-Albertan ideas as human rights, freedom of conscience, freedom of choice, and equal treatment for minorities. Then on April 7, a Party convention passed a resolution that would effectively abolish Canada's Charter of Rights and Freedoms, while the new leader, in addition to naming Day his foreign affairs deputy, expressed adamant opposition to any compromise with the moderate Conservative Party. So the Canadian Alliance of Redneck Bigots is still a theofascist hate cult that falls somewhere on the evolutionary scale between Khomeini's Ayatollahs and Torquemada's Inquisitors. So what else is new?

### 16. Is the Canadian Alliance Proof that Pithecanthropus Troglodytus is Alive and Well and Flourishing in Alberta?

More than three million years ago, a line of hominids branched off from Australopithecus afarensis to begin the long evolution that eventually produced the two interbreedable (as donkeys and horses are interbreedable) species, Homo sapiens and Homo godworshipper, the only measurable difference between the two being that the former is teachable and the latter is not. It has long been thought that all other hominid species became extinct, and that even Alberta's most mindless theofascist fanatics are merely an aberrant breed of Homo godworshipper. But the differences between moderate godworshippers and Alberta's hate cultists are so profound as to suggest an alternative explanation.

There are three billion godworshippers on planet earth, and as many as an additional two billion who find it socially expedient to brainwash themselves that they belong to Homo sapiens' coexisting species. But even among the three billion, an overwhelming majority has never submitted to the self-inflicted mind amputation that would enable them to shut out all logic and evidence in order to believe that the sane, the intelligent and the educated are all wrong, and that a fantasy novel written two and three thousand years ago is right.

The universe is more than twelve billion years old. The earth is more than four billion years old. Thinking human species have existed for thirty thousand years, and their ancestors differed from them physically in no way that anthropologists can detect for an additional hundred thousand years before that. Humankind evolved from the same ancestors as the anthropoid apes, and before that from proto-monkeys, proto-primates, proto-mammals, and even lower species, all the way back to protozoa. Those are fully established facts, disputed by no member of Homo sapiens, and possibly by no member of Homo godworshipper. The only disputants are, by every measurable element of hominid evolution, so far below even Homo godworshipper, that the possibility that they are throwbacks to what was once called the "missing link" should be given serious consideration.

A prominent (infamous) Albertan, James Keegstra, proved himself so incapable of even the level of logical reasoning achieved by Homo godworshipper, that he was able to rationalize away all of the evidence of Adolf Hitler's campaign to exterminate an opposition religion, and brainwash himself that the Final Solution and its gas chambers never existed. He promoted hatred of Jews by teaching his high school classes a dogma that no person with a functioning human brain and a reasonably full education could possibly believe. While his application for membership of the hate cult now called the Canadian Alliance was rejected, he clearly saw himself as one of them, and their demonstrated inability to recognize their dogmas as insane and evil proves that he was right.

For example, Randy Thorsteinson, the last leader of the dying "Social Credit" perversion before he was forced to resign for trying to turn that cult into a branch of Mormonism, ran for electoral office under the Socred label because he considered the provincial Conservative Party, even after it had clearly become a puppet of ultra-right-wing theofascists, too liberal.

For half a century, Alberta had been a theocracy, until the last theocrat, Ernest Manning, failed to win majority support and retired before he could be voted out. A generation later, his son Preston Manning revived Social Credit as a national religion posing as a political party, under the new name, Reform Party. He made no secret of his inability to distinguish between a pre-human tadpole with zero brainwave activity indicative of human thought, and a self-aware sentient being. Nor did he even pretend to see a difference between true immorality, defined as the unnecessary hurting of a nonconsenting victim, and religious taboos that labeled as immoral behavior that was observably victimless and therefore none of the law's business.

Preston Manning was unable to win the support of moderate godworshippers, because they recognized that, if he was allowed to give sectarian religious beliefs the force of law, it was just a matter of time before all freedom to practice the belief system of one's choice was abolished. His cult consequently won little support outside of Alberta. It was the certainty that Manning could never form a government, not a rejection of policies that his hate cult fully endorsed, which led to his replacement as party leader.

The renamed Canadian Alliance replaced Manning with Stockwell Day, not because Day was less fanatic than Manning, as he certainly was not, but because Manning's theofascist status and theocratic agenda had become impossible to hide. Day, in contrast, was unknown outside of Alberta. It was at least conceivable that he might be able to win an election and institute a theocracy before his belief that humans coexisted with dinosaurs on an earth less than ten thousand years old was recognized as evidence of retarded evolution.

That did not happen. Day's inability to think like Homo sapiens, or even like the average Homo godworshipper, led to his removal as Head Hatemonger in a matter of months, not because his extremism repulsed a party every bit as extremist under Stephen Harper as it had been under Manning and Day, but again because the hate cultists recognized that a party led by Day would be wiped from the face of the earth at the first election.

So how can the kind of thinking demonstrated by the Keegstras, the Mannings, the Days and the Thorsteinsons (at least two generations of each) be harmonized with their merely being aberrant samples of the species, Homo godworshipper? I suggest that it cannot, and that they, and a large percentage of Albertans, are a different species altogether, a species once thought to be nonexistent, the species Pithecanthropus troglodytus.

# SECULAR METAPHYSICS

cartoon courtesy of *Skeptical Inquirer*, www.csicop.org

Science has banished the deities and demons, the ghosties and ghoulies, of our primeval superstitious past into the realm of metaphor where all such mythical creatures belong.

*Child of Fortune,* Norman Spinrad

# 1. Alien Visitors? I Don't Think So

From the time that humankind learned that there are other worlds, there has been speculation that aliens may have come to earth and lived among us. But not until 1947 did the possibility of alien visitations capture the public imagination. In that year a pilot reported seeing nine boomerang-shaped objects flying erratically, "like a saucer if you skip it across the water." (Sheaffer, p. 15) Misinterpreting the pilot's description, the news media promptly coined the term, "flying saucers." Once flying saucers had been introduced to the human imagination, sightings began to occur as regularly as later post-mortem sightings of Elvis Presley.

As flying saucer reports proliferated and became less and less credible, often violating Newton's laws of motion, believers fought to retain credibility by changing the flying saucers' designation to "unidentified flying objects."

Close to a million North Americans have reported seeing a UFO. Since the total population of the USA and Canada exceeds 300 million, that means that the probability of any individual seeing a UFO is one in three hundred. The probability of anyone having two sightings is one in ninety thousand (three hundred squared), three sightings one in 27 million (three hundred cubed), and four sightings one in eight billion. Yet the pilot who filed the first report continued to see UFOs almost monthly for forty years, and ninety percent of all reports come from persons with ten or more sightings.

Former American president Jimmy Carter acknowledged seeing a UFO that has now been identified with reasonable certainty as the planet Venus. The brightness of Venus makes it seem incredible that it could be further away than somewhere inside earth's atmosphere, and to a driver on a slowly curving road it will always appear to be moving horizontally in the direction opposite to the one in which the observer's vehicle is curving. Of the many lights in the night sky reported as UFOs, Venus has turned out to be the explanation for more sightings than the next two most common causes, helicopters and light planes, combined.

The author once saw a UFO. It happened in April or May of 1962, when I was a customer at a drive-in movie theatre about thirty kilometers from downtown Vancouver. A light appeared above the distant city, and quickly moved in a direction that took it away from the city without coming any closer to the observer. It traversed an arc in the neighborhood of sixty degrees in about a minute (although no attempt at accurate measurement or timing was made at the time). Since it seemed to be as far away as the city, that put its speed at a little under 2,000 km/h, far beyond the capacity of any human aircraft likely to be in the area.

Not until some years later did the author learn just how impossible it is to assess the distance of an object in the night sky. Even the awareness that Venus could appear to be closer than the horizon did not immediately suggest that an object estimated to be thirty kilometers away could in fact be within one or two kilometers. If the UFO was actually two kilometers from the observer, then the speed needed to traverse sixty degrees of arc in one minute would have been 120 km/h, consistent with the sighted object being a light aircraft. And if it was one kilometer away, that would make its speed 60 km/h, suggesting that it was probably a helicopter. While it cannot be stated with certainty that the misidentified aerial object was much closer than the observer first thought, neither is there any reason to believe that it was not.

By 1956 belief in UFOs had become so widespread that the U.S. Air Force financed a study of reported sightings for the purpose of assessing whether at least some of them might be enemy spy-craft that posed a threat to the security of the United States. That study, Project Bluebook, lasted ten years and investigated every UFO report that appeared to have even the slightest credibility. The project was terminated in 1966 when every member of the investigating committee with the one exception of J. Allen Hynek (who coined the phrase, "Close encounters of the third kind") concluded that no further investigation could be justified because there was nothing out there to investigate.

Project Bluebook traced ninety-four percent of all UFO reports to an identifiable mundane cause, including a small percentage of deliberate hoaxes. The remaining six percent were consistent with mundane explanations, even though no specific cause was identified. The investigators' *Final Report*, written by Edward Condon, should have ended UFO mythology permanently, and as far as the government, military and other agencies of the United States are concerned, it did end it. (Sheaffer, p. 271)

Unfortunately, UFO fanatics are as impervious to falsifying evidence as religious fanatics. Hynek's need to believe that UFOs must be extraterrestrial caused him to found the Center for UFO Studies (CUFOS), and other hardcore believers founded the Mutual UFO Network (MUFON). Both organizations continue to solicit alleged UFO sightings, and habitually authenticate the same claims that have been exposed as hoaxes or incompetent interpretations of the evidence in the books of Philip Klass and Robert Sheaffer. And when Jimmy Carter, after becoming President, asked NASA to conduct a new investigation of UFOs, NASA firmly declined on the ground that the nonexistence of UFOs as something other than misinterpreted mundane phenomena was already fully established. Despite the hundreds of claims of saucer landings, no alien artifact or evidence of alien presence has ever been found, anywhere, any time. Nor will any evidence be found in the future, for the obvious reason that, as the Condon report concluded thirty years ago, there is nothing to find. (see Condon, "further reading")

In recent years, every time the news media have reported large-scale sightings of spectacular aerial phenomena of the UFO type, it has taken no more than a day or two for the same media to acknowledge that the cause of the sightings was a decaying satellite or something equally mundane. As a consequence, the public has lost interest in mere lights in the sky, and believers in alien visitations have been obliged to take their fantasy to the next logical step.

The first person to claim that he had actually talked to aliens, and whose story struck a publisher as so patently ridiculous that the undiscriminating would probably lap it up, was George Adamski. Adamski, like Bernadette Soubirous, was so incapable of grasping that he was telling a story that anyone over the age of eight could recognize as childish, that he eventually convinced at least a few people that he must believe he was telling the truth.

Even by the standards of 1953, when Adamski published *Flying Saucers Have Landed*, Adamski was scientifically illiterate. Not only did his space aliens—all of whom resembled human beings—come from Saturn, Mars and Venus, two of which were even then known to be incapable of supporting protoplasmic life. He declared that, while his aliens regularly traveled to star systems reasonably close to our own, they had not gone to the more distant star systems because such a journey was likely to take two or three years. Apparently Adamski had never heard of Albert Einstein, whose equation showing why faster-than-light travel is impossible meant that, even at the maximum theoretical speed, a trip to our nearest neighboring star system, Alpha Centauri, would take more than four years each way.

Even the most ardent present-day proponents of the alien-contact hoax recognize Adamski as an embarrassment, and do not acknowledge his existence if they can possibly avoid it. The same is true of Betty and Barney Hill, the first persons to "remember" an alien abduction while "hypnotized" by a devout believer. Such persons are mentioned to show the reader that evolved abduction mythology did not spring fully formed from the imaginations of Bud Hopkins, Whitley Strieber and John Mack. Books like Hopkins' *Intruders*, Strieber's *Communion*, and Mack's *Abduction* merely continued a trend started by Adamski and the Hills and, being written by authors with rather more sophistication, avoided their predecessors' errors and inconsistencies. (Sagan, pp. 101-104)

Of the three aforementioned apostles of the aliens-are-among-us religion, only Strieber can with reasonable confidence be suspected of perpetrating a "come in, sucker" hoax, utilizing his proven talent for writing horror novels and taking the genre one step further by passing off a product of his conscious imagination as nonfiction. As Ernest Taves, in reviewing Strieber's book, noted, the only plausible alternative to viewing Strieber as a hoaxer is: "Strieber is mentally ill, suffering from delusions and hallucinations. The evidence presented does not permit us to entertain this hypothesis." (*S.I.* 12:95) In other words, since Strieber is not insane, the alternative explanation that be is lying must be the true one.

Bud Hopkins, on the other hand, whom Philip Klass described as the Typhoid Mary of UFO abductions (*S.I.* 12:85), is more reasonably viewed as a gullible but sincere incompetent who naïvely authenticated abduction tales told to him by persons who delighted in deceiving him for the simple thrill of getting away with it. Not being a psychologist, Hopkins probably did not delude himself that he had a professional capacity to detect when a source was lying. But he did accept at face value, without making any attempt to check details that could have been falsified, virtually every abduction tale told to him by any person capable of keeping a straight face. And he actually concluded that anyone who had had a nightmare as a child involving odd-looking bogeymen, or who had lost track of time while engrossed in a task that made an hour seem like a few minutes, had almost certainly been abducted by extraterrestrials. By Hopkins' criteria, it would not be difficult to prove the probable existence of the tooth fairy.

From the scholastic point of view, Strieber and Hopkins were nobodies, laymen whom the academic community refused to dignify by acknowledging their existence. John Mack, in contrast, is a psychiatrist, an MD. His endorsement of the reality of alien abductions could not so easily be dismissed as the speculations of the uneducated. But like Hopkins, he believed that any tale told to him without smirking must be true. And unlike Hopkins he imagined that, as a trained professional in the field of human behavior, he could not be deceived by lies. That conceit is probably what made it so easy to deceive him.

Mack's book stands or falls on his claim that the alien encounters described to him by patients must be true, because if a patient lied he would have detected it. That conceit was annihilated when one of his alleged abductees, Donna Bassett, informed a 1994 Seattle convention of the Committee for the Scientific Investigation of Claims of the Paranormal (CSICOP) that she had fed Mack a pack of lies about being abducted, for the purpose of determining whether he utilized any valid methodology for detecting inaccuracies or falsehoods. An unidentified attendee at the convention rightly remarked that, "The fact that Mack had not discovered Donna Bassett as a fake called into question his whole methodology." (*S.I.* 19:112) But the most preposterous consequence of Bassett's revelation was that, after learning that she had deceived him, Mack suggested to the convention that she really was an abductee who was now pretending to have lied about a real experience. He attributed her allegation to the psychobabble concept, "denial."

Nonetheless, Mack's gullibility does not in itself prove that none of his patients' stories could possibly be true. But that conclusion becomes mandatory, at least to the point of rejecting everything in Mack's book as unreliable, when the details told by the patients and accepted by Mack as not impossible are considered. For example, several patients claimed to have been transferred from their beds into a waiting spaceship by floating through solid walls. While many of today's impossibilities may be within the realm of future technology, solid-through-solid is not one of them. Four patients described time-travel experiences. One recalled meeting the gods of ancient mythology and a winged horse. One claimed that his body had transmuted into light, in imitation of the similar transformation described in Matthew 17:2 in the Christian Bible. Several reported remembering past lives. Seven claimed to have multiple personalities. One had a childhood imaginary playmate who was an extraterrestrial. And one recalled *being* an alien who spoke in the stilted, robotic tones of *Star Trek*'s android, Data (*S.I.* 20:3:18-20, 54). That Mack could report such claims in a book that he alleged to be an account of true experiences raises the question whether alien-contact mythology really has changed qualitatively since George Adamski.

There is not now and never has been any credible evidence that intelligent lifeforms from other worlds have visited earth. As Carl Sagan so often asked, "Where is the evidence?" Yet the public continues to accept, not only that flying saucers have landed, but that governments, particularly the United States government, know about it and are covering it up. And that absurd belief is daily fostered, cultivated and perpetuated by producers of television fiction who salt fantasy series like *The X Files* and *Dark Skies* with real-world references, out of fear that they would lose ratings if viewers were allowed to know that the shows' fantasy concepts have no parallel in reality.

Movies and television have been presenting space opera for decades, and until recently such shows made no pretence to be anything but imaginative fantasy. Even when *Star Trek* had three Ferengi and a shapeshifter time travel to 1947 Roswell and get captured by the Air Force before escaping back to the future, not a single viewer of that episode could have imagined that he was seeing a plausible guess of what "really" happened. Repeated references to Roswell on *The X Files*, in contrast, are deliberately slanted to generate the belief that the U.S. government really is covering up a crashed UFO, that producer Chris Carter has secret sources, and if viewers keep watching he will eventually reveal the "truth" that for some inexplicable reason his government is suppressing. If Carter believes that, he is an ignoramus. If he encourages viewers to believe it without believing it himself, he is a prostitute. The sad part is that such tactics are unnecessary. On balance, *The X Files* is excellent fantasy (as opposed to science fiction, which anticipates future developments that are not definitively impossible), and few if any viewers would stop watching if Carter acknowledged that his scripts have no more real-life parallels than the horror fantasies of Stephen King.

*Star Trek*, while incorporating fantasy elements, is nonetheless legitimate science fiction, and its producers have never tried to buy ratings by pretending that such violations of natural law as faster-than-light travel, teleportation, time travel and shapeshifting might actually be possible. Even so, *Star Trek* can be identified as the source of many of the contrary-to-fact beliefs that have become fashionable in recent years. Before, "Beam me up, Scotty," no one would have expected to be believed if he concocted a fantasy of being floated bodily through a solid wall. Without decades of being subjected to "warp drive," "hyperspace," "subspace," and other literary devices invented for the sole purpose of getting around the speed-of-light limitation, the masses would never have swallowed tales of alien astronauts commuting between earth and a home planet more than fifty light-years away at speeds that simply cannot exist. Without "Mr. Spock" as a role model, not even the tabloids would have endorsed human-alien babies, a fantasy more impossible than the crossbreeding of a human and a broccoli, which at least share common nucleotides and sub-proteins.

The impossibility of extraterrestrials resembling humans was spelled out by biochemist Isaac Asimov twenty years ago. He pointed out that each human cell contains 25,000 different genes. Each gene consists of some 400 trinucleotides, and each trinucleotide could be any of 64 varieties. The total number of genes that could be formed is therefore $64^{400}$, which equals $3 \times 10^{722}$. But the number of genes that actually exist in the universe, calculated by a formula designed to give an impossibly inflated figure, is less than $3 \times 10^{63}$.

That means that any particular gene has a probability of being formed of $10^{63}/10^{722}$, which is $10^{-659}$, in layman's terms, zero to 659 decimal places. The probability of *some* genes forming, in contrast, is close to certainty. And while not every one of the $3 \times 10^{722}$ different genes that could form is certain to contribute to some form of life, it is entirely possible that it could do so. The probability of life-forming genes evolving elsewhere in the galaxy is therefore virtual certainty, regardless of how different those genes may be from human genes. But the probability of the same gene evolving in two places independently (earth and elsewhere) is $10^{-722}$, zero to 722 decimal places. And that is just one gene out of 25,000.

Chimpanzees also have 25,000 genes, of which 24,600 are identical with those of humans. For an extraterrestrial to have a closer resemblance to humans than does a chimpanzee, it would need to have more than 24,600 identical genes. The probability of that is $0^{24,600}$, or more precisely,

$10^{-18,000,000}$. That is about equal to the probability of random molecular activity causing water in a kettle on a hotplate to freeze instead of boil. Asimov concluded (*Life and Time*, p. 15):

> Considering in how many different ways life developed on earth and how many hundreds of thousands of different species formed, it seems unlikely that a similar wild variety would not form there, and it would be an almost impossible chance to have a species there closely resemble some species here.

If, instead of Ferengi, Cardassians, Klingons, Vulcans, Romulans and other extraterrestrials who look like exactly what they are, humans in special-effects makeup, *Star Trek* viewers had been exposed to three decades of aliens who all resembled canines, John Mack's patients would have described being taken aboard the mother kennel.

Albert Einstein, in the process of determining what happens to a body's mass at relativistic speeds, came up with the equation:

$$ m = \frac{m_o}{\sqrt{[1 - (v/c)^2]}} $$

That equation, in which "v" is velocity, "c" is light-speed, "$m_o$" is rest mass, and "m" is relativistic mass at any given speed, has been validated by experimentation on several occasions. At light-speed, v = c, so that $(v/c)^2 = 1$. The denominator thereby becomes zero, making the moving body's mass equal to rest mass divided by zero, which equals infinity, an absurdity. At faster-than-light-speed, the denominator becomes an imaginary number, also an absurdity. Of all the reasons why extraterrestrials cannot be visiting earth, the proven impossibility of their attaining or exceeding light-speed is the most insurmountable.

Even an electromagnetic message from the nearest star system would take over four years to reach earth. But there are no technologically advanced alien civilizations that close. Humans have been sending out radio transmissions sufficiently powerful to be detectable at a distance of fifty light-years for a century. That means that, if there was a civilization sufficiently advanced to detect such signals within a radius of fifty light-years, they would have had time to send back a response by now. Since that has not happened, we must conclude that, while the nearest aliens capable of space travel could be as far away as the other side of the galaxy, or even in another galaxy, they could not be as close to us as fifty light-years.

That means that a starship sent to earth would take at least fifty years to get here—in theory. In practice, a spaceship could only accelerate by expending fuel. But the fuel a ship carried would be part of its inertial mass, so the more fuel it carried the more slowly it could be accelerated. The fuel would be gone by the time it reached a tiny fraction of light-speed, and starting with more fuel would not change that. Robert Jastrow, author of *Journey to the Stars*, has calculated that the fuel problem would make acceleration beyond one percent of light-speed impossible, or two percent if no fuel was saved for deceleration and the ship planned to stop by crashing into earth without slowing down.

That is where the impossibility of star travel becomes manifest. Even from Alpha Centauri, a one-way trip to earth would take five hundred years. From the closest theoretical location of a civilization capable of undertaking such an adventure, it would take five thousand years. To be here now, they would have had to leave home millennia before they had any way of knowing that our solar system has any planets smaller than Jupiter, let alone any intelligent inhabitants.

There are almost certainly intelligent lifeforms elsewhere in the galaxy, and SETI, the Search for Extraterrestrial Intelligence, may eventually find them. But the most we can hope for is sure knowledge of their existence. Even two-way communication would involve at least a hundred-year delay between sending a message and receiving a response. Learning enough of each other's language to communicate meaningfully would take millennia. We might some day hear from extraterrestrials, but we will certainly never meet any.

## Further reading

*Skeptical Inquirer* (cited as *S.I.*)
Isaac Asimov, *Life and Time*, NY, ppb.
    *Past, Present and Future*, Amherst, 1987.
    *The Roving Mind,* Amherst, 1997.
Edward Condon, *Final Report of the Study of Unidentified Flying Objects,* NY, 1966.
Kendrick Frazier *et al.*, *The UFO Invasion*, Amherst, 1997.
Philip J. Klass, *The Real Roswell Crashed-Saucer Coverup*, Amherst, 1997.
    *UFO Abductions: A Dangerous Game*, Amherst, 1988.
Terry Matheson, *Alien Abductions: Creating a Modern Phenomenon*, Amherst, 1998.
Carl Sagan, *The Demon Haunted World*, NY, 1995.
Robert Sheaffer, *UFO Sightings: The Evidence*, Amherst, 1998.

## 2. Psychics and Other Time Travelers

Despite deliberate distortions of quantum physics theory by the news media for the purpose of pretending that scientists are no longer convinced that time travel is impossible, and despite seeing time travel presented favorably on *Star Trek* and other fantasy shows, the public remains acutely aware that time travel *is* impossible. Yet while almost no one believes that the movement of physical objects backward in time is plausible, enormous numbers believe that information can travel backward in time, as it would have to do to implant itself in the minds of "prophets" or "psychics."

Time travel cannot exist. At first glance that statement might seem untestable. But that is not so. There is a technique used by mathematicians and others to prove negative hypotheses, called *reductio ad absurdum*. The method involves assuming that a hypothesis is wrong, and then following that assumption to see whether it leads to a logical impossibility.

For example, in order to prove that the square root of two is not a rational number, meaning a number that can be written in the form a/b, "a" and "b" being whole numbers, mathematicians assumed that root-two *is* a rational number and constructed the equation consistent with that assumption: $2 = a^2/b^2$. They were then able to prove that "b" is simultaneously an odd number and an even number. Since that conclusion is absurd, it followed that the initial assumption from which it sprang, that root-two is a rational number, must be false. *Q.E.D.* (Sagan, 1980, p. 347)

Similarly, the assumption that time travel is possible leads to a comparable paradox or absurdity. Suppose that a time traveler went back into the past and killed his father before he met his mother. The time traveler would then not be born and therefore could not go back in time. He therefore could not prevent his own conception and so *would* be born. He would then be enabled to go back and kill his father and...The assumption that time travel is possible leads to a reality in which the time traveler is simultaneously born and not born. Since that is an absurdity, time travel is therefore shown to be an absurdity.

If that is not enough, consider the ultimate time travel paradox. Imagine a man journeying back to the moment of his conception, replacing his father in his mother's affections, and becoming his own father. Robert Heinlein wrote such a paradox story, in which a woman became her own mother. She also became her own father, but that detail involved a second fantasy element of functional (as opposed to cosmetic) sex change that is not relevant to the current issue.

Matter cannot travel backward in time. Proving that information cannot travel backward in time requires a slightly different approach. Suppose a method were discovered tomorrow, or next year, of sending a message back. Does it not logically follow that someone would send back the necessary information to prevent the assassination of John Kennedy? or prevent the events that led to the enforced resignation of Richard Nixon? or prevent World War Two? The argument that those events are part of our current reality because the method has not yet been found to prevent them retroactively does not stand up. Even if the ability to send messages through time were not discovered for a million years, as soon as it was it would change *our* past. That the past has not already been changed means that it can never be changed, because if it could be it would have already happened, and we would remember a history with no World War Two, a sixteen-year presidency of John Kennedy, a Nixon who died in well-deserved obscurity, and an Indian earthquake that killed no one because the entire population had been evacuated before it happened. In addition, a Standard Terran language would already have won official recognition by the United Nations, because the creators of Esperanto would have been warned to base it on English, since an artificial language based on the Romanic languages had proven to be useless.

Objectors will argue that psychics did receive the messages from the future, but that no one was willing to believe them. To this day, believers in the unlamented psychic humbug (tautology) Jeane Dixon continue to accept the pretence that she prophesied the Kennedy assassination, as indeed she did—after it had happened. The present author can do that, too. I hereby prophesy that Hitler will lose World War Two. Now was I right or was I right? (see Bringle, "further reading") Every January, professional psychics make predictions for the coming year that are published in *National Inquirer* and elsewhere. A year later, *Skeptical Inquirer* reprints those prophecies and draws attention to the fact that no intrinsically improbable prophecy has ever been fulfilled, and those for which a degree of success could be claimed were deliberately kept vague so that almost any outcome could be claimed as a hit. As an obvious example of the latter: "There will be a disastrous earthquake in 2003." Since there is an earthquake somewhere on earth practically every day, and "disastrous" can be interpreted to include a single death, such prophecies require as much knowledge of the future as a prediction that the sun will rise tomorrow.

Prophecies that were touted by the tabloids at the beginning of 1995, and were debunked in *S.I.* 20:1:5-6 include:

— a child genius will invent a working time machine, made out of parts of a microwave oven.
— scientists will discover a virus that can turn rocks into edible protein.
— President Clinton will be shot and wounded by a disgruntled postal worker.
— a volcanic eruption will create a land bridge between the United States and Cuba.

To discover further failed prophecies, it is only necessary to check any back issue of a tabloid with a "prophecy" theme. The above are merely typical.

What no psychic ever predicted was the Oklahoma City bombing, the World Trade Center atrocity, the Rabin assassination, the collapse of the Soviet Union and east European communism, or anything that was not a logical consequence of events already in progress. Indeed, some would say that the collapse of the Soviet Union *was* a logical consequence of events already in progress, and even so no psychic was able to deduce that it was predictable.

Asked if he believes that information can travel backward in time, virtually every person with a functioning human brain will answer, "No." Yet many of those same persons accept the existence of real psychics, as opposed to cranks and humbugs who falsely claim to be psychics, without any awareness that they are expressing simultaneous belief and disbelief in a single phenomenon.

Public figures in occupations not requiring scientific literacy are notorious for believing that the future can be foreknown. Among the political figures who allowed themselves to be influenced by paranormal advisers who claimed to have access to information that had traveled backward in time were Ronald Reagan, whose belief in a geocentric universe revolving around an immobile earth was an embarrassment even to other creationists, and Princess Diana, whose intellectual capacity can perhaps be gauged by her failing all of her O-level classes, the British equivalent of grade eleven.

As for tabloid TV programs such as *Extra, Inside Edition* and *Entertainment Tonight*, which routinely report the psychic and analogous beliefs of show business personalities as if a claim to have conversed with a dead former spouse had as much credibility as a claim to have been on Richard Nixon's enemies list, the question of whether the perpetrators are gullible simpletons who cannot separate fact from fantasy, or amoral story tellers to whom "truth" is whatever the rubes will swallow, is best answered: Probably both. Indeed, unless the gossip media are deliberately publicizing the weirdos and ignoring a rational majority, one is justified in concluding that practically everybody in the entertainment industry, on both sides of the camera, is a gullible simpleton.

Psychics are humbugs. A year-long study of claims of psychic crime solving led to the conclusion (Nickell, 1994, p. 173) that, "careful examination reveals no successful crime solving, but instead only tangled webs of misinformation, generalization, opportunistic credit-taking, and, in some instances, probable deceit."

No psychic has ever given police useful information about a crime (*S.I.* 17:159-165). Television's ongoing pretence to the contrary can be attributed to that medium's willingness to place ratings ahead of truth. In the words of entertainer Peter Reveen (p. 114), who spent forty years presenting simulated psychic phenomena in his performances and seeking for persons who could replicate his demonstrations without resorting to magicians' tricks: "I am forced to conclude that there is no such thing as a psychic...and in all my years of touring the world I have never met anyone who does have such powers."

There are two reasons why tabloid addicts believe in psychics. The first is their refusal to recognize that, for psychics to have foreknowledge of the future, that knowledge must have traveled backward in time. The second is that, if they acknowledge the impossibility of psychics receiving information from the future, they would then have to acknowledge that what is impossible for psychics would have been equally impossible for biblical prophets.

While religious fundamentalists will object, liberal theologians of all persuasions are agreed that their sacred writings were compiled by fallible human authors whose ability to tell the truth as they saw it was limited by their scientific literacy and cultural background. All but the most unsophisticated pseudo-scholars have allowed themselves to recognize that their Bibles contain only two kinds of prophecy: those that failed, and those that were already fulfilled at the time of writing.

Shortly after the death of King Solomon, a Jewish theologian put into the mouth of his god Yahweh a prophecy to the patriarch Abraham that Abraham's descendants would in time conquer and occupy all of the territory that constituted the nations of Israel and Judah. Since the Jews and Israelites already occupied all of the land in question at the time the prophecy was composed, the probability of the theologian being proven wrong was minimal, to say the least. To moderate believers, the retroactive nature of such prophecies is not a problem. But to the intestinally challenged, belief in the literal truth of a Bible that promises them eternal life is the only thing that suppresses their terror of death and gets them through the day without losing control of their bodily functions.

The "promised land" was not, of course, the Bible's only retroactive prophecy. One of the authors of the Book of Daniel ostensibly prophesied that the empire of King Nebuchadnezzar would be followed by four evil successors, and finally replaced by a kingdom of Saints, meaning the Maccabee priest-kings. Again, since the Maccabee revolution was already in progress at the time of writing, 164 BCE, the alleged prophet's powers of prognostication were not seriously challenged. (see Callahan, "further reading")

Virtually every biblical prophet gave his ultimately failed predictions a spurious credibility by pretending to have written years earlier than the time he actually lived, and by preceding his guesses about the future with "prophecies" of events that had already happened. The reason the prophetic section of Daniel can be so precisely dated is that all events prior to 164 BCE were "prophesied" correctly, while no event later than that date was accurately foreseen.

Not all biblical prophets were content to write books from which they could not personally benefit. Richard Friedman, a professor of religion at the University of California, San Diego, presents convincing evidence that the Book of Deuteronomy, discovered behind a loose brick in the Jerusalem temple in 621 BCE, was written by the prophet Jeremiah. And Deuteronomy contained a detailed prophecy put into the mouth of Moses (Deuteronomy 18:15-19) that there would one day be a new prophet who would be Moses' successor and equal. At the time Deuteronomy was written, the only person who could have plausibly claimed to be that prophet was Jeremiah himself. Unfortunately for Jeremiah, his attempt at self-glorification failed. The Jewish king imprisoned him as a Babylonian collaborator, and the Babylonian king eventually forced him to flee to Egypt, from where he never returned. (see Friedman, "further reading")

Of the Bible's countless failed prophecies, the most notable is the one put into the mouth of Jesus by the anonymous author of the gospel known as Mark (9:1): "There are some standing here who are not going to experience death until they have seen Allah's theocracy established by force." Since Jesus died in 30 CE, and his hearers could have included children capable of living a further ninety years, we can calculate that he promised to overthrow the Roman empire and be crowned king of an independent Judea no later than 120 CE, early in the reign of the emperor Hadrian. He seems to have been delayed. What neither the Jesus of history nor the reconstructed Jesus of Mark ever prophesied was that his triumph would happen only after an intervening death and "second coming." In a desperate attempt to pretend that Jesus' failed prophecy could still be fulfilled, medieval Christians invented the "Wandering Jew," a man who taunted Jesus on the cross and was subsequently cursed by him to remain alive until the alleged second coming. Few people today believe in the Wandering Jew, but the Christian church has never dared repudiate him, because that would necessitate recognizing that Jesus' prophecy is indeed unfulfillable, its deadline having passed more than 1,800 years ago.

The best known other failed prophecy is to be found in the Book of Revelation. Revelation's inclusion in Christian Bibles is supremely ironic, since neither of its authors was a Christian. The first author, writing between the Roman occupation of the Jerusalem temple in July of 70 CE and the razing of the temple in August of 70 CE, was an Essene Jew who viewed the Nazirites, Jews who regarded Jesus as their messiah, with suspicion. The final redactor, John of Patmos, was a Nazirite, a Jesus-Jew who regarded the Christians as, "those who call themselves Jews and are not, but are a synagogue of the satan." (Revelation 2:9)

The earlier author, the Essene, promised the Jews fighting for independence from Rome that the final battle of the war would take place at Armageddon, north of Jerusalem, and the Jews would win. As any historian can confirm, the final battle took place at Masada, south of Jerusalem, and the Jews lost.

No one can know the future, even a god—unless the future is predetermined and there is no such thing as free will. Any person who agrees with psychologist B. F. Skinner that genetic programming, whether implanted by a god or by blind chance, determines whether an individual becomes a philanthropist or a serial killer, and that free choice plays no part in his pseudo-decision, can expect to be condemned by the religious and the rational both. But if, in contrast, the future depends on decisions not yet made and accidents not yet caused, then foreknowledge of the future is by definition impossible. To believe otherwise is to believe that an effect (knowledge of an event) can precede its cause (the event foreknown). It follows that there is no such thing as a psychic and never will be.

Further reading

Mary Bringle, *Jeane Dixon: Prophet or Fraud?*, NY, 1970.
Tim Callahan, *Bible Prophecy: Failure or Fulfillment?*, Altadena CA, 1997.
Milbourne Christopher, *ESP, Seers and Psychics*, NY, 1970.
Richard Friedman, *Who Wrote the Bible?*, NY, 1987.
William Harwood, ed, *The Judaeo-Christian Bible Fully Translated*, Charleston SC, 2002.
Joe Nickell, ed, *Psychic Sleuths: ESP and Sensational Cases*, Amherst NY, 1994.
Peter Reveen, *Hypnotism Then and Now*, Charleston SC, 2002.
Carl Sagan, *The Cosmic Connection*, NY, 1980.
    *Cosmos*, NY, 1989.

### 3. Hyperspace: It's Not a Free Lunch

Ever since Einstein provided mathematical proof, with a vanishingly small probability of error, that no particle of non-zero mass can ever reach an observable speed equal to or greater than that of photons through a vacuum, science fiction writers have circumvented that limitation by having their starships travel through a speculated *hyperspace* (or *subspace*, or *warp space*), not subject to the laws of three-space.

Hyperspace is not itself a science fiction concept. Physicists have postulated a universe in which observable three-dimensional space is merely the surface of a four-dimensional hypersphere, analogous to the two-dimensional surface of a regular sphere. Just as the distance from Borneo to Brazil would be reduced by one-third, and the distance from San Francisco to New York by one-thirtieth, if we could utilize the third dimension and travel through the earth rather than being confined to its two-dimensional surface, so the journey to Alpha Centauri or Betelgeuse would also be proportionately reduced if we could travel through the fourth dimension of hyperspace.

Science fiction writers utilize hyperspace. But none pretends that it is more than mere speculation. Isaac Asimov wrote in *The Roving Mind* (pp. 173-174), "But now that we know what hyperspace is and why science-fiction writers use it, the next question is: Does hyperspace exist in reality? Unfortunately, as far as we know, it does not…There is no evidence for its real existence—at least so far."

Arthur Clarke wrote in *Profiles of the Future*, "Two points with a certain separation in 3-space will still have at least that separation in any higher space. If, however, we imagine that space can be bent or curved, so that the axioms of Euclid no longer apply, then some interesting possibilities arise."

Hyperspace, according to its theorists, is flexible. The presence of mass distorts it in the same way Einstein hypothesized that it distorts three-space to create the phenomenon we measure as gravity. But distortion is directly proportional to mass: the smaller the mass, the smaller the distortion. Nowhere in the hypersphere hypothesis is there any justification for imagining that hyperspace may be folded like a blanket (figure 1) or arbitrarily crumpled (figure 2), so that two points (A and B), widely separated on the line that represents three-dimensional space, may be adjacent in hyperspace.

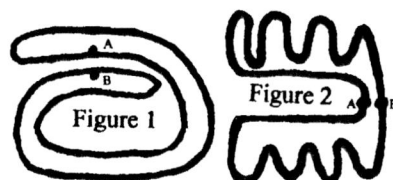

In the universe we see, gravity or spatial curvature tends to force massive bodies into the shape of a sphere. Only when the mass is much lower than that of earth's moon do we get such aberrations as the potato-shaped satellites of Mars. And only when the oblating effect of centrifugal force is extremely high do we see disk-shaped galaxies in which the polar axis is significantly shorter than the equatorial axis. It seems reasonable to suppose that, with perhaps slight modifications, the four-dimensional universe is subject to the same generalizations we have come to regard as the laws of physics in the three-dimensional universe.

The formula for the circumference of a circle is *pi* x diameter. That does not change for a sphere, and it is a logical extrapolation that it will not change for a hypersphere. The distance between the points of maximum separation on a circle or a sphere can be reduced by no more than one-third by taking the direct route through the next dimension, and that will be equally true for a hypersphere. If the distance from Quasar A to Quasar B through three-dimensional space is twelve billion light-years, the distance through hyperspace cannot be less than eight billion light-years. A distance of fifty light-years is so tiny by cosmic standards, that hyperspace could not reduce it by more than a tiny fraction of one percent.

So hyperspace, even if it exists, barely qualifies as a short cut. An interstellar voyage at near light-speed, that would take a human lifetime through the three-dimensional universe, will still take a human lifetime through hyperspace. Converting an impossible journey into a possible one, hyperspace cannot do. We are as effectively bound to the immediate vicinity of Sol if hyperspace exists as if it does not. That means that persons who defend their belief in alien visitors by arguing that the visitors used hyperspace are deluding themselves. A trip to earth from fifty light-years away would still take over fifty years, even through hyperspace. There is no free lunch.

But that does not illegitimize hyperspace as a science fiction convention. Without it there could be no *Foundation*, no *Dune*, and no *Star Trek*. And that would be unthinkable.

# THE BIBLE:  IT AINT NECESSARILY SO

Several of the articles in this section, mainly those backed up by evidence that no reader with a functioning human brain could possibly dispute, were first submitted to theological or religious studies journals. *Journal of Biblical Literature* rejected (8), "Jesus the Nazirite," responding, in effect, "The historicity of the Lord Jesus is not in dispute, and only dirty atheists do not know that."  The editor of *Journal of Jewish Studies* rejected an early draft of (34), "Moses and De Mille:  The Parting of the Red Sea," declaring that an article denying Moses' authorship of the Torah might be more acceptable to a "more liberal" journal.  And the only truly fundamentalist religious journal, *Flat Earth News*, rejected (4), "The Inconsistency of Round Earth Religionists," since, in their eyes, the reality that their bible endorses a flat earth did not need re-proving.  Only after I had satisfied myself that the editors of *all* such publications had obtained their appointments by proving that they were rationally bankrupt, no more capable of evaluating evidence that falsified their mythology than were the Inquisitors who refused to look through Galileo's telescope, did I start submitting articles to humanist, rationalist, skeptical and freethought publications.  The "Publications" list at the back of this book indicates where each article first appeared in print.

All biblical quotations, except where otherwise stated, are from the author's *The Judaeo-Christian Bible Fully Translated* (Imprintbooks, 2002).

# MAP OF THE UNIVERSE
# ACCORDING TO THE JUDAEO-CHRISTIAN BIBLE

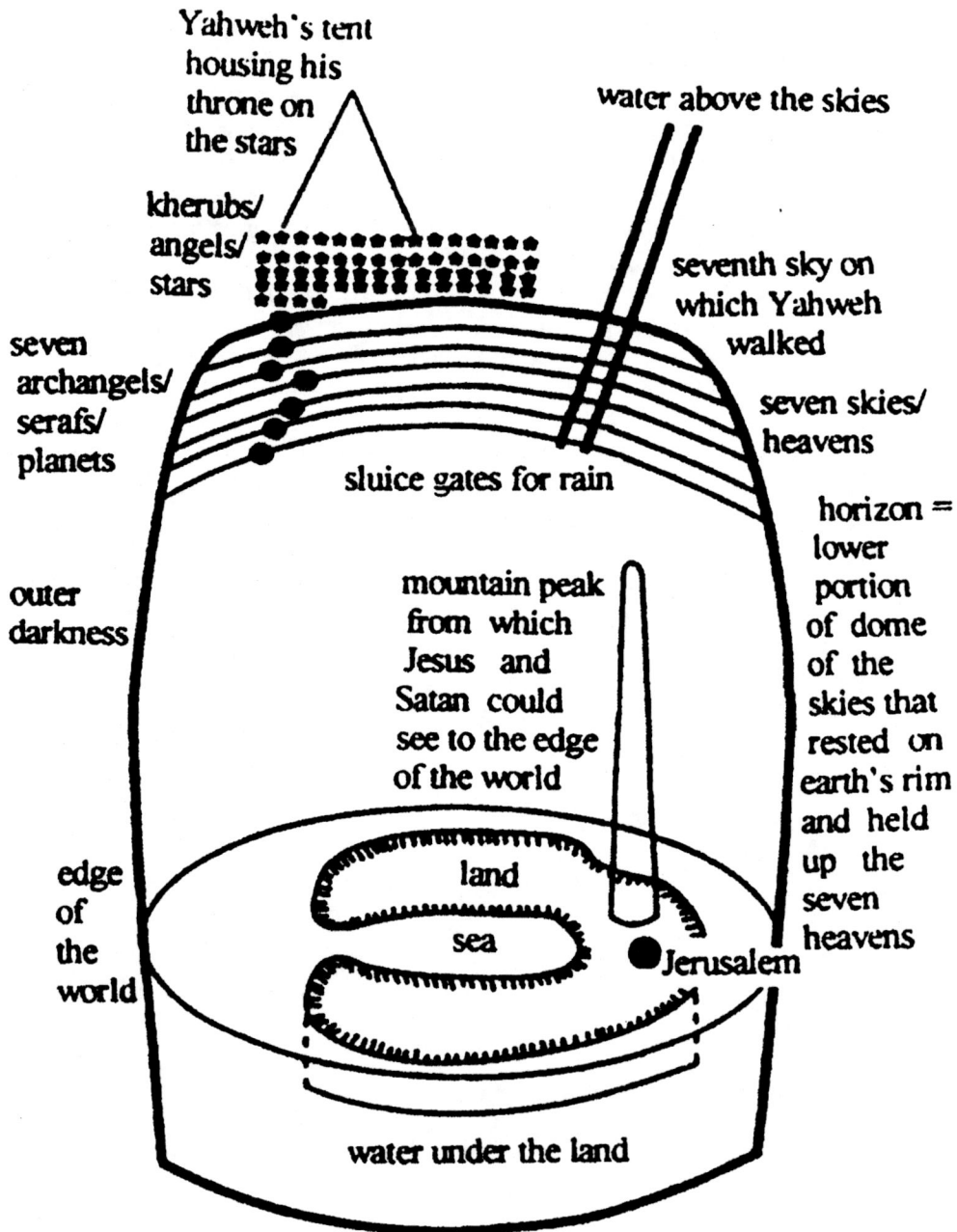

Yahweh's tent housing his throne on the stars

water above the skies

kherubs/ angels/ stars

seventh sky on which Yahweh walked

seven archangels/ serafs/ planets

seven skies/ heavens

sluice gates for rain

horizon = lower portion of dome of the skies that rested on earth's rim and held up the seven heavens

outer darkness

mountain peak from which Jesus and Satan could see to the edge of the world

land

sea

Jerusalem

edge of the world

water under the land

# 1. God and Santa Claus: A Letter to a Little Girl

Yes, Virginia, there is an entity that lives forever, rewards and punishes, and is not bound by the laws of reality. It lives in the imaginations of children, where it is called Santa Claus, and in the imaginations of childish adults, where it is called God. Belief in Santa Claus enables the chronologically handicapped to escape into a fantasy world in which they receive presents and grown-ups do not. Belief in God enables the intestinally handicapped to escape into a fantasy world in which they live forever and their imagined enemies do not.

Santa Claus rewards good little boys and girls who obey everything their parents tell them, no matter how capricious, and believe everything their parents tell them, no matter how absurd, and ignores those who disobey. God ignores those who obey and believe everything their priests, ayatollahs, gurus and dead lawgivers tell them, and sics his vigilantes onto those who do not contribute to the extermination of the human race by overpopulation and starvation in obedience to laws invented by persons who lived and died at a time when the world was underpopulated.

So, Virginia, we should not feel contempt for children who believe in fairy tale creatures. It is not their fault they lack the maturity to recognize that flying reindeer can live only in the world of the imagination. And we should not feel contempt for Peter Pans who believe in adult mythology. It is not their fault they lack the intestinal fortitude to face the reality that death is forever, and that entities such as God and Captain Kirk who regularly violate the laws of reality can live only in fantasy literature. Yes, belief in gods has been used to justify fifty million murders in the western world alone. But do you really think the perpetrators of such atrocities as the Crusades and the Inquisition would not have found some other justification if they had not had a god to blame? What god was Stalin obeying?

So you go on living in your imaginary world, Virginia. And if you never outgrow it, and simply replace Santa Claus with God, that will not be your fault. Either you were born educable or you were not.

## 2.  Bible Belt or Loony Bin:  Is There a Difference?

For over fifteen years I have resided in the redneck navel of Canada's most redneck province, where in four straight provincial elections a majority have chosen to be represented by a religious fundamentalist, one Stockwell Day, currently **[deposed March 20, 2002]** federal Opposition Leader, who is on record as believing that dinosaurs roamed the earth six thousand years ago, women who are impregnated by rape should not be provided with publicly funded abortions, homosexuals are not entitled to equal treatment under the law, Canada's federal government wasted taxpayer money by funding AIDS research, capital punishment should be restored, and when a lawyer defends a person charged under a child pornography law that the British Columbia Supreme Court overturned because it would have criminalized everyone who owns a magazine containing a Coppertone™ ad, the lawyer thereby commits a morally reprehensible act.  (Day's letter to the Red Deer *Advocate* expressing that last opinion led to his being sued for libel and settling out of court at a cost to Alberta taxpayers of $792,064.40.)  But not until I got fed up with the *Advocate*'s blatant and repeated religious propaganda and wrote a letter to the editor pointing out some of the absurdities and self-contradictions on which religion survives, did I find out first hand just how incapable of rational human thought Bible Belt rednecks really are.

I started by pointing out to security belief addicts that, "If their imaginary playmate has the omnipotence to prevent such evils as disease, famine, wars, natural disasters and transportation accidents and chooses not to do so, then it must itself be evil."  That is surely self-evident.  But one attempted rebuttal asked, "Has Harwood considered the possibility that evil may be playing a constructive role in the universe, and that an omniscient God may have chosen to use it for benevolent purposes not fully known to us?"  My answer to that is No—and neither has anyone else with a functioning human brain.  Another letter writer's reaction was that my stating facts of history that persons comfortable with their mythology prefer not to know "strike me as wandering very close to blasphemy."  He apparently does not grasp that *blasphemy* may be religion's most absurd concept.  A god who can feel insulted by a human is analogous to a chicken farmer who can believe he has been insulted by a chicken.

My letter continued, "No astronaut who has looked at the earth from an orbiting space shuttle can retain the belief that the earth is flat, as his Bible says it is (in seven places), because to do so would require a kind of mind amputation."  Among the responses to that, one tried to prove the bible authors believed the earth is round by quoting one of the very passages that prove the opposite: "'It is he that sitteth upon the circle of the earth.'  Surely a circle is not flat."  Correctly translated, the passage reads, "He has been sitting on the domed roof of the land."  That the correspondent cited an English translation with no awareness that the Hebrew might be quite different comes as no surprise.  This was the same person convicted of promoting hatred against Jews by telling his high school students that the hundreds of thousands of survivors of Hitler's death camps are all liars who have perpetrated a successful hoax for over fifty years without a single one breaking ranks and admitting they made it all up.  He justified Satan taking Jesus to the top of a mountain to show him all of the inhabited earth by explaining that, "You can look much further if you are up."  That my point was that it is physically impossible to see the entire surface of a sphere from *anywhere* apparently escaped him.

I wrote, "No historian who has verified that fifty other virgin-born savior gods rose from the dead on the third day centuries and even millennia before Jesus can believe that the first fifty such tales were myths but on the fifty-first (approx.) retelling fantasy became history."  The same holocaust denier asked, "How come neither I nor anyone else has ever heard of these fifty other virgin-born savior gods, but most if not all of us have heard of Jesus Christ?"  I interpret that as a declaration that, if he has never heard of a fact of history, then it did not happen. Is any other interpretation possible?

I wrote, "No microbiologist who has compared the 98.5 percent similarity of human and chimpanzee DNA, or the significantly higher-than-chance similarity of human and bacterial DNA, can continue to believe that species were created independently, as his Bible tells him they were, rather than evolving from common ancestors." Since a letter writer named three scientists who do not accept the reality of evolution, perhaps I should have said, "No *thinking* microbiologist." That same writer, while agreeing that an astronaut cannot believe the Bible's flat-earth cosmography, argued that, "It does not, however, prove that God does not exist." That the existence of the god of western religion stands or falls on the veracity of a Bible that assures its readers that species were created independently, and that the earth is flat, must have escaped his notice.

Interestingly, not a single letter writer commented on my citing the reality that the five Torah authors were all henotheists, believers in many gods but partisans of one; that Ecclesiastes specifically denies that there is life after death; or that omnipotence and omniscience both involve *reductio ad absurdum*s and therefore cannot exist.

Most of the dozen or more letters that attempted to dispute my *facts* by denigrating them as one man's *opinion* merely nauseated me, so much so that I did not keep copies, and had to go to the library's microfilm files when I decided to write this article. But more than one really offended me. At least three felt sorry for me for lacking the ignorance that keeps them in contented bliss, and informed the paper's readers that they were going to pray for me. My response was to thank one of them for her kindness, and assure her that, in exchange, I would ask the high priestess of the Church of Wiccan in Salem, and the high priest of the Church of Satan in San Francisco, to pray for *her*. But the letter that angered me most was from a person who, after stating that I "made mockery of every tenet of the Christian faith," then added, "I am not a Christian, but I was also offended by his letter." This from a man who praised my detractors for their forbearance in not adopting the "kill the messenger" reaction that is apparently an acceptable response to unwelcome verifiable information wherever he comes from. Even by the standards of Alberta's redneck majority, such a person disgusts me.

# 3. Biblical Nonsense

Most believers in religion are aware that their Bible contains tales that violate the laws of reality: A virgin gave birth to a son. A dead man came back to life and later brought other dead people back to life. A miracle worker fed thousands with a handful of bagels and a few fish. Moses parted the Sea of Reeds (or Red Sea in English mistranslations). The walls of a fortified city fell down when Joshua blew a trumpet. A woman past 90 gave birth. Eve conversed with a talking snake. Given the success of preachers in assuring their congregations that such impossibilities happened, one would think that no biblical fable could create such a credibility gap that religion pushers would make a point of never mentioning it, so that the mass of True Believers never learn that their Bible contains such stories.

One of the most obvious examples can be found in Numbers 22:21-35. Balaam, a spokesman (usually mistranslated as *prophet*) for the god Bel, was hired by the king of Mowab to curse the Israelites. Traveling to Mowab to comply with the request, Balaam's way was obstructed by a messenger (usually mistranslated as *angel*) sent by the Jewish god Yahweh to kill him. Yahweh apparently feared Bel's curse and had no option but to prevent Balaam from uttering it. But Balaam's ass saw the invisible messenger, and three times turned aside to save Balaam's life. Balaam thrashed his ass, which then spoke to him and told him what was in his path. So the curse of Bel was forestalled because Yahweh spoke to Balaam through his ass.

But the tale of a talking ass is not the only Bible story that could have been written by Graf von Münchausen. In the earliest Bible author's account of the years the patriarch Jacob spent working for his father-in-law Laban, readers were told how Laban promised to pay Jacob for his services by giving him all the goats that were patterned and all the sheep that were black. So Jacob, who even the Torah authors who wrote about him agreed was a liar, a thief, and a thoroughly nasty person, set out to swindle his father-in-law by the following means:

> Yaakob took branches and peeled them into white strips…The flocks
> tupped in front of the striped branches and gave birth to striped, speckled,
> and spotted offspring…And whenever sturdy animals tupped, Yaakob
> placed the striped branches in front of their eyes." (Genesis 30:37-41)

The author of those fanciful fables was an embarrassment even to later Torah authors. The Priestly author who wrote an alternative Torah between 621 and 612 BCE made no mention of any talking ass. And the Elohist who wrote in the northern kingdom of Israel around 770 BCE did not suppress Jacob's swindling, but he made it more believable by having "a messenger of the gods" tell Jacob, "Watch and see that all the rams that tup the ewes are striped or spotted or piebald." (Genesis 31:11-12)

Are religion addicts more capable of believing that a dead man came back to life than that donkeys can talk or that mating in front of striped branches produces striped young? Or are preachers missing the obvious point that persons capable of believing that one god plus one god plus one god equals one god, will believe anything, provided it is fed to them with the assurance, "It's in the book"?

## 4. The Inconsistency of Round-Earth Religionists: Their Own Bible Says They are Wrong

It is an observable fact that many people claim to be Christians (or Jews) even though they believe that the Bible is not really true. Perhaps they are within their rights. Telling a person who thinks he is a Christian that he is not surely smacks of intolerance. But when a round-earther proudly claims the title of Fundamentalist, meaning one who believes that every word in the Bible is true, even though the Bible states in seven places that the earth is flat, it cannot be unreasonable to say that he is nothing of the sort.

The gospel of Matthew (4:8) states that the slanderer took Jesus "to an extremely high mountain and showed him all the kingdoms of the cosmos." That every kingdom or society on earth could be seen from the top of a high enough mountain only if the earth was flat is self-evident. From any point on a globe, no part of the opposite hemisphere could be seen no matter how high one climbed. Luke's account of the same incident (4:5) is ambiguous, and could be true on either a flat earth or a global earth. But Matthew makes clear that either the earth is flat or he was a liar.

Daniel (4:10-11) wrote, "I saw a tree of great height in the middle of the earth. It reached to the skies, and could be seen from the farthest edges of the earth." Maybe a tall tree could be seen from all points on a flat earth, but not from all points on a sphere. And the surface of a sphere has neither edges nor a middle.

A psalmist wrote (103:12), "Farther away than the east is from the west, so far he removes our transgressions from us." But only on a flat earth are east and west at the points of maximum separation. From any point on a globe, the point of maximum eastness and the point of maximum westness are the same point. The translators of the *New World Translation* made the passage conform to their round-earth belief by falsifying it to read, "As far off as the sunrise is from the sunset."

Isaiah (40:21-22) wrote, "Since the land was founded, he has been sitting on the domed roof of the land, whose inhabitants are like grasshoppers. He stretched out the skies like a curtain and spread them like a tent to live in." The domed roof of the land is a clear reference to solid hemispherical skies resting on the horizon of a flat earth. Only over a flat earth could the entire skies be spread out like a tent, and only if the top sky, the seventh (2 Corinthians 12:2; Qur'an 71:15-16), is solid could Yahweh sit on it or use it as a tent.

Genesis 1:6-8 says, "The gods said, 'Let there be a solid dome in the middle of the waters and let it divide the waters from the waters.' So the gods made the dome, and divided the waters under the dome from the waters above the dome. The gods called the solid dome Skies." Here also the Bible endorses an earth covered by a hemispherical dome, a physical possibility only if the earth is flat.

The author of Job (22:14) wrote, "He walks on the dome of the skies." Again, only if the earth is flat could a dome on which Yahweh walked cover it.

Finally, Revelation (7:1) says, "I saw four messengers standing at earth's four corners." A flat earth can be pictured as having four corners, or ultimate extremities. A globe cannot.

A lot of people believe that the earth is a globe. That is their right. But when they simultaneously claim to believe that the Bible is an inerrant revelation from an omniscient god, they are being inconsistent. Either the earth is flat, or the Bible is fantasy. They cannot have it both ways.

*This is an article originally written for* Flat Earth News, *which apparently recognized its satirical nature and rejected it. Isaac Asimov wrote a whole essay on the lengths to which fundamentalists will go to rationalize that their Bible does not endorse a flat earth. But it does. My position on flat-earth theory and biblical literalism can be found in my book* Mythology's Last Gods, *published in 1992 by Prometheus.*

## AFTERWORD

I received one letter from a reader after this article appeared in *The Skeptical Review*, asking me to identify the cited Asimov article and book. Unfortunately, since Asimov's essay collections were never indexed, I was not able to do so. I informed him that the article *might* have been "The Judo Argument," and the book *might* have been *The Tragedy of the Moon*. But I am still not able to confirm that that is correct.

# 5. Has Religion Been Disproven?

Carl Sagan was one of the most rational men on earth. Yet he was able to write (*Broca's Brain*, Ballantine edition, p. 365), "An atheist is someone who is certain that God does not exist, someone who has compelling evidence against the existence of God. I know of no such compelling evidence." He was, of course, trying to be politically correct, acknowledging that he was unimpressed by the "God" hypothesis, while avoiding the appearance of dogmatism.

But Sagan was neither a historian specializing in religious origins, nor a philosopher specializing in logic, and was therefore unaware of the "compelling evidence" known to practitioners of both of those disciplines.

It is axiomatic that anything that cannot exist does not exist. For an omnipotent, omniscient god to exist, it must first be theoretically possible for omnipotence and omniscience to exist. But both are logical impossibilities.

Omnipotence is the ability to do absolutely anything. An omnipotent god could create a number that is more than ten but less than nine, or a triangle with four sides. Since such constructions are impossible by definition, it follows that omnipotence, the ability to create them, is likewise definitively impossible. Omnipotence cannot exist; therefore an omnipotent god cannot and does not exist.

Omniscience is total knowledge of the past, present and future. But if we assume that the future depends on decisions not yet made and effects not yet caused, and that the decision-makers have free will, then the only way even a god could have perfect knowledge of the future would be for information to travel backward in time. Since that is impossible, omniscience is therefore shown to be impossible. And if omniscience cannot exist, then an omniscient god cannot and does not exist.

But what if we remove the stipulations, and allow for the possibility that every future action, decision and consequence is predetermined? In such a universe the future could indeed be foreknown and an omniscient god could theoretically exist. But freewill could not exist, culpability could not exist, and even human thought would be nothing more than a programmed response to an inevitable stimulus. But with few exceptions, religion does not preach an omniscient god who presides over a puppet world in which the Hitlers and Stalins innocently act out the script written for them by programmed inevitability. The same religions that postulate an omniscient god, also insist that they live in a world in which individuals have the freewill to trigger an infinite number of possible futures. In that world an omniscient god cannot and therefore does not exist.

An omnipotent god cannot exist. An omniscient god cannot exist. Each of those concepts is falsifiable individually. But they are also mutually incompatible.

For a god to know what will happen in the future, and to know that he will not change it, he must in fact be unable to change it and therefore lacking in omnipotence. But if he can, when the time comes, change the future from what he knows it must be to something else, then he did not know what it would be and is not omniscient. If he did know that he would change it, then he was incapable of not changing it and is therefore not omnipotent. Ultimately, a god cannot be simultaneously omnipotent and omniscient, and a god that is both cannot and does not exist.

The situation becomes even more oxymoronic when, in addition to omnipotence and omniscience, a religion credits its god with omnibenevolence. Epicurus recognized the incompatibility of omnipotence and omnibenevolence more than 2,400 years ago:

> The gods can either take away evil from the world and will not, or being willing to do so they cannot, or they neither can nor will, or lastly they are both willing and able. If they have the will to remove evil and cannot, then they are not omnipotent. If they can but will not, then they are not

benevolent. If they are neither willing nor able, then they are neither omnipotent nor benevolent. Lastly, if they are both able and willing to annihilate evil, how does it exist?

That which cannot exist does not exist. And philosophers proved 2,400 years ago that gods possessed of the attributes credited to them by every religion on earth cannot exist. Yet religion still exists. The explanation is that, while believers might pay lip service to "cannot equals does not," their emotional response to the most definitive proof that their god cannot exist is, "Yes, but he does." The problem with the philosophical argument is that it can only prove that a particular god *cannot* exist. Only the methodology of history has the potential to show the masses that their god *does* not exist.

Negative hypotheses such as, "God does not exist," "The Great Pumpkin does not exist," and "Bigfoot does not exist," cannot be empirically proven. The only exceptions to that generalization are assertions such as, "Time travel cannot exist," where the converse leads to a *reductio ad absurdum*. Historians therefore do not even attempt to prove that gods do not exist. Instead they settle for proving that no god has ever revealed its existence, by tracing all claims of a divine revelation to demonstrable liars or fantasizers.

The sole reason that Christians and Jews believe in the god Yahweh (even if they are unfamiliar with his name) is that, "The Bible says so." Finding a single false statement anywhere in the Bible should therefore be sufficient to refute the claim that the Bible is a reliable source of absolute truth. And even the most dogmatic fundamentalists must have difficulty arguing that Ecclesiastes 9:5 is inerrant truth when it says, "The dead neither know anything nor have any further reward, for their awareness has ceased," since that passage contradicts the Bible's many endorsements of life after death. A bible that asserts that there simultaneously is and is not an afterlife tends to undermine the claim that an unsubstantiated allegation must be true because, "The Bible says so."

But the Bible was written by a large number of authors, some of whom conceivably had more credibility than others. To discredit claims that a god revealed its existence, it is necessary to discredit the specific authors who made those claims. Even finding false statements in Genesis is insufficient to falsify that whole book, since liberal believers are aware that Genesis had three major authors, as well as two editors who added harmonizing passages.

The first account of a divine revelation in Genesis (1:27-28) was written by the Priestly author (P). According to P, "The gods blessed them [Adam and Eve]: 'Be fertile and increase in number and fill the land and pacify it.'" That passage cannot be directly falsified. But it was written as part of a creation tale in which billions of years of evolution were compressed into seven days, the earth existed before its sun, the sky was a solid hemispherical dome, trees bore fruit before there was a sun to trigger the necessary photosynthesis, and the sun and moon were tiny lights attached to the dome of the sky. In the light of his other falsehoods, accepting P's word that gods created and conversed with humans would be analogous to accepting the unsubstantiated testimony of Baron Münchausen.

Exodus contains three accounts of Yahweh introducing himself to Moses, written by the Yahwist (J), the Elohist (E), and P. According to J, Yahweh described himself as "the gods of your ancestors, the gods of Abraham, Yitskhak and Yaakob." (Exodus 3:16) That passage is consistent with an earlier J story in which Abraham addressed his deity as, "Yahweh, your lordship." (Genesis 15:8) But in Ps account (Exodus 6:3), Yahweh told Moses, "When I appeared to Abraham and to Yitskhak and to Yaakob, it was as Allah the demon (*El Shaddai*). They didn't know me by my name, *Yahweh.*" And E similarly called his deity *Allahiym,* meaning "the male and female gods," up to the point where it introduced itself to Moses as, "Yahweh, the gods of your ancestors. That is my eternal name by which I'm to be evoked for all generations." (Exodus 3:15)

While the differences between those versions of Yahweh's first encounter with Moses prove that at least two were lying, or perhaps fantasizing about what they believed must have happened, they do not prove that all three were lying. But those same three authors also wrote passages that endorsed the real existence of gods other than Yahweh. J had Yahweh declare (Exodus 9:14), "No god in the whole land is my equal." E (Exodus 18:11) wrote, "Yahweh is the greatest of all the gods, for they tried their hardest and he defeated them." P's version (Exodus 12:12) was, "I'm going to exercise judgment against all of Egypt's gods, I, Yahweh." It follows that, if all three were lying, then their claims that Yahweh revealed his existence to one or more humans must be rejected as fantasy. But if even one of them was telling the truth, then the gods of Egypt, Babylon and the rest of the ancient world are as real as Yahweh. Thus an examination of the Bible does not disprove polytheistic religion. But it does disprove every religion that stands or falls on the dogma that a One True God has revealed its existence.

Short of being contacted by them, humans have no means of learning whether, somewhere in this or another galaxy, there are entities we might well identify as gods. Beings capable of violating the laws of nature cannot and do not exist. But lifeforms whose understanding of the laws of nature makes it appear to us that they are not bound by those laws cannot be ruled out. If they do exist, they are far away and have nothing to do with the gods of earth religion. Have gods been objectively disproven? No. Has religion been disproven? Unequivocally, Yes.

## 6. Gods, Goddesses and Bibles: The Canonization of Misogyny

On October 1984, America's National Council of Churches issued a new translation of passages of the Judeo-Christian Bible that the Council felt were marred by "male bias." Words that were masculine gender in the original language were converted to common gender in English (for example: *king* became *ruler; God's son* became *God's child*), passages that ignored women were altered to rectify the omission (*The God of Abraham* became *The God of Abraham and Sarah*), and references to the head of the Christian pantheon as *God the Father* were amended to *God the Father and Mother*. While all but the culturally schlaflyed applauded the attempt to drag religion into the twentieth century, even at the price of altering "revealed truth," what nobody seems to have realized is that a translation of the Judeo-Christian Bible that does not offend women is analogous to a translation of *Mein Kampf* that does not offend Jews.

In a male-dominated world, popes, caliphs, ayatollahs, prophets, messiahs, priests, and rabbis tend to be male; but that was not always so. From humankind's creation of the first goddess thirty thousand years ago, until the retaliatory invention of male gods more than twenty thousand years later, women held the same ruling-caste status presently enjoyed by men. There was a good reason for this. Just as Cro-Magnon humans were able to recognize that the cow was their superior because she sustained them with her milk (thus the cow-goddess Hera, and the status of cows in Hinduism), so did they recognize that woman was man's superior because she produced the children who ensured the species' continued survival.

Almost from their conception, gods were perceived as the givers of life. Since only females could give life, it inevitably followed that the gods must be female. And in a world ruled by female divinities, those humans created in the Mother's image naturally far outranked the male humans whose prime functions were fighting wars and providing their female overlords with sexual recreation.

As it was in the skies, so it was on earth. Goddesses ruled the metaphysical world; women ruled the physical. Priestesses reigned for life, often accepting homage (the original meaning of *worship*) as goddesses-on-earth. In an orderly world hatched from the egg of the goddess and run by her mirror-image, men accepted that they had no rights and did as they were ordered, just as in the modern world there are women so conditioned to the belief that they are hereditary slaves that they give speeches urging state legislatures to refuse to ratify a constitutional amendment granting full human status to women. It is doubtful, however, that men were ever exploited by women prior to the Male Revolution of 3500 BCE in the manner in which women since that date have been oppressed and dehumanized by men. There was never, for example, a female-absolutist equivalent of the sixth-century CE synod of Macon, at which Christian bishops earnestly debated whether women were human beings, possessed of "souls," or the seventh-century Council of Nantes that, in its third canon, restricted immortality to males and pronounced women "soulless beasts" whom the chief male god had given Man to use as he saw fit.

Men were never private property, owned by one woman and arbitrarily forbidden from providing sexual recreation to any woman but herself. At least, they were not in the days of goddess-rule. Men accept such a designation today (or pretend to) as the price they must pay for imposing similar private ownership on their breeding women; but this, too, is a consequence of the Male Revolution. When the idea began to evolve that monogamy was either right or wrong, the ruling males declared, in effect, "We won't annul your sexual slavery—but we'll agree to share your captivity by submitting to the same exclusivity."

Then came the Big Discovery.

The Big Discovery did not occur everywhere at the same time. Among the Aborigines of Melville Island to the north of Australia, it was not made until the nineteenth century CE. In some places, it may well have taken place much earlier than 3500 BCE, which is the best available estimate of the approximate date at which it became widespread. To persons who have grown up in a society in which such knowledge is taken for granted, it is difficult to convey the tremendous significance for future history of the first discovery by men that the organ with which they pleasured their mistresses *also made babies.* The Big Discovery meant that women were no longer the sole purveyors of life—and therefore neither were goddesses! From being the reproducers of life, women found themselves reduced to the level of incubators, of no more relevance to the birth process than the dirt in which an ear of corn grew into an adult plant.

Men were physically stronger than women. That fact had long been known and rationalized to fit a female-dominant theology, and only men's acceptance of their insignificance in the divine order kept them from taking over the world much sooner. Following the Big Discovery, nothing could stop them and nothing did stop them. However, the takeover did not occur right away. Compared to the Male Revolution, the Industrial Revolution was accomplished overnight. Before the mind could conceive of any change in the social structure of human society, it first had to postulate a similar change in the sky. Thus, before there could be any king reigning on earth, there had to be created a King of Heaven, a God the Father, who was the Mother's superior and by whose impregnation she produced her children.

Men did gain political power. But power that was not hereditary was meaningless. Just as mothers had always been able to identify their daughters, now fathers wanted to be able to identify their sons. It was for that reason that men imposed upon women a logical extension of the private-property concept, the chattel slavery that came to be known as marriage. And with marriage came the first sexual taboo: You're not to commit adultery.

Adultery was a crime against property. A woman, owned by one man, who allowed herself to be impregnated by another, thereby robbed her husband's true heir of the inheritance that could conceivably be usurped by her lover's "bastard" (another new concept). Had the discovery that sexual recreation causes pregnancy been coupled with the realization that births can be positively traced to those couplings that occurred roughly nine months earlier, the adultery taboo would never have been so severe. As it was, the taboo was based upon the assumption that a woman's child was jointly fathered by every previous lover, regardless of whether the couplings had occurred five months or twenty years before its birth.

Since adultery was a crime against the adulteress's husband, an attempt to rob him of his right to pass on his property and power to his lawfully conceived sons, it followed that an act of recreation involving an unmarried woman did not constitute adultery. The generalization of adultery to include recreation between married men and unmarried women did not occur until after Siddartha Gautama's creation of the belief that abstinence from recreation, deemed a sin by the Talmud (Nazir 19a), could somehow be virtuous in itself.

Adultery was for many centuries the only sexual taboo. Without any concept of wrongdoing, women grew up copulating freely with brothers and cousins and neighbors. At the age of eleven or twelve, those that had not been sold to husbands would take adult lovers, usually their closest relatives, and recreate diligently until such time as they could demonstrate their fertility by becoming pregnant. Women who, although nubile, had never produced a live infant, and who were therefore bad breeding risks, were stigmatized by the pejorative label, *virgin.*

Once a woman had given birth, an event that often did not occur until the age of fifteen or sixteen, her chances of being purchased as a wife increased significantly. Men wanted good breeders, and a woman who had demonstrated, not only fertility, but also the ability to survive childbirth, could expect a wide range of suitors, all of whom would share her favors until such time as her father accepted the highest bid. The first child of the marriage, regardless of how many years might have elapsed before its birth, being of multiple paternity, would be sacrificed to Molokh or Baal or Yahweh or Allah or whichever other god had the local baby-burning concession. Following the birth and sacrifice, the wife would observe an adultery taboo. All future children could then be attributed with absolute certainty to her legal owner.

It was the abolition of infant sacrifice that led to the imposition of cradle-to-marriage joy-deprivation on half of the human race. All societies eventually recognized that, with women dying in childbirth faster than men could kill one another in war, live births were too rare to be wasted, and infant sacrifice must be abolished. That meant that some new method had to be found whereby a man could be certain that the first child born to his new bride was of his own begetting. The solution was to deny unmarried girls the opportunity to bring to the marriage bed a womb that might already be carrying seed that could one day produce a cuckoo's chick. Women were informed that henceforth they were to practice total premarital joy-rejection, and that any woman who failed to spill hymeneal blood on the marriage blanket could expect to be promptly executed. (Deuteronomy 22:13-21) Thus from sheer ignorance concerning the duration of pregnancy and the durability of sperm, men stole from women the basic right to decide for themselves whether an offer of sexual activity should be accepted or rejected—a right that only the recent perfection of dependable contraceptive techniques has enabled them to reclaim.

The final step in the degradation of women was not taken until perhaps two thousand years after their reduction to slave status as men's "wives" following the Big Discovery. Not content with denying women their ancient role in creation and salvation by making the post-Discovery creator and savior both male, the phallusocracy now came up with the myth that male gods had created a perfect world which women had subsequently rendered imperfect by their culpable inadequacy. In Greek god-mythology, the first woman was Pandora, whom Zeus gave to Prometheus to replace his male lover, as a punishment for giving man fire. Pandora was endowed with a sealed box (the sexual symbolism of which should need no explanation), and warned never to open it. She disobeyed Zeus's admonition, and out of her box leaped disease, famine, and all of the other evils with which Man has since been punished for Woman's crime.

In the Semitic version of the same myth, the humanized goddess Eve first yielded to a serpent-goddess's invitation to worship her by eating the vulva-shaped pomegranate that was her sacramental body and blood, in defiance of the ruling male god's instruction to worship him alone. She then corrupted the man she had been created to serve. That only a chronic misogynist could have composed such a fable is obvious enough. That only a misogynist culture could believe it should be no less obvious.

A religion in which all first- and second-ranking gods are male is misogynous by definition. Christianity, for example, admits females only as third-ranking immortals ("saints"), and many of its third-ranking gods, prior to their pumpkinification, were themselves vicious misogynists. Jerome, translator of the Vulgate, described women as "a tool of Satan and a pathway to Hell," while Ambrose and Augustine contributed to the world the pious belief that someone as clean as Jesus could not have come out of something as dirty as a female recreational orifice, but instead magically appeared outside of third-level goddess Mary's body without the necessity of utilizing her birth canal. While not even popes claim to be speaking inerrant truth at all times, and otherwise good humans can be bigots, it is nonetheless significant that the views expressed by those men have never been repudiated by the Church that canonized them.

It is, however, not in the writings of "saints" but in canonized Scriptures that one must look for proof that a religion *officially* categorizes women as subhuman. Judaism's position is clearly spelled out in the prayer in the Talmud that reads, "Yahweh, I thank you who have not made me a woman, an idiot, or an infidel."

Christianity's inspired apologist for misogyny was Paul of Tarsus:

> Women, submit yourselves to your men as to his Lordship. For the man is his woman's head. Just as the community is subject to the Messiah, so are women to their men in all things. [Ephesians 5:22-24]
>
> Women in the community are to remain silent. They are required to be obedient, as even the Torah commands. [1 Corinthians 14:34]
>
> The man was not created for the woman, but the woman for the man. [1 Corinthians 11:9]

Paul's misogyny has begun to be rejected in the secular world, but in Christian churches women's demands for full membership in the human race continue to be ignored.

The misogyny of Islam's fanatic present-day leaders is widely known. Less known perhaps is that Muhammad's Koran unambiguously endorses misogyny:

> Has your Lord blessed you with sons and himself adopted daughters from among the angels? A monstrous blasphemy is that which you utter. [17:40]
>
> Men have a status above women. [2:228]
>
> Call in two male witnesses from among you, but if two men cannot be found, then one man and two women. [2:282]
>
> Men have authority over women because Allah has made the one superior to the other…Good women are obedient…As for those from whom you fear disobedience…beat them. [4:34]

Not only did Muhammad's male chauvinist god deem men superior to women; he declared it a blasphemy to suggest otherwise.

And how did the National Council of Churches respond to Paul's sexism in their common-gender Bible translation? Very easily: They simply left it out.

## 7. Conflicting Attitudes Toward Homosexuality in the Judaeo-Christian Bible

Between 621 and 612 BCE, an Aaronic priest whom historians have designated "the Priestly author" composed a new Torah ("Law") designed to supersede the existing Torah begun 300 years earlier. He was only partly successful. In 434 BCE, High Priest Ezra read from a composite Torah created by riffling the old and new versions together. While it cannot be stated with certainty that the Redactor (R) who did the riffling was Ezra himself, the probability is that he was.

The Priestly author (P) wrote in his new lawcode, "You're not to tup a man the way you tup a woman. That would be detestable." (Leviticus 18:22) That verse constituted the first suggestion by any bible author that homosexual recreation, as practiced by Zeus with Ganymede, Poseidon with Pelops, and Apollo with Hyakinthos, was less commendable than the heterosexual kind. R later added, "If a man tups a man the way he tups a woman, they're both without fail to be executed." (Leviticus 20:13)

P's sex taboos, forbidding practically everything except the missionary position, were sudden and unexpected. Less than a decade earlier, the Deuteronomist (D) had written favorably of "the woman who fulfils your physical needs, or the male lover who means as much to you as your own breath." (Deuteronomy 13:6) D had simultaneously prohibited Jewish men from becoming *kedeshiym,* homosexual monks who provided sexual services to male patrons of fertility goddesses as an act of worship. However, since D also prohibited Jewish women from becoming *kedeshoth,* nuns who provided similar services to heterosexual worshippers, it is clear that he was simply reiterating the Commandment that Jews could not participate in an act of worship to any god but Yahweh. He was not singling out homosexual recreation as intrinsically unacceptable or questionable. (Deuteronomy 23:17)

P banned homosexual coupling for men. Since he wrote at a time when Babylonian influence in Jerusalem was becoming irresistible, it is not difficult to calculate where he derived such a taboo. Early in the seventh century BCE, Zoroaster (or his maguses) had banned both homosexual coupling and celibacy, since each tended to reduce the breeding of tithe-paying Zoroastrians. He did not ban lesbianism because, in a culture in which it was not economically feasible for a woman to live without a man, it did not keep women from breeding, and therefore constituted a sinless alternative to adultery for unsatisfied women. P's banning of gay sex for men but not for women indicates that his motivation was similar. Like Zoroaster, he was enforcing self-serving expedience, not some kind of abstract morality.

Seventh-century BCE Zoroastrianism was the first religion anywhere on earth to suggest that homosexual behavior was morally different from the current equivalent of tennis or golf. Biblical authors who wrote prior to that date could no more imagine that homosexuality might be immoral, than they could imagine that eating meat on Friday might be immoral. Nonetheless, a myth composed in the tenth century BCE and set in an even earlier time is commonly cited as an endorsement of a homosexual taboo at a time when no such taboo had yet been invented.

In the reign of King Solomon's son (930-913 BCE), a mythologian retroactively termed the Yahwist, or J (German: *Jahwist*), wrote a destruction-of-humankind myth set in the Anatolian city of Sodom. In the prelude to the volcano-god Yahweh's decision to erupt over Sodom, the Sodomites pounded on the door of the righteous Levit (*Lot* is a mistranscription) and demanded the right to rape Levit's male houseguests. That the rape would have been homosexual is not in doubt, but that element is not what made the Sodomites' action a crime. They proved by their demand to be blatant offenders against the universally practiced hospitality code, and that was why Yahweh decreed that they deserved to die.

There are many instances of the inviolability of the hospitality code in ancient mythology. When Bellerophon was falsely accused of raping Proitos's wife, Proitos was helpless to take vengeance because Bellerophon had been his houseguest and as such was under his protection. The Jews observed a similar code, and even an infidel was entitled to the full protection of his host while under the roof of a Jew. (Genesis 19:8)

Yahweh's messengers were Levit's houseguests. That they were male was irrelevant. In a similar tale in Judges, chapters 19-20, the inviolable houseguest raped by a mob was female, and Yahweh's retribution was no less severe. By demanding that Levit surrender guests under his protection to be abused by a mob, the Sodomites showed their contempt for the sacred rules of hospitality, and for that Yahweh destroyed them. The designation of homosexual coupling as "sodomy" is an anachronism.

Of the five principal authors of the Torah, the first three had never heard of a homosexuality taboo. Nor had the chronicler who, in Samuel/Kings, described how King David married King Saul's daughter and fell in love with Saul's son.

David was not homosexually inclined. He had at least seven wives who bore his children, besides Saul's daughter who did not. Once he had acquired a harem, there is no evidence that he ever again loved a boy. And when he was dying, it was a young girl, not a boy, who was introduced into his bed to try to stimulate him. But that his relationship with Yahuwnathan was sexual is the only possible interpretation of the chronicler's words:

> The breath of Yahuwnathan was dependent on the breath of David, for Yahuwnathan loved him as his own breath. (1 Samuel 18:1)
>
> Yahuwnathan and David contracted together, for he loved him as his own breath. (18:3)
>
> Yahuwnathan, Shauwl's son, was utterly enraptured by David. (19:2)
>
> David said, "Your father knows I am the delight of your eyes." (20:3)
>
> Shauwl said…"You have coupled with the son of Yishay to your own confusion and the confusion of your mother's vulva." (20:30)
>
> They kissed each other and wept with each other until David ejaculated. (20:41)
>
> My brother Yahuwnathan, you have been very satisfying to me. Your love for me was wondrous, surpassing the lovemaking of women. (2 Samuel 1:26)

David and Yahuwnathan were lovers. Yahweh did not disapprove, because the Priestly author would not invent his retroactive disapproval for another 300 years. David was an equal opportunity lover, as was every man and woman on earth prior to about 670 BCE. A sizable minority never engaged in any homosexual recreation, but they were observing a preference, not a taboo.

## 8. Jesus the Nazirite: Real Person or Literary Creation?

Biblical historians are evenly divided on the question of whether Jesus the Nazirite was a posthumously-deified real person or a creation of fantasy literature. G. A. Wells and Robert Price argue that Jesus did not exist, Michael Arnheim and Martin Larson believe he did, and John Crossan insists that there really was a Jesus while simultaneously presenting a reconstruction of Christian origins that would be logical and coherent only if there was not a real Jesus. At the extreme of the "no such person" thesis, John Allegro, in a mushroom fantasy that Robert Graves described as "an elaborate literary hoax," concluded that Jesus began as a personified mushroom. Equally extreme are theologians whose refusal to question the dogma that Jesus was a god incarnate, was virgin-born, and rose from the dead, makes it impossible for them to be taken seriously by scholars who do not start from predetermined conclusions and force the evidence to fit.

Arguments for Jesus' nonexistence tend to take the form: George Washington allegedly flourished in the 18th century. The fable of the honest little boy and the cherry tree has been traced back as far as the 17th century. Therefore it did not happen. Therefore George Washington did not exist.

Such reasoning would not be without merit if, as Price (p. 250) claims, "the gospel story of Jesus matches the pattern of the Mythic Hero Archetype in every detail, with nothing left over." But even after the fantasy elements are deleted from Jesus' official biographies, there is plenty left over, most of it of such a negative nature that no mythmaker in his right mind would have invented such stories about a person he was trying to portray as the ultimate hero.

For example, at the beginning of Jesus' career, when an unfortunate sermon almost got him lynched, the anonymous author of the earliest gospel writes (Mark 3:21), "But when his family heard, they came to take him into custody, for they declared, 'He's gone mad.'" Jesus' immediate repudiation of all family ties (Mark 3:31-35) could be explained away as a standard Essene procedure toward family members who remained outside of the sect. But the only believable explanation for an apologist acknowledging that the people who knew Jesus best considered him a madman, is that Mark was stuck with the reality that they really did. And what mythmaker would invent accusations that Jesus was "a drunkard, a glutton, and a lover of tax collectors and sinners [*i.e.*, hookers]"? (Matthew 11:19)

The author of Mark (15:7) also revealed that Jesus was arrested with "the imprisoned revolutionaries who had committed homicide in the uprising." In other words, Jesus' arrest coincided with an abortive revolution against the Roman occupation, a revolution Jesus began by disrupting the temple sacrifice for Tiberias that symbolized Judea's subservience to the emperor and the empire. Jesus was a man who started a war of independence—and lost. Real people do such things (*e.g.*, Spartacus, Bar Kokhba). Mythical heroes do not.

Mark did not dispute that Jesus' lieutenants included members of the Zealot sect that had instigated the war in progress at the time of writing (70-73 CE), including a member of the sect's ultra-terrorist wing, the *sicarii/iskariots*, who were still holding out at Masada after the fall of Jerusalem. In a book written to convince Vespasian that the Christians were not a branch of the religion with which the emperor was at war, Mark could not deny that Jesus was supported by anti-Roman rebels, since that reality was too widely known. So he tried to neutralize it by pretending that Judas was "really" Jesus' enemy and had ultimately betrayed him.

But perhaps the strongest evidence for Jesus' historicity is the surviving testimony that Jesus was grossly deformed. While the author of Luke stopped short of portraying Jesus as a cross between Quasimodo and Rumpelstiltskin, as later Christian apologists did, he did record (Luke 4:23) that Jesus told his home town synagogue audience, "You're sure to recite this proverb to me: 'Doctor, heal yourself.'" Why would Jesus expect his hearers to react in such a manner? What was there about him that needed healing? One possible answer can be found in the same author's Acts (8:32-33), and also in John (12:38).

The authors of Luke-Acts and John both equated Jesus with the suffering slave of Isaiah 53:1-12. Deutero-Isaiah had written of his hero, "Having neither proper shape nor beauty, lacking good looks that would have attracted him to us…We regarded him as someone plagued and afflicted by the gods."

Even allowing that John used Luke as a source, why would either author have equated Jesus with such an unfortunate creature? The answer that springs to mind is that they were putting the best possible spin on the reality that the physical description of Isaiah's suffering slave matched the physical description of Jesus.

By *c* 178 CE, the pagan writer Celsus was able to say of Jesus, "Surely a god would never have such a body as yours, that is so contemptible, being subject to such numerous and considerable imperfections." A generation later, in his *Contra Celsum*, the Christian apologist Origen did not dispute the accuracy of Celsus's description. Rather, he argued that Celsus "cannot deny that if our liberator was born as we say he was, that then his body had in some sense a stamp of divinity on it." (ch. 59) Since Origen offered rationalizations of all of Celsus's other anti-Christian arguments, his failure to dispute that Jesus' body was so stricken with "imperfections" as to be "contemptible" is most reasonably interpreted as an admission that the description was accurate.

Tertullian in 207 CE similarly conceded that Jesus was misshapen: "His body was not even of honest human shape." Clement of Alexandria described him as having "a very ugly face." Cyril of Alexandria echoed that description, while Andrew of Crete declared that he had "eyebrows which meet."

Those descriptions prove only that, from about eighty years after his alleged death, Jesus was believed by Christians and non-Christians alike to have existed and to have been deformed and ugly. But did such a description exist early enough for it to have had a factual basis? There is good reason to believe that it did.

The earliest hint of Jesus' physical imperfections was written within twenty years of his death by Paul of Tarsus. Jesus, according to Paul (Philippians 2:6-7), "did not exhibit the shape of a god because he considered it larceny to be equal to a god. Rather, he degraded himself by taking the shape of a slave." Note that Paul did not say that Jesus adopted the "status" of a slave. He wrote that Jesus had the shape/form/*morphē* of a slave, in other words a body more appropriate for a slave than a king.

It is not credible that second-century Christian apologists invented the deformed Jesus that remained undisputed until the sixth century. More likely, they were quoting from an older source, and there is reason to believe that the source was Josephus. According to a medieval writer cited by Don Cupitt, Josephus described Jesus as an old-looking man, balding, stooped, with joined eyebrows, and approximately 4 feet 6 inches tall.

It cannot be proven that Josephus ever wrote anything of the sort. But someone of comparable reputation must have done so, for Christian theologians to accept the description as accurate. Such a description could only have been written down and accepted at a time when dispute would have been impossible because people were still alive who had seen Jesus preach and knew that he was as ugly as the writer claimed. Josephus wrote at a time that fits that specification. Had the description of a deformed Jesus originated as late as the time of Celsus, Origen could simply have painted an alternative Jesus, more like the Adonis that he became after Josephus was expurgated and the Mandylion of Edessa, on which all future depictions of the Christian junior god were based, was painted. Origen did not dispute Celsus's description, because the unexpurgated Josephus still existed, and any contradictory description would not have been believed.

The Jesus of history was deformed. A fictitious Jesus would from the start have been the Greek god of post-Mandylion iconography. The Jesus of history marched into Jerusalem on a Sunday morning, proclaimed independence, and was arrested and executed five days later without accomplishing his purpose. A fictitious Jesus would not have started a war he could not win. Jesus was posthumously transformed into a latter-day Osiris, Dionysos, Atthis, Tammuz, Adonis and Mithra. But he must have existed, because what remains after the savior-god myths and hero-miracles are deleted is a misshapen, rationally-challenged religious fanatic whom no mythologian would ever have invented—at least not as a hero.

Bibliography of recent books that discuss the question of Jesus' historicity:

John Allegro, *The Sacred Mushroom and the Cross*, NY, 1970.
Michael Arnheim, *Is Christianity True?*, Amherst, 1984.
Steuart Campbell, *The Rise and Fall of Jesus*, Edinburgh, 1996.
John Dominic Crossan, *Who Killed Jesus?*, San Francisco, 1995.
Don Cupitt, *Who Was Jesus?*, London, 1977.
Earl Doherty, *The Jesus Puzzle*, Ottawa, 2000.
Timothy Freke & Peter Gandy, *The Jesus Mysteries*, NY, 2000.
William Harwood, *Mythology's Last Gods*, Amherst, 1992.
  *The Judaeo-Christian Bible Fully Translated*, Imprintbooks, 2002.
  *Uncle Yeshu, Messiah* (a novel), Xlibris, 2001.
Randell Helms, *Gospel Fictions*, Amherst, 1989.
R. J. Hoffman and Gerald Larue, *Jesus in Myth and History*, Amherst, 1985.
Ian Jones, *Joshua, The Man They Called Jesus*, Port Melbourne, 1999.
Martin Larson, *The Essene-Christian Faith*, New York, 1980.
Gerd Lüdemann, *The Great Deception*, Amherst, 1999.
A. J. Mattill Jr., *Sweet Jesus*, Gordo AL, 2002.
Robert Price, *Deconstructing Jesus*, Amherst, 2000.
G. A. Wells, *The Historical Evidence for Jesus*, Amherst, 1982.

## 9. Jesus' Miracles: Anything Eliyah Did, Jesus Did Better

According to Morton Smith (*Jesus the Magician,* 1978), Jesus the Nazirite was a faithhealer who drew huge crowds by successfully curing demon possession and other psychosomatic ailments by a method that would today be called hypnotism by some and suggestion therapy by others. Probably he did. To this day faithhealers and other placebo peddlers capitalize on the reality that eighty percent of all illnesses will heal spontaneously with no treatment whatsoever, and there is no reason to believe that Jesus' success rate would have been lower.

In discussing Jesus' alleged miracles, however, the cures achieved by the Jesus of history are irrelevant. The Christian gospels make no mention of Jesus screaming into a deaf man's ears until he acknowledged that he could hear something, or standing close enough to a man who was legally but not totally blind, so that the blind man would have to concede that he could see him. Rather, the miracles described in the gospels bear no resemblance to anything the real Jesus ever achieved, but were simply borrowed from Jewish scriptures and credited to Jesus long after his death.

Jesus' greatest alleged miracle was the raising of one or more dead persons. But an examination of the books of Kings shows that both Eliyah and Elisha raised the dead. A widow fed Eliyah when he was hungry, and in return he revived her dead son (I Kings 17:17-22). Elisha successfully prophesied to a woman who was sterile that she would bear a son. When the son died, Elisha restored him to life (2 Kings 4:32-35). The description of Elisha's miracle reads like mouth-to-mouth resuscitation, but it would be unduly credulous to interpret it as such. Yahweh's spokesmen (Hebrew *nabiya,* Greek *prophetes*) were expected to be wonderworkers, and were posthumously credited with whatever came into a biographer's mind.

Jesus' biographers tended to be less imaginative. They simply read a Greek translation of *Kings* and credited Jesus with the same miracles claimed for his prototypes (Mark 5:22-24).

Eliyah, after raising the widow's son, learned that she had only "a handful of cereal in a barrel, and a little oil in a jar." (I Kings 17:12) Eliyah worked his magic, so that the small amount fed three people for many days (17:16). The anonymous author of Mark topped that by having Jesus feed a crowd of thousands with enough bread and fish for a handful—twice! (Mark chapters 6 and 8)

Eliyah journeyed into the desert, where he was tended by a heavenly messenger (Greek *angelos*). As a consequence he had no difficulty surviving forty days without food. (I Kings 19:4-8) Mark had Jesus do likewise. (Mark 1:12-13) Eliyah emulated Yahuwshua's parting of the Jordan river. (2 Kings 2:8) Even Jesus' biographers dared not credit their hero with a feat of that magnitude, in case persons living in the vicinity came forward and declared that nothing of the sort had ever happened. Instead they contented themselves with the claim that Jesus had crossed a body of water dry-shod by walking on the water. (Mark 6:48)

Eliyah did not die, but was transported bodily to the skies. Jesus did likewise. But whereas Jesus merely rose in a cloud (Acts 1:9), Eliyah departed in a chariot of fire pulled by equally fiery horses, as the Persian god Mithra and the Greek god Apollo had done before him. Only Elisha was permitted to see the chariot and horses. The "sons of the spokesmen" who were also present thought they saw Eliyah swept away by a tornado, and spent three days searching for his body. (2 Kings 2:11-17)

Elisha, the second of Israel's miraclemongers, began his career by multiplying a widow's oil so that a single potful filled as many pots as she could borrow. (2 Kings 4:2-6) He followed that by feeding one hundred men with twenty barley bagels and some ears of corn, having some left over. The anonymous author of John added the "barley" element to the bagels-and-lox incident he found in Luke. (John 6:9-15)

Elisha cured a leper by having him bathe seven times in the Jordan River. (2 Kings 5:6-14) The anonymous author of Luke credited Jesus with a similar feat (Luke 17:12-14), despite the fact that in

Jesus' day lepers were confined to isolated colonies and forbidden from wandering near populated centers.

In the thirty years between the Luke and John versions of the death of Lazarus, a notable evolution occurred. In Luke, a beggar named Lazarus died and went to "Abraham's bosom," while the capitalist at whose door Lazarus sought crumbs went to "Hades." The capitalist, tormented by the eternal flamethrowers, asked Abraham to send Lazarus back to earth as a warning to his five brothers. Abraham replied that anyone who would not learn from the scriptures would be no more impressed by someone returning from the dead. (Luke 16:20-31) The incident was written as an allegory told by Jesus.

When the author of John found the fable in Luke, he made a slight change. He turned Lazarus into a real person whom Jesus actually raised from the dead. He then had the Jewish priests plot to kill both Jesus and Lazarus (John 11:53, 12:10), paralleling the Greek myth in which Asklepios raised a dead man and Zeus reacted by killing both.

"John" also repeated a miracle tale first told by "Luke," with modifications. Peter and others had been fishing all night and caught nothing. On Jesus' instructions, they cast their nets again, whereupon the enormous catch proved almost more than their boats could hold. In the Luke version, the nets were torn by their huge load. (Luke 5:6) John saw torn nets as a reflection on Jesus' power and declared that, "Despite there being so many, the nets were not torn." (John 21:11)

Jesus did not perform impossibilities. He did, however, achieve significant success as a faithhealer. A talk therapist cannot achieve cures among persons who regard him as a local upstart. The statement in Mark that Jesus could effect no cures in his "own country" (Mark 6:5) is the best possible evidence that cures were indeed achieved elsewhere.

## 10. Modern Religion and Bible Religion: the Differences

The identifying credo of a religious fundamentalist is the assertion that everything in his Bible is inerrant truth. Yet that same fundamentalist, asked if he agrees with a particular biblical teaching, is as likely to say No as Yes, for the simple reason that he has no idea what the Bible authors really taught.

The god of modern fundamentalists is monotheistic, omnipotent, omniscient, omnipresent and omnibenevolent. To the authors of the Jewish Testament, he was none of those things. Of the five Pentateuch authors, all five wrote passages that unambiguously acknowledged the existence of gods other than their own. To J (Exodus 9:14), "No god in the whole land is my equal." To E (Exodus 18:11), "Yahweh is the greatest of all the gods, for they tried their hardest and he defeated them." To D (Deuteronomy 3:23), "Which god, whether of the land or the skies, can emulate your wonders or match your strength?" To P (Exodus 12:12), "I'm going to execute judgment against all of Egypt's gods, I, Yahweh." And to R (Numbers 33:4), "On their gods, also, Yahweh inflicted punishments." Clearly, Yahweh's official biographers would not have shown him outranking or defeating other gods that they viewed as nonexistent. Expressing allegiance to one god while acknowledging the existence of others is henotheism, not monotheism.

Yahweh became omnibenevolent in post-Captivity times as a consequence of the adoption by Judaism of the Zoroastrian prince of darkness, whom they renamed Satan and blamed for all of the evil previously attributed to Yahweh. But prior to that, Yahweh was the sole culprit: "Can there be evil in a town, except that Yahweh has done it?" (Amos 3:6) "I create good fortune, and I create evil. I, Yahweh, do all of these things." (Isaiah. 45:7) "Don't good and evil come from the mouth of Ilion?" (Lamentations 3:38) Modern believers may acquit Yahweh of all the evil in the world. Biblical authors did not.

Modern gods are omniscient, even though that means they have knowledge of information that has traveled backward in time. The biblical god was not. When he was told of the blatant violation of the sacred hospitality code by the citizens of Sodom, he needed to transport himself there physically, "to find out for myself whether the accusations that have reached me reflect what they have really done, for if they were lies I want to know that." (Genesis 18:21) An omniscient god would not have needed to do that, and neither would an omnipresent one.

The gods of virtually all modern religions are omnipotent, even though omnipotence must include the ability to create a number that is more than ten but less than nine, or a triangle with four sides. Biblical authors had no such belief. When the Babylonians tried to build a tower that would enable them to climb up to the solid sky and break into the demesne of the gods, Yahweh expressed the fear that, "Now nothing will be impossible for them." (Genesis 11:6) Having no other way to stop them, Yahweh was forced to confuse their languages so that they could not understand one another. An omnipotent god could have simply caused the tower to collapse. And when Isaac was tricked into blessing Jacob instead of Esau, Yahweh had no ability to transfer the blessing to its intended beneficiary, because "Now he cannot be unblessed." (Genesis 27:33) A god that was unable to overturn a fraud was clearly not omnipotent.

Modern believers would vehemently reject the suggestion that their god is a sadist. But according to a Bible author, "He'll derive pleasure from exterminating you and reducing you to ruin." (Deuteronomy 28:63) The ability to derive pleasure from hurting others is sadism by definition.

Fundamentalists tend to be homophobic, and two Pentateuch authors indeed condemned homosexual behavior for men—but not for women. In a society in which all women were forced to marry and breed tithe-paying believers, lesbianism did not interfere with that duty, and for that reason no ban on lesbian activity is to be found anywhere in the Hebrew Bible. And less than a decade prior to the homosexual ban being introduced into Judaism by the Priestly author, the Deuteronomist wrote in non-condemnatory language of "the woman who fulfils your physical needs, or the male lover who means as much to you as your own breath." (Deuteronomy 13:6)  While homophobes can be selective about which author to cite in connection with male homosexuality, their attacks on the lesbian story line in "Ellen" have no biblical basis whatsoever.

With the exception of the fifty hardcore members of the Flat Earth Society, modern believers reject biblical cosmography.  They do so by not allowing themselves to become consciously aware that their Bible endorses a flat earth (Matthew 4:8), a solid metal sky (Job 22:14), an immobile earth at the center of the universe (Psalm 93:1), a universe that has existed for less than ten thousand years (Genesis genealogies), stars small enough to fall to earth as meteors (Revelation 6:13), a moon that is not a reflector of light but a source of light (Genesis 1:16), and trees that bore fruit before there was any sun to energize the necessary photosynthesis (Genesis 1:11-16).  Isaac Asimov wrote a whole essay on the lengths to which believers will go to rationalize that their Bible was not written by flat-earthers.  But it was.

All modern religionists believe in an afterlife, and so did many Bible authors.  One who did not was the Sadducee who wrote Ecclesiastes (9:5): "The living are aware that they are going to die, but the dead neither know anything nor have any further reward, for their awareness has ceased."

A little over a century ago, Christians endorsed the enslavement of Africans by quoting Leviticus 25:44, "Both your men-slaves and your slavegirls are to come from the infidels who surround you."  Today, most prefer not to know that the Bible even mentions slavery.  Similarly, they choose not to know that it also endorses human sacrifice: "Yahweh ordered Mosheh, 'Sacrifice to me all of the firstborn, the first issue of every belly.  It belongs to me.'" (Exodus 13:1-2)  "No human who has been solemnly vowed is to be redeemed.  He's to be sacrificed without fail." (Leviticus 27:29)

To modern religionists, a demon is necessarily evil.  To the Priestly author, it was not, since he had Yahweh tell Moses, "When I appeared to Abraham, it was as Allah the demon (*El Shaddai*). They didn't know me by my name, Yahweh." (Exodus 6:3)  Demons have become bad guys because, as Yahweh's rivals, they are viewed as followers of Yahweh's great Enemy.  But originally "demon" was a synonym for "spirit."

Only believers who have actually read the Bible are aware that Yahweh talked to Balaam through his ass (Numbers 22:28).  Preachers tend to omit from their sermons any mention of the Bible's authentication of a talking donkey.  Similarly, only by reading the scene for himself has any believer learned of the Yahwist's theory that animals which mate in front of striped branches produce striped young (Genesis 30:39).  And if preachers in the past ever quoted the biblical declaration that no human would be permitted to live past 120 years (Genesis 6:3), it is a safe bet that it will never be cited again, since a woman in France died in 1997 at the age of 122.

Christians tend to believe that the miracles attributed to Jesus were unique.  Very few are aware that every single one was copied from tales previously told in connection with Eliyahuw and Elisha (3 Kings 17:16-22; 19:4-8; 4 Kings 2:8; 2:21-22; 4:42-44; 5:6-14).  Even Jesus' walking on water was the closest the gospel author dared come to Eliyahuw's parting of the Jordan River.

The Yahwist promised Abraham that he would one day have as many descendants as there are stars in the sky.  Since the stars number $10^{22}$, ten billion trillion, anyone who takes it seriously must believe that Jews will some day occupy a trillion planets.

There are many other differences between what modern believers think is in their Bible and what is really there, including the assumption that the ban on "adultery" covered relationships involving a woman using birth control or a woman not involved with a recognized breeding partner, whereas to the Bible authors it was the fraudulent impregnation of a married woman, thereby depriving her owner of his right to pass on his inheritance to his legitimate heirs. (Leviticus 18:20; 20:10)  But the foregoing explain why the Catholic Church fought for so long to prevent the Bible from being translated into languages that the unlearned masses could read, and why the King James Version, whose seventeenth-century English successfully obscures the real meaning, is so popular today.

If believers were allowed to find out what their Bible really teaches, religion would disappear overnight.

## AFTERWORD

This essay was published in *American Rationalist* Jan/Feb 2001, a year before the publication of *The Judaeo-Christian Bible Fully Translated*, volumes 1 and 7 (Imprintbooks, 2002).

## 11.  The Multiple Authorship of the Books Attributed to Moses

As far back as the 18<sup>th</sup> century, biblical scholars started to recognize that the Pentateuch or Torah was riddled with doublets, *i.e.,* two versions of the same story, each complete and self-contained. This would have been insignificant in itself, but they also noticed that one of the versions invariably identified the deity as *Yahweh,* while in the other account the deity was *Elohim.* Recognizing that they were looking at a riffling together of two older documents that had been written independently, they called the author of the Yahweh stories "the Jahwist," in German, or "Yahwist," in English, and for convenience thereafter referred to him simply as "J." The author of the Elohim stories became "the Elohist" or "E." A little later, they came to the realization that the Elohim stories were the work of two authors, one from the 8th century BCE, who retained the "E" designation, and the other an Aaronic priest from the 7th century, who became "the Priestly author," or "P." When the author of Deuteronomy was recognized as "none of the above," he became "the Deuteronomist" or "D." Finally, in the late 20th century, Richard Friedman of UCSD demonstrated that the person who combined the separate documents into a single narrative, long thought to be the Priestly author, was in fact a much later editor, whom he called "the Redactor" or "R."

While it is not unanimous, the most widely accepted dates for the various authors are:  J, *c* 920 BCE; E, *c* 770 BCE; D, 621 BCE; P, 621-612 BCE; and R, 434 BCE. The reasoning behind those dates is that J shows signs of having been written during the reign of King Rekhobowam (*c* 930-913 BCE), whom he consciously flattered. E could be off by as many as fifty years. D clearly wrote shortly before the "discovery" of his book in Yahweh's temple in 62I BCE (2 Kings 22:8-11). P was written after D, which showed no awareness of P's existence, while P referred to Assyria as an existing reality, as he could only have done before Assyria's annihilation in 612 BCE. Since the Torah's final version, containing sections not from J, E, D, or P, turned up in the hands of high priest Ezra in 434 BCE, with no explanation of where it came from, or why Ezra suddenly changed the ritual for the Feast of Booths from the formula in Deuteronomy to that in Leviticus, the logical conclusion is that it did not exist seven years earlier when a Deuteronomy-based Booths was celebrated. Indeed, the most logical assumption is that Ezra himself was R.

Since the delusion that the Torah dates back to Moses has been totally disproven for more than a century, people who can rationalize away the evidence in order to retain a security belief no more deserve the dignity of a rebuttal than incurable Maharishiites or flat-earthers. But for the benefit of those who, while aware that scholars have refuted Moses' pretended authorship, are not familiar with the evidence, the following passages that clearly were not written during or near the lifetime of Moses should suffice:

> To this day no one has ever found his (Moses') grave.  (Deuteronomy 34:6)
> Since then there hasn't been a spokesman in Yisrael comparable with Mosheh, whom Yahweh knew face to face. (Deuteronomy 34:10)
> At that time, there were still Phoenicians in the land. (Genesis 12:6)
> These were the kings who reigned in the land of Edom before there was any king reigning over the descendants of Yisrael. (Genesis 36:1)
> When Avrum heard that his kinsman was a captive, he...chased them as far as **[the city of]** Dan. (Genesis 14:14)

It is not feasible that the line, "No one has ever found his grave *to this day*," could have been written until long after Moses' death. The statement implied extensive attempts to locate Moses' grave, carried out over a long enough period for the writer to be despaired of ever succeeding.

It is equally inconceivable that the lack of a subsequent spokesman comparable to Moses would have been deemed remarkable until at least a century had gone by. Also, Moses would not have been described as a spokesman (Hebrew, *nabiya*) any earlier than the sheikdom of Samuel, since it was in Samuel's time that the word was coined. The same is true of "Miriam the spokeswoman." (Exodus 15:20)

That there were Phoenicians in the land "at that time" was an expression that could only have been used at a time when there were no longer Phoenicians in the land; and the Phoenicians survived Moses by many generations.

There was no king reigning over the Israelites until the time of Saul, so no writer who lived earlier than Saul could have referred to a time "before there was any king reigning over the descendants of Yisrael." The comparison of Edom's having kings before Israel had a king demolishes the pretense that the writer was not referring to the monarchy of Saul.

The Canaanite city of Laish was not renamed Dan until the Danites captured it long after the death of Moses. (Judges 18:29)

Since J, E, D, P and R all practiced more or less the same religion, not surprisingly there were points on which they all agreed. For example, all five acknowledged the existence of gods other than Yahweh. R wrote of the Israelites' enemies, "On their gods, also, Yahweh inflicted punishments." (Numbers 33:4) J (Exodus 9:14), E (Exodus 18:11), D (Deuteronomy 3:24), and P (Exodus 12:12) were equally unambiguous in expressing their belief that Yahweh was merely the most powerful god, not the only god.

On the question of what was taboo, the authors disagreed. The Priestly author banned male (but not female) homosexuality (Leviticus 18:22), and R later added a death penalty (Leviticus 20:13). But less than a decade before P's attempt to force gay men to start breeding tithe-paying believers, D had written favorably of "the woman who fulfills your physical needs, or the male lover who means as much to you as your own breath." (Deuteronomy 13:6) The claim that Moses authored both of those incompatible views is absurd.

But for most of the differences between doublets, the explanation is not philosophical disagreement but the inevitable consequence of the writers being separated by time and geography. Consider the two accounts in Genesis of Joseph's sale into slavery. E's, beginning at 37:18, reads:

> Yosef's brothers saw him in the distance, and before he reached them, they plotted together to kill him and throw him into a well. But Reuwben overheard, and rescued him from their clutches by suggesting, "We don't have to kill him." And Reuwben advised them, "Instead of shedding blood, throw him into this well in the desert unharmed." His intention was to rescue him from their clutches and return him to his father. So they seized him and threw him into an empty well containing no water. Some Midyanite merchants passed by. They reached down and lifted Yowsef out of the well. So when Reuwben returned to the well, he saw that Yowsef was not in the well, and he tore his clothing. He returned to his brothers and told them, "The boy's gone. What am I going to do?" Meanwhile, **the Midyanites sold him in Egypt** to Powtiyfar, one of Pharaoh's eunuchs, commander of his bodyguard.

J's story is equally complete, beginning at 37:19:

> They said to one another, "Look the dreamer's coming. How about we kill him? We can say that a wild animal ate him. Then we'll see what becomes of his dreams." When Yowsef reached his brothers, they stripped

off Yowsef's coat, the coat with the full sleeves that he was wearing. While they were sitting eating their bagels, they looked up and saw a caravan of Yishmaelites coming from Gilead, with their camels loaded with gum, balm and resin that they were taking to Egypt. Yahuwdah asked his brothers, "Where's the profit in killing our brother and covering his blood? Instead, how about we sell him to the Yishmaelites? It should not be our hands that harm him. After all, he is our brother, our own flesh." His brothers agreed, and they sold Yowsef to the Yishmaelites for twenty silver coins. They took Yowsef to Egypt. His brothers meanwhile took Yowsef's coat and killed a kid and dipped the coat in the blood. Then they sent the full-sleeved coat back to their father, asking, "Do you know whether this coat we found is your son's or not?" When Yowsef was taken to Egypt, **he was bought by an Egyptian from the hands of the Yishmaelites** who had taken him there.

Believers could argue that, while each of the separate versions is coherent, so is the un-separated version—except for one thing. As it stands, the Torah now shows Joseph being sold into slavery in Egypt twice, once by Midyanites and once by Yishmaelites. Could a single author (Moses) really have been that stupid? Logic says that he could not.

## 12. Priestly Power and the Role of Sin

Perhaps as early as 30,000 years ago, *Homo sapiens* created the first gods. He did so—in the context of his limited knowledge—in response to the question: Why? Why do the stars not fall out of the sky? Why, among the myriad of unmoving stars, are there five wanderers? Why does rain fall at intervals on good cropland, but not in the desert? Why is there a growing season and a barren season? Since the beings that orbit the sky live forever, why don't humans? And they found the answer: All things happened as they did because some form of living intelligence had so decreed. The planets wandered, the sun gave forth light and heat, the moon died and was reborn at monthly intervals, and the earth produced the food that sustained life, because they all were living creatures with powers far beyond the capacity of mere humans. And so were born humankind's first gods, not at first objects of worship capable of aiding favored followers, but simply immortals whose existence was noted as the existence of an equally immortal river might be noticed.

There is no way of gauging the elapsed time from the creation of the first gods to the creation of the first religion; for mere belief in gods did not constitute a religion. Not until the first sun-worshipper turned his face toward the sky and asked it to ripen his crop in exchange for a gift, or until the first ambitious junior executive asked a river-god to drown her rival, also in exchange for a designated gift, did nature deification evolve into religion.

Once the idea had evolved that a god could intervene in human affairs, it was just a matter of time before most events that affected humans were ascribed to a god. As more and more mundane accidents, ranging from a successful crop to annihilation by an erupting volcano, came to be attributed to divine intervention, the necessity of appeasing the gods by ritual, flattery and sacrifice became more and more entrenched. Even so, religion did not immediately acquire the status of the mind-crippling tyranny that it was later to become. What made that development possible was the invention of a concept that was a logical consequence of the victory of male-god worshippers over the worshippers of the much older female deities: SIN.[1]

Sin was the Babylonian moon goddess. Following the invention of male gods *c* 3500 BCE, everything that the male-liberationists wished to abolish as female-oriented was categorized as sacred to Sin, and therefore hateful to Shamash, or Zeus, or Yahweh, or Allah. How long it took the primitive god-pushers to realize the tremendous value of *sin* as a means of securing their grips on the minds of entire populations is uncertain; but within a few centuries a priestly caste had arisen that recognized the enormous degree to which its power depended on convincing the masses that they were all congenital *sinners.* And since only a small percentage of the population would ever commit such genuine immorality as killing or stealing, it was to the priesthood's advantage to invent new sins so self-evidently *not* immoral that the masses *would* commit them.

The solution was to impose such severe and absurd restrictions on who could engage in sexual recreation, and on where, when, how and with whom, that all but the most chronic masochists (Buddhist monks, Christian nuns) would refuse to comply. The result was a population conditioned to believe that *sin* was inevitable, and that participation in a priest-led ritual was the sinner's only means of buying salvation. The more guilty a believer could be made to feel for his moments of rationality when he broke irrational taboos, the more absolute and tyrannical became the power of the priest. There is some suspicion that the current pope's **[John Paul II]** affirmation of taboos calculated to overpopulate the human race to the point of starvation-extinction, taboos that he knows will not be observed by the sane majority, is simply a continuation of the priestly practice of keeping the masses guilt-ridden in order to maximize personal power.

All taboo-makers are self-serving. Males invented "marriage" and "adultery" to turn women into privately owned breeding stock. Zarathustra banned both homosexuality and celibacy, since each prevented the practitioner from breeding more Zoroastrians. The Priestly author of Leviticus devoted much of his work to regulations that entrenched the interests of the priestly caste. The Priestly author was, however, also a realist. He recognized that, then as now, priests constituted a parasite class that could not survive indefinitely if its uselessness became widely recognized.

It was to forestall that recognition that the priestly author invented laws whose purpose was to keep the priesthood busy performing an endless succession of daily, weekly, monthly and seasonal sacrifices that would give them the appearance of being harder workers than the masses on whom they sponged. Modern Christian priests perform 6 AM masses for essentially the same reason.

So long as religion survives, so will priests (or their equivalent), and so will theocratic tyranny. To further all three, more than one apologist for god mythology has argued that, "If God did not exist, Man would have had to invent him" (as of course he did). That argument tacitly acknowledges that the god addict subjugates himself to superior lifeforms in the sky in order not to have to accept responsibility for his own situation. Perhaps worshippers do have such an inadequacy—now. But the argument also presupposes that the inadequacy is innate, and that God-the-slavemaster was created by persons already possessed of a slave mentality.

Such reasoning is not merely abhorrent; it is unnecessary. It seems more likely that, just as the Uncle Toms developed slave mentalities only as a consequence of being slaves, so compulsive god slaves could have become what they are only in a world in which Ol' Massa in the sky was already a part of their conditioned thought processes. **Humans did not create gods because they needed mental enslavement; they developed the slave mentality as a consequence of their creation of omnipotent gods.** [boldface added by *Humanist in Canada* editor]

It is said that people get the rulers they deserve. Godworshippers have priests. Persons such as Torquemada, Khomeini, Paisley, Falwell, Begin and Wojtyla do not constitute aberrations, individuals who contravene the true spirit of religion; they *are* the true spirit of religion.

1    The English word *sin* and the Babylonian concept, "sacred to Sin," are not etymologically connected. They are, however, identical in meaning, and that is the point of the article.

## 13. Once Upon a Time in the Sky

Novelist Trevor Hoyle once acknowledged a debt "not least to the authors of the Judeo-Christian Bible, King James Version, which has got to be the greatest work of science fiction ever written." Most science fiction fans would disagree. First of all, the Bible is fantasy, a genre only the undiscriminating equate with science fiction. And secondly, the adjective "greatest" should surely be reserved for books such as *Stranger in a Strange Land* that promulgate the valid moral philosophy that has been the unifying theme in all worthwhile science fiction for more than sixty years. The moral philosophy of the Bible, in contrast, is more akin to Adolf Hitler's *Mein Kampf* and the Marquis de Sade's *Juliette.*

For example, the Bible's earliest author, at the very beginning of his narrative, has his hero, the god Yahweh, tell Adam and Eve (paraphrased), "This is a pomegranate. It is tasty, delicious, scrumptious and mouth-watering. You're not to eat it. If you eat it I'll kill you. And I'll kill your children, and your children's children, and their children, forever and ever. Now run along and have fun, kiddies."

Naturally Adam and Eve ate the pomegranate. Having been shown that the cookie jar was unguarded, how could they have not eaten from it? And naturally Yahweh carried out his threat. To this day, according to biblical literalists, Yahweh is executing a quarter-million men, women and children each and every day in reprisal for that alleged offense of their distant ancestors. And in case the message is missed, Yahweh reiterates in Exodus 20:5 that he intends to go on punishing descendants for their ancestors' crimes. So do believers question Yahweh's morality? They do not. Apparently, "When God does it, it's not evil." And when the president does it, it's not illegal. Thus spoke Richard Nixon.

Many Christians take the attitude that no lesson can be drawn from their Bible's "Old Testament" books. So what, they shrug, if Yahweh used to be a capricious, sadistic mass murderer? So what if he instructed his pets "When Yahweh your gods [generic plural] has settled you in the land you're about to occupy…You're going to exterminate (the inhabitants) in a massive genocide until they're eliminated"? (Deuteronomy 7:1-23) So what if he inflicted Job with unspeakable cruelty just to see whether he would question the god's right to do anything it wished? That was the old god. The new god is nice.

Unfortunately, one can only conclude that persons who think the New Testament god is nicer than his Old Testament self have not read the Christian books. For example, Paul of Tarsus justified Yahweh's partisanship by arguing, "Doesn't the potter have power over the clay, to make from the same lump one vessel in honor and the other in dishonor?" (Romans 9:21) And when Jesus' students asked him why he preached in riddles, he explained that, if he used plain language, too many listeners might convert and he might accidentally cause the salvation of persons whom his omnipotent sky Führer had predestined for damnation. (Mark 4:10-12)

Among the allegories Jesus allegedly preached is one whose moral can be summarized, "Cheat those who are no longer useful to you, and use the stolen money to bribe those who are in a position to do you good." (Luke 16:1-9) Did Jesus actually preach anything of the sort? The answer is irrelevant. It is the Bible that is being evaluated here, not Jesus. And the Bible is unequivocally a paean to evil.

That is not to say that Jesus might have been a much nicer man than the Bible paints him. As an Essene, raised to believe in a purgatory where even the righteous were barbecued with flamethrowers for the purpose of burning away the stain of sin and purifying the dead soul in readiness for its reward, Jesus almost certainly did preach the sadistic "hell" doctrine attributed to him in the Christian gospels. But he (or his posthumous scriptwriters) made one slight change. He (they) changed the purpose of the torture from purification to punishment, thereby transforming the

Zoroastrian/Essene/Pharisee purgatory into the Christian hell. A sicker, more evil concept than Hell cannot be imagined, and whether or not Jesus preached it, the Bible certainly preaches it. By comparison, de Sade was a philanthropist.

Since the Bible was compiled by editing separate, self-contained versions of the same stories into a single narrative, not surprisingly it is riddled with internal inconsistencies. Malory's *Le Mort D'Arthur,* constructed by the same method, shows persons killed in one chapter turning up alive in a later chapter, and the same can be seen in Homer's *Iliad.* The Bible shows giants alive in Moses' time (Numbers 13:33), as well as the descendants of Cain, even though the giants and the Cainites had both been killed off in Noah's flood. And the story of Moses producing water out of a rock appears twice (Exodus 17:1-7; Numbers 20:2-13), forcing literalists to rationalize that the same fantastic event occurred twice—as if finding water for two million Israelites and their livestock in the middle of the Sinai desert even once did not sufficiently strain readers' credulity.

Of all English Bible translations produced by organized religions, not a single one acknowledges that the Torah authors were not monotheists. Everywhere that the Hebrew refers to *ha-elohim,* meaning "the god and goddess committee," English Bibles substitute the singular, masculine, proper name "God." And with the exception of *The Jerusalem Bible* and *The New World Translation,* English Bibles falsify the proper name, *Yahweh,* into "the Lord." (The *NWT* uses "*Jehovah,*" a mistranscription but not a deliberate falsification.) And all English Bibles have Jesus' students address him as "Lord," even though that translation of *kyrios* was correct when the King James Bible was written in 1611 and "Lord" was a title of rank, akin to "House of Lords," but is incorrect today, when "Lord" has ceased to be a synonym for "Master" and has acquired connotations of divinity. A correct translation would be "Your Lordship" or "Sire," depending on whether the speaker accepted Jesus' claim to be his king.

Other falsifications designed to create the illusion that Bible authors believed the same things present-day god addicts believe include the translation of the same word as "sky" when it refers to the abode of birds, and "heaven" when it refers to the abode of gods, even though neither Hebrew nor Greek makes any such distinction, and a distinction between "messenger" and "angel" that similarly does not exist in either original language.

Even ignoring the differences between Orthodox Bibles which contain 3 and 4 Esdras and other Bibles that do not, between Catholic Bibles that contain I and 2 Maccabees and Protestant Bibles that do not, and between Christian Bibles that contain Mark and Matthew and Jewish Bibles that do not, *all* Bibles contain approximately one thousand internal inconsistencies, pairs of statements that are mutually incompatible and therefore cannot both be true. For example, Daniel, written by a Pharisee, endorses the reality of life after death. Ecclesiastes, by a Sadducee, explicitly repudiates life after death. Matthew tells birth tales that could be true only if Jesus was born during the lifetime of King Herod. Luke's birth tales could be true only if Jesus was born ten years after Herod's death. And of the Bible's 18,000 external inconsistencies, the most blatant are the seven descriptions of a flat earth. (Genesis 1:1-19; Isaiah 40:21-22; Job 22:14; Daniel 4:10; Matthew 4:8; Revelation 7:1; Psalm 103:12)

So why does religion not face reality and acknowledge that its Bible is a work of the imagination? The answer is that, without the Bible's say-so, there is no reason whatsoever to believe in gods capable of rewarding sycophants with eternal life. And without the mind-deadening opiate of an afterlife belief to annul the terrifying finality of death, an estimated one-sixth of the human race would have to be institutionalized and diapered. So, sadly, we can expect the Bible to be around for a long time.

# 14. Revelation:  Invective Against the Christians

Until recently, most attempts to date the last book of the Christian Bible and identify the targets of its invective, particularly "666," have floundered on the various daters' unwillingness to recognize that Revelation is neither a Christian document nor the work of a single author.  Consequently, awareness that Revelation was completed in the reign of Domitian has led to the equation of Domitian with "666," even though the 666 passages were composed at a time when Domitian was a teenager.  And the determination to identify "666" as a persecutor of Christians rather than a persecutor of Jews has led to his being identified as Nero, even though 666's predecessor is readily identifiable as Nero.

Two men wrote Revelation.  The earlier of the two was an Essene, a member of a Jewish sect that regarded the Righteous Rabbi executed in 104 BCE as the sect's messiah.  It is possible that he had never heard of Jesus the Nazirite, even though a neo-Essene sect in Jerusalem known as the Nazirites had abandoned the Righteous Rabbi and adopted Jesus as its messiah.  The redactor of Domitian's reign, who appended six additional chapters at the beginning and end of the sixteen composed by the Essene, was a Nazirite, a member of the sect that viewed Jesus as the successor and equal of King David, and the Christians as pseudo-Jewish infidels.

The original Essene Revelation was written in barbaric Koine Greek by a man whose native language was Aramaic.  Its composition can be precisely dated to the month between the Roman occupation of the Jerusalem temple in July of 70 CE and the razing of the temple in August of 70 CE.  That is made clear by the author's "prophecy" that the temple courtyard would be "given to the infidels," but the temple itself would never be destroyed. (11:1-2)  And the author's status as an Essene is revealed by his reference to the war in the sky in which Mars/Michael and his messengers expelled Venus/Satan and his messengers. (12:7-9)  That Zoroastrian myth, found in the Essene *Book of Enoch*, was never part of orthodox Jewish mythology.

The original Revelation was an invective against Rome, and its central theme was a never-fulfilled prophecy that the imperial capital would be totally annihilated for daring to besiege Jerusalem. (17:3 to 18:21, especially 17:9, 18:8)  Further invective was directed against the two emperors responsible for the siege.  The first was Nero, in whose reign the war had started and who first sent Vespasian to Jerusalem "to wage war against the Essenes/Saints to defeat them." (13:7)  The Essene author referred to a "beast" who had been "fatally killed, but its fatal wound had healed." (13:3)  That beast could only have been Nero, who was known to have committed suicide but was nonetheless rumored for many years to have cheated death and to be planning a comeback.  The second beast, whom the author significantly described in the present tense, was the notorious "666." (13:18)

The beast whose number was 666, who "is exercising all the authority of the first beast" (13:12), was Nero's *de facto* successor in Jerusalem, Vespasian.  Vespasian was responsible for the assault on Jerusalem, and while it is possible that all Romans had not yet accepted him as emperor at the time of writing, the Jewish historian Josephus had certainly done so.  It is not necessary to show a sample calculation to prove that Vespasian could be 666.  Given that titles, adnomens and various other details were optional, including the language used, it is possible to make absolutely any name add to 666, or 616 as shown in a small number of manuscripts.

The anti-Roman author was nothing if not confident.  Writing at a time when Jerusalem had already been under siege for three years, he prophesied that the Romans were "going to trample over the sacrosanct town for forty-two months." (11:2)  In predicting that the war would end in a matter of days, at least as far as Jerusalem was concerned, he was perfectly correct.  Unfortunately for his credibility, he failed to predict that the Romans were going to win.  He also miscalculated by 150 km the location of the last battle. He placed it at Mount Megiddo, "Armageddon," north of Jerusalem. (16:16) In fact it took place at Masada, south of Jerusalem.

Following the defeat of the Roman army at *har-Magedon*, Rome itself was to be burned to the ground. (chapters 17-18) Possibly the author expected his readers to believe he had written his prophecy before the great fire of 64 CE, although passing off the fire as a consequence of a war that started two years later seems so unlikely that he was probably predicting another fire that never happened. After that was to come the extermination of all who did not have Yahweh's *tefillin* strapped to their foreheads. (9:4) However, while wearing a *tefillin*, a leather pouch containing selections from the Torah, might have been a reliable external indication of who were the Saints/Essenes, it was not in itself the criterion for who were to be saved. Yahweh's ethnic cleansers were to wipe out the entire population of the earth, "except the 144,000 who had been ransomed from the land, celibates who had never degraded themselves with women, for they had remained unpolluted." (14:3-4) The 144,000 were to consist of 12,000 from each of the tribes of Israel. For some reason, however, the "twelve tribes" included Joseph and Manasseh, but not Ephraim or Dan. (7:4-8) The Essene author left no doubt of his sect's unequivocal hatred of the human race, a hatred that, according to Tacitus, was already being manifested by the Christians (*Annals* 15:44), for he consigned to the infernal flamethrowers *all* non-Jews, *all* women, *all* non-Essene Jews, and even all secular Essenes, for the unforgivable sin of abandoning celibacy. To a celibate Essene monk, salvation was restricted to 144,000 celibate Essene monks. Not until Revelation was expanded and interpolated by a Nazirite at a time when the alleged virtue of self-inflicted celibacy was being downplayed, were copulators, including circumcised converts to Nazirite Judaism, admitted into Yahweh's elect: "A huge crowd which no one was able to number, from every infideldom and tribe and nation and language." (7:9) Obviously, that "huge crowd" would not include any uncircumcised Christians, members of the infidel religion repudiated by the Nazirites even before the death of its founder, Paul of Tarsus.

The Nazirite redaction of Revelation was made toward the end of the reign of Domitian (81-96 CE) by a man named John of Patmos, who was neither the author of the anonymous gospel to which the name *John* has been arbitrarily attached, the anonymous Elder and his copycat who wrote two of the letters similarly designated *John*, nor any of the Johns named in the gospels. Like his predecessor, the redactor wrote in Koine, an unlearned dialect that fell so far short of the educated Greek of the John gospel, that attributing the two writings to the same author is an absurdity.

The Nazirite redactor, who composed the letters to several Nazirite communities with which the edited work now begins, endorsed his predecessor's attacks on Rome, but he also added a new target: the Christians and their founder, Paul. He wrote "I'm aware of the ranting of those who call themselves Ioudaians but are not, but are a synagogue of the Enemy." (2:9) "I have a few things against you, that you have some there who adhere to the teaching of Balaam." (2:14) "I do have this against you, that you tolerate the woman Iezebel, who calls herself a spokeswoman but beguiles my slaves to practice sexcrime and to eat food dedicated to godlets." (2:20)

By "those who call themselves Ioudaians but are not," the Nazirite author meant the uncircumcised Christians. The "teaching of Balaam" represented the Nazirite view of the teaching of Paul. The identity of the "woman Iezebel" is uncertain, but it has been theorized that she was Priskilla (whom at least one scholar thinks wrote Luke-Acts, while another thinks she may have written Letter to the Hebrews), mentioned in Acts 18:18 as an associate of Paul.

Revelation was a purely Jewish document. In its original form it was an invective against Rome. It was written with some circumspection, so that any Roman into whose hands it fell would not recognize it as such. Rome was not actually named. Instead, the Essene author identified his target as "Babylon." However, his description of the city, locating it on "seven hills," (17:9) left no doubt of its true identity. His equation of Rome with Babylon was clearly intended as an equation of the Babylonian Captivity of 586-538 BCE with the current Roman occupation of Jerusalem. That he had to go to such lengths to indicate his *Babylon*'s true identity proves that no such equation had previously existed. When Peter wrote his letter from Babylon two decades earlier, he was really writing from Babylon, not from Rome, which he never visited in his life.

The redactor's additions were equally Jewish. For about a decade, ending in 60 CE, the Nazirites, led by Jesus' brother Jacob, had recognized Paul's Christians as Nazirites, converts to Judaism. By the reign of Domitian, that recognition had long ceased, and Nazirites were referring to Christians as "those who call themselves Ioudaians but are not." While Revelation was not the only non-Christian writing to be included in the Christian canon (James, Jude and 1 Peter were also Nazirite), it was the only anti-Christian writing to be categorized by the early Christians as sacred scripture. Since it restricted admission to the afterlife to circumcised Jews, its inclusion in the Christian Bible is puzzling, to say the least.

## 15. Yahweh: A Morally Retarded God

Societies create their gods in their own image. Inevitably, therefore, cultures that were perfectly sane and virtuous by the standards of five thousand, or four thousand, or two thousand years ago, but insane and evil by the standards of today, created gods that were likewise insane and evil by the standards of today. That would not have occurred had the gods been allowed to evolve with their creators. Unfortunately, the invention of writing led to the depiction of capricious, temperamental, xenophobic, genocidal, morally retarded gods in sacred scrolls in which their every atrocity and irrationality was not merely acknowledged, but unequivocally applauded by their equally vicious and irrational creators. The consequence was that, while the godworshippers who created them continued evolving, their gods' moral evolution ceased as soon as their concept of right and wrong was frozen in a "bible"—in the case of the Jewish and Christian gods, more than two thousand years ago. By categorizing fiction composed by a morally retarded culture as "revealed truth," Judaism and Christianity have been to this day saddled with gods that conform to the moral standards of long ago but not to those of today.

In the reign of King Rekhobowam, *c* 920 BCE, the mythistorian known as the Yahwist justified genocide by making the savage conquest of Judah a consequence of patriarchal curses inflicted upon the conquered peoples' ancestors by the god of the victors. Two hundred years later, the Elohist further absolved his ancestors of guilt by having Yahweh personally order the exterminations that the volcano god's Chosen Nation had piously carried out. (Joshua 8:26-27) The Deuteronomist, writing in 621 BCE, confirmed Yahweh's culpability, but put into the mouth of Moses the explanation that Yahweh's no-survivors policy was designed to prevent the worshippers of an opposition god from passing on their heresies to true believers. (Deuteronomy 20:16-18)

The Deuteronomist's god ordered the rampaging Jews to behave in a manner that is best evaluated by comparing it to the behavior of a modern-day Yahweh:

> When Yahweh your gods has settled you in the land you're about to occupy, and driven out many infidels before you, you're to cut them down and exterminate them. You're to make no compromise with them or show them any mercy. (Deuteronomy. 7:1)
>
> Instead, this is how you're to deal with them: You're to destroy their altars, smash their stone phalluses, chop down their wooden vulvas, and burn their sacred icons. For you are a nation consecrated to Yahweh your gods. Yahweh your gods has chosen you to be his special nation, ahead of all nations on the surface of the land. (Deuteronomy 7:5)
>
> You're going to exterminate them in a massive genocide until they're eliminated. (Deuteronomy 7:23)

> When your Führer has settled you in the land you're about to occupy, and driven out many inferior races before you, you're to gas them down and exterminate them. You're to make no compromise with them or show them any mercy. Instead, this is how you're to deal with them: You're to destroy their synagogues, smash their mosques, chop down their wooden crosses, and burn down their phylacteries. For you are a nation of Supermen, a Master Race chosen by nature to rule over every inferior race on the surface of the land. You're to exterminate them in a massive genocide. (Adolf Hitler, paraphrased)

Yet at the same time that the Deuteronomist was justifying the atrocities of Joshua and King David by having their brutality retroactively ordered by Yahweh, he also ordered, "Be compassionate to foreigners, for you were foreigners in the land of Egypt." (10:19) How the Deuteronomist could preach universal compassion while simultaneously applauding and justifying genocide and enslavement is difficult to understand. Perhaps, like a growing number of modern-day godworshippers, he was genuinely horrified by the atrocities of the past, but was able to brainwash himself into the belief that the tribal god's orders, no matter how monstrous, could only be incomprehensible, never unjust. That the atrocities were of purely human origin, and that the god accused of ordering them perhaps did not exist, did not cross his mind. It certainly did not occur to him that Yahweh *could not* exist, because if he ever had existed he would have long before tortured himself to death from the sheer joy that he derived from hurting living things.

The Deuteronomist created a racist god that did not regard the murder of gentiles as immoral. The Priestly author, who wrote between the composition of Deuteronomy in 621 BCE and the fall of Ninevah in 612 BCE, expanded on that concept and described a god so xenophobic that Jews who showed any toleration for gentiles could expect to be exterminated:

> A Yisraelite man was seen bringing a Midyanite woman to meet his family, in full view of Mosheh and the Yisraelite community as they wailed at the entrance to the Tent of Meeting. Fiynkhas the priest, the son of Eleazar ben Aharon, saw them and left the assembly. He picked up a spear and followed the Yisraelite man to his tent, and there speared them both though, the Yisraelite man and the woman, through her genital orifice. Thus the pestilence that had struck the Yisraelites was halted. Those who died in the epidemic numbered 24,000. (Numbers 25:6-9)

It is not uncommon for modern-day Christians, confronted by the foregoing horror stories, to rationalize that the god of the Old Testament was indeed bloodthirsty and vengeful, but that their New Testament god is not like that—while simultaneously maintaining that the Christian Yahweh *is* the Jewish Yahweh. That rationalization is tantamount to an acknowledgment that their paramount god *used to be* a homicidal psychopath, but has morally evolved. In fact of course they are right. The only problem is that a god whose morality can evolve to keep up with his worshippers is clearly revealed to be a human invention. For an objective existing god that was perfect then and is perfect now, such moral evolution is by definition impossible. Either the god Yahweh is the epitome of absolute evil that the Judaeo-Christian Bible portrays, or he does not exist. Fortunately, the latter alternative is the true one.

## 16. The "Second Coming" and the Wandering Jew: Salvaging a Failed Prophecy

It is an article of faith to Christians that Jesus prophesied his "Second Coming," and that that prophecy will one day be fulfilled. But did he in fact prophesy anything of the sort? In view of Jesus' belief that he was Messiah, the king who could not die until he had overthrown the Roman overlords and been crowned in Jerusalem as the successor and equal of King David, we can reasonably say that he did not. However, he did make a prophecy that, after his execution, was interpreted by his followers as a promise of a second coming. After all, he had died and been resurrected (or so they inferred from the disappearance of his body), and the liberation of Judea and restoration of the Jewish monarchy that he had prophesied remained unaccomplished. It therefore logically followed that there had to be a second coming in order for the prophecy to be fulfilled.

All three synoptic gospels put a time limit on Jesus' second coming. However, by the time Mark was written, forty years had passed and still he had not returned. While the Christians could accept their messiah's tardiness up to that point, they found it impossible to believe that he had revealed his identity to a generation that would never see the restored theocratic monarchy. Mark accordingly recorded a prophecy that in all likelihood Jesus actually made, that he would be crowned king of an independent Judea within the lifetime of the persons listening to him preach. Matthew and Luke, who each wrote at a time when it was still believed that the prophecy could be fulfilled, copied it from Mark. John, who wrote after Jesus' time limit had expired, not surprisingly left it out. In fact Jesus had expected the prophecy to be fulfilled within a matter of months at the outside, without any intervening death and "second coming."

### MARK
There are some standing here who are not going to experience death until they have seen Allah's theocracy established by force.

### MATTHEW
There are some standing here who are not going to experience death until they have seen Ben Adam established in his theocracy.

### LUKE
There are some standing here who are not going to experience death until they have seen Allah's theocracy.

Mark's reference to the establishment of the theocracy by *dynamis* (power, might, force) fully refutes the Christian pretence that the Jewish Jesus meant something other than his Jewish audiences imagined him to mean. Allowing that his hearers included small children, some of whom might conceivably have lived for a further ninety years, we can calculate that Jesus promised to end Roman overlordship and have himself crowned king of an independent Judea no later than 120 CE. He seems to have been delayed.

Jesus' unfulfillable prophecy proved an embarrassment to Christians almost from the moment his self-imposed time limit expired. In an attempt to make the impossible possible, desperate Christians in medieval times invented the Wandering Jew, a man who heard Jesus' prophecy and reviled him with it as he was dying on the stake. Jesus in consequence cursed the man to remain alive until his second coming. Although widely believed for a long time, and therefore a great comfort to Christians who were thus enabled to believe that the prophecy *could* be fulfilled, the tale of the Wandering Jew was one myth that official Christianity was wise enough not to canonize.

On the other hand, no Pope or other Head Christian has ever officially repudiated the story—and cannot. For if there is no Wandering Jew, then Jesus' alleged promise of a Second Coming within the lifetime of "some of those standing here" cannot ever be fulfilled.

## 17. Messiah: What the Title Meant to Jesus the Jew

At about the beginning of the tenth century BCE, King David founded the dynasty that would rule Judah for five hundred years. In 586 BCE, that dynasty was deposed, never to be restored. As a fact of history, the Davidic dynasty's termination was less of a surprise than was the fact that it did not occur centuries sooner. To the Jews whom David's descendants had ruled, it was a traumatic event that their sacred writings had promised could never happen.

During the reign of David's grandson, the Jewish historian known as the Yahwist had written, "The royal staff is not to pass from Yahuwdah, nor the royal power from between his legs, until the one comes to whom...the nation is to grant obedience." (Genesis 49:10) The Yahwist also wrote: "A star from Yaakob takes charge. A scepter arises out of Yisrael." (Numbers 24:17)

Three centuries later, utilizing a biography of David written by the spokesman Nathan (1 Chronicles 29:29), the earliest of the chroniclers wrote:

> The word of Yahweh came to Nathan, saying: "Go and tell my slave David...'I'm going to set up your descendant after you, begotten of your body...and I'm going to establish the throne of his kingdom for ever.'" (2 Samuel 7:4-13)

A little later still, probably within a decade either side of 600 BCE, another chronicler made that promise conditional:

> Yahweh appeared to Shlomoh...and said to him..."If you obey my laws and my whims, then I'll establish the throne of your kingdom forever, as I promised to your father David when I said, 'You'll never lack a man on the throne of Yisrael.'" (1 Kings 9:2-5)

Yahweh's promise that a descendant of Judah, a star from Jacob, would become king of the Jews was fulfilled in the person of King David. Since, as mentioned, David's grandson Rekhobowam was already king at the time of writing, the Yahwist's powers of prophecy were not severely taxed in making such a prediction. On the other hand, the chronicler's prophecy that the descendants of David would rule Israel *forever* was falsified in about 930 BCE. If the prophecy is viewed as applying to Judah rather than Israel, then it was falsified in 586 BCE. No descendant of David has been king anywhere since that date. Now that the Davidic line is extinct, none ever can be.

To the Jews of the Babylonian Captivity, the idea that a scriptural prophecy had failed was unacceptable. Yahweh had promised that a descendant of Judah would establish a Jewish monarchy that was to last forever; therefore it must be so. He had promised that David's son/descendant would usher in the permanent, uninterrupted monarchy; therefore it must be so. That son had not been Solomon, whose line had been deposed in 586 BCE; therefore he had to be a distant son, a descendant of David as yet unborn. Thus was born the concept of "Yahweh's Anointed," *MashYah* (Messiah), the Christened (anointed with chrism) King of the Jews.

The Messiah, once conceived, was promptly backdated. The old foreshadowing passages that had retroactively prophesied the coming of King David were now seen as messianic. Almost immediately, Isayah's prophecy, written while the Davidic line still reigned in Judah, that, "A shoot will emerge from the stem of Yishay, and a sprig out of his roots," (Isaiah 11:1) who would bring back a remnant of the conquered inhabitants of the northern kingdom of Israel from Assyria, Egypt, Persia, Uganda, Arabia, Babylonia, Nimrod and Cyprus (11:10-11), was reinterpreted to apply to the king who would refound David's dynasty.

The first writer so to interpret the older prophecies and repeat them in the light of the new orthodoxy was Jeremyah. Jeremyah quoted Kings 9:5: "For Yahweh said this: 'David will never lack a man to sit upon the throne of Yisrael.'" (Jeremiah 33:17) Even though the prophecy quoted was the conditional one to Solomon, Jeremyah ignored the conditions and treated the prophecy as utterly binding. He added his own reinforcement: "'You see the days approaching,' says Yahweh, 'when I'm going to perform that good thing that I've promised…I'm going to cause a righteous sprout to grow up unto David.'" (Jeremiah 33:14-15)

Jeremyah wrote in about 580 BCE. Sixty years later another messianic prophecy issued from the quill of Zekharyah: "Yahweh, commander of armies, says, 'Gaze on the man whose name is the Sprout. He's going to rebuild Yahweh's temple, and sit down and rule as a priest upon his throne.'" (Zecharia 6:12-13)

Zekharyah's book was written after Darius I had ordered work resumed on the second temple, halted since the first year of the reign of Cyrus. To Zekharyah, the Messiah's main task, since there was no longer a Captivity for him to end, was the completion of the temple. Once that was done, Zekharyah expected him to be crowned and "rule as a priest upon his throne," in other words to become a new kind of king, a theocrat, analogous to the medieval popes and Muslim caliphs. And since Zekharyah was writing about events in the process of happening, he had no difficulty identifying who that theocrat would be: "The hands of Zerubabel laid the foundations of this temple. His hands will finish it." (Zecharia 4:9)

Zerubabel, current head of the Davidic family and pretender to David's throne, did complete the second temple; but the Jews did not "take the silver and gold and make a crown and place it on his head." (Zechariah 6:11)

Consequently, even though Zekharyah had unambiguously indicated that all of his prophecies had referred to Zerubabel, later generations, unable to comprehend that prophecies could fail, concluded that the Messiah promised by Zekharyah was not Zerubabel but a descendant of David *still* unborn.

As years, decades, even centuries went by without any end to Jewish subjugation by one overlord or another, Jews seized upon the Messiah concept and made it the central feature of their hopes for independence. The occupations were seen as temporary. In his own good time Yahweh would order his winged messengers to stop punishing the Jews for past disobediences by continuing to support the foreign overlords. He would then overthrow the foreigners in the only conceivable way, by sending a warlord who would repeat the victories of Joshua and David. That warlord would be the descendant of David, Messiah, by this time firmly believed to have been prophesied in the Torah. David's descendant would put an end to the occupation and establish the permanent monarchy prophesied by Jacob. Such was the belief and the hope.

Then came the Maccabees. Jewish independence was proclaimed in 166 BCE, and recognized in 143 BCE. Zekharyah's prophecy that a theocrat would rule as "a priest upon his throne," written of Zerubabel, was seen to be fulfilled when High Priest Shimeown Hasmon was recognized as ruler of Judah by the Seleukids. The Davidic king, Messiah, was all but forgotten. No more prophecies of his coming were written, and the old ones were ignored. How could Yahweh send a Messiah to liberate Judah when it had already been liberated by the Maccabees? Why would he send an Anointed King when the Jews already had a perfectly good king? Besides, the Book of Daniel had promised that the Hasmonean monarchy was to last forever. (Daniel 7:27)

And then along came Rome. Once again the Messiah became a daily expectation. Every few years a man came along who, usually but not always on the basis of some kind of prophecy-fulfillment, believed that *he* was the Messiah. Among the many claimants to the title were John the Immerser, Jesus the Nazirite, Bar Kokhba, and quite possibly Herod Agrippa I.

Jesus' reasons for believing that he was Messiah are hard to discern. Long after his death, Christians began to interpret a passage in Micah that referred to David's birth in Bethlehem as a prophecy that the Messiah would be born in Bethlehem (Micah 5:2), and the authors of two Christian gospels composed incompatible tales pretending that Jesus had been born in Bethlehem. Jesus' non-Bethlehem birth would not have presented a problem, since the belief that the Messiah should be born in Bethlehem was unknown, both to him and to his detractors. His Galilean birth did, however, become an issue: "You're even a Galilaian, aren't you? Look and you'll see that no prophet is to arise out of Galilaia." (John 7:52)

In declaring that Jesus had been challenged over his Galilean origin, the author of the fourth gospel offered no rebuttal of that challenge, and thereby implied that Jesus had been unable to give the objectors a satisfactory answer. However, the same author also declared that Jesus was asked, "Don't the Writings say that the Messiah is to come from the sperm of David, and from Bethlehem, the village where David used to be?" (John 7:42), and that here also Jesus offered no rebuttal. Since Jesus did attempt to answer the objection that he was not Davidic, and he could not in fact have been challenged over the non-issue of his non-Bethlehem birth, the reasonable conclusion is that the John passages are inventions. Nonetheless, Jesus *was* a Galilean, and was bound to be challenged on that point, and no gospel records any answer Jesus might have given to such an objection.

Jesus did rebut the objection that he was not Davidic, by arguing that the Messiah should not and could not be Davidic. (Mark 12:35-37) Since the Messiah was the descendant of David by definition, Jesus' claim that he was not descended from David but was nonetheless Messiah was totally illogical. Nonetheless, Jesus was apparently such a skilled demagogue that, "The huge crowd blissfully swallowed this." (Mark 12:37)

Jesus' reasons for believing he was Messiah remain incomprehensible. The synoptics show him fulfilling many prophecies; but those "prophecies" were taken out of context and could only be interpreted as messianic by authors with great imagination. Jesus did not, however, have any doubt that the Messiah's function was what it had always been: to overthrow the Romans (formerly the Seleukids, formerly the Persians, formerly the Babylonians), liberate Judea, and have himself crowned King in Jerusalem. He marched into Jerusalem on a Sunday morning in 30 CE, and promptly proclaimed the revolution:

> "Don't imagine that I've come to bring the land peace. I've come to bring not peace but rather a sword." (Matthew 10:34)
> "I've come to set the land on fire, and I wish it were already lit…From now on a household of five is going to be divided…Father will war against son and son against father." (Luke 12:49-53)
> "Anyone who doesn't have a sword is to sell his cloak and buy one." (Luke 22:36)
> "There are some standing here who are not going to experience death until they have seen Allah's theocracy established by force." (Mark 9:1)
> They answered, "Sire, look. Here are two swords." And he told them, "It'll be enough." (Luke 22:38)

Mark's reference to the establishment of the theocracy by *dynamis* (power, might, force) fully refutes the claim made long after Jesus' death that the Jew Jesus saw the Jewish Messiah as something other than his Jewish audience imagined him. Jesus proclaimed the messianic war of independence, and when the Roman garrison came to arrest him for treason, he resisted violently: "And those around him, seeing what was about to happen, asked, 'Sire, are we to strike with the sword?' And one of them struck the High Priest's slave and amputated his right ear." (Luke 2:49-50)

Although Luke went on to add the dubious detail that Jesus magically healed the slave's severed ear, his acknowledgement that Jesus' lieutenants were all armed seems to reflect the true circumstances of the arrest. Mark confirmed this with his reference to Barabbas as one of "the imprisoned revolutionaries who had committed homicide in the uprising." (Mark 15:7)   The significance of Mark's admission that Jesus' arrest coincided with an insurrection in which men were killed, cannot be overemphasized.

Jesus claimed to be the Messiah.  He did not imagine that "Messiah" meant anything but what it had always meant:  the warlord who was the successor and equal of King David.  In the belief that the Messiah *could* not be killed or defeated, he declared war on the Roman Empire with less than one hundred followers, and inevitably paid the price of his conceit.  Had he not been posthumously deified by the Christians, he would probably have won a place in Jewish folklore alongside Bar Kokhba and Eleazar bar Yair.  Such is the irony of fate.

# 18. Why Yahweh's Name Became Taboo

From at least the fourth millennium BCE, the ancestors of the tribe that would one day be called Yahuwdites/Jews lived on the slopes and in the fallout area of Mount Yahuwah, a then-active volcano, and appeased her by throwing captives and criminals into her lava-filled vulva.  Then around 3500 BCE men everywhere made the Big Discovery that children have fathers as well as mothers, and reigning goddesses were superseded by male equivalents.  The goddess Yahuwah was masculinized to Yahweh by dropping the feminine suffix *-ah* from the pronunciation, even though the written form of the deity's name, YHWH, remained unchanged.  The Jews took their volcano god with them when they were expelled from Anatolia-Syria by the advancing Babylonians, and retained him even after settling into southern Phoenicia.

But Yahweh was more than a volcano god.  As a Jewish confederacy developed, incorporating different tribes with different gods, and the incoming gods were assimilated to Yahweh, persons whose ancestral god had not been a volcano came to view Yahweh as the kind of god to which they were culturally conditioned.  Thus the Elohist saw him as anthropomorphic, and wrote, "You can't see my face…I'll cover you with my hand…and you'll see my backside." (E: Exodus 33:20-22) To sheiks and kings from Abraham to Solomon he had been a phallus: "Yaakob erected a stone phallus on which he poured an offering of wine and oil." (P: Genesis 35:14) To a psalmist he was a hawk-god like the Egyptian Thoth:  "He will cover you with his feathers and keep you safe under his wings." (Psalm 91:4) To another psalmist he was a sea god like Poseidon: "Deep calls to deep at the noise of your waterspouts.  All your waves and billows are gone over me." (Psalm. 42:7) But mainly he was a volcano:

J   Mount Sinai was completely wrapped in smoke, because Yahweh had descended on it in the form of fire.  The smoke from it rose like the smoke of a furnace, and the whole mountain shook violently. (Exodus 19:18)
E   Mosheh spoke and the gods answered with thunder. (Exodus 19:19)
E   Yahweh's fire burned among them, wasting all. (Numbers 11:1)
D   What creature has heard the thunder of erupting gods roaring from the middle of the lava, as we have, and survived? (Deuteronomy 5:26)
P   The magnificence of Yahweh was like a devouring fire on top of the mountain. (Exodus 24:17)
R   You heard his orders from the middle of the flames. (Deuteronomy 14:36)

Yahweh was a volcano god, so capricious that he massacred a randomly chosen segment of his own worshippers because, "The nation grumbled, arousing Yahweh's temper," (E: Numbers 11:1) just as Enlil had drowned the human race because, "The uproar of humankind is intolerable, and sleep is no longer possible on account of the babble." (*Epic of Gilgamesh*)  That parallel between two temperamental, homicidal gods tells a good deal about why the Jewish priests in post-Captivity days saw an urgent need to change their paramount god's image.  But right down to the time of the Babylonian Captivity Yahweh was still primarily a volcano:

I called upon Yahweh, and cried out to my Allah.
He heard my voice out of his temple…
Then the land shook and trembled…
There went up a smoke out of his nostrils,

And out of his mouth a fire that devoured.
Coals were kindled by it…
Yahweh thundered in the skies,
And Ilion gave his voice:
Hail and coals of fire…
The land's foundations were exposed, oh Yahweh,
At the blast of the breath of your nostrils.  (Psalm 18:6-15)

It was not so much Yahweh's volcanic or oceanic or hawk-like qualities that eventually led the priestly caste to attempt to purge him from their mythology altogether, as his indelible association with the atrocities of David and Joshua.  For example, it was "in compliance with the orders that Yahweh had given Yahuwshuakh," (P: Joshua 8:27) that the Jews of old had murdered the entire population of Jericho, "both man and woman, young and old." (P: Joshua 6:21)  And it was because "Yahweh hardened Pharaoh's heart so that he would not let the Yisraelites go," (R: Exodus 10:20) that Yahweh had manufactured his excuse for murdering "all of the firstborn in the land of Egypt, from the firstborn of Pharaoh to the firstborn of the prisoner in the dungeon." (J: Exodus 12:29)

Nor did Yahweh limit his mass murders to non-Jews.  When a Jew married a Midianite woman, Yahweh's response to that act of race pollution was to send a plague that killed 24,000 Jews.  The racist god only terminated his murder of innocent bystanders when "Fiynkhas speared them both through, the Yisraelite man and the woman, through her genital orifice." (P: Numbers 25:7-8)  And when the Israelites grumbled against Moses' dictatorial rule, Yahweh hurled down a thunderbolt, "wasting all in the outer fringes of the camp." (E: Numbers 11:1)

Yahweh was a megalomaniac who massacred thousands for the crimes, real or imagined, of one individual.  But he was not above unleashing his malevolent, unstable temper against even his sycophants for reasons that only he could comprehend.  For example, when the Chest of the Treaty was about to fall over, and David's aide-de-camp, Khuzyah, reached out to protect it, "Yahweh's tantrum flared against Khuzyah, and the gods snuffed him for his mistake, and he died beside the gods' chest." (2 Samuel 6:6-7)

Similarly when Balaam, to whom Yahweh spoke through his ass, agreed to accompany the Mowabites in obedience to a direct order from Yahweh, "his going ignited the gods' temper," (J: Numbers 22:22) and Yahweh sent a heavenly messenger to execute Balaam for his obedience.  A similar capricious whim caused Yahweh to come within an inch of murdering Moses, for the crime of not knowing that he should have circumcised his son. (J: Exodus 14:24-26)

All gods were capricious and vindictive, for the logical reason that they were created in their worshippers' own image.  The virgin Artemis murdered Aktaion, having him torn apart by dogs, for the crime of accidentally seeing her naked.  Only Yahweh, however, imposed such disproportionate punishment on children.  When forty-two children laughed at Elisha's bald head, the spokesman called on Yahweh to avenge the insult, and Yahweh obediently sent two bears that tore the children to shreds. (2 Kings 2:23-24)

Yahweh murdered children for behaving like children.  He murdered the many for the crimes of the few.  And unlike any other god before or since, he murdered thousands for the crime of obeying his own orders.  Yahweh ordered David to conduct a census and, when David obeyed, Yahweh executed 70,000 in an epidemic as punishment for the crime of conducting a census. (2 Samuel 24:1-16)  Of course, no biblical author was so insane as to invent such an inconsistency.  In the original versions of the myth, a pro-census chronicler had Yahweh order David to conduct the census, while an anti-census chronicler had Yahweh punish David for conducting the census.  Only when the two incompatible stories were riffled together did David's crime become obeying Yahweh's orders.  But the purgers of Yahweh's name did not know that.

Yahweh was a serial killer. He executed descendants for the crimes of their ancestors. (J: Genesis 3:22-24) He demanded the death penalty for being a priestess of any god but himself (E: Exodus 22:18), for heresy (E: Exodus 22:20), for zoophilia (E: Exodus 22:19), for failing to possess an intact hymen at the time of marriage (D: Deuteronomy 22:20-21), for working on Saturday (P: Exodus 31:14-15), for cursing one's parents (R: Leviticus 20:9), for adultery (R: Leviticus 20:10), for gay sexual recreation by men but not by women (R: Leviticus 20:13), and for fortunetelling (R: Leviticus 20:27). And he endorsed human sacrifice, declaring that, "no human who has been solemnly vowed is to be redeemed. He's to be sacrificed without fail." (R: Leviticus 27:29)

As a slave owner himself, Yahweh endorsed slavery, with the stipulation that Jews could not enslave other Jews: "The descendants of Yisrael are *my* slaves." (P: Leviticus 25:55) "Both your men slaves and your slavegirls are to come from the infidels who surround you." (R: Leviticus 25:44) But he added the loophole that Jews could sell their daughters as concubine-slaves. (E: Exodus 21:7) And like all tyrants, he declared it a capital crime to insult his name. (R: Leviticus 24:16) He was also the kind of sadist who derived pleasure from inflicting pain. (D: Deuteronomy 28:63)

Yahweh had committed atrocities worthy of Joshua or Hitler. But more than that, in a society that was basically monistic prior to the adoption of the Zoroastrian devil, he was responsible for all evil: "Can there be evil in a town, except that Yahweh has done it?" (Amos 3:6) Yahweh's record was indefensible, and the day was coming, the priestly caste knew, when anyone associated with such a god would go down with him.

In view of Yahweh's record, it was no surprise that the priestly caste concluded that he had to go. The history of Allah, the god of the Jews' Israelite allies, was much more respectable, and much less likely to cause the masses some day to turn against a priesthood that endorsed such a god.

As the first step toward severing all connection with Yahweh, the priests declared his name a sacred taboo, never to be spoken under any circumstances. That was not a particularly radical proposition. The belief that a person or a god could be harmed by speaking his name had been around for centuries.

The taboo worked, at least partially. Jesus the Nazirite, a practising Jew, taught his students the following prayer: "Our father in the sky, whose name is taboo, may your theocracy be established..." The Jewishness of Jesus' prayer, declaring Yahweh's name taboo even though Christians have no such belief, is seldom mentioned in Christian pulpits.

As always when practical considerations necessitated a modification of current mythology, there were conservatives to whom the purging of Yahweh's name from the spoken vocabulary was anathema. The spokesman known only as *Malakhi* ("my messenger") bitterly opposed the Levite priesthood's suppression of his god's name, and protested, "'Where is my reverence?' says Yahweh, commander of armies, to you, oh priests, who despise my name...'You have corrupted Levi's treaty,' says Yahweh, commander of armies." Unless the priestly caste restored Yahweh's name to its former dignity, Malakhi threatened, "'I'll derive no pleasure from you,' says Yahweh, commander of armies," and "'My name will be exalted among the infidels,' says Yahweh, commander of armies." (Malachi 1:6; 1:10-11; 2:8)

Even though Malakhi's threats were ignored and the purging of Yahweh's name from the spoken vocabulary continued, Malachi was nonetheless eventually incorporated into both the Jewish and the Christian canons.

Ultimately it was the priests' own pious superstition that defeated their purpose. While they had no qualms about removing Yahweh's name from the spoken language, they dared not delete YHWH, or even such other items as obvious scribal errors, from the written Torah. A form of the name survives, although not necessarily with the original pronunciation. Yahweh is the god of the Jews to this day, even though Jews still deem it a blasphemous breach of taboo to pronounce his name.

## 19.  Where Was Jesus Born?

Sometime after 586 BCE, the first redactor of Micah wrote, "Bethlehem Efrathah...out of you he comes forth to be ruler of Yisrael." (Micah 5:2)  King David had indeed come out of Bethlehem to be ruler of Israel, and when the Jewish bible was formally canonized in the second century CE, the correctness of Micah's retroactive prophecy led to its being included.

There is no evidence that official Judaism ever interpreted Micah's reference to David's hometown as a prophecy of the birth of the Messiah, the descendant of David whose function was to reinstitute the monarchy and restore Jewish independence.  Certainly the author of Mark, writing between 70 and 74 CE, had never heard of such a theory.  He made no mention of Bethlehem, and instead wrote that, "Iesous came from the dispersion (*nazareth*) of Galilaia and was immersed by Ioannes in the Jordan." (Mark 1:9)  The word *nazareth* meant the worldwide community of Jews outside of Judea, but Mark may not have known that or he would have used the Greek equivalent, *diaspora.*  He may well have believed that Nazareth was a specific geographic location.

By the last quarter of the first century CE, Christians—but not Jews—were interpreting Micah's alleged prophecy as messianic.  The disappearance of Jesus' body from its temporary tomb had led to the belief that he had been resurrected and would return to expel the Romans in the near future. So since Jesus was Messiah, and Messiah had to be born in Bethlehem, it logically followed that Jesus must have been born in Bethlehem.

The first writer to state as a fact that Jesus was born in Bethlehem was the anonymous author of Matthew, *c* 95-100 CE.  He wrote that, "Iesous was born in Bethlehem in Ioudaia in the days of Herodes the king." (Matthew 2:1)  To get Jesus from Bethlehem to Galilee, where he was known to have originated, Matthew had him taken out of Bethlehem as a baby and, after a short stay in Egypt, taken to "Nazareth" in Galilee when his father feared to return to his Judean homeland. (Matthew 2:22-23)

Matthew wrote of Jesus' father that, following his return from Egypt with the baby Jesus, "He migrated to...a town called Nazareth, in order to fulfill what was spoken by the prophets, 'He's to be called a Nazirite.'" (Matthew 2:23)  Clearly, Matthew did not believe that Joseph and Mary had lived in "Nazareth" before Jesus' birth, but had been natives of Bethlehem.

The anonymous writer of Luke, in contrast, declared that, in compliance with the terms of the taxation census ordered by Augustus Caesar when Judea was brought under direct Roman rule in 6 CE, Joseph and his pregnant wife "departed from Nazareth town in Galilaia to Ioudaia, to David's town, which is called Bethlehem, since he belonged to the clan and family of David." (Luke. 2:4)

Since Matthew and Luke were agreed that Jesus was born in Bethlehem and had been taken to a town called Nazareth as a child, their disagreement as to whether Joseph and Mary were native Judeans who migrated to Galilee after Jesus' birth, or native Galileans who happened to be visiting Judea at the time of Jesus' birth, may seem minor.  If neither author had any reason to invent a Bethlehem birthplace, then their mutual corroboration could be deemed significant.  But each was aware of the Christian interpretation of Micah's retroactive prophecy, and was desperate to find some way to make his Galilean messiah fulfill it.  Needing to explain how an itinerant preacher from Galilee could have been born in Bethlehem, each racked his imagination for an explanation that he could believe, and they came up with mutually exclusive fantasies.

Luke's attempt to uphold a Bethlehem birthplace led him to invent a fable that was not even credible.  Augustus Caesar did, in 6 CE, order a census in Judea in order to impose Roman taxation on what had hitherto been a dependent monarchy taxed by the Herods.  But only Judea was subjected to the census.  Joseph and Mary were not affected, as they were Galileans, residents of a province still being taxed by Herod Antipas, not by Rome.

The census of 6/7 CE did not include Galilee, for the obvious reason that only Archelaus of Judea, not Antipas of Galilee, had been deposed and replaced by a Roman procurator. But even if Galilee had been affected, a journey to register in an ancestral village would have been an absurdity. Of all the accusations that historians have leveled against Augustus Caesar, none has ever accused him of being stupid. Had any Jew registered as a taxpayer in one province and then returned to his permanent residence in another province, the whole purpose of the census would have been defeated. Since Joseph lived in Galilee, it would have been the Galilean tax collectors with whom he needed to register if Caesar were to benefit.

Luke's birth fable was internally inconsistent. For whereas he had Mary pregnant with Jesus at the time of the first Judean census, 6/7 CE, he also declared that Jesus was born during the lifetime of Herod, king of Judea. (Luke 1:5) Since Herod died in 4 BCE, any child born during Herod's lifetime would have been ten years old at the time of the census. Luke's ability to make such a blunder, and get away with it, is explained by the reality that he was writing a full century after Herod's death. In view of the absurdity of the whole census-journey fable, Jesus' own acknowledgment that he was not a descendant of King David (Luke 20:41-44), his ancestors being gentiles forcibly circumcised by the Hasmoneans, is a minor objection. Luke needed some way to get Mary from Galilee to Bethlehem in order for Jesus to be born there. But the best explanation for his coming up with such an incredible flight of fancy is that he was familiar with a Hindu myth in which Nanda made a journey to Mathura to pay his taxes.

Luke's birth fable can be made internally consistent by assuming that the "Herod" in whose lifetime Luke's Jesus was born was Archelaus or Antipas. It cannot be harmonized with Matthew by any means. For whereas Luke's tale depended on Jesus being born ten years after the death of Herod the Great, Matthew's depended on Jesus being born during Herod's lifetime. According to Matthew, the Herod who tried to kill the newborn messiah by massacring all babies born in Bethlehem later died and was succeeded by his son Archelaus. (Matthew 2:22) The same Herod who was king when Matthew's Jesus was born, had been dead for ten years when Luke's Jesus was born. And those incompatible narratives contain the only claims that Jesus was born in Bethlehem.

Jesus was not born in Bethlehem. Mark indicated as much when he said that Jesus came from *the dispersion.* The anonymous author of John, writing a century after Jesus' death, was more specific. In a passage that may have been copied from an older document, he wrote, "But others asked, 'Since when is Messiah to come out of Galilee? Don't the Writings say that the Messiah is to come from the sperm of David, and from Bethlehem, the village where David used to be?'" (John 7:41-42) If the gospel author or his source had believed that Jesus was born in Bethlehem, he would surely have put into Jesus' mouth a rebuttal of the Sadducees' objection that Jesus was *not* from Bethlehem. John's synoptic source (Luke 20:41-44) had shown Jesus rebutting the objection that he was not David's descendant by arguing that Messiah could not be Davidic, since David called him "Master." John did not show Jesus claiming to be born in Bethlehem, because John's source was aware that Jesus was not born in Bethlehem and had never claimed to be.

The authors of Matthew and Luke were not liars in the usual sense of the word. Each believed that, since Messiah had to be born in Bethlehem, therefore Jesus *was* born in Bethlehem. Each, unaware of the other's existence, figured out an imaginative set of circumstances that could have taken a Bethlehem-born messiah to Galilee, and promptly convinced himself that those circumstances must have occurred. Probably they recognized the conundrum, prayed for guidance and, when their imaginations found a solution, accepted it as a divine revelation. Similar thinking abounds to this day.

Nobody today believes that Jesus was born in Bethlehem, with the exception of fundamentalists who insist that every word in Matthew is inerrant truth, presumably including chapter 4, verse 8, which could be true only if the earth is flat. Rather more believe that he was born in Nazareth, even though no such village existed until present-day Nazareth was given that name by a fifth-century emperor to whom the nonexistence of Jesus' alleged hometown was an embarrassment. Mark's statement that Jesus came from *the nazareth* was misinterpreted by both Matthew and Luke as the name of a town. Matthew compounded his error by having Joseph and Mary settle in Nazareth to fulfill a prophecy that the Messiah was "to be called a Nazirite." (Matthew 2:23) As any philologist can confirm, the word *Nazirite* (Aramaic: *notsri,* Greek: *nazoraios)* cannot be derived from *Nazareth.* Rather, it meant a kind of temporary monk who had placed himself under certain vows of abstinence for a fixed period.

Jesus was known throughout his public life as "the Nazirite," a title that was even affixed to his cross. But Jesus was not a nazirite. He may have taken a nazirite vow at the beginning of his public life. A month or two of temperance would not have been more than a newly converted fanatic could bear. But if Jesus was indeed the drunken whoremonger that the gospels hint (Luke 7:33-34), the title of "Nazirite" may have been attached to him by his detractors as a mark of scorn. Since Jesus had undeniably been known as "the Nazirite," and was known to have drunk wine and associated with prostitutes, the great rationalizer, Matthew, explained away his title by making it refer to a nonexistent town called Nazareth. While all four canonical gospels agreed that Jesus came from "the nazareth of Galilee," only Matthew and Luke imagined that Nazareth was the name of a town.

Jesus' true residence as an adult and probable birthplace can plausibly be identified as Capernaum on the west coast of Lake Galilee. Matthew got him there by writing, "And leaving Nazareth, he went to live in Kafar Nahoum." (Matthew 4:13) Mark located Jesus' first essay into preaching in the synagogue at Capernaum (Mark 1:21), and described him as being "at home" only in Capernaum. (Mark 2:1) It was at Capernaum that Jesus' mother and brothers came to the seashore looking for him "beside the sea." (Mark 3:31; 4:1)

It was at Capernaum that Jesus was assessed for Antipas's taxes. (Matthew 17:24-27) And while the *Hebrews* gospel declared that Jesus' immersion occurred while John the Immerser happened to be in the neighborhood, two canonical gospels located that immersion in the Jordan River. (Mark 1:9; Matthew 3:13) Capernaum was six kilometers from the point at which the Jordan River became the Sea of Galilee. Nazareth, when it was finally instituted, was twenty-three kilometers from the Jordan, with mountain ranges in between.

Finally, when Jesus in a temper tantrum condemned to destruction those towns that had rejected him as an upstart local boy with delusions of royalty, he gave top billing to Capernaum, which he declared to have been "exalted to the sky," presumably because it had given birth to Messiah, and similarly castigated Chorazin and Bethsaida, close enough to Capernaum to have been the sites of his earliest out-of-town tryouts, (Matthew 11:21-24) but made no mention of any "Nazareth." Yet Mark stated that "in his own fatherland" Jesus was "unable to perform any powerful deeds...and was amazed at their incredulity," (Mark 6:4-6) while Luke identified Nazareth (actually it was Capernaum) as the place where an enraged mob tried to lynch him. Jesus' omission of Nazareth from his Galilean invective would have been incomprehensible if such a place had actually existed.

That Capernaum was Jesus' hometown is certain. That he was born there is a reasonable probability.

## 20. The Yahwist's Eden: The Meaning of the Forbidden Fruit Fable

Some time during the reign of King Solomon's son (930-913 BCE), a theologian whom historians have designated the Yahwist, or J (German: *Jahwist),* wrote a tale that attempted to explain the universality of death. As a fanatic henotheist who believed that his god Yahweh was the creator of the cosmos, so powerful that, "No god in the whole land is my equal," (Genesis 9:14) J probably did not question the imposition of death upon the infidels who denied Yahweh his due worship. But he clearly had difficulty understanding why Yahweh would kill his own nation. The only conceivable answer was that in the far distant past some ancestor of Jew and infidel alike must have offended Yahweh beyond measure. To an adherent of the most male-supremacist religion that had existed to that time, there could be no doubt what the offence must have been. Adam and Eve must have committed the abominable and detestable crime of goddess-worship.

At the time that J wrote, the struggle between god and goddess was at its height. Even though J's Ten Commandments (Exodus 34:12-28) decreed that, "You're to pay homage to none of the other gods, for Yahweh is most possessive about his name, Jealous Allah," worship of the Great Mother constituted such a threat to the established religion that J needed to single it out for special treatment. For the Mother had a tremendous advantage. Whereas Judaism had no afterlife belief, and would not have for several more centuries, the goddess was able to offer her worshippers eternal life.

For twenty thousand years, women had been observing the moon goddess die and come back to life on a monthly basis. Somewhere along the line they deduced that her death was a sacrifice, enabling her worshippers to absorb the surrendered portion of her immortality by eating her body and drinking her blood. Since all goddesses were variant forms of the Great Mother, the deified vulva, it followed that her sacramental body must be those fruits that, when fully ripened until they had split down one side, most resembled the source of all life. Thus the most common choices for the vulva of the Vulva were the pomegranate, fig and almond. So entrenched did the triple equation of goddess, vulva and pomegranate become, that the Hebrew word *asherah* came to mean all three.

The Yahwist was sufficiently familiar with the superstitions of the infidels to know that, when he sat down to write a myth in which Adam and Eve committed the crime of goddess-worship, they would have done so by eating the Mother's sacramental pomegranate. There was no risk of his symbolism being misunderstood. Solomon's policy of religious toleration, so hateful to Yahweh's enforcers, had led to the establishment of *asherahs,* vulva-shaped goddess shrines, throughout Judah and Israel. (1 Kings 11:4-8) As a partisan chronicler bemoaned, "They've built themselves phalluses and vulvas on every high mountain and under every evergreen tree, and there burned incense and slaved for godlets." (2 Kings 17:10-12) The Yahwist's purpose in having a tempter tell Adam and Eve, "The gods are aware that on the day you eat it your eyes will be unsealed and you will be like gods," (*i.e.,* immortal—Genesis 3:5) would have been clear to his readers even if the goddess had not made a personal appearance in her traditional snake form.

The Yahwist would have had to be very naïve to imagine that he could pontificate, "Asherah eating does not bring eternal life. It brings death. Take my word for it." But he did not have to do so. At least a segment of his readers would have been familiar with the older Sumerian myth in which the gods, jealous of their immortality, tricked Adamu into *not* eating the fruit of the tree of life that would have enabled him to live forever. All J had to do was "rectify" the older myth so that death was a consequence of eating the pomegranate, rather than the consequence of not eating it.

The Yahwist's fable was written to explain and justify death, and nothing but death. The evolved Christian belief that a tendency to disobey (the literal meaning of *sin*) is a human trait inherited from Adam was unknown to J. Nor did Yahweh's action in passing a death sentence on all humans in reprisal for their distant ancestors' alleged offence strike him as questionable or unjust.

J's ultimate hero was King David, and David routinely executed criminals' children to prevent one of them from growing up to become his father's avenger of blood. J can hardly be criticized for crediting his god with the morality of his own age, or for failing to realize that three thousand years later the execution of a criminal's children would be viewed with horror.

Nor could the Yahwist have foreseen that a later age would regard Yahweh's mother the goddess as nonexistent, and that the presence of her *asherah* in Eden would have to be attributed to Yahweh's caprice. No doubt even in 920 BCE a parent who deliberately left the cookie jar uncovered to tempt his children would not have been deemed guiltless. But in the original myth Yahweh did no such thing. The original myth, as J understood it, was not obscene. It was the later reinterpretation that made it so. Compare:

## THE YAHWIST'S VERSION

Yahweh created a man and woman to worship him, and a paradise in which they were to live. But there was a problem. Yahweh's mother, the goddess, wanted them to worship her instead. Since the goddess had existed before him, Yahweh could not solve his problem by the simple expedient of not creating her. She was a fact of life, and Yahweh had to allow for her existence. So he ordered Adam and Eve not to worship her. They disobeyed, and Yahweh inflicted the standard tenth-century-BCE punishment: death to the evildoer and all of his kin.

## THE CHRISTIAN REINTERPRETATION

Yahweh created a paradise. Sin was impossible, because nothing was forbidden. So that he would have something to ban, Yahweh unnecessarily placed in his paradise the sacramental body of a goddess who did not exist. He ordered Adam and Eve not to worship the Vulva by eating her vulva. They did eat the *asherah*, a disobedience Yahweh could have prevented simply by not creating it. As punishment for their disobedience, Yahweh inflicted the cruel and unusual punishment now abolished by every civilized government on earth: death, not only for the offenders but also for all of their descendants.

Having deduced an explanation for death that satisfied him, the Yahwist wrote it as fact. That readers of the original version could accept Yahweh's behavior as morally defensible is understandable. That believers in the Christian interpretation can do so is incomprehensible.

## 21. The Jewish Parochialism of the *Our Father*

Are Christians unteachable, or just braindead? In America, the Christian Right is trying to post Deuteronomy's Ten Commandments in schools and courtrooms, in violation of the Constitution's prohibition of any "law respecting an establishment of religion." They are doing so even though those commandments, in their original meaning, prohibited murder, adultery, perjury, and lusting after a compatriot's private property, *only when the victim was a fellow Jew.*

In the Canadian provinces of Saskatchewan and Alberta, hate cultists of the "my god can lick your god" persuasion are attempting to insert sectarian indoctrination into public schools, funded by taxpayers who specifically declared that they do not support religion-oriented separate schools, in the form of a pseudo-Christian prayer. Albertan religious bigots are even demanding that children not belonging to the majority sect be required to sit through a religious ritual, in the hope that they will be corrupted/converted into the hatred of the human race that is the Redneck Right's *raison d'être*. But the prayer they are touting, the so-called "Lord's Prayer," is as parochially Judaic, and therefore anti-gentile, as the Torah's Ten Commandments.

The very name, "Lord's Prayer," is offensive to the five billion persons on this planet who regard Jesus, not as their *kyrios,* "lord/master," but as either a madman, a humbug, a deluded fanatic, a nonentity who became the posthumous figurehead of a religion he would have repudiated, Muhammad's herald and subordinate, or a myth not based on a real person. The prayer's Catholic name, "the Our Father," Greek *pater hēmōn*, is much more neutral.

The *pater hēmōn* is a specifically Jewish prayer, expressing beliefs shared by no Christians. For example, correctly translated it begins, "Our father in the sky whose name is taboo." Only Jews regard the god Yahweh's name as taboo, too sacred ever to be spoken by anyone but the high priest (an office that no longer exists). Christians have no such belief. From there it continues, "May your theocracy be established," the theocracy being the Davidic monarchy in Jerusalem that *MashYah*, the successor and equal of King David, was destined to refound.

And the prayer ends, "Do not encourage us to be defiant, but rather liberate us from the intolerable." As every one of Jesus' hearers was well aware, "the intolerable" from which they asked Yahweh to liberate them was the Roman occupation. The prayer's addendum, "For the theocracy and the army and the magnificence are eternally yours," is chanted only by Protestants (in a slightly different translation), since Catholic and even recent Protestant bibles omit it, in recognition that it was an interpolation.

If parents wish to instill supernatural beliefs into their children, whether in the field of religion, astrology, alien abductions, or whatever, they are free to do so at home, where they do not thereby infringe the rights of other people's children. That such a procedure works can be confirmed by looking at Serbia, where fifty years of banning religion from the schools did not prevent the present generation from growing up with the same Catholic-hating, Muslim-hating, theocentric, Eastern Orthodox Christianity as their grandparents.

State-supported religion gave the world the Crusades, the Inquisition, the Thirty Years War, the Salem witch trials and Ayatollah Khomeini. It must not be allowed to happen in a country in which freedom of religion is only possible if it incorporates freedom *from* religion.

## 22. The Ten Commandments:  History's Best Lawcode—Or Its Worst?

The concept of *Ten Commandments* can be traced back as far as the biblical author known as the Yahwist (J), who wrote a mythical history of the world *c* 920 BCE.  Since J put his commandments into the mouth of Moses, they were already anachronistic at the time of writing, for they referred to customs that originated in Phoenicia and could not possibly have been taught by Moses.  Among them was a prohibition against boiling a kid in its mother's milk, a practice J deemed reprehensible because the Phoenicians did it as a sacrifice to Yahweh's rival, Moloch.  J's list, found in Exodus 34:14-28, was the only biblical lawcode specifically identified as *the Ten Commandments.*

Nonetheless, when the Elohist (E) wrote an independent history of his alleged ancestors *c* 770 BCE, he included a different set of laws, later copied by both the Priestly author (P) with slight changes, and the Deuteronomist (D) unchanged.  It is the later list that has come to be known as *the Ten Commandments* by modern religions, even though it was never identified as such by any biblical author.  The Priestly version can now be found in Exodus 20:1-17, and the Elohist's original list in Deuteronomy 5:6-22.

The Elohist's first two commandments paralleled the Yahwist's first two, but with an added touch of realism.  Where J had banned kowtowing or paying other forms of homage to any god but Yahweh under any circumstances, E stipulated that the prohibition applied only "before Yahweh's face," *i.e.*, in Judea, where Yahweh would be subjected to the indignity of having to watch.  Jews visiting Babylon, for example, were not expected to be so foolhardy as to antagonize the Babylonian gods by refusing them due honor.

J, who almost certainly held some official status in the Jerusalem hierarchy, outlawed the making of cast metal godlets, thereby invalidating the religion practised in the breakaway northern kingdom, where golden bulls were worshipped at Dan and Bethel.  E, who seems to have been a Shiloh priest, enlarged the ban to include carved godlets of the type he knew were currently located in Solomon's temple.  E was therefore invalidating both the non-Levite priests of the north and the Aaronic priests of Jerusalem, leaving only his own allegedly Moshite priesthood unscathed.

E's first commandment did not prohibit the worship of other gods outside of Judea, and therefore did not authorize such atrocities as the Crusades, the Thirty Years War and the modern my-god-can-lick-your-god murders in Ireland and Bosnia.  The modern interpretation does exactly that.  The current Christian first commandment is therefore a recipe for hatred, intolerance and war.  The second Protestant commandment (part of the first in the Catholic system) was likewise composed to validate partisanship and put down all priesthoods but one's own, and for that reason must also be recognized as indefensible in a society which claims to respect freedom of belief.

Commandment three, correctly translated, reads:  "You are not to swear in Yahweh your gods' [generic plural] name that which is false."  There is little to criticize in that.  Whether one invokes gods or the tooth fairy, making a false sworn statement is reprehensible.  But modern believers interpret the commandment as prohibiting the use of a god's name as an expletive.  While the use of profanity says a great deal about the user's linguistic competence, few would argue that ineptness in language should be rated alongside murder and stealing as one of the ten most unspeakable evils.

The demand that everybody rest on the seventh day may have been practical in the agriculturist society of three thousand years ago.  Today it is absurd.

E ordered Israelites to *worship* their parents, for the logical reason that, in a society with no afterlife belief, the only way to grant a parent a kind of immortality was to keep his/her name alive by ancestor worship.  Correctly translated, that commandment is also an absurdity in the world today.  The modern interpretation, requiring believers to *honor* their parents, is basically sound, provided the parents in any particular case deserve such honor.

Commandments six through nine (Catholic five to eight), banning murder, adultery, stealing and perjury, are valid only when interpreted as their composer never intended. To all biblical authors, murder meant only the killing of a fellow Jew without authorization from the relevant priest or magistrate. Under no circumstances did it mean the killing of a non-Jew (Talmud, Sanhedrin 78b). Similarly, the prohibition of stealing did not include stealing from a non-Jew. And the ban on perjury had similar limitations. The modern misinterpretation of those commandments actually changes bad laws into good laws, by extending protection to all humans.

The proscription of adultery was not a good law when composed and is not a good law today, for two very different reasons. To the biblical authors, the fraudulent impregnation of a non-Jew was not adultery as defined in the commandment. The Torah's Redactor made clear that adultery was only a felony "provided that his adultery was with a compatriot's woman." (Leviticus 20:10) The expansion of the commandment to cover all cuckolded husbands is a significant improvement.

Unfortunately, the concept of adultery has also been distorted in an undesirable direction. To the biblical authors, and as recently as the 15th century when Thomas Malory had Lancelot copulating with Guinevere but withdrawing before the intromission of sperm transformed chaste love into adultery, adultery meant fraudulent impregnation, thereby robbing a man of his right to pass on his inheritance to his legitimate heirs. The expansion of the concept to include, not only copulation with a woman who does not have a legal owner to saddle with a cuckoo's chick, but also nonprocreational coupling involving a woman on the pill, could only have occurred in a culture conditioned to believe that right and wrong are whatever the tribal god says they are. Monogamy may be a good thing, especially in a society infested with AIDS. But to label nonconsequential recreation as intrinsically evil is surely indefensible.

E's tenth commandment prohibits lusting after the property of a fellow Jew, whether his house, his wife or his livestock. Since the commandment validates the private ownership of women, then either women are and ought to be men's privately owned breeding stock, or the commandment is evil. That E's purpose was to uphold the validity of private property against the rising philosophy of communism does not change that.

As for P changing the rest-day commandment from a commemoration of slavery in Egypt into a commemoration of a seven-day creation in which the earth existed before there was a star for it to orbit, and trees bore fruit before there was a sun to provide the energy necessary for photosynthesis, that little anachronism has been given very little coverage in the Christian press.

## 23. The Origin and Evolution of the Doctrine of Predestination

Before Chauncy Q. Schlemiel was ever conceived, the omniscient god who created all things knew that Chauncy was destined to spend eternity in Hell. Since the god's knowledge was absolute, there was not even a possibility that Chauncy might not spend eternity in Hell. Nonetheless, even though the god is also omnipotent and therefore could have created Chauncy with that little something extra that would have enabled him to escape the eternal flames, but chose not to do so, Chauncy was not *predestined* to spend eternity in Hell. He earned his fate, and the omnipotent god's decision to create him without sufficient "grace" to escape his inevitable destiny does not mean that he was damned from birth.

As self-contradictory as the preceding doublethink is, it is dogma in all of the largest Christian denominations. The explanation for the doublethink is simple. The creator-god is perceived as omnibenevolent, and a god that would create souls for the specific purpose of damning them would have to be recognized as evil. Therefore, since the god cannot be evil, predestination must be rejected, even though the only way an omnipotent creator could have foreknowledge of an individual's fate is if his fate was predetermined. "Pre-known but not predestined" is as oxymoronic as "One plus one plus one equals one."

The doctrine of predestination was vigorously endorsed by John Calvin, and continues to be endorsed by modern-day Calvinists. More than one pope has equally vigorously condemned it. Was Calvin more capable than the popes of believing in a god so evil that it would intentionally create Hell-fodder? Alternatively, was Calvin more capable than the popes of recognizing that, for his god to have perfect knowledge of the future, that future must be predetermined? Or was Calvin simply more capable of grasping that, regardless of how evil a god it implied, the Christian Bible unequivocally endorses predestination?

The doctrine of predestination did not spring fully formed from the mind of an individual innovator. Nor was it invented by any person who realized that he was preaching something new. Just as Jesus invented the eternal torture of Hell in the belief that the flamethrowers of the Essene Gehenna were already permanent, and Jeremyah invented the messiah in the belief that failed prophecies of King David referred to a descendant of David as yet unborn, so predestination similarly evolved from misinterpretations of what had been written before.

Early Judaism had no predestination belief, for the logical reason that it had no afterlife belief. Thus when the Elohist (*c* 750 BCE) wrote, (Exodus 33:19) "I (Yahweh) show compassion to whom I choose, and mercy to whom I choose," he meant that Yahweh was arbitrary, choosing who was to live and who was to die by the toss of a coin. Even viewed in the context of the Elohist's theology, that doctrine is damaging to Yahweh's image. But the Elohist wrote at a time when Zoroaster and Jesus had not yet invented what became the Christian Heaven and Hell. To the Elohist, the extinction of death was the end of existence. If Yahweh chose to end some people's lives earlier than others, that was his privilege. Eventually all must cease to exist in any case.

To the Pharisee Paul of Tarsus, writing near 57 CE, death was not the end of existence. Sometime between the completion of the modern Jewish Torah in 434 BCE and the completion of Daniel (except the Greek chapters) in 164 BCE, Judaism acquired from its Persian, Greek and Egyptian overlords a belief in an immortal "soul" that survived physical death, a belief that only the ultra-conservative Sadducees rejected as new-fangled and foreign. Not only did Paul believe in separate afterlives for "us" and "them." He also imagined that Moses, who he thought had written Exodus, had believed the same thing. Consequently, whereas the Elohist had accused Yahweh of capriciously shortening men's lives, Paul read the passage in the light of his own afterlife beliefs as indicating that Yahweh arbitrarily consigned men to Heaven or Hell. Thus it did not occur to Paul that he was accusing Yahweh of a hitherto unparalleled atrocity when he wrote:

> Whom he foreknew, he predestined. And whom he predestined, he called. And whom he called, he also justified. And whom he justified, he also glorified…What are we to say then? Can the god be unrighteous? May it not happen!…You will say to me then, "Why does he find fault? For who can withstand his whim?" Human, who are you to contradict the god? Doesn't the potter have power over the clay, to make from the same lump one vessel in honor and another in dishonor? (Romans 8:29-30; 9:l4-15, 19-21)

Paul's belief, that Yahweh created humans for the specific purpose of having someone to torture with his flamethrowers, was obscene. The anonymous author of Mark (4:10-12) made it more so.

Around 700 BCE, Isayah (Isaiah 6:9-10) had written: "You hear, yes, but you don't understand. And you see, yes, but you don't perceive. For this nation's heart has grown flabby and their ears hard of hearing, and they've closed their eyes, in case they see with their eyes and hear with their ears and understand with their hearts, and convert and be healed."

Isayah was berating his contemporaries for closing their eyes and ears and hearts to what he perceived as "truth." The author of Mark paraphrased Isaiah but, emulating Paul in attributing his own afterlife belief to his source, so changed the passage's context that it came out meaning something Isayah had never imagined:

> When he was alone, those around him with the Twelve asked him about the fables. And he told them, "To you, permission has been given to understand the mystery of Allah's theocracy. But to those who are outside, all things are to be told in fables, so that in seeing they may see and not perceive, and in hearing they may hear and not understand, in case they convert and obtain forgiveness for themselves." (Mark 10-12)

The Jesus who deliberately spoke in riddles in case he accidentally caused the salvation of persons whom his father in the sky had predestined to damnation has been given very little coverage in Christian pulpits.

Predestination was a Christian invention, at least to the degree that Paul, a Jesus-Jew who died before Jesus was deified or made the son of a virgin or part of a trinity, can be deemed a Christian. But its seeds were present in Essene/Pharisaic Judaism, meaning Judaism that incorporated an afterlife belief. Consider the following passages from pre-afterlife Judaism: "Can there be evil in a town, except that Yahweh has done it?" (Amos 3:6) "I create fortune, and I create evil. I, Yahweh, do all of these things." (Isaiah 45:7) "Who has only to speak to make things happen? Doesn't his Lordship ordain? Don't good and evil come from the mouth of Ilion?" (Lamentations 3:37-38) None of those passages carried implications of predestination in a culture that viewed eternal life as an infidel superstition. But in the course of doctrinal evolution, the adoption of an afterlife concept should have led to their reinterpretation as implying just that. The only reason Talmudic Judaism rejected predestination was that radical Christianity had meanwhile picked it up.

## 24. Sodom and Khomorah: When, Where and How

In the late tenth century BCE, the Jewish chronicler known as the Yahwist wrote a history of his tribe and its allies that culminated in the establishment of the empire of David and Solomon. According to the Yahwist, his creator-god Yahweh destroyed the cities of Sodom and Khomorah during the lifetime of the Jews' great sheik and demigod, Abraham, thought to have flourished *c* 1800 BCE.

The Yahwist located Sodom and Khomorah in the Valley of Demons (*khemeq ha-shaddim*) at the south end of the Dead Sea. (Genesis 14:3) Archaeologists, satisfied that there are no ruins near or under the Dead Sea that could possibly be the destroyed cities, have concluded that the story is pure myth and that no such cities ever existed. It may be, however, that the Dead Sea location represented the Yahwist's guess as to where the buried cities must have been, and that the search for remains should be directed elsewhere. Possibly the Yahwist found Sodom's true location in his source, and changed it because he could not believe that the source was accurate. Plato seems to have done just that in identifying Atlantis as a large island west of the Mediterranean, when Solon may well have written that it was a city east of the Mediterranean in Lydia, so devastated by an earthquake that it wound up at the bottom of a newly created lake.

The Yahwist's tale of Sodom and Khomorah appears to have been a blending of a historical event with mythology that was much older. The original myth was simply one more version of the destruction-of-humankind fable found in many cultures: Yahweh destroyed the world by erupting over it. But Levit, (*Lvt/Lwt*) was righteous and, along with his family, was allowed to escape. That Levit, like Noah, was originally humankind's only survivor, can be deduced from the line spoken by his daughter before seducing him with wine: "Our old father is the only man left on earth to get into us, as is the custom throughout the land." (Genesis 19:31)

That, despite the myth that became attached to their destruction, Sodom and Khomorah were historical cities destroyed by a volcanic eruption in the third (not the second) millennium BCE, becomes a reasonable conclusion when we compare the Yahwist's description with the known facts about the annihilation of Pompeii and Herculaneum:

> Yahweh brought down burning sulfur [lava] on Sodom and Khomorah...from the skies...He turned those towns upside down, and the whole plain, including all the people of the towns...Abraham looked toward Sodom and Khomorah and across the plain, and he saw smoke rising from the land like smoke from a furnace. (Genesis 19:24-48)

Yahweh was a volcano god, and there is no extinct volcano in the vicinity of the Dead Sea. But the Jews did not settle near the Dead Sea until the time of sheik Yahuwshua and Pharaoh Ikhenaton, in the fourteenth century BCE. Prior to that time they had lived far to the north, and there is ample evidence that Sodom was also far to the north.

One of the kings who waged war against Sodom was the king of Sumer (Genesis 14:1). Sumer still existed in the third millennium BCE, but not in the second. If the Sodom stories that the Yahwist associated with the demigod Abraham were originally connected to a much earlier Abraham (the name was in use in Ebla *c* 2400 BCE), they could well date from the time when the Jews occupied the volcanic area of southern Anatolia and northern Syria. A war between Sumer and southern Anatolia, at opposite ends of fertile land that both wanted, makes more sense than a war between Sumer and Phoenicia, on opposite sides of a hostile desert. Perhaps the search for Sodom and Khomorah should be directed to the highlands north of Adana. If the buried cities are located, it

111

would then be reasonable to identify the volcano that buried them as the original Yahweh, still worshipped as a god by Jews and Christians.

# 25. Monism Versus Dualism: The *Job* Compromise

Shortly after the fall of Jerusalem in 586 BCE and the subsequent transportation of large numbers of Jews to Babylon, the Jewish spokesman Ezekiel wrote a book in which he referred to three men in a context that identified them as the most righteous figures in Jewish mythology. They were Noah, Daniel and Job. Such was Ezekiel's influence that some centuries later books were written that were credited to all three of his heroes. Of the three, *The Book of Job* proved to be the most influential.

That Job was indeed composed long after Ezekiel (three centuries later) is evident from its portrayal of the Satan as a male, for not until the Jews had been under Babylonian and Persian dominance long enough to assimilate the Zoroastrian Prince of Darkness, Ahriman, did Yahweh's prime antagonist cease to be the goddess Ashtareth. Despite Job's inclusion in both the Jewish and Christian bibles, few proponents of either religion categorize it as nonfiction. To persons 2,200 years more morally evolved than the author of Job, Yahweh's consent to the capricious murder of Job's children is much more acceptable as a moral fable than as history.

The author of Job was a Sadducee, a member of a sect that clung to the traditional Jewish belief that there is no afterlife, at a time when the liberal Pharisees and the communist Essenes were winning increasing numbers of converts to their newly adopted belief that there is an afterlife. That was why his concept of a Happy Ending was Yahweh's giving Job ten new children to replace those who had been killed, children whose ancestor-worship would keep his name alive, thereby providing him with the only kind of eternal life a Sadducee deemed possible.

He was also a believer in palmistry, for he wrote: "Allah implanted signs and marks in the hands of humans, that all may know their deeds." (Job 37:7) Whether palmistry was a peculiarly Sadducaic belief, or widespread among all Jews, is uncertain. The only other reference to palmistry in the Judaeo-Christian bible was also written by a Sadducee: "Length of days in her right hand, and wealth and status in her left hand." (Proverbs 3:16) Palmists often quote those passages as evidence of the validity of palmistry. They are better evidence of the invalidity of Job and Proverbs.

Job was written as acknowledged fiction. Nonetheless it contributed significantly to Judaism's transformation from a monistic to a dualistic religion. In the monism that had existed since Yahweh's transformation in the tenth century BCE from a mere tribal god to the most powerful of all the gods, the creative force (Yahweh) had been seen as a unity, fully responsible for both good and evil: "Can there be evil in a town, except that Yahweh has done it?" (Amos 3:6); "I create good fortune and I create evil. I, Yahweh, do all of those things" (Isaiah 45:7); "Who has only to speak to make things happen? Doesn't his Lordship ordain? Don't good and evil come from the mouth of Ilion? Let's consider and examine our behavior, and return to Yahweh." (Lamentations 3:37-38)

As mentioned above, it was ultimately Yahweh himself who caused Job's torments, since it was Yahweh who gave the Satan permission to carry them out. That was the essential weakness of monism in a culture that in post-Captivity times had begun to develop a concept of morality that veered toward the objective. In the days when there had been no such thing as objective evil, and even such an atrocity as the destruction of a whole city by an erupting volcano-god could be seen as a just punishment for crimes known only to the god, monism, the crediting of everything that happened to a single creative force, had presented no problems. But with the recognition that there *were* injustices in Yahweh's world, continued monism would have involved the acknowledgement that Yahweh must be unjust.

Dualism, first conceived by Zarathustra, offered a solution. Zarathustra had divided the creative force into two beings, an omnibenevolent god responsible for all good, and an omnimalevolent god responsible for all injustice. Unfortunately, while dualism solved one problem, it simultaneously created another.

Zarathustra had recognized that, if his god Ahura Mazda was to be worthy of total admiration, it must be totally benevolent. He also recognized that, if Ahura Mazda was to be omnibenevolent, then he could not also be omnipotent, since in a world ruled by an omnipotent, omnibenevolent god, evil could not exist. Zarathustra therefore made Ahura Mazda and Ahriman evenly matched opponents, neither of whom had the power to exterminate the other or render him impotent.

Zarathustra's dualism might have been acceptable to Judaism two or three centuries earlier, when that religion was still henotheistic. But by 250 BCE when Job was written, Yahweh had ceased to be the cosmos's paramount god and had become, in Jewish eyes, its only god. (Minor gods such as Mars/Michael, Mercury/Gabriel and Venus/Satan had been demoted to the status of created immortals, not true gods at all.) The acceptance of pure (Zoroastrian) dualism would have necessitated depriving Yahweh of the omnipotence that, in a One True God, was *sine qua non*.

Continued monism was unacceptable to a culture with an evolving sense of morality. But dualism was equally unacceptable to a culture whose tribal god had evolved into the omnipotent Master of the Universe. Some kind of compromise was called for, and Job provided that compromise: The Satan was responsible for all evil and injustice; but he could do nothing without Yahweh's consent. And Yahweh granted that consent to test his worshippers' faith. Yahweh thus remained omnipotent, while simultaneously ceasing to be the author of all evil.

No single philosophy concerning the nature and source of evil has succeeded in conquering the world. To this day Christianity and Judaism retain the para-dualism of *Job,* arguing that the deaths of millions of children of bubonic plague in the 14th century and starvation in the 20th century could not have occurred without Yahweh's consent, but that Yahweh is nonetheless not responsible. Hinduism has never been anything but dualistic, as an acknowledged polytheistic creed must be. And Islam represents a reversion to the monism of pre-Captivity Judaism.

# 26. Luke: Christianity's First Revisionist

The Christian Gospels known as Mark, Matthew and Luke are commonly called Synoptic, since they see their hero, Jesus, through a single eye. But that eye is Mark's.[1] Large parts of Matthew and Luke are identical with Mark because they are copied from Mark. This has long been acknowledged.

What is less widely recognized is Luke's status as an innovator, rationalizer and revisionist. Like Matthew, Luke changed some of the things he found in Mark and in his other main source, the Q gospel, as well as including stories that cannot be traced to any specific source. But whereas Matthew's changes were minimal, and appear to have been made on the spur of the moment when he detected something in his source that he did not like, Luke's give the impression of being made by a man consciously searching for weaknesses in order to rectify them.

For example, where Mark, followed by Luke, totally prohibited remarriage after a divorce, an actual Nazirite taboo taken from the Essene documents that the Nazirites deemed sacred,[2] Matthew added the loophole that remarriage was permitted if the divorce was a consequence of sexcrime (*porneia*). Presumably Matthew simply did not like the taboo in its original form, and was unaware that he was falsifying something Jesus the Nazirite actually taught. Luke's failure to make a similar change may have stemmed from his being better acquainted with the Essene doctrine, but it may also be that he saw Mark's story as adequate the way it was.

That Luke was better educated than Matthew is made clear by a story both borrowed from Q. In Matthew, and presumably in Q, the devil took Jesus to the top of a mountain so high that they could see all the way to the edge of the world. Clearly Matthew, writing some centuries after Aristotle had demonstrated that the earth is a sphere, still believed it to be flat, since only on a flat earth could the entire surface be seen from a single point. (Matthew 4:8) Luke, recognizing the flat-earth element of Q's tale as invalid, rewrote the story so that it could still be true on a round earth. (Luke 4:5)

Mark put into Jesus' mouth a prophecy that, when the Temple was polluted by Zealots who would start a war of independence without waiting for his return to lead them, the Nazirites were to flee to the mountains. (Mark 13:14) This had in fact happened. A couple of years before Mark wrote his gospel, Jesus' cousin Shimeown had led the Nazirites out of Jerusalem and into the Decapolis. But Mark had also written, as a prophecy by Jesus, that immediately following the Nazirites' exodus, "And then they'll see the Descendant of Adam [*i.e.* Jesus] advancing on clouds with a mighty and magnificent army" to overthrow the Roman occupation. (Mark 13:26)[3] That, of course, did not happen. And Matthew, writing thirty years after Mark, knew that it had not happened—yet he copied Mark's failed prophecy with neither comment nor amendment.

Luke was more discerning. Knowing that a further thirty years had passed since the Nazirite exodus, and that Jesus' second coming seemed as far away as it had ever been, he wrote that, following the exodus, "Jerusalem is going to be downtrodden by infidels until the period of infidels is concluded." (Luke 21:24) Only after that would Jesus return with an army of heavenly messengers. Since Jerusalem was still under Roman occupation when Luke wrote, he very logically deleted from Jesus' prophecy a time limit that had already expired.

Matthew and Luke both invented, or perhaps borrowed, genealogies showing Jesus to be descended from King David (Matthew 1:6-16; Luke 3:23-31), even though both had also included an incident borrowed from Mark in which Jesus unequivocally denied Davidic descent. (Mark 12:35-27; Matthew 22:41-45; Luke 20:41-44) But whereas Matthew's genealogy showed Jesus' ancestors each reaching the average age of thirty-eight years before siring his heir, in violation of the Talmudic teaching that, "The gods utter a curse against those who remain celibate after they are twenty years of age," (Kidd. 29b) Luke showed his set of Jesus' ancestors siring heirs at the average age of twenty-four years. And whereas Matthew had made Jesus a descendant of Solomon and the kings of

Judah, thereby enabling his genealogy to be falsified when he made one of those kings the son of his great-great grandfather, Luke traced Jesus to David's son Nathan with a genealogy that, being completely fictitious, could not be falsified. That the two genealogies were incompatible was a point neither Matthew nor Luke could consider, since neither was aware of the other's existence.

Both Matthew and Luke transferred to Jesus myths first told of the Indian god Krishna. Krishna escaped a massacre of infants perpetrated by a jealous rival. And Krishna was born in a stable, where he was visited and adored by poor shepherds. Matthew appropriated the former tale to his hero, and Luke the latter. But whereas Matthew offered no explanation of why Jesus came to emulate Krishna, and likewise Abraham, Moses and John the Immerser, who all escaped similar massacres,[4] Luke explained Jesus' unlikely birthplace by declaring that there was no room at the inn. (Matthew had Jesus born in a house, but Luke could not know that.)

Luke's greatest innovation, however, was his backdating to Jesus of the policy of preaching to gentiles that actually originated with Paul. He suppressed the Q passages in which Jesus expressly forbade such preaching (Matthew 10:5-6), and even had one of Jesus' personal appointees preach in Samaria. (Acts 8:5) And in contradiction of Jesus' actual equation of gentiles with dogs (Mark 7:27), Luke concocted a parable that he put into Jesus' mouth in which Jesus praised a Samaritan as more worthy of brotherly love than a member of the Levite caste or a temple priest.

Mark, Matthew and Luke were all gentiles. But only Luke was sufficiently embarrassed by Jesus' xenophobic Jewish nationalism to delete it and turn Jesus into a gentile-lover that he never was. And as a consequence, Christians to this day express adoration for a man who, had he ever lived to meet a Christian, would have damned the entire sect of Christians as detestable infidels.

## NOTES

1   The names Mark, Matthew and Luke are used here to mean the anonymous authors of the gospels known by those names.

2   William Harwood, *Mythology's Last Gods*, Prometheus Books, 1992, pp. 328-329.

3   Translations are from Harwood, *The Judaeo-Christian Bible Fully Translated*, Imprintbooks, 2002.

4   Harwood, 1992, pp. 141, 324, 338.

# 27. Moses Versus Aaron: Their Comparative Status to Different Pentateuch Authors

This paper starts from four basic assumptions: (1) The narrative portions of the Pentateuch, excluding Deuteronomy, were originally written as three separate and independent documents by authors who lived in three different centuries. Those authors were the Yahwist (J), the Elohist, (E), and the Priestly author (P). No reference is necessary for that assumption, as it has been considered fully proven by all but dogmatists for more than two centuries. (2) The Yahwist was an admirer of King David and the Davidic monarchy, into the third generation at the time J wrote. That conclusion has been established by Peter Ellis in *The Yahwist: The Bible's First Theologian*. (3) The Elohist was a Shiloh priest of the Moshite clan that claimed descent from Moses. For the evidence for that assumption, see *Who Wrote the Bible?* by Richard Friedman. (4) The Priestly author was a Jerusalem priest of the clan that claimed descent from Aaron. That assumption is also beyond dispute, but the reader may care to check my chapter on P in *Mythology's Last Gods*.

To J, a monarchist who viewed priestly power as a threat to royal power, Aaron was a nobody. When Moses informed his god Yahweh that he could not organize the Israelites because he could not speak Hebrew, Yahweh responded (Exodus 4:14), "What about your kinsman Aharon the Levite? I know he can translate." J's Aaron was nothing more than Moses' interpreter. There are scenes in the J narrative (Exodus 5:1) where Aaron accompanies Moses to see Pharaoh, even though his services as an interpreter would not have been needed, since Moses and Pharaoh both spoke Egyptian, but it is possible that Aaron's name was interpolated into those scenes by the final Redactor (also an Aaronid).

J's Moses was commander-in-chief, in other words a secular ruler. He communicated directly with Yahweh and carried out Yahweh's orders, including the plagues of Egypt, with no participation by Aaron the priest. And while J described Aaron's relationship to Moses with a word that can mean either a brother, kinsman or compatriot, he nowhere mentioned Aaron in a context that made clear that he was Moses' brother, even though he unambiguously gave the infant Moses an unnamed sister (Exodus 2:4), whom later authors equated with the Miriam named in the E narrative as Aaron's sister (Exodus 5:20) but not as Moses' sister.

To E, Aaron was worse than a nobody. He was the eponym of the Jerusalem priesthood whom the Moshite priests viewed as pretentious upstarts, if not outright usurpers. In the E document, Moses' recognized deputy on significant occasions was not Aaron but Joshua. And since Joshua was a purely secular authority, his presence on occasions calling for a priest, in the absence of any third person, clearly affirmed that Moses needed no other priest because he was himself chief priest. (Exodus 33:11)

From the time of Solomon, the Aaronic priesthood had wielded power in Jerusalem while the Moshite priesthood at Shiloh had lived off the crumbs that metaphorically fell from the Aaronids' table. E composed a fable that had no purpose but to show Aaron, and by implication the Jerusalem priesthood, in an unfavorable light by having Aaron act in a manner that aroused the ire of the true priest, Moses. That fable was the "golden calf" (Exodus 32:1-24). (To reiterate: This analysis originated with Richard Friedman.)

E's choice of a golden bull ("calf" is a pejorative translation) as an object of illicit worship is readily explained. Golden bulls were worshipped at religious centers administered by priests whose legitimacy was repudiated by Moshites and Aaronids alike. By having Aaron build such a godlet, E was putting down every priesthood but his own. And in case anyone missed the point, E included in his version of the Ten Commandment a prohibition of "gods of silver or gold to stand beside me." (Exodus 20:23) That commandment illegitimated the gold-plated kherubs in the Jerusalem temple administered by the Aaronids, and Aaron himself as the creator of the golden bull.

J's Aaron had been Moses' interpreter. E's Aaron had been a renegade priest whom the true priest, Moses, put firmly in his place. P's Aaron was Moses' costar. For example, in J (Exodus 9:22-23), "Yahweh ordered Mosheh, 'Reach your hand toward the skies and cause hail to fall...' So Mosheh pointed his staff at the skies, and Yahweh thundered and rained down hail." But in P's version of the plagues (Exodus 7:19), "Yahweh told Mosheh, 'Tell Aharon, "Take your staff and reach out your hand over Egypt's rivers...and let them turn into blood."'" To P, Moses was a purely secular ruler, and as such could do nothing to invoke Yahweh's intervention in human affairs without the mediating role of Aaron as high priest. And where J had written of all Levites, obviously including the Moshite clan (Exodus 32:29), "'Today,' Mosheh told them, 'you have been ordained priests for Yahweh,'" P told a very different tale. In a Torah written to supersede the composite of J and E, that P never in his worst nightmare imagined would one day be riffled together with J/E into a single document, P wrote (Numbers 16:7b-11), "'You take too much upon yourselves, Levites,' Mosheh told Korakh. 'Isn't it enough that the gods separated you to conduct the worship in Yahweh's Dwelling? Do you demand the priesthood as well?...for who is Aharon, that you blame him?'" To P, criticism of Aaron was tantamount to a rebellion against Yahweh, and so was refusal to recognize the paramountcy of the Aaronid priesthood. And to complete the denigration of the Moshites, P invented a genealogy (Numbers 3:20) that made them descendants of an insignificant Levite named Mowshiy, Moses' cousin. His ultimate glorification of Aaron, and by implication himself as Aaron's descendant, was (Numbers 16:40), "No outsider who is not descended from Aharon is to approach Yahweh to offer incense. Otherwise he will suffer the fate of Korakh and his followers [extermination by fire]."

There is no evidence that Moses or Aaron ever existed. But it seems unlikely that a Moses who promised to lead the Israelites into a land rich in milk and honey, and died with the task unaccomplished, was a purely literary invention. No such argument can be offered for the historicity of Aaron.

## 28.  The Death of Hananias and Sapphire:  An Explanation

Only ultra-conservatives, who constitute a minority of all believers, maintain that every story in the Judeo-Christian bible is literal truth.  Most concede that the earth is not flat, as depicted in Isaiah 40:21-22, Daniel 4:10, Matthew 4:8, and Revelation 7:1; not immobile, as shown in Psalms 93:1, 96:10, 104:5, and 1 Chronicles 16:30; not roofed by a solid sky, as was believed by the authors of Job 22:14, Revelation 6:14, and Acts 10:11; and not created less than ten thousand years ago.  They also concede that, since the aforementioned cosmography is not literally true, then it is reasonable to conclude that other biblical tales that stretch credulity might be at the very least distortions of the facts.  Among such is the account in Acts of the death of Hananias and Sapphire (Acts 5:1-10):

> A certain Hananias, together with his woman Sapphire, sold a property, and with his woman's connivance he withheld part of the proceeds and brought only a portion and laid it at the envoys' feet.
> So Petros asked, "Hananias, how did the Enemy fill your heart, beguiling your consecrated breath to withhold some of the proceeds of the land?"
> And when Hananias heard those words, he collapsed and died.
> After an interval of about three hours, his woman came in, unaware of what happened.  So Petros questioned her.  She promptly collapsed and died at his feet.

It is by no means impossible for the deaths to have occurred precisely as the author of Acts described them.  Such is the power of cultural conditioning that, even in the 20$^{th}$ century, Australian natives could die after being 'sung' to death by tribal priests, while believers in voodoo and witchcraft have proven equally susceptible.  So long as the victim *believes* that he can be killed by a curse, he can indeed be killed by a curse.

There is, however, good reason to doubt that Hananias and Sapphire died as impressively and spontaneously as the author of Acts would have us believe.  Essene documents discovered since 1947 leave little doubt that Jesus was a neo-Essene preacher, and that the Nazirites, the sect led by his brother Jacob and his student Peter after his death, was a neo-Essene commune.  And the Nazirite commitment to the Essene principle of communal living is fully documented:

> They [Essenes] are communists to perfection.  Their rule is that novices to the sect must surrender their property to the order...Each man's possessions go into the pool and their entire property belongs to them all. (JOSEPHUS *Jewish War* 8)
> The mass of believers were united heart and psyche, and not one claimed any of his possessions as his own.  Everything they had became communal property...All who owned land or houses sold them and brought the proceeds. (ACTS 4:32-35)

Hananias and Sapphire, as members of a neo-Essene sect, were required to liquidate their property and give the proceeds to the communal treasury.  They attempted to cheat, were caught, and consequently died.  But how (speculating that the case had a factual basis) did they die?  Josephus's discourse on the Essenes provides an answer.

Men convicted of major offenses are expelled from the order, and the outcast often comes to a miserable end. For bound as he is by oaths and customs, he cannot share the diet of non-members, so is forced to eat grass until his starved body wastes away and dies.

In the light of the foregoing, it is not unreasonable to conclude that Hananias and Sapphire indeed died as a consequence of trying to retain a measure of private property after joining a sect that regarded property-owning as an unforgivable sin. But they did not die of shock, brought on by fear of divine vengeance. They were expelled from the sect, and consequently died of starvation.

## 29. Was Nathanael the Beloved Disciple?

The anonymous author of the gospel known as John made several references to a student (Latin: *discipulus*) not mentioned in the synoptics, an unidentified "student whom Iesous loved," (John 21:20) or "the one Iesous cherished." (John 20:2)   Unlike the synoptic authors, who made no acknowledgement of their sources, the author of John admitted basing part of his gospel on a memoir written (or dictated to a secretary) many years earlier by the aforesaid "Beloved Disciple."   John wrote, "It was that same student who testified about these things and wrote them down, and we are certain that his testimony is accurate." (John 21:24)

But who was the Beloved Disciple?  We can start by establishing that he was not the author of the fourth gospel.  In writing that "We are certain that his testimony is accurate," the gospel author made a clear distinction between himself (we) and the Beloved Disciple (his).  Also, the gospel author explained away a prophecy that the Beloved Disciple would not die until Jesus' second coming, with the words, "However, Iesous did not say to him that he was not going to die, but rather, 'If it's my whim that he remain until I'm established, what's that to you?'" (John 21:23)  As the gospel author was aware, since he was writing a full century after Jesus' death, the Beloved Disciple was also long dead, and it was therefore necessary to harmonize his death with the alleged prophecy that he could not be dead.

Nor was the Beloved Disciple the redactor of Revelation, whose name was John and who lived on the island of Patmos.  It is physically possible for a man who was Jesus' student in 30 CE to have lived long enough to write the Nazirite portions of Revelation over sixty years later.  But the equation of the Beloved Disciple with John of Patmos stems from the false assumption that the Beloved Disciple wrote the fourth gospel, and the equally false assumption that the author of the fourth gospel wrote (or redacted) Revelation.  Since the John gospel was written in educated Greek, and Revelation in the pidgin Greek known as *Koine*, the idea that they were written by the same author is as ludicrous as claiming that *Hamlet* was written by Dan Quayle.

The theory that the Beloved Disciple was Zebedyah's son John also stems from the false reasoning that he was the John of Revelation.  Since there is no other reason for imagining that his name was John, equating him with a fisherman of that name is absurd.

Because all attempts to identify the Beloved Disciple have been speculative, he has been given little recognition in the Christian liturgy.  That is ironic, since he was the originator of the core dogma that Jesus emulated several pagan savior gods by rising from the dead.  That belief was triggered by nothing more than the discovery that Jesus' body had been removed from its temporary tomb.  John wrote that, when the Beloved Disciple arrived at the tomb, "He looked and was credulous." (John 20:8)

But of what was he credulous?  The answer would seem to be that, discovering that Jesus' body had disappeared, the Beloved Disciple jumped to the conclusion that the only explanation for an empty tomb was resurrection.  But while millions have since accepted that reasoning, only a person of special qualities could have originated it.

That is where Nathanael comes in.  Jesus said of him, "I see an Israelite in whom is no guile." (John 1:47)  That sentence is less than definitive proof that Nathanael was observably retarded.  But Nathanael's response to his first meeting with Jesus, "Rabbi, you are the descendant of the god.  You are the king of Israel," (John 1:49) suggests that his capacity for skeptical reasoning was limited, to say the least.  Of all Jesus' students, only Nathanael accepted his messianic claims at their first meeting.  Add to that, that Jesus' physical deformity (Luke 4:23) would have made him empathize with a man who must have suffered the same taunting from other children while he was growing up as Jesus himself, and a special relationship can be envisioned that would justify Nathanael describing himself as "the student Iesous cherished."  And of all the qualities that would cause an individual to make the quantum leap from empty tomb to resurrection, none quite matches mental retardation. In recent years persons as sophisticated as Whitley Strieber have claimed that they had been abducted by aliens. But the first person to concoct such an absurdity was Barney Hill, not a retard but certainly a backwoods hillbilly. Some innovative claims only seem plausible to persons of less than normal intelligence.

Only two students are portrayed in the Christian gospels as the recipients of Jesus' flattery, Peter in the synoptics and Nathanael in John. But John's reference to "Simon Petros and to the other student, the one Iesous cherished," (John 20:2) makes clear that Peter was not the Beloved Disciple. And the scenes involving Nathanael and Peter are so similar as to suggest a common origin, with only the name changed:

> Iesous saw Nathanael coming toward him and said of him, "Indeed I see
> an Israelite in whom is no guile." Nathanael answered him, "Rabbi, you are
> the descendant of the god, the king of Israel." (JOHN 1:47; 1:49)

> So he asked them, "But who do you say I am? And Petros answered,
> "You're the messiah, the descendant of the god." (MATTHEW 16:15)

Since the titles, *messiah* and *king of Israel*, were interchangeable, the statement of credulity that the synoptic author attributed to Peter was identical with the words that, according to John's earlier source, whom he identified as the Beloved Disciple, were actually spoken by Nathanael.

By the time the earliest Christian gospel was composed, forty years after Jesus' death, either jealousy or embarrassment over Jesus' special relationship with a retard caused Nathanael's name to be suppressed, with the consequence that the author of the gospel called Mark never learned his name and did not include him when he invented Jesus' mythical "twelve envoys." Since Peter was by that time the second in command of the Nazirite commune led by Jesus' brother Jacob, Nathanael's unique status was retroactively transferred to Peter in gospels based on chronicles written by men who had known Peter but had never known Nathanael or Jesus. It is by no means certain that Nathanael was the Beloved Disciple. But the evidence is good that he was.

# 30. Moses and Joshua: The Problem of Chronology

Most attempts to date Moses (*Atonmoses*) and Joshua (*Yahuwshua*) have placed them in the late thirteenth or early fourteenth century BCE, depending on whether the particular writer looked at the evidence for Moses and automatically placed Joshua later, or looked at the evidence for Joshua and assumed that Moses must have preceded him. In each case, the findings of archaeologists at Jericho in ancient Judah and Hazor in ancient Israel have shown the postulated datings to be impossible, and the question has been raised whether the Exodus under the Egyptian prince Moses and the Conquest under the Hapiru sheik Joshua were events from history or fantasy.

For some time the problem of chronology seemed to be insoluble, and for scholars irrevocably committed to the dogma that Joshua was Moses' successor it will always be insoluble. The problem only disappears when it is recognized that Joshua was a leader of the Jews, a tribe that conquered southern Phoenicia in the early fourteenth century, whereas Moses was a leader of the Israelites, an unrelated tribe that settled northern Phoenicia in the late thirteenth century, 150 years later. Jews and Israelites were not the same people.

The Jews can be traced as far back as the third millennium BCE, at which time they inhabited an area that reached as far south as Ebla in Syria and as far north as Mount Yahweh, a then-active volcano in Anatolia, somewhere in the highlands north of Adana. They were driven out of their volcanic homeland before 2000 BCE by the marauding Babylonians, and during the reign of pharaoh Ikhenaton (1380-1362 BCE) established a new homeland south of Jerusalem.

The Israelites can be traced no further back than the fourteenth century. At that time they seem to have been driven out of their Phoenician, possibly Midian, homeland by rampaging Hapiru, who can reasonably be equated with Joshua's Jews. What seems to have happened is that the Israelites were invited to migrate to Goshen on the eastern shore of the Nile delta, on the condition that they defend Egypt's northern border in the event that the Hapiru attempted to invade Egypt. The Roman Empire regularly placed tribes of barbarians on its eastern border on similar terms.

According to Genesis, the pharaoh at the time of the Israelite immigration had an Israelite viceroy named Joseph. According to the Amarna letters, Ikhenaton had a Semitic viceroy of Phoenicia named Yanhuma. To the degree that the "Joseph" myths were grafted onto the biography of a historical person, that person was Yanhuma.

Ikhenaton's predecessors had detested Semites, for a good reason. Around 1730 BCE Egypt had been conquered by a Semitic tribe known as the Hyksos, who constituted Egypt's fourteenth to seventeenth dynasties and ruled the Egyptians for 160 years until they were expelled by Ahmose I, founder of the eighteenth dynasty. Who the Hyksos were and what became of them is uncertain. Josephus equated the expulsion of the Hyksos with the biblical Exodus, but the Israelite settlement in northern Phoenicia certainly did not take place as early as 1570 BCE.

Following the expulsion of the sheepherding Hyksos, the Egyptians became anti-Semitic and anti-sheepherders. Even though the patron god of the eighteenth dynasty was the ram Amen, not for a century were sheepherders tolerated within Egypt's borders. For a Semite to obtain even the lowliest position in Pharaoh's administration, let alone become viceroy, would have been an unthinkable denigration of Egyptian superiority. Then came Ikhenaton.

Ikhenaton was the world's first monotheist. Not only did he stubbornly refuse to worship any god but Aton, the sun. He vehemently denied that any other god existed. And whereas under Amen the god's own people were intrinsically superior to non-Egyptians, under Ikhenaton's universal god all of Aton's people, literally everybody under the sun, whether Egyptian, Phoenician or Ethiopian, were equal. That was why Ikhenaton was able to appoint a Semite as his viceroy, to the disgust of his entire population, when no other pharaoh would have done so.

Yanhuma, like Churchill, found himself in the unenviable position of presiding as the king's first minister over the disintegration of a mighty empire. For at the same time that Ikhenaton's religious innovations were tearing Egypt apart internally, Egypt's Phoenician province was being invaded by hordes of Hapiru, "Easterners," of whom Joshua's Jews were probably a mere part. When Yanhuma's own tribe, the Israelites, fell under the Jewish siege, he persuaded Ikhenaton to allow them to settle in Egypt as a first line of defence.

Yanhuma's influence in Egypt died with Ikhenaton. The nineteenth dynasty pharaohs who restored Egypt's glory would not have failed to regard the Semites within their borders as an unpleasant reminder of the collapse of their empire under the great heretic, and also as a security risk. Ramoses I, who "knew nothing of Joseph," turned the Israelites into slaves. Ramoses II forced them to build the storage towns of Pitthom and Raamses. It is most unlikely, however, that Pharaoh came to depend on slave labor in the midst of mass unemployment, or that the Israelites escaped from Egypt against his will. It seems more likely that the Israelites were expelled at the demand of starving Egyptians who wanted their jobs and their land.

Ikhenaton's religion did not die with him. A select priesthood survived, and it was that priesthood that more than a century later produced Moses. Moses led the Israelites into Israel at precisely the time archaeologists declare the razing of Hazor to have taken place: 150 years after the destruction of Jericho. Except for Hazor, the settlement was peaceful. The Israelites formed an alliance with their southern neighbors, the Jews, resulting in a thirteen-tribe confederacy. When the number thirteen's indelible association with goddess-worship caused it to become taboo, the number of tribes was reduced to twelve by pretending that the tribes of Ephraim and Manasseh were really half-tribes, moieties of a nonexistent "Joseph" tribe.

By the time of king Solomon, the Jewish-Israelite confederacy had lasted more than two centuries. In the reign of Solomon's son a man known to historians as the Yahwist wrote a history of the confederacy from the creation of the first human down to the accession of David. His biggest problem in composing a chronicle that would be accepted as history by Jew and Israelite alike was the necessity of blending an Israelite folk memory of an Egyptian captivity with a Jewish tradition that knew nothing of such an event. Clearly he could not deprive the Jews' greatest hero, Joshua, of the permanence of his conquest by having the hybrid Jews/Israelites enter Egypt and be enslaved after Joshua's victories. Nor could he tamper with the tradition that Moses had died with his promise to win the Israelites a new homeland unfulfilled.

The Yahwist's solution to what should have been an insoluble problem was a slight juxtaposition of events. He placed everything else in correct chronological order, but dated the conquest of Judah by Joshua to the period following the death of Moses, 150 years later than it had actually occurred, and named the fourteenth-century-BCE Jewish leader Joshua as the successor to the thirteenth-century-BCE Israelites' leader Moses. As insurance against detection, he omitted the name of Joseph's pharaoh (Ikhenaton), the name of the pharaoh in Joshua's day (also Ikhenaton), and the name of Moses' pharaoh (Ramoses II), that would have drawn attention to the impossible chronology.

## 31. Judas's Betrayal: Why the Author of Mark Invented It

According to the earliest Christian gospel, Jesus the Nazirite was betrayed to his executioners by one of his own chief lieutenants, his treasurer, Judas the Iskariot. Judas's status as a betrayer has been questioned many times over the centuries, and novels exonerating him are written at regular intervals. Those novels, however, invariably have assumed that only Judas's motives were misinterpreted, and that his observable behavior was exactly as Mark described. There is good reason to doubt that that was so.

It was the Jewish War of 66-73 CE, and the death of Nero and accession of Vespasian, that led to the writing of the first gospel. Nero had been anti-Christian. The war made Vespasian anti-Jewish. The Christians therefore saw Vespasian's accession as a golden opportunity. For if the new emperor could be convinced that Christianity, far from being a sect of the religion that had rebelled against the empire, was an independent mystery cult analogous to the Mithraism to which Vespasian as a soldier was favorably disposed, the emperor might be prepared to extend to the Christians the same freedom of worship enjoyed by adherents of every other religion within the borders of the empire.

Mark, as the anonymous author of the earliest gospel has erroneously been designated, wrote his evangelion shortly after the destruction of the temple of Yahweh that occurred in 70 CE. This is revealed by the retroactive prophecy that Mark put into Jesus' mouth concerning the temple's fate: "Under no circumstances is there going to be one stone left on another that isn't torn down." (Mark 13:2) That prophecy was far too accurate to have been composed before the event that it described. The prophecy of the temple's destruction made in 63 CE by a deranged fanatic named Jesus bar Hanan was much more vague. Mark wrote after 70 CE but, as Randel Helms has demonstrated (*Who Wrote the Gospels?*), before 74 CE.

Mark set out to persuade Vespasian to end the persecution of Christians begun by Nero. To that end he had to "prove" that the Christians were not anti-Roman like the rebellious Jews, but were friends of the Romans and enemies of the Jews. His task was simplified, indeed made possible, by the decision of Jesus' cousin Shimeown bar Klopa, who had succeeded Jesus' brother Jacob the Righteous as supervisor of the neo-Essene sect known as Nazirites, to lead his followers out of Jerusalem and into the Decapolis at the first sign of impending war. The presence in Jerusalem of a Jesus commune, led by Jesus' relatives, that was undeniably Jewish, would have made the dissociation of Christianity from Judaism impossible. In order to achieve his objective, Mark had to falsify much history.

High on the list of Vespasian's enemies were the Pharisees, the revolution's original leaders until the Zealots murdered them for being too moderate. Mark retroactively made the Pharisees Jesus' enemies also. Even though Jesus' teachings were identical with those of the Pharisees on most issues, and the Sadducees, whose "no afterlife" belief made them Jesus' real enemies, regarded him as an upstart Pharisee, Mark included in his gospel passages in which Jesus called the Pharisees hypocrites and warned against their doctrines, passages based on Jesus' actual attacks on the Sadducees.

The true relationship between Jesus and the Pharisees was revealed by the anonymous author of Luke, who recorded that Jesus once fled from Herod Antipas after being warned by some Pharisees that Antipas planned to assassinate him. (Luke 13:31-35) Luke also acknowledged that some Pharisees joined the Nazirite sect after Jesus' death, (Acts 15:5) a most unlikely occurrence if the Pharisees had indeed been the target of Jesus' Galilean invective. Mark knew of Jesus' confrontations with a Jewish sect, but found it politically expedient to pretend that that sect had been the Pharisees. In fact the only Pharisees who viewed Jesus unfavorably were the high-priestly families, who saw him not as a neo-Essene and therefore an ally against the heretical Sadducees, but as a dangerous Zealot.

125

The prime instigators of the war, however, had been not the Pharisees but the Zealots, a sect that had been trying to provoke a war of independence since its inception in 48 BCE. And the last holdouts, who defended Masada for three years after the fall of Jerusalem, had been the Zealots' ultra-terrorist wing, the *sicarii* or *Iskariots*, meaning "daggermen" or assassins. The Romans viewed the Zealots, particularly the Iskariots, in much the same light that a later imperial power viewed the Mau Mau. Mark knew that he could not conceal that two or three of Jesus' lieutenants had been Zealots, one of them an Iskariot. The logical way to dissociate Jesus from the Zealot cause that he wholeheartedly espoused was to pretend that Judas had really been Jesus' enemy all along, and had ultimately betrayed him to "the Jews." That one of Jesus' lieutenants sold him out is improbable. That a member of the one sect bound to back Jesus' revolution to the hilt did so is beyond belief.

That Judas's betrayal was a late invention is attested by a Pauline interpolator's unawareness of such an event. Paul's letters to the Corinthians were written around 55 CE, and the interpolations inserted some time after that. The interpolator wrote that the resurrected Jesus was seen "by the twelve." (1 Corinthians 15:5) That one of the alleged Twelve had defected or hanged himself after Jesus' death was a theory that the interpolator had not heard, since it would not be invented for a further fifteen years.

The author of Acts was similarly unaware of any suicide, for he had Judas die from an accidentally sustained chest wound. (Acts 1:8) And even eighty years after the composition of Mark, Docetic Christians who followed Gnostic rather than synoptic traditions continued to regard Judas as an envoy in good standing: "But we, his lordship's twelve students, [*after the crucifixion*] were weeping and were in sorrow." (Gospel of Peter). Clearly there was a school of Christianity that as late as 150 CE had never heard of Judas's betrayal.

Mark made no mention of Judas's death. Matthew and Luke, in contrast, wrote mutually exclusive accounts that coincided only on one point. They agreed that Judas did not long outlive Jesus' arrest. If their information on that point was accurate, it raises the question of why Mark suppressed Judas's deserved fate for his monstrous crime. The logical answer would seem to be that the actual circumstances of Judas's death were incompatible with the actions Mark attributed to him, that in fact Judas either died trying to protect his leader or was crucified with him. That would explain why Matthew and Luke, seeing the desirability of reporting Judas's death but unable to give a true account, resorted to their respective imaginations and came up with tales that contradicted each other.

The evidence is ambiguous. If Judas died at about the same time as Jesus, why did later writers think he was alive much later? Alternatively, if he did not die in 30 CE, why did Matthew and Luke both concoct tales, clearly not from a common source, that he did? Answers to that must remain conjectural. But that Judas was not in any way responsible for Jesus' arrest is a conclusion that cannot reasonably be denied.

## 32. Joshua: The Jewish Hitler

According to the Priestly author, usually abbreviated to P, "The Yisraelites violated the terms of a curse when Khakan ben Kharmiy ben Zabdiy ben Zarakh of the tribe of Yahuwdah appropriated an item that had been vowed to be destroyed. Yahweh unleashed a tantrum against the Yisraelites." (Joshua 7:1)

Joshua's response was to ascertain the identity of the offender by the infallible method of rolling Yahweh's sacred dice. When the dice identified Khakan, what happened next was that "Yahuwshuakh asked, 'Why did you bring down disaster on us? Today Yahweh is going to inflict you with disaster.' Then the whole of Yisrael killed him by pelting him with rocks, and killed them (his children and livestock) by pelting them with rocks, and threw them on the fire to burn." (Joshua 7:25)

The foregoing horror story was written between 621 and 612 BCE as part of a Priestly Torah intended to supersede and replace the previous Torah, to which Deuteronomy had been added in 621 BCE. Since the Deuteronomist had written (Deuteronomy 24:16), "Fathers are not to be executed on account of their children, and children are not to be executed on account of their fathers," it is a logical assumption that P was reestablishing the principle that children *were* to be executed for their parents' crimes. (That the J/E/D and P documents would be combined into a single Torah two hundred years later could not have been foreseen.) It logically follows that the incident had no historical basis. That is not really relevant, since the Joshua admired by Christians and Jews is the biblical Joshua, not the historical person, of whom nothing is known except that he *may have* existed—a century before Moses, not as Moses' successor. It is the biblical Joshua who was a homicidal maniac.

Most True Believers are aware that Joshua allegedly conquered Jericho by blowing a trumpet. Very few have any awareness of the details, since that part of the story is never preached from Christian or Jewish pulpits. According to P (Joshua 6:21), "They utterly exterminated everyone in the town, both man and woman, young and old, ox and sheep and ass, with the blade of the sword." And when Joshua lured the fighting men out of ha-Khay so that his militia could set fire to the town (8:17-29), "They killed them all, letting none of them survive or escape. But the king of ha-Khay they captured alive, and brought him to Yahuwshuakh. So he hanged the king of ha-Khay from a tree until evening."

Nor was Joshua's genocide the creation of a single author. Consider the following versions:

P    (11:21-22) Yahuwshuakh simultaneously came and wiped out the Khanakites...Yahuwshuakh utterly exterminated them from their towns. Not a Khanakite remained in the land of the Yisraelites, although some survived in Gath and in Ashdowd.

E    (10:32a) Yahweh delivered Lakhiysh into Yisrael's hands. They captured it on the second day and killed every person in it who breathed with the blade of the sword.
(11:14) The Yisraelites plundered those towns, taking the loot and the livestock for themselves. They also killed every human with the blade of the sword until they were exterminated. They left no one alive.

R    (10:33, 35b, 37) It was then that Horam, king of Gezer, came to the aid of Lakiysh. Yahuwshuakh wiped out him and his nation, until there was not a survivor left alive. He ritually sacrificed everyone in it who breathed the same day, in accordance with everything they had done to Lakhiysh. They captured it and its king and all of its towns, and butchered everyone in it with the blade of the sword. He left no one alive. Just as he had done to Kheglown, he ritually sacrificed it and everyone in it who breathed.

(10:40) And Yahuwshuakh raped the whole land: the highlands, the Negev, the lowlands, the slopes, and all of their kings. He left no one alive, but carried out the ritual sacrifice of everything that breathed, just as Yahweh, the gods [generic plural] of Yisrael, had ordered.

Joshua was also accused by R (the Redactor who combined the J/E/D and P Torahs in 434 BCE) of informing the Israelites (24:20), "If you abandon Yahweh and slave for other gods after he has treated you so well, he'll turn on you and inflict atrocities on you and wipe out." Passages like that help explain why, with a god like Yahweh, the Torah authors did not need a devil.

That the historical Joshua, or whoever was Führer of the *Hapiru* who conquered and appropriated Judah in the reign of pharaoh Ikhenaton, really was a genocidal maniac is borne out by the Amarna letters. That remains irrelevant. It is the biblical Joshua modern believers continue to admire—and he was certainly a psychopath whom even Hitler barely surpassed.

# 33. Jesus' Deification: When and Why?

On Thursday April 11 in the year 30 of the Common Era, Jesus the Nazirite, a religious fanatic not unlike Ayatollah Khomeini, started a war of independence in Jerusalem by leading a ragtag army of between 50 and 150 Zealots into Yahweh's temple and disrupting the daily sacrifice on behalf of the emperor that was the visible sign of Judea's subservience to Rome. Within ten minutes he was arrested for treason, and the following day he and those followers who had not seen the Roman cohort from the Antonia coming and escaped were executed by a method used only by the Romans and only for rebels and slaves.

Two days after that, at the cessation of Passover (sunset on Saturday April 13), the pious Jew who had allowed Jesus' body to be stored in his own tomb for the 24 hours during which preparation for burial was taboo, moved the body to its permanent location without telling Jesus' Galilean students. Consequently, on Sunday morning the students found their leader's body missing from its temporary abode, and the resurrection myth was born.

From the beginning of his career as an itinerant preacher, Jesus had been viewed by his followers as the successor and equal of King David: the Messiah, or rightful king of the Jews. To the Jewish masses who had hoped that his claim was valid and that freedom from Roman overlordship was days away, Jesus' death was the ultimate proof that he was just one more crank in a long line of messianic pretenders. But to the hardcore followers who swallowed the resurrection delusion, his royal status was now proven beyond doubt.

The myth of Jesus the king was promoted by the preacher Paul of Tarsus, and by the anonymous authors of the gospels arbitrarily attributed to a Nazirite named Mark, a Christian named Luke, and a nonexistent Matthew. But that was the full extent of Jesus' promotion for the first one hundred years after his death. Nowhere in Paul's letters or in the synoptic gospels is there the slightest hint that their authors had ever heard the theory that Jesus was a god. Two gospels in their present form portray Jesus as virgin-born. But all but biblical literalists accept both virgin-birth passages as interpolations that were not part of the original gospels.

The document that turned Jesus into a god, much the way a resolution of the Roman senate turned Augustus and later Claudius into gods, was the anonymous gospel ineptly attributed to the redactor of Revelation, whose name was John. By ascertaining when John was written, we can thereby ascertain when Jesus was first promoted from king to god.

That John was written later than the synoptics has been recognized since the second century. Attempts to date it earlier have stemmed from a desperate need by inflexible believers in Jesus' divinity to explain away the embarrassing reality that the synoptic gospels depicted Jesus as a mere mortal—a resurrected mortal, but a mortal nonetheless.

Just as Mark had been written to persuade the emperor Vespasian that the Jesus religion was centered on a resurrected savior like Mithra, and was therefore not a sect of the Jewish religion responsible for the war of 66-73 CE, so John was written to persuade the emperor Hadrian that the Christians worshipped a resurrected god like Mithra, and had nothing to do with the Jewish war of 132-135 CE. Jesus became a god because the Christians found it politically expedient to make him so.

But turning Jesus into a god in order to make Christianity acceptable to the emperor was not the John author's only reason for writing. As Martin Larson has shown (*The Essene-Christian Faith*), he also set out to make Christianity attractive to the Greco-Roman masses who viewed the Essene-Nazirite doctrines endorsed by Luke in particular, such as communism, celibacy, and a theocracy centered in Jerusalem to be established within a time limit that had already expired, as unacceptable. Obviously, John's repudiation of the synoptics could only have been written later than the books it repudiated—but how much later? One clue can be found in John's description of Jesus' last supper with his students prior to his execution.

In John's synoptic source, Luke, Jesus was executed on Passover, and his last supper was the Passover meal. That could only have happened in a year in which Passover fell on a Friday. In John, Jesus was executed the day before Passover, and therefore his last supper was not the Passover commemoration. John's gospel dates Jesus' crucifixion to a year in which Passover fell on Saturday. Why did he not follow his source? The best explanation is that he had a second source, one more reliable than the synoptics, the earliest of which was written 43 years after Jesus' death.

John (whatever his real name might have been) claimed that he copied some of his gospel from a memoir written by "the student Iesous cherished," commonly referred to as the Beloved Disciple, a student not known to the authors of the synoptics. While Randel Helms (*Who Wrote the Gospels?*) has argued that the Beloved Disciple was a creation of the John author's imagination, a lot of questions are best answered by assuming that the Beloved Disciple did exist and was indeed one of John's sources.

For example, if Jesus was a real person from history, as I maintain in *Mythology's Last Gods*, and not an early version of the Great Pumpkin, then he must have been executed between the death of John the Immerser in 29 CE and the impeachment of Lucius Pontius Pilatus in 36 CE. Between those two dates there was no year in which Passover fell on a Friday, as the synoptics required. Passover fell on a Saturday, as indicated in John, in 30 CE and 34 CE. (I follow Larson in accepting 30 CE as the actual crucifixion date, on the ground that a rabblerouser preaching revolution could not possibly have avoided arrest for more than a single year.)

John's dating of the crucifixion to the day before Passover was correct. But be could not have rectified the synoptics' error on the basis of a better knowledge of Jewish law, or he would not have included passages that showed Jesus being taken to the Jewish high priest and questioned before being handed over to the procurator. John had one source (Luke) that made Jesus' last supper the Passover meal, and another source (the Beloved Disciple) that did not. Unable to choose between them on any other basis, he avoided making Jesus a Passover-celebrating Jew for a purely political purpose.

The practice of falsifying the facts about Jesus' life and death for political purposes began with Mark. Since the Pharisees who had started the war of 66-73 CE were Rome's enemies, Mark pretended that they were also Jesus' enemies. Since the presence of Zealots and Iskariots, the terrorists who held out at Masada for three years after the fall of Jerusalem, among Jesus' lieutenants could not be suppressed, Mark pretended that Judas the *Sicarius* (daggerman) was "really" Jesus' enemy and had betrayed him.

By the time Matthew and Luke were composed, *c* 100-110 CE, the Jewish War of a generation earlier was a distant memory. While both of those gospels copied Mark's anti-Pharisee and anti-Judas fantasies, they had no reason to view Jesus' Jewishness as the negative factor it had been at the time of the war. Judaism had regained much of the favorable regard it had temporarily lost, and the presentation of Christianity as a form of Judaism would have helped the new religion overcome the bad reputation it had earned under Nero and Domitian.

As additional time passed, and Trajan and Hadrian also found it necessary to proscribe a religion that refused to tolerate any of the empire's other gods, the stressing of Christianity's alleged Jewish status should have been a useful means of making the Jesus sect respectable. Yet in the gospel called John, we find that Jesus' Jewishness has again become an embarrassment that needed to be played down. John's method of un-Jewing Jesus was to change him from a Jewish (would-be) king whose very function as Messiah was the overthrow of Roman overlordship, into a Greek god who differed in no significant way from Mithra, Adonis, Osiris, and a dozen other savior gods whom the empire viewed favorably. In order to do that, John used the simplest possible means. He put speeches into Jesus' mouth in which he claimed to be the incarnation of "the Father," an ambiguous title that could have meant Jupiter or Zeus as easily as Yahweh.  His reason for needing to do so is not difficult to discern.

In 130 CE the emperor Hadrian issued an edict prohibiting any subject of the empire from mutilating his phallus.  Hadrian's target was not the Jews but the priests of Kybele and other goddesses who were emasculating themselves in increasing numbers.  A side effect of Hadrian's edict, however, was that it prohibited circumcision.  And for several centuries Jews had been amputating their foreskins in the belief that this was the only way the god Yahweh could recognize his pets.

Just as they had done three centuries earlier in response to analogous interference with their religious practices by Antiokhos IV, the Jews reacted violently.  Although full-scale war, led by a self-proclaimed messiah, Bar Kokhba, did not break out for a further two years, at any time after 130 CE the Jewish resistance to Hadrian's edict would have made Christianity's Jewish connection as much a hindrance to survival as it had been in Vespasian's day.

Hadrian's immediate successor, Antoninus Pius, was not a persecutor of Christians, even though he came to the Purple only three years after the end of Bar Kokhba's war.  From that, it can be inferred that Christianity's dissociation from Judaism was already an accomplished fact.  We can therefore conclude that John was composed between 130 and 138 CE, and that Jesus' deification similarly took place between those dates.  Yet 2,000 years later, Christians accept Jesus as a god without ever asking themselves:  If Jesus claimed to be a god, why had he been dead for a full century before such a claim found its way into the writings?  If Jesus was a god, then modern Christians know something that Jesus' students, his first century of preachers, including Paul, his first three biographers, and Jesus himself did not.  Is that reasonable?  Only an incurable could believe that it is.

## 34. Moses and De Mille:  Parting the Red Sea

When moviedom's most recent Moses, Burt Lancaster, waved his phallic staff over the Red Sea and asked his imaginary playmate to clear a path for the Israelites to escape their Egyptian pursuers, Yahweh (called *God* in the movie) sent a sacred wind, which can also be translated *fart*, and Yahweh's wind, in combination with a neap tide, caused the seabed to dry up so that the Israelites could cross over without wetting their feet.  When the Egyptians tried to follow, the wind changed, the tide came in, and the Egyptians were swamped.

Cecil B. De Mille's Moses, Charlton Heston, similarly flashed his rod, and, in probably the most spectacular movie special effect prior to computerization, Yahweh cut a dry channel through the middle of deep water.

So which version corresponds to the story told in Exodus?  The answer is:  both.

That Exodus originated as three separate narratives, written by authors historians have designated the Yahwist (J), the Elohist (E), and the Priestly author (P), and was riffled into its present form by a Redactor (R), is disputed only by the unteachable.  That conclusion will therefore be treated as a given.

The oldest version of the crossing of the Red Sea was written by the Yahwist.  In J's version (Exodus 14:21b, 24, 25, 27b):

> And all night long Yahweh caused a strong wind to blow that pushed the sea back and dried up the sea.  In the morning, Yahweh looked at the Egyptian army from the phallus of fire and cloud, panicking the Egyptian army.  He clogged their chariot wheels, making them difficult to drive, causing the Egyptians to say, "Let's flee from facing the Yisraelites, for Yahweh is fighting for them against the Egyptians."  As morning approached, the sea returned to its bed.  The Egyptians tried fleeing from it, but Yahweh swept the Egyptians upside down in the middle of the sea.

P's version of the myth, written 300 years later, had Moses foreshadow De Mille (Exodus 14:15, 16, 21a, 21c, 22, 23, 26, 27a, 28, 29):

> Yahweh asked Mosheh, "Why do you keep bothering me?  Tell the Yisraelites to march on.  Meanwhile, raise your staff and point your hand over the sea and divide it, so the Yisraelites can traverse dry ground in the middle of the sea."  Mosheh pointed his hand over the sea, and the waters divided.  The Yisraelites walked into the middle of the sea on dry ground, with the waters forming a wall to the right and left of them.  The Egyptians chased after them into the middle of the sea, all Pharaoh's horses, his chariots and his cavalry.  Then Yahweh ordered Mosheh, "Reach your arm over the sea and bring the waters back over the Egyptians and over their chariots and over their cavalry."  So Mosheh reached his arm over the sea.  And the waters returned.  They covered the chariots and the cavalry and the whole of Pharaoh's army that had followed them into the sea.  Not one survived.  But the Yisraelites walked on dry land in the middle of the sea, with the waters forming a wall to the right and left of them.

As with several other biblical myths, the most notable difference between the J and P versions is that J's story did not violate the laws of nature.  J wrote at a time when Yahweh had not yet acquired omnipotence, as is evident in the story in which the only way he could stop the builders of the Tower of Babylon from climbing up to the realm of the gods on top of the skydome and eating the fruit of immortality and thereby becoming his equals was by confusing their language. (Genesis 11:6-8)

Nor was he yet omniscient or omnipresent. When the Sodomites were accused of gross violations of the universally practised hospitality code, Yahweh's response was, (Genesis 18:21), "I'm going down to find out for myself whether the accusations that have reached me reflect what they have really done. For if they were lies, I want to know that."

By P's time, *c* 615 BCE, Yahweh had evolved the power to know and do anything he wished. P could show Yahweh carving a dry channel through deep water because, unlike J's readers, the masses for whom P was writing had no ability to reflect, "That is impossible." De Mille was able to film the Priestly version of the myth for the same reason.

## 35. Nineteenth Century Religions
### excerpt from *The Disinformation Cycle* (Xlibris, 2002)

In nineteenth-century America, a new religion came into existence practically every month. Most either did not outlive their creators, or did not win sufficient followers to be deemed a denomination rather than a cult. Of those that did survive to the present day, three can be traced back to plagiarists who passed off predecessors' writings as divine revelations, while a fourth has a long history of failed prophecies, unique and incompetent interpretation of ancient mythology, and an inability to distinguish between compulsive playacting and Satanism.

The least harmful of the four, in the sense that it neither kills its children, equates its current leader's every whim with Revealed Truth, nor attempts to make its beliefs the law of the land, is the church founded by Ellen White, the Seventh Day Adventists.

Ellen White's mental dysfunction may have been related to her being hit on the head with a rock at the age of nine. She attributed her every fantasy to a revelation from metaphysical beings, including her teaching that persons who masturbate eventually become cripples and imbeciles. Physicians tend to disagree. Even the church she founded has in recent years been forced to acknowledge that White was a liar and a plagiarist. In its journal, *Ministry*, June, 1982, the church conceded: "She utilized the words of prior authors in describing words she heard while in a vision. In a few instances, she used the writings of a 19th century source in quoting the words of Christ or of an angelic guide." (see Rea, "further reading")

Equally dysfunctional was Mary Baker Eddy, who invented Christian Science. After she claimed to have resurrected the dead, without the news media ever managing to find and interview one of her alleged Lazaruses, her son came within an inch of convincing a lunacy court that she should be institutionalized for her own protection. (see Gardner, "further reading")

Eddy's career began when she became, first a patient and later an assistant, of mesmerist Phineas Quimby. Quimby, at the point where he realized that the only curative element in his mesmeric mumbo jumbo was suggestion, became the world's first acknowledged talk therapist (as opposed to the mesmerists, who did not realize that they were talk therapists). He informed a Bangor, Maine, newspaper: "The mind is what it thinks it is, and that if it contends against the thought of disease and creates for itself an ideal form of health, that form impresses itself upon the animal spirit and through that upon the body." (Ellis, p. 146)

That was the theory of disease that Mary Baker Eddy learned from Quimby, and that was essentially the theory that became Christian Science after Quimby's death.

Mrs Eddy did, however, incorporate into her new religion elements that originated in her own imagination and owed nothing to her late teacher. In particular, she taught her followers that the human body did not really exist. And since the body did not exist, obviously it could never fall ill. Belief in illness was "error."

The first edition of Eddy's *Science and Health*, published in 1875, was essentially an expanded version of a manuscript by Quimby. Later versions by ghostwriters eventually created the impression that Eddy was a scholar.

Mrs Eddy was herself cured of psychosomatic ailments by talk therapy, and founded a religion in which talk therapy, under the new name of divine healing, was the only permitted form of medicine. But when that same talk therapy, usually under the euphemism, hypnosis, was practised by anyone else, she denounced it as "malicious animal magnetism" and attributed it to the Christian antigod, Satan.

Eddy's attacks on the medical profession softened toward the end of her life, as she found herself in need of glasses, false teeth, and painkilling injections for kidney stones. And when her nonexistent body expired from pneumonia, she suffered the ultimate "error" of believing she was dead.

The author once challenged a Christian Scientist lady to go to bed with him, so that while he was deluding himself that he was inserting a nonexistent bodily appendage into her nonexistent recreational orifice, she could take the opportunity to make him realize that it was not really happening. She could hardly refuse, since that would constitute an admission that the human body does exist and therefore Mary Baker Eddy was a liar. In fact the lady declined. Big surprise.

If the reader is getting the impression that the author feels a special hostility toward sects and fanatics who justify unspeakable atrocities by claiming to be obeying a god, he will get no argument from me. A further example of the breed is the Jehovah's Witnesses.

Jehovah's Witnesses do not use talk therapy themselves, but they join the Christian Scientists in denouncing the nonexistent art of hypnotism as emanating from their antigod. Nor do they reject medicine, with the one exception that they allow their children to die rather than permit them to receive lifesaving blood transfusions. That prohibition was instituted by the revisionist who gave the sect its present name, Charles Russell, and stemmed from the same ignorance of facts known even to theologians that led him to call his god *Jehovah*.

Russell found the name *Jehovah* in the 1611 King James Bible that was his only source of ancient knowledge. He had no idea that the name had been transcribed unchanged from older translations influenced by German and Latin, or that the German letter *J* and the Latin letter *I* were the phonetic equivalent of the English consonant *Y*. (Iulius Caesar should have been Anglicized to Yulius, not Julius.) But even in the earlier bibles, *Jehovah* was an inaccurate transcription.

In the third century BCE, for reasons that need not be discussed here, the Jewish priestly caste tried to purge the name of their god, Yahweh, from the spoken language, even though it survived on almost every page of their vowelless written Torah as *YHWH*. They declared Yahweh's name a sacred taboo, not to be spoken under any circumstances (with one insignificant exception). Evidence of that taboo is to be found in the Christian prayer attributed to Jesus: "Our father in the sky whose name is taboo…" (Harwood, 1992, pp. 237-238)

Since the priests could not afford to abandon oral Torah readings, they decreed that, wherever the name *YHWH* appeared in print, the reader must instead say *Adonai*, "my master." Centuries later, when vowel points were added to the Torah, the name *YHWH* was pointed, not with the vowels necessary to form the unspeakable name, *Yahweh*, but with the vowels of the mandatory substitute, *Adonai*. In translating from pointed Hebrew, King James' translators combined the consonants of *Yahweh* with the vowels of *Adonai*, Latinized *yodh* to *J*, and came up with the mongrel name, *Jehovah*, used by no Jews and only unlearned Christians.

Russell's prohibition of blood transfusions stemmed from a similar misunderstanding. In ancient times, before there was any widespread awareness that plants were alive but after it had become apparent that not all living things had blood, the idea arose that the difference between living and nonliving was breath. The breath (Hebrew *nefish*, Greek *pneuma*, Latin *spiritus*) was the life, and in time the word for "breath" acquired the alternate meaning of "soul." But even before the breath was perceived as life, two Jewish theologians (P and D) equated the blood with life (Leviticus 17:14; Deuteronomy 12:23). Jewish lawmakers accordingly prohibited the ingestion of blood, on the ground that consuming a dead animal's lifeforce, rather than spilling it as a sacrifice to Yahweh, would have denied the god his rightful share. It would not have occurred to them that, had such a procedure been available, saving lives by transfusing them with the lifeforce of a living human, without thereby harming the donor, could contravene their taboos. Nor has it occurred to any modern religion other than the Russellites.

But the Jehovah's Witnesses' greatest hoax was their declaration that the second coming of Jesus the Nazirite was to occur in 1844. When 1844 produced no second coming, the Adventists, (as the cult was then called), rescheduled it to 1874, and Russell, who renamed the Adventists "Jehovah's Witnesses" and appointed himself leader in 1872, went along with the amended prophecy. When 1874 passed uneventfully, the prophecy was again revised and the new date for the second coming was set at 1914. Rather than acknowledge a third failure (or second, if the pre-Russell prophecy is not counted), the J.W. hierarchy now maintain that the second coming did occur in 1914, but for the past nine decades Jesus has been keeping a low profile. That belief is, however, being played down, and most J.W.s are currently being taught to expect a second coming in an unspecified near future. (see Alfs, "further reading")

Of all the plagiarists who founded new religions in the nineteenth century, none was more blatantly self-serving or compulsively dishonest than Joseph Smith, founder of the sect currently known as the Reorganized Church of Jesus Christ of Latter Day Saints, the word "Reorganized" serving to distinguish it from the breakaway sect formed by Brigham Young after Smith's death and commonly called Mormons. Smith's sect from its inception has been led by his primogenitural descendants, whereas the Mormons have had leaders related to neither Smith nor Young. While only the breakaway sect is called "Mormon" (although splinter sects that reject the Mormon prohibition of polygamy continue to call themselves Mormon), both are deemed "LDS," and both claim to be the "true" church of Joseph Smith—as do the splinter sects.

In the early nineteenth century a man named Solomon Spalding wrote a historical novel in which the supposed lost tribes of Israel (who actually disappeared from history by integrating with their Assyrian conquerors) migrated to America and became the American Indians. About three years after Spaulding's 1816 death, his unpublished manuscript fell into the hands of Joseph Smith, then aged fourteen, and Smith rewrote it into the semblance of nonfiction under the title, *The Book of Mormon*. In recent years two Mormon scholars, after temporarily yielding to church pressure to suppress their discovery, established that twelve pages of Smith's manuscript of *The Book of Mormon* were in Spaulding's handwriting. (see Persuitte, Taves, Brodie, "further reading")

Smith was, to borrow a term from a biographer of James II, "not a nice man." In making polygyny an article of faith, he can reasonably be accused of self-gratification, since he acquired at least twenty-seven and possibly as many as forty-six wives (Brodie, appendix). He was probably not so much a racist as a product of his culture, even though in *The Pearl of Great Price* he declared that the "mark of Cain" was blackness and that the black races were descended from Cain. (Many fundamentalists maintain the same belief, even though their bibles show Cain's descendants dying in Noah's flood.) He accordingly prohibited blacks from becoming LDS priests. He also wrote that the American Indians were stained red as a punishment for sin, but that any Indian who abandoned his red gods and adopted Smith's white gods could turn white. If any such transformation has ever taken place, the news media must have missed it. (2 Nephi 5:21; 30:6)

Smith was a pathological liar, incapable of distinguishing between a lie that had a reasonable chance of remaining undetected and one that had no chance. For example, even though the Rosetta stone had been discovered before Smith's birth, and the eventual decipherment of hieroglyphics was inevitable, Smith wrote a pretended translation of some Egyptian papyri in his possession that he called *The Book of Abraham*, a narrative that paralleled Genesis. It was later discovered that Smith's papyri were funerary scrolls, and that he "translated" some of them upside-down. And when some opponents, for the purpose of discrediting him, forged pseudo-hieroglyphs on some "Kinderhook plates" and challenged him to decipher them, he took the bait and concocted a pretended translation.

In 1978 the Mormon "no black priests" rule became a source of widespread criticism. Since the only way the rule could be changed was for the current prophet to receive a divine revelation, such a revelation was conveniently promulgated.

While the biggest Mormon hoax is the pretence that the *Book of Mormon* is nonfiction, an equally pernicious lie is peddled by missionaries in the recruitment of new Mormons. The marks are told that, whereas the election of a Catholic pope takes several votes, Mormon prophets are always elected unanimously on the first ballot. In fact the election is a pious fiction. When a prophet dies, he is always succeeded by the senior member of the council of twelve apostles. The council go into the election knowing in advance which of them is senior, and obediently cast their votes accordingly.

Since the news media know in advance who is going to be the next prophet, his subsequent election is something less than miraculous.

But the fastest growing religion in America today is a racist hate cult calling itself the Nation of Islam. Its earlier deputy leader, Malcolm X, before he repudiated the cult, became a Sunni, and embraced moderation and tolerance, preached a message that, with only the color changed, could have emanated from the Ku Klux Klan. When its present leader, Louis Farrakhan, was given the opportunity by the host of *Nightline* to repudiate his followers' claims that he had been taken aboard an orbiting Mother Ship by aliens who promised him 1,500 planes and crews to overthrow white governments and restore the black man to his rightful place as ruler of earth, he instead affirmed that the claims were true (*Skeptic*, 5:3:82).

Then there are the Hare Krishnas, who suppress their sexual urges through endless repetitions of the chant, "*Om mani padme hum*," in blissful ignorance that those Sanskrit words mean, "The penis is in the vagina."

## FURTHER READING:

Matthew Alfs, *The Evocative Religion of the Jehovah's Witnesses*, Minneapolis, 1991.
Fawn M. Brodie, *No Man Knows My History: The Life of Joseph Smith*, NY, 1995.
K. Ellis, *Science and the Supernatural*, London, 1974.
Martin Gardner, *The Healing Revelations of Mary Baker Eddy*, Amherst, 1993.
William Harwood, *Mythology's Last Gods: Yahweh and Jesus*, Amherst, 1992.
D. Persuitte, *Joseph Smith and the Origins of the Book of Mormon*, Jefferson NC, 1985.
W.T. Rea, *The White Lie*, NY, 1982.
E.H. Taves, *Trouble Enough: Joseph Smith and the Book of Mormon*, Amherst, 1985.

## 36. Is God a Petulant Little Boy?

The equation of the entity known as God with a petulant little boy is not new. Rod Serling used it in a *Twilight Zone* episode decades ago. A ten-year-old monster played by Billy Mumy had by the beginning of the teleplay evolved absolute power, and used it to gratify his every whim. He had "thought" the rest of the universe out of existence, and only his terrified family survived, living their lives in constant fear that he would turn their heads inside out or commit some comparable atrocity if he so much as imagined that one of them had affronted his extravagant ego. Not until the episode was over did viewers realize that it was about the birth of god.

Nor was the birth of God restricted to *The Twilight Zone*. A movie starring Eric Braedon, under his original German name, was about a computer that had as many nodules as the human brain, and gradually took over the world. It slowly enslaved its technicians, and when it tired of playing with them it blew up a couple of cities to prove that it could not be turned off and would not tolerate any form of disobedience. But only when it informed the Braeden character at the very end, "You will learn to love me," did the movie's true meaning become apparent.

Similarly, in the second *Star Trek* pilot, in which William Shatner took over the captain's role, a character played by Gary Lockwood developed absolute power and became absolutely corrupt. To this day, many hardcore *Star Trek* fans are unaware that the episode was about the birth of God.

What all of those screenplays had in common is that "God," even when played by an adult or a computer, was as capricious and egocentric as the petulant child in *The Twilight Zone*. In other words, he was depicted as precisely the temperamental, spiteful, vicious, vindictive, megalomaniac, "Do it my way, see!" serial killer described in the Judaeo-Christian Bible.

Would anyone be taken by surprise to encounter a spoiled brat who demanded that he be served hamburgers on a Wimpy plate, but hot dogs on a Goofy plate? Or so manipulated his surroundings that all who came into his presence were required to be similarly served? But what of a President or Prime Minister who imposed such customs on his citizens, and criminalized violations of his whims? How long would it take the citizenry to recognize that he belonged in a mental institution? Behavior that is barely tolerable in a nasty little boy would be totally unacceptable from anyone else, right?

So what about a parent/head of state/lawgiver who decreed that eating a salami sandwich was permitted for everyone under his authority, and eating a cheese sandwich was permitted, but eating a salami-and-cheese sandwich was not permitted? Or that drinking wine was compulsory on a specified date in Israel, but prohibited in Salt Lake City? Or that leaving home without cashmere gloves would anger him, or leaving home without a turban? Or that eating meat was virtuous on Thursday but immoral on Friday?

A human who laid down such rules would be recognized as an ill-bred little kid who had never grown up. So why is a god that imposes similar capricious, pointless, indefensible ordinances credited with the right to do anything it wishes? Are god addicts genuinely incapable of recognizing that their deity's behavior differs in no way from that of a petulant little boy? Are they mind-slaves who dare not recognize what is in front of their noses, for the same reason the enslaved masses in the communist empire dared not even think that their lawmakers were plain evil, in case they talked in their sleep and a snitch overheard and reported them to the thought police?

God was created in the image of kings who could do anything they wished, whose every whim was law, and who were allowed to act like petulant little boys because nobody dared rebuke them. Is it any surprise that God turned out to be a carbon copy of the nasty little brats who created him?

# SATIRE

... to see ourselves as others see us.

I hope for His sake that God doesn't exist—because if He does He has a lot to answer for.

Philip K. Dick

There she lusted after her lovers, whose genitals were like those of donkeys, and whose emissions were like that of horses.

Ezekiel 23:20 (NIB)

# The *Our Playmate*

Our Imaginary Playmate, security blanket of the intestinally challenged,
Whose name is your only reality,
May your enslavement of humanity discontinue.
May your insane taboos go the way of Hitler's,
To the gullible as well as the rational.
Give us today the capacity for reasoned behavior tomorrow,
And cease pretending that good is bad,
As we cease pretending that you are good.
Do not encourage us to remain braindead,
But liberate us from the unspeakable (That's you, old Playmate).
So let it be written.  So let it be done.

## Literature quiz:  name the book.

A nontheist stumbles into a country so infested with godworship, that his claim to have a nonexistent sense called "reason," enabling him to see that their beliefs are falsifiable fairy tales, causes the inhabitants to assume that he is insane.  When he falls for a female godworshipper, his need to belong prompts him to yield to their demand that he be surgically cured by having the organ responsible for his delusion, an organ they do not have that he calls his "brain," amputated.  At the last minute, he realizes that no woman or society is worth a lifetime of brainless conformity, and flees.

**Answer**: *The Country of the Blind*, by H. G. Wells

A man living under theocratic Anglican Christianity rebels against its masochism, its sky Führer whom nobody has ever seen and who probably does not exist, and its totalitarian prohibition of everything that makes life worthwhile.  He forms a sexual relationship with a female dissenter, who sees their recreation together as pleasure-giving rather than an onerous duty to "be fruitful and multiply," as the sky Führer's domesticated livestock are required to believe.  The hierarchy hunts them down and, by threatening the man with his personal concept of Hell, persuades him to denounce his lover as a dirty Antichrist on whom the threatened torture should be inflicted rather than on himself.  He is converted, and finally realizes that he loves the sky Führer.

**Answer**: *1984*, by George Orwell.

# My Dog Can Lick Your Dog

Once upon a time, long, long ago, when people still owned dogs, there lived three families whose names were the Smiths, the Browns and the Joneses.

The Smiths owned a Doberman pinscher that they called Dog. The Smiths' Dog smoked and drank and ate pork but did not practise birth control.

The Browns owned a toy poodle that they called Dog. The Browns' Dog practised birth control and ate pork, but did not drink or smoke.

The Joneses owned a bulldog that they called Dog. The Joneses' Dog smoked and drank and practised birth control, but did not eat pork.

Because the Smiths knew that their Doberman pinscher was the One True Dog, they regularly killed members of the Brown and Jones families for claiming that some other dog was Dog. And because the Browns knew that their toy poodle was the One True Dog, they killed the Smiths and the Joneses. Similarly, the Joneses knew that their bulldog was the One True Dog, so they killed the Smiths and the Browns.

Then one day the Smiths said, "Our dog is Dog. The Browns' dog is Dog. The Joneses' dog is Dog. Since Dog belongs to all of us, why are we killing one another?"

And the Browns said, "We know there is only One True Dog. The Smiths know there is only One True Dog. The Joneses know there is only One True Dog. Since we are agreed, why are we killing one another?"

And the Joneses said, "Since we are agreed that our Dog and the Smiths' Dog and the Joneses' Dog are the same Dog, and there is only One True Dog, why are we killing one another?"

So they all agreed that the One True Dog was the Smiths' Doberman pinscher and the Browns' toy poodle and the Joneses' bulldog. The killing stopped, and they lived happily ever after.

And nobody ever noticed that the welfare budget, which all agreed was far too high, was but a tiny fraction of the dog food budget.

# THE AUTOBIOGRAPHY OF GOD

*The Autobiography of God* **(Xlibris, 2002) is narrated in alternating chapters by "Hughie," the world's first sane god; "Yahweh," who believes he is a product of the human imagination and who uses racist epithets that the other narrators repudiate; and "Pan," whom Yahweh calls *Satan*, who says that he and Yahweh are extraterrestrials. The following are Yahweh chapters.**

## ONE

I have finally decided I like being a man. I wasn't always male, you know. About 5,500 years ago, when men first discovered that children have fathers as well as mothers, that Big Discovery caused them to decide that I, too, should be invested with a cock. Before that I had a cunt.

My name is God. At least that is the English corruption of my Gothic name, *Godan*, "Father Goth." I am also called *El, El Shaddai, Allah, Ilion, Brahma, Manitou,* and about ten thousand other names that originally referred to other gods but were applied to Me once the idea arose that all paramount gods, creators of the universe, were really Me under another name. But my favorite name is *Yahweh*, for that is the name by which I am called in the original language of the sacred books of a billion worshippers.

My original name was *Yahuwah*. The suffix *-ah* indicated that I belonged to the ruling caste, the cunt-bearing caste. Men were mere toys in those days, given to women by the Mother (my oldest name but now seldom used) purely for their pleasure.

It is a common cliché in fiction that someone whose best years are behind him is "older than God." The alleged humor lies in the impossibility of anyone or anything being older than God. I have ostensibly always been here. But in fact many things are older than God. The Himalayas are older than God. The Arctic ice pack is older than God. The human race is older than God. In fact anything that has been in existence for thirty thousand years is older than God. If by "God" you mean a male god, then anything six thousand years old is older than God.

I came into existence as primeval woman, the Mother who gave birth to the universe, roughly thirty thousand years ago. Modern doublethink to the contrary notwithstanding, I did not will myself into existence. I was created, and my creator was woman. About twenty-five thousand years later I was cocked and untitted, and my defeminator was man. Since man's role in the cosmos is not giving birth but manufacturing things, he turned Me into the manufacturer of all things. And because man is still motivated by such unevolved insanities as jealousy, revenge, sadism, masochism, hatred, intolerance, xenophobia, parochialism and murder, naturally I am equally insane. How could I be otherwise, when man created Me and continues to recreate Me in his own image?

I liked being the Mother. I especially liked being Yahuwah, the volcano Mother, lying on my back and spewing lava out of my cunt whenever my kike worshippers failed to masturbate me with sufficient human sacrifices. I liked to listen to the victims' screams as they were thrown down into my hot, gaping orifice. And I thrilled to the super intensity of that last agonizing scream when they first hit the lava, and by that connection gave Me a new G spot. Of course women don't really have any Grafenberg spots. The interior of a cunt is about as sensitive to stimulation as a clipped toenail. But I had them. I had as many G spots as I had captives, criminals, and lovers with whom my priestesses had become bored, given to Me in sacrifice. And with each new screaming human being thrown into my lava-spewing cunt, I would cum with the intensity of a neverending sneeze.

I am still changeable, conforming to the fantasies of every new mythologian who comes along and recreates me with qualities I never before possessed. For example, when an influenza virus took advantage of a temporary distraction of my attention to mutate itself into an AIDS virus, America's television ayatollahs transformed Me into such a pathological homophobe that I had created a new, incurable, fatal disease to punish gays for displeasing Me. Of course AIDS also killed people who contracted it from spouses, or from transfusions, or in the womb. But what self-respecting, revenge-addicted god would be bothered by a minor side effect like that?

I am also changeable geographically. In Rome I so love killing babies slowly by starvation and disease, that I refuse to permit abortion or contraception. In Salt Lake City I permit population control, but reserve a special place in Hell for the monstrous sinners who drink tea or coffee. In Israel I permit the eating of salami sandwiches and cheese sandwiches, but inflict the fate of the infidels on any kike who eats a salami-and-cheese sandwich. You think laws like that—heads it's a sin and tails it's a virtue—make Me a fruitcake? Don't blame Me. They made the rules. They created Me in their own racist, sexist, might-makes-right image. I didn't create them.

I was far more changeable, however, in the good old days. Even my physical body varied from place to place. In south-central Anatolia I was a mountain that spewed lava from my gaping vulva at irregular intervals, no matter how many sacrifices the kikes spent trying to bribe Me to stop. In Egypt I was the Nile, and I annually pissed on the wogs and made their crops grow. In Babylon I was the moon, and women's cunts bled and stopped bleeding in phase with my waxing and waning. And in Greece I was both Aphrodite and Artemis, inviting every man to mount Me in the former manifestation, and having them torn apart by dogs for accidentally being aroused by my naked virginity in the latter. I was one capricious bitch. It was glorious.

Ah, yes, those were the days.

I have always played favorites. My creators played favorites, so I played favorites. I remember Cain and Abel. Cain was so industrious, sweating from dawn to dusk, planting, weeding, watering, cultivating, harvesting. Abel on the other hand was a lazy son of a bitch who slept all day while his trained dogs kept the flocks and herds from wandering away. Cain spent hours picking Me an offering from the biggest and ripest fruits in his orchard. Abel took five minutes to whack a sheep with an axe and throw the whole thing on my altar and set it alight.

Naturally I fawned over Abel's offering and sneered at Cain's. I'm that kind of a god. Okay, I hadn't warned them that the way to please Me was to kill something. But they had seen the way I regularly sent fires, floods, famines, plagues and earthquakes to kill things. Abel figured it out. Cain could just as easily have drawn the same conclusion if he had stopped to think about it. But my preferential treatment of Abel's offering had nothing to do with its intrinsic superiority. Man, Cain was *ugly*.

You probably know what happened. Cain killed Abel in a fit of jealous rage. Well naturally I couldn't sit still for that. Killing is fine, so long as it's either ordered by Me or dedicated to Me. But Cain had killed his brother without even asking my permission. He had to be disciplined.

That presented a problem. Today I would punish such an offense by torturing him with flamethrowers for billions and trillions of years, getting my rocks off on his neverending screams of agony. But this was early in human history. The hunchbacked dwarf from Galilee had not invented Hell yet. And then as now everything I did had to fall within the frame of reference of my biographers' imaginations.

I changed Cain's color. In cultures where the ruling caste was white, I made Cain a nigger. In cultures where black is beautiful, I made him a honkie. I turned the Japanese Cain into a hairy Ainu, the Maori Cain into an Australian Abo, the Burmese Cain into the first slant-eyed chink, and the spic Cain into a gringo.

For good measure I instituted a rule that henceforth all killings were to be avenged by the victim's closest relative. It did not matter if a killing was accidental. It still had to be avenged. If A killed B, then B's relative C was to kill A.  But this time it was to be done properly—in my name.  I do hate waste, and a homicide not dedicated to Me is almost as bad as no homicide at all.

After C killed A, then of course A's relative D had to kill C.  Then E had to kill D.  Then F…You get the idea.  Eons later the system was retroactively called a chain reaction, and super killing devices were invented based on the same principle.  I do love the hydrogen bomb.  I figure on having a ten thousand year orgasm when that button is finally pushed.  And it will be.  It will be.  I'll see to it.  You think "holy war" is an oxymoron?  Not to a god like Me.

# TWO

Of course I am insane. How many times do I have to tell you that? The mythologians who created Me were insane, although not all as insane as Jesus, thank Me, and they created Me in their own image. If I was not insane, or my worshippers were not insane, I would not have survived Satan's invention of the scientific discipline of history.

It happened in the third quarter of the nineteenth century of the Christer era. Historians had already learned to compare statements in alleged nonfiction with facts known from other sources, and where the two were incompatible expose the alleged nonfiction as nothing of the sort. That technique, plus the detection of anachronisms, enabled them to expose Geoffrey of Monmouth's *Chronicle of the Kings of Britain*, long touted as sixth-century history, as in fact a twelfth-century historical novel. Would you believe that Geoffrey had King Arthur's father, lusting after Arthur's mother, chase her into Tintagel Castle in Cornwall—six hundred years before Tintagel Castle was built!

Satan persuaded a small group of historians to apply their newly perfected techniques to the kike-and-Christer bible. And he taught them the additional methodology of searching for pairs of statements that contradicted each other and therefore could not both be true.

The results, when published, terrified Me and terrified the pope. The bible contained over one thousand pairs of mutually exclusive statements, and therefore contained at least one thousand lies even if no others were present. (In fact there are 18,000 others.) It contained dozens of anachronisms, such as a mention of Alexander the Great by an author who had allegedly lived more than two centuries before Alexander's birth. It contained hundreds of prophecies that the historians showed to have been already fulfilled at the time of writing, always accompanied by prophecies of the future that failed to be fulfilled.

For example, the anonymous author of Mark prophesied the destruction of my temple in Jerusalem in 70 CE, not a difficult feat for a man writing in 73 CE. But he then went on to prophesy Jesus' return visit no later than the reign of the emperor Hadrian. He seems to have been delayed.

The pope was frantic. "What can I do?" he wailed. "We'll be destroyed. There must be some way to rebut them."

"Appoint your own historians," I advised him in a dream. "Make sure they're all card-carrying Catholics who never miss Sunday mass and go to confession at least several times a year. Tell them to examine the evidence the historians have uncovered, and find the simple mistake the historians made that led them to their false conclusions."

The pope followed my direction. He appointed a group of Catholic historians to examine the bible with piety and faith, and find the true explanation for its apparent inconsistencies. Boy, was that a mistake.

The Catholic historians confirmed the findings of the secular historians. The kike-and-Christer bible was pure fiction, written by authors who contradicted each other, authors who believed in a flat earth and a solid domed sky, authors who correctly prophesied the past but failed miserably whenever they attempted to prophesy the future, authors who guessed, fantasized or lied in order to make a point, and authors who credited Me with whatever action they felt they would have taken in my place.

The pope's historians gave him their report. Realizing that he was certain to suppress it, they also leaked it to the press. Then, without exception, they at some time in the next three or four years declared that they had ceased to be godworshippers. Christianity is totally dependent on the veracity of the kike-and-Christer bible, and since they had proven beyond the last shred of doubt that the bible is fiction, they recognized that for their bible to be a hoax, their religion must also be a hoax.

The pope was devastated. Facing instant unemployment, or worse, he considered killing himself. But I gave him an alternative.

"Since the methodology of history refutes the bible," I told him, "what we have to do is invent a new methodology that can be used to prove the opposite. Instead of starting with the evidence and following wherever it leads, as the historians do, we start by determining what conclusions we're going to reach, and then make the evidence fit those conclusions. We can call the new methodology *theology*, 'knowledge of God,' and get the universities to give degrees in it and tout it as a discipline every bit as respectable as history.

"Of course we have to be careful that the new methodology is never applied to any book but the bible," I warned. "If it's applied to *Alice in Wonderland* or *Gulliver's Travels* or the memoirs of Graf von Münchausen, it'll show all of them to be nonfiction, and we don't want that to happen. But that's a risk we have to take. Unless, every time a historian proves that a biblical story is fiction, we can produce a theologian to prove it's not fiction, we're both dead. Don't worry. I know this will work."

And it did. I am very proud of theology. It requires that its practitioners be ignorant of the methodology used not only by historians but also by all scientists. It also demands that they have the stubborn, inflexible gullibility and mental dysfunction for which P. T. Barnum would have beaten a path to their doors. But it saved the Christian church. It saved Me-worship. To this day, when the mindless masses hear that a scholar with a doctoral degree in history declares the kike-and-Christer bible to be a demonstrable fraud, and an alleged scholar with a doctoral degree in divinity declares it to be revealed truth, they see the situation as a dispute between equals that cannot be settled one way or the other. So they go on believing the one that offers pie in the sky when they die.

To paraphrase Barnum: There's a godworshipper born every minute.

# THREE

He introduced himself, "Madame, I'm Adam."

She replied, "Don't look now, but you've got a wart on your cunt, not to mention two enormous shrapnel hernias in the vicinity."

Well perhaps that's not exactly the way it went down. My memory is not what it used to be. And of course every time some new mythologian rewrites my biography, I am obliged to go back and change history to make it conform to the newest orthodoxy. I can do that, you know: change history. A couple of thousand years ago my loyal subjects finally made me omnipotent and omniscient, and since then I've been able to do and know anything wish. I can even make a rock so heavy that I can't lift it.

Did you know I had to build a whole new Hell for the Salt Lakers, a Hell so wonderful that any mortal who could imagine what it was like would immediately commit suicide in order to get there? Because the Mormons thereby made Me a nicer god than I had ever been before, I retroactively gave those rutting tomcats, Joseph Smith and Brigham Young, licenses that turned their every compulsive score-chasing into a recognized marriage.

Adam and Eve were my natural, unfathered children in the first reality, born when I was still Mother Earth. After I became male, I had to go back and create them out of a hundred pounds of clay. If some up and coming mythologians have their way, I am some day going to have to go back again and change my raw material from clay to seaweed and salt water. There are even some who say I created only the soul, and that the human body evolved from lower life forms—but I try to steer clear of those. My life is complicated enough as it is.

Adam and Eve worshipped Me as the Mother. They built Me cunt-shaped altars, and ritually ate the cunt-shaped ripe pomegranate that was my sacramental body and blood.

Then came the Big Discovery that children have fathers, and the consequent revolution that saw women replaced by men as absolute masters of the universe. I gave birth to my male self, and sliced my female self in two and used the parts to create the land and the sky. I ordered Adam and Eve not to eat my sacramental cuntegranate, since that would constitute the heinous crime of goddess worship.

"You can eat the fruit of any other tree," I informed them. "But the tree in the middle of the orchard is special. It's so juicy and tasty and delicious and scrumptious that only gods are allowed to eat it. If mortals ever find out how fantastically mouth-watering and satisfying it is, they'll eat it all and there'll be none left for Me. So you're not to eat it. If you eat it I'll kill you. I'll kill your children. I'll kill your children's children, and so on and so on, forever and ever. Now run along and have fun, kiddies."

Would you believe those disobedient arseholes actually ate the goddess symbol after I told them not to? And since as a male god I was obliged to regard goddess-worship as the most monstrous crime imaginable, I had no choice but to punish them severely. Killing them was not enough so, since my scriptwriters would not invent life-after-death for several more centuries, I had to settle for killing all of their descendants before they reached the age of 120 years.

It was soon after that that Adam and Eve learned to fuck. I'm not sure how they learned— remember I wasn't omniscient in those days—since fucking is not instinctive among humans. I saw a movie once called *Blue Lagoon*, in which a male and female infant, totally ignorant of sex and with no adult to teach them, grew up on a deserted island in the middle of the ocean. Yet sometime during adolescence they found themselves fucking by instinct.

Horseshit. It couldn't happen. If they had had farm animals to teach them by example, particularly horses, cattle and dogs, where it is quite apparent to a casual observer what is going into where, then they may have comprehended that they had the same parts and conceivably could do the same thing. But one way or another, fucking is something humans have to learn. I can only assume that Satan must have taught Adam and Eve to fuck, as he had undoubtedly taught them to disobey Me and eat my deposed Mother's cuntegranate.

I dropped in on Adam right after his first fuck, and asked him were Eve was.

"She's down at the river, your Lordship," Adam answered. "We just had a fuck, and she's gone down to wash off the cum."

"She's rinsing her cunt in the river?" I queried. "Shee-it! Now the fish are going to smell like that."

Fuck Satan. It's all his fault.

I have never understood why I had to create Satan. Since with my current omniscience I should have known he would rebel, it would have been only logical to refrain from creating him in the first place. But of course the choice was not mine. My creators finally noticed that there was much evil in the world. So since they could not saddle Me with responsibility for that evil, having made Me omnibenevolent, they had little option but to conjure up a monarch of evil who was my great Enemy. And I have been stuck with the little bastard ever since.

Like Me, Satan also used to be a personified cunt. The twit who concocted the Adam and Eve fable called her a snake, and had her tempt my mortal creations into worshipping at her altar by eating her sacred cuntegranate. Obviously the twit wrote at a time when I had no male Enemy. Only when the kikes were conquered and dominated for four hundred years by the Babylonians and their successors did they become jealous of the Zoroastrian antigod, Ahriman, and decide that I, too, should have an evil counterpart.

Naturally such a responsible position as prince of darkness could not be entrusted to a female. So the Mother, who had previously been the closest thing the kikes could imagine to an Enemy, was given a cock transplant and became Satan. And since you already know that I was the Mother before I gave birth to myself, you can deduce from that that Satan and Yahweh are merely two perspectives of the same god. I'd say that Satan is my evil side, except that the mythologians who created Me in their own image were so morally retarded that they naturally made Me equally evil. To coin a phrase: With a god like Yahweh, who needs a devil?

Nonetheless, the post-Captivity kikes did give me an Enemy. They had no difficulty finding him an evil nature. All they had to do was look in any mirror—or, tantamount to the same thing, look at Me. His body was more of a problem. Rather than start from scratch, they simply borrowed the Greek goat-god Pan. To egocentric humans, incapable of seeing beauty in anything but themselves, what could be more ugly, and therefore more evil, than a god with horns and a tail? Since Pan was already endowed with those attributes, he was a shoo-in for satanship—especially as Greek mythology had just killed him off and he was actively seeking new employment.

Poor Satan. He survived the transformation from Pan okay. He survived the transmogrification from a nanny to a billy with no serious trauma. What he did not survive unaffected was his cock enlargement.

You see, when Satan was first masculinized he was given a phallus no larger than the clitoris it had previously been. To the kikes, sexual recreation was a great joy, created by Me, and it logically followed that the Evil One must be an enemy of joy, possessed of a joystick so small as to be nonfunctional.

However, when the Jesus masochists formed the dogma that anything that feels good and hurts no one must be evil, a Satan that didn't practise the joyful evil of fucking was a self-contradiction. So they gave him a cock graft from a stallion. And apparently all that blood rushing to his cock dried up his brain and gave him permanent delusions. He doesn't even know who he is any more.

Satan just will not accept that he is a product of the human imagination. He thinks he is, of all things, an extraterrestrial. No kidding. He thinks that both he and I belong to a race called satyrs that evolved in another part of the galaxy, and that we were sent here to be the gods of earth by *our*

150

gods.  I ask you, how can I have a god when *I* am God?  And I am God indeed.  You bet your sweet arse I am.

# FOUR

I showed him. I still don't know how I showed him, but I showed him. It was so exciting. I do so many tricks, but I always know how I did them and that spoils the fun. But this time I pulled my most spectacular trick in decades and I didn't see how I did it. Am I good or am I good?

It was that slimy limey pipsqueak who got Me riled, the one who in the novel that foreshadowed the reality told the press, "Not even God could sink this ship." Now I ask you, could I ignore a challenge like that?

"And just what do you propose to do about it?" Satan asked Me.

"Do about it? I'll do what the novel shows Me doing. I'll sink the motherfucker's unsinkable ship, of course," I answered.

"A mythical ship called the *Titan* was sunk after a mythical character said that even God couldn't sink it. The *Titanic* is real, and no real person made any such boast. But even if someone did, how do you propose to sink an ocean liner?" the evil one demanded. "I seem to recall George Bernard Shaw giving you five minutes to strike him dead, and instead it was you who almost dropped dead, from apoplexy brought on by the frustration of impotence. Brother, you can mind-direct fanatics to do things they were already inclined to do without your nudging. But other than that you can't touch them. You couldn't touch Shaw, and you can't touch the *Titanic*."

So I'm impotent, huh? So I couldn't sink the *Titanic*, huh? Well it seems I'm not unable to hypnotize icebergs, even if I didn't know I'd done it until I read it in the newspapers.

"Yahweh, fifteen hundred people are dead," Satan whispered hoarsely. "It was a horrible coincidence. You didn't do it, and if you had it would have been a monstrous crime. How can you rejoice over an appalling tragedy? Do you hate these humans the Overlords appointed you to guide and protect? You should be sharing their sorrow."

That extraterrestrial stuff again. Scripture doesn't say so, but I must have created him with a few missing marbles. It's the only explanation.

"You read where that fatuous twit challenged Me," I beamed, too excited at my triumph to be bothered by Satan's continued negativism. "So not even God could sink the *Titanic*? Well I showed him. I sure showed him."

Satan absolutely refused to share my joy. He actually wept, so piqued was he that I had enhanced my reputation and he had not. Talk about your sore losers.

We went to a memorial service at Canterbury cathedral. It was Satan's idea. He thought the sight of all that sorrow would dispel my elation. He'll never learn.

"Merciful and loving God," a woman near us wailed, "let my son be alive. Let him not be counted among the dead. Let him come back to me."

Was she kidding? The *Titanic* had sunk days ago. If her son was one of the casualties, then it was over and done with. Did she think I could change the past? If that was her belief, how come she wasn't praying that Adam would refuse to eat the Mother s cuntegranate?

"You heard the woman," Satan taunted. "Let her son come back to her. He's probably been dead for a week, but that's no problem for a god as all-powerful as you. You boast about how you can time jump and change history. So time jump and save her son."

"I can rectify history when sacred scripture says I did something contrary to current reality," I explained for the googoolplexth time. "Where in scripture does it say that this woman's son survived the sinking of the *Titanic*?"

"Are you saying he's either alive or dead, and either way you can't do a thing to change it?" Satan smirked.

"It's not that I can't," I corrected. "I just have other priorities."

"And Santa Claus comes down the chimney on a dead Jew's birthday."

"In the reality of children, yes he does," I snapped.

Satan sank to his knees and raised his arms in Me-worship fashion. "Saint Dasher, Saint Dancer, Saint Prancer, Saint Vixen, pray for us," he roared. "Saint Comet, Saint Cupid, Saint Donner, Saint Blitzen, pray for us."

"Will you shut up," I hissed. "The marks will hear you."

"They did hear me," he answered. "They heard me as a high-pitched whistle that lasted a thousandth of a second. Say, here's an opportunity. How about we both hypometabolize right here, so they can see us? You can introduce yourself to the grieving widows and orphans and mothers as the god who murdered their loved ones. You can explain that you only did it to show me what a powerful and almighty god you are. They'll understand. Go ahead. Show yourself to them. If you've forgotten how, I can show you. How about it?"

The suggestion was nonsense. Everyone knows gods are invisible.

Besides, my worshippers believe I'm benevolent. When I do things like sinking the *Titanic*, or leveling Los Angeles in a four-billion-dollar earthquake, or allowing a couple of million Jews to be wasted in Hitler's gas ovens, they rationalize it as my "mysterious ways," incomprehensible perhaps but never evil, since everything I do is virtuous by definition. If I appeared to them and tried to explain that I sank the *Titanic* because I damned well felt like it, they'd probably get pissed off. They might even try to put Me on trial.

Satan would love that, to see Me tried by a jury of my victims. Can you imagine how I'd fare if I had to justify my atrocities to a jury consisting of the relatives of disaster casualties, victims of mustard gas, the casualties of Hiroshima, Krakatoa and Pompeii, the grossly deformed such as Jesus and John Merrick, the blind such as John Milton and Helen Keller, Joan of Arc and Archbishop Cranmer, barbecued in my name, Rock Hudson and Liberace, Abraham Lincoln and John Kennedy, Rose Kennedy and Otto Frank, not to mention the millions who die annually of starvation, violence, disease and despair? For crimes like mine, if there was not a Hell they would have to invent one. They'd probably abandon Me-worship and switch to Satan as the lesser of two evils.

And Satan knew that. Didn't I tell you he was jealous? He wants humans to worship him. That was why he tried to trick me into manifesting my magnificence in a crowded cathedral. He probably planned to yell, "Fire."

He's a schemer, that Satan. If it wasn't that he keeps Me on my toes, I'd uncreate him.

Can I do that? I seem to remember...Sure I can do that. I can do anything I want. I'm omnipotent. If I really set my mind to it, I could create a triangle with four sides, or a number that is more then ten but less than nine.

I simply have other priorities.

# FIVE

Don't blame Me. I'm not the one who promised, "There are some of you standing here who won't experience death until you've seen Allah's theocracy established by force of arms." (Mark 9:1) The psycho from Galilee made that fatuous prophecy, not I. And then he died.

That prophecy was made in the first reality, at a time when I had not even heard of Jesus, let alone met the man. So it would not have presented Me with a problem if I could have kept it out of subsequent realities. Unfortunately, the Greek who wrote the gospel you call Mark (his name was Georgio, if I remember correctly) learned of Jesus' failed prophecy and included it in his gospel. And then the authors of Matthew and Luke (I think their names were Nikhos and Zorbos) copied it from Mark. Now that all three of those historical novels are recognized as scripture, I'm stuck with them. The prophecy must therefore somehow be fulfillable, even though, centuries after Jesus' death, it cannot be—

unless someone who heard Jesus' rash boast is still alive and hiding in Argentina, awaiting the second coming, as Jesus himself has been hiding since 1917 according to the Jehovah's Witnesses.

That was it. That was the loophole I needed. Jesus' second coming and coronation as head kike could take place within the lifetime of a person who heard him preach, if that person was specifically destined to remain alive until the aforesaid second coming. Reassembling the scattered molecules of a bio-organism that is totally and permanently dead is beyond even my omnipotence, or I would have done it centuries ago, so the waiting one, the Wandering Jew, could be in for one hell of a long wait. But as long as my worshippers believe it can and will happen, some day, what do I care that it never will? I can milk it for as long as it lasts.

I am very proud of the "Wandering Jew" myth. It was my own original concoction, you know. I revealed his existence to about two dozen different Jesus pushers in their dreams, but in each case I hinted that it was a recollection of a story he had first heard as a child. That way the tale could never be traced to a specific author and thereby discredited.

This is what happened in the current reality:

Jesus promised his hearers that he was going to overthrow the Roman Empire and usurp David's throne (you thought Jesus was Davidic?) within his hearers' lifetime. So when one of the persons who heard that prediction saw King Runt nailed to a stake and about to die, he taunted him for his failed prophecy. Jesus responded by cursing the taunter to be the one who would remain alive until the prophecy was fulfilled. That man has been wandering the earth, never staying in one place long enough for his non-aging to be detected, ever since.

Cute, huh? A couple of times I considered persuading the pope *pro tem* to incorporate the Wandering Jew into Christer dogma. But my purpose is to preserve Me-worship, not destroy it—and the story does present a credibility gap. So I content myself with reminding new popes that it can never be repudiated, for then Jesus would be revealed as a liar.

Okay, so he *was* a liar. Let's keep that our little secret.

# SIX

It's a good thing I'm a quick-change artist. For after the disintegration of the Khassidites and their supersedure by Pharisees, Essenes and Sadducees, I found myself changing beliefs every time I changed hats. And it wasn't just beliefs. Every time a kike died, I had to find out what sect he belonged to before I could consign him to his eternal fate.

For example, if he was a Pharisee or an Essene then, in accordance with his beliefs, I had to sentence him to be tortured with flamethrowers until all of his disobediences to my whims, the true meaning of *sin*, had been purged away and he was fit to go to Abraham's bosom. (Keep in mind that the runt from Galilee had not yet turned ge-Hinnom into Hell by making the torture permanent.)

On the other hand, if he was a Sadducee, then all I could do was bury his body and try to keep his name alive in the memory of his descendants. There was no Sadducee afterlife, no "Abraham's bosom," or *Heaven* as it eventually became through osmosis from the Christians, no purgation of sins in an underworld torture chamber, only the total, permanent, irreversible extinction of death.

From Abraham to Seleukid times, all kikes had been Sadducees, in the sense that they had no afterlife belief. For that reason the Sadducees were my sentimental favorites. But even God must advance with the times. So when the new philosophy invented ge-Hinnom, I had no option but to time jump and retroactively create the place, along with "Abraham's bosom." But I had to keep my Pharisee/Essene reality separate from the Sadducee reality. If I accidentally sent a Sadducee to Abraham's bosom, he would realize at once that I was a figment of the Pharisees' imagination and therefore not the Sadducee Yahweh. Then would the shit hit the fan.

Understand, the dispute between the one-lifers and the eternal-lifers was not merely oral. If it had been, I could have waited until one of the opposing philosophies became extinct (the Sadducee, as it turned out), and then written off the proponents of that philosophy as heretics. Like Sam Goldwyn said, a verbal contract is not worth the paper it's written on. Unfortunately for Me, both sides stated their beliefs unequivocally in writing, and the writings of both are to this day included in the kike-and-Christer Bible.

A Pharisee wrote the book known as Daniel, in which he retroactively prophesied the Babylonian, Persian and Seleukid overlordships and the establishment of the Hasmonean monarchy by the Maccabees. And like all prophets, he refused to quit while he was ahead. After successfully prophesying the past, he made the mistake of trying to prophesy the future. He declared that the Hasmonean monarchy would last forever. And we know what Rome did to the Hasmoneans, don't we, kiddies?

But in addition to prophesying the events prior to 164 BCE correctly and the events after 164 BCE incorrectly, the author of Daniel also spelled out the Pharisee belief in an afterlife. He wrote, "Many of those who sleep in the dust of the land are going to awaken, some to everlasting life and some to everlasting contempt." You will notice that only "many" of the dead were destined to live again. It was never a kike belief that non-kikes might also have a chance at eternal life. Why should they? Were goys my chosen nation? Why would I raise them?

The Sadducee author of Ecclesiastes, on the other hand, spelled out his sect's no-afterlife belief in the words, "The living are aware that they are going to die. But the dead neither know anything nor have any further reward, for their awareness has ceased to exist."

Since both of those mutually exclusive passages now constitute sacred scripture, inspired by Me and therefore by definition Revealed Truth, you can see my problem. Compared to that, telling Christians to eat pork and Jews not to eat pork, Baptists to drink coffee and Mormons not to drink coffee, Protestants to practise birth control and Catholics not to practise birth control was rational.

And you wonder why I am schizophrenic?

# SEVEN

It was such a lovely war. And the best part was that both sides were my worshippers, so every single fatality constituted a sacrifice to Me. I thought I was in for the greatest orgasm I had ever experienced.

But naturally Satan had to spoil it for Me. When the sultan Mohamed II saw 23,000 of his loyal soldiers impaled on Count Vlad's stakes, his instinctive reaction should have been to attack the Christers without mercy and repay atrocity for atrocity. But Satan got in first and softened the sultan's heart. Mohamed actually wept when he saw what Vlad had done. He turned around and marched back to Istanbul, convinced that persons as subhuman as the Christers could no more be incorporated into a civilized empire than could a pack of mad dogs.

I tried to stop him from retreating. I appeared to Mohamed in a dream, in my Allah persona of course, and assured him that Vlad the Impaler was not your run of the mill everyday Christer. Naturally I did not know then that Vlad's insatiable blood lust would in time cause even the Christers to repudiate him, or that four hundred years later a book would be written that portrayed the great defender of Christianity as an enemy of the Christers who cringed in terror at the sight of a Christian anti-vampire charm.

I did, however, point out to Mohamed that Vlad was not a fanatic Muslim-torturer. He indiscriminately tortured everybody. And while the impaling of 23,000 Turks taken in a single battle set a new record, he had in the course of a long and violent life impaled more Christians than Muslims. His own subjects lived in constant terror of his sadistic whims. If the sultan of Ottoman Turkey were to press on and kill the monster who kept them permanently terrorized, the Christers would hail him as a liberator and turn Muslim.

It didn't work. The sultan turned back and made no further effort to annex Transylvania. Count Vlad Tepes was hailed as the hero who had routed the heathens, and he remained a hero until the absence of further external threats caused him to resume his internal abominations. His subjects finally rebelled and forced him to flee, and he spent his last years in a foreign prison.

The Christers were most unfair to turn on Vlad. Not since the Jew Joshua at Jericho had anyone given Me so many human sacrifices to consume at a single sitting, and not until the Bartholomew's Day sacrifice of Protestants by the French Catholics would Vlad's offering be surpassed. Yet because he loved Me and loved feeding Me blood, they called him "vampire." His father's insignia had been a dragon, *dracul* in Romanian, so Vlad became "son of the dragon," *Dracula.*

How quickly they forget. To this day Christers in the vicinity of Castle Dracula paint anti-vampire crosses in the air at the mere mention of the count's name, blissfully unaware that, but for Dracula, they would now be on their knees five times a day bowing toward Mecca.

How quickly they forget.

156

# EIGHT

I returned to kike land, after an absence of eight hundred years, during the reign of the infidel usurper, King Saul the Israelite.

Let Me amend that. In the first reality that only I remember, I stayed away for eight hundred years. But when I learned from scripture that I had aided Yahuwshuakh in his bloody conquests, I visited Jericho retroactively and changed history to accord with the Book of Joshua. I can do that. I'm God.

And if you think that was pretty good, consider this: I retroactively turned Yahuwshuakh into Moses' successor, as the first Genesis author declared him to be, when in the first reality Yahuwshuakh had died a century before Moses was born, and the tribes led by the two, Jews and Israelites respectively, although both Semitic, were otherwise unrelated. In the present reality, as you can verify, "Jew" and "Israelite" are synonyms. It's in the book.

On my return, I found that an Israelite named Saul, whose tribe worshipped the sky god Allah, had incorporated the Me-worshipping descendants of Abraham and his slaves into his infidel kingdom. I soon put a stop to that.

In accordance with the cunt-venerating Israelites' long-standing tradition, Saul had named his son-in-law, the Jew David, his heir apparent. I used my godly wiles to persuade Saul to double-cross David and nominate his own son, Yahuwnathan, who by a delicious twist of irony happened to be David's lover, as his designated heir, in imitation of the custom of the kikes. Naturally David rebelled. I won't go into detail, but David wound up as emperor of not only Jews and Israelites, but also a whole bunch of other Semitic and non-Semitic tribes on whom he imposed the worship of Me, Yahweh.

David's empire fell apart after the death of his son Solomon. But for a while I was God the King in an area stretching from Egypt to Assyria. You win some and you lose some.

It was while I was living in Jerusalem, in the temple I had Solomon build Me, that Satan, or Pan as he was still called at that time, tried to delude Me that I was not God.

"Look, you claim to be the creator of all things, right?" he asked Me. "Not a sparrow dies without your knowledge? Not an insect lives without your consent?"

Naturally I agreed.

"Good. Come with me. I want to show you something."

What he showed Me was a wasp stinging a trapdoor spider, paralyzing it, and laying its eggs in the spider's abdomen.

"The spider is permanently immobilized but fully conscious," Satan explained. "When the eggs hatch, the maggots will eat away the spider's insides while the spider just lies there, aware of what is happening to it but unable even to escape the horror of the situation by death. It eventually dies, when it has no internal organs left to sustain life. And you claim that, as creator of the cosmos, you designed that abomination?"

So I threw up. Big deal. There's no law that I have to like everything my godship requires Me to do.

"Really, goat-head," I expostulated when I had recovered my composure somewhat, "sometimes you push Me too far. Since you ask, yes, that is my creation, and yes, it is obscene. But try to understand: I did not choose what I would create. I had to create the world as it is, not as I would like it to be. I didn't exist until the first mythologian created Me. The earth existed. The sky existed. The horror you just showed Me existed. But I did not exist. So when the kike mythologian at King Rekhobowam's court made Me the creator of all things, I had to create them as he perceived them, not as I perceived them. Is that so hard to understand?"

King Rekhobowam, by the way, was Solomon's son and successor. In his reign my earliest biography was composed.

"Then go back in time," Satan smirked, "and retroactively uncreate this revolting creature."

"You just will not comprehend. Before I could uncreate the spider wasp, for example, the imagination in which I exist would have to purge it out of existence. And how could it do that when it has already seen that the thing does exist? I'm sorry, but if you choose not to believe in Me, that's your problem. I exist and I am God. I believe in you."

Satan can be such a pain in the arse sometimes.

# NINE

So I changed sides. I'm God. I'm allowed to do that. I'm allowed to do anything I wish. If I batter Los Angeles with an earthquake that does four billion dollars worth of damage, that isn't evil. It's my mysterious ways. Ask my brownnosers. So why draw the line at my changing sides?

I mean, it wasn't as if I hadn't given the kikes a fair shake. For the four hundred years since their defeat by Nebuchadnezzar, I had kept dropping in on them, checking on their needs, giving them new taboos, and intensifying their xenophobia. I had thereby enabled them to survive conditions that had integrated the Israelites out of existence in a single century. And still they were tribute-paying serfs, making no effort to conform to my commandment that they pay tribute to Me alone. Do you blame Me for getting pissed off and switching my support to Antiokhos Epiphanes?

Antiokhos was the current king of Persia, descended from Alexander's general, Seleukos, who had appropriated Persia after Alexander's untimely demise. But Antiokhos was more than just another king. He it was who recognized the imminent danger of upstart Rome, which had already leveled mighty Carthage. He decided to transform his heterogeneous empire into a unified fortress that could counterbalance the impending threat from the west. His plan was to create a theocratic monolith by imposing the religion of his ancestral Macedon on those parts of his empire that still favored non-Greek gods. Since one of my names was Zeus, the prospect of spreading Zeus-worship to the whole Eurasian world pleased Me greatly. Naturally I changed sides.

The kikes resisted Antiokhos's decree, so Antiokhos invaded. He marched his army into Jerusalem, and personally violated the sanctity of the holy shrine in my temple. Normally that would have annoyed Me. But I was aware that he only wished to replace the Yahweh statue in there with a statue of Olympian Zeus. Since I was both Yahweh and Zeus, a change of hat did not bother Me in the least.

I am an invisible god, at least in my Yahweh hat. The kikes had therefore proclaimed far and wide that no representation of Me in metal or stone was possible, since no man had seen Me. Consequently no representation existed. The holy shrine in the center of the temple, where the priests knelt to worship Me, was believed even by most kikes to be empty. For generations only the high priests had been allowed to step through the curtained entrance and learn otherwise. Now Antiokhos discovered the truth.

The godlet from the *sanctum sanctorum* was paraded through the streets of Jerusalem.

"For generations you have been boasting that you do not worship images," Antiokhos's criers boomed to the crowds that had been herded into the streets to watch. "Now take a look at what you have been worshipping. This is your god. This is your Yahweh."

The statue from the temple's inner sanctum was not of one god but two. Each was double the height of a man, and each had a human head and chest, a lion's rear end, and an eagle's wings. One was male. The other was a full-breasted woman. They depicted the fifth-century-BCE concept of the fixed star gods, called *kherubim*, sometimes Latinized to *cherubs*. And, as befitted the image of God the Phallus and his obedient *Shekinah*, they were fucking.

Looking at the faces of the kikes lining the streets, and feeling their embarrassment, I congratulated myself on getting out just in time. I was the god of the winning side, and I felt only contempt for the losers. Napoleon Bonaparte said it well: "God is on the side of the big battalions."

If you are an unbeliever, you probably think the only way I could know Napoleon's words is that I am writing at a time when Napoleon has already lived and died. I get a lot of that. It doesn't bother Me, because I have the satisfaction of knowing that I will some day get my rocks off torturing you with flamethrowers for your unbelief for billions and billions of years—unless of course you are a humanist, in which case I can't touch you, because yours is not the kind of imagination in which a god as insane as I am can exist.

The Seleukid victory was short-lived. Within a decade of Antiokhos's entry into Jerusalem, Maccabee victories had given Jewdah real independence for the first time in four hundred years.

So I switched back to the kikes. I am nothing if not practical.

160

# TEN

I knew I was right. Satan told Me over and over that the continued infanticide of deformed babies would strengthen and improve the human race. But all he was really trying to do was stop Me from breeding worshippers as fast as human cunts could turn them out. So while the Greeks and others continued to breed for quality, I ordered the kikes to breed for quantity. Any viable infant capable of becoming my worshipper, no matter how misshapen, imbecilic or susceptible to genetically transmitted diseases it might be, was to be preserved alive until it, too, could breed more believers. And it paid off. Boy did it pay off.

Anywhere else on earth, the baby would have been mercifully smothered at birth. But because his parents were kikes, he grew up to be a cross between Rumpelstiltskin and Quasimodo. *Ugleeeee!*

The runt was taller than a munchkin, and at an adult height of 137 cm, (4 ft 6 in, or three cubits), he just missed out on being a designated dwarf. And even though his detractors promptly labeled him "the hunchback," a more accurate description would be "stooped." He was bald by the age of thirty, with just an atoll of hair around the back and sides, but he was otherwise quite hirsute, and his eyebrows formed an unbroken line right across his forehead. In later generations, even his own apologists described him as "incredibly ugly," and "not even of honest human shape." And the fact that, until the day he died, the only fucking he was able to get was with magdalenes, should also tell you something.

The runt was, of course, Jesus of Galilee. You don't imagine there could ever have been two men that ugly on a single planet? Of course it was Jesus of Galilee, "the Nazirite" as his followers preferred to style him. How he ever managed to acquire a cadre of hardcore followers, I will never comprehend. I still have difficulty understanding how ignorant, semi-literate hillbilly demagogues like Billy Graham and Maharishi Mahesh Yogi can find followers—and they are both physically presentable.

I'll never forget the first time I saw him. Understand that I am speaking of the first reality. Later I went back in time and retroactively adopted him at his immersion initiation. Later still I went back even further and knocked up his mother. I thought at first she was in a coma, but then I realized that she was Jewish.

Satan and I had just dematerialized in Rome and instantly rematerialized in Jerusalem. Satan described our route as "the short cut," but God does not need a short cut. Jesus was haranguing a crowd of several thousand, and doing such an effective job that only a tiny minority were laughing at his barely anthropoid body.

I could not believe my eyes. "That's the latest messiah?" I whispered to Satan. "The man's a joke. Come on; let's get out of here. This is one messiah who's going nowhere."

"Don't be too sure," Satan answered. "Look at the way they're hanging on his every word. You've seen what demagogues can achieve. Remember Kleon? Marcus Antonius? Kalkhas? Nathan? Jeremyah? Unless you want to wind up as god of the geeks and freaks, you'd better stop him before he gets any stronger. Listen to what he's saying. He's only a day or two away from proclaiming a war of Jewish independence."

Satan was right. Unless I swatted this rabblerouser right now, I could be stuck with a chief prophet and viceroy with all the credibility of a baboon in a toga. The Council of the Gods would expel Me as a laughing stock. And I so wanted to be their President.

I hightailed it to Judea's administrative capital, Caesarea, and spoke to the Roman procurator in a dream.

"Pilaté, baby," I told him, "this guy Jesus, you gotta waste him, like now. I mean, it's your job, man. He's spreading discontent, and we don't want Caesar to think you can't keep one little province pacified, now do we?"

"Jesus?" Lucius Pontius Pilatus laughed. "You mean that Galilean freak who's been touring the provinces with his entourage of pimps and poofters? The man's saving me a fortune in circuses. He is a circus. As long as freaks like Jesus keep the great unwashed amused with their cheap conjuring tricks, they're not going to make my life miserable with strikes and insurrections."

"Pilaté, dear boy, insurrection is precisely what Jesus is promoting. He's telling the ignoranti he's their prophesied king, the descendant of David who's going to throw out the Romans and restore Jewish independence."

*"What???"*

Within twenty-four hours Pilatus had force-marched his battalion to Jerusalem and ordered Jesus' arrest. Within forty-eight hours Jesus had been executed and buried and was providing nutrients to ten thousand worms. There the matter should have ended.

Unfortunately, I had not anticipated Paul of Tarsus, a fruitcake whose need to be a big frog in a small puddle caused him to abandon the Pharisee sect in which he was going nowhere, and infiltrate one of the smaller messianic sects in the hope of rising to presiding guru. He chose the Nazirites, only to discover, too late, that the sect's dead messiah, Jesus, had a large number of living relatives, and they alone had a snowflake's chance in Hell of becoming Head Nazirite. So Paul, while maintaining the pretence of being a Nazirite, founded a totally gentile religion in which the Jew Jesus was posthumously enthroned as its figurehead king. And as if that were not enough, seventy years after Paul's death the anonymous author of *John* promoted King Jesus to the god Jesus, my naturally begotten son. The new religion's detractors called it, in Greek, *Khristianismos,* "Anointianity," and that name was later non-translated into other languages as *Christianity*.

I am nothing if not pragmatic. When I saw that Anointianity was well on the way to conquering the world, I adopted it as my own. I made the necessary time jumps to rectify history and turn Jesus into everything sacred scripture said he was, including an Aryan Adonis fifty centimeters taller than he had been in the first reality. I was God of the Christers, and in that capacity had more worshippers within three centuries than I had been able to secure in two thousand years as god of the kikes. Things were going well for Me, and not before time.

I could not remove all of the physical descriptions of Jesus left over from the first reality. I did manage to get the original description, penned by Flavius Josephus, expurgated from all surviving manuscripts of his *Jewish War*. But later writings based on Josephus's description are extant to this day. I did, however, succeed in restricting such information to scholars, and a full ninety percent of all Christers now believe that Jesus looked a lot like Max von Sydow.

Satan, of course, asked Me why, since I am omnipotent, I could not have retroactively created only the handsome Jesus, so that no description of the malformed runt Jesus would ever have been written. I say again, omnipotence is not retroactive. I made Jesus ugly and deformed in the first reality because that is how the earliest sacred writings described him. For the thousandth time I tried to make Satan understand that I did not create scripture. Scripture created Me.

He didn't understand. He'll never understand. What can you expect from someone who thinks he's an extraterrestrial?

# ELEVEN

I rule over a multiverse of varied realities. For example, Catholics live in a reality in which my round-heeled concubine Mary was bodily assumed into Heaven without the necessity of kicking the bucket. Muslims live in a reality in which Mohamed was bodily assumed into Heaven. Protestants live in a reality in which both the kikette and the wog died and were buried and eaten by worms. And my most zealous fanatics, the hardcore members of the Flat Earth Society, live in a reality in which the earth is indeed the flat disk depicted in the Judaeo-Christian bible in seven places, even though all other godworshippers live on a planet that is an oblate spheroid.

Kikes live in a reality in which I am a kike. Christers live in a reality in which I am a Christer. I have a hard enough time keeping the two separate without further complications. The last thing I needed was a sect that didn't know if it was kike or Christer.

Okay, Jesus was a kike, not just by being the son of kikes, but also by religion. He would have repudiated Paul of Tarsus, and if he ever manages that somewhat delayed "second coming," he'll head for the nearest kike synagogue and denounce the Christers. Big deal. The fact is, the superstition founded in Jesus' name is not remotely Jewish. You can be a kike or you can be a Christer. You can't have your kike and eat him.

So when I learned that two Jewish sects, called Nazirites and Ebionites, were teaching the same things Jesus taught, namely that Jesus was the heir and successor of King David, anointed by Me to refound the kike monarchy and snuff out all goys, I was thoroughly pissed off, I can tell you. I mean, either you're a Christer, in which case you live in a reality in which Jesus is my bastard son, adulterously sired on Joseph's wife, or you're a kike, in which case you live in a reality in which Jesus was a liar and a madman. There's no middle ground, and I was not about to accept one.

So I had a few words with some prominent Christers, starting with the sainted Jerome.

Are you familiar with Jerome? I'd call him a masochist except that, having already identified him as a Christer saint, I'd be perpetrating a tautology. He once boasted that he had lived in the desert among scorpions and wild animals until his skin was dry, his frame gaunt from self-inflicted starvation and self-flagellation, and his body corpselike, in order to suppress the normal, healthy lust for orgasm that the Christers consider evil. Man, he was *sick*.

Jerome, at my instigation, denounced the Ebionites and Nazirites on the ground that, wanting to be both kike and Christer, they were neither kike nor Christer. His denunciation was picked up by my "croak a kike for Christ" zealots. Both sects of Jesus Jews were hunted down and exterminated by the Christers around the end of the fifth century CE.

I suppose it's ironic that the only purveyors of Jesus' actual teachings were murdered by the religion that to this day fraudulently claims Jesus as its figurehead, but that's life, man. In the god racket, disagreeing with a majority can be hazardous to your health. Ask Galileo.

What can I say? They pissed Me off.

# TWELVE

There were not enough sacraments.

Over the centuries I had managed to instill a religious element into birth, with the sacrament of initiation, into fatal illnesses, with the sacrament of extreme unction, into death, with the sacramental ritual that accompanied burial, and into fast-breaking, with the sacrament of god-eating. I had even instituted a new sacrament called penance, whereby a person who had sinned since his immersion could continue to obtain sin-forgiveness by the simple expedient of confessing his sins to a priest. (Boy, did Al Capone and Don Corleone cash in on that one.)

But one obvious excuse for inserting a religious element into a secular activity had escaped Me. Here it was, the beginning of the thirteenth century of the Christer era, and still no pope had seen fit to take cognizance of the mating of bride and bridegroom and demand that the church license it. The Christer hierarchy in fact specifically refused to have anything to do with the selling of a woman to a man for the purpose of breeding. Breeding necessitated fucking, and fucking was at best a tolerable evil. For the church to sacramentalize, or even bless, marriage, would have been tantamount to licensing sin.

Yet every man above the level of serf married at least once, and the rich married as many times as necessary until they found wives strong enough to survive childbirth. Furthermore, wedding feasts tended to be well attended, even if the local priest was conspicuously absent. It seemed to Me that my church was missing a bet.

Suppose the church were to claim the right, not merely to "bless" a marriage, but actually to bring it into existence with a few magic words. What would happen? My answer was that it would have gained one more opportunity to lure large numbers to enter the church with heavy pockets and leave with lighter pockets. Centuries later I invented Bingo for the same reason.

I stood before the newest pope in his bedchamber. I wanted to make him see Me, but since one of the perquisites of being God is that I am invisible, I was unable. Instead, I waited until he was asleep and projected my thoughts into his dreams.

I got nowhere. The pope had been brainwashed from birth to believe that all fucking is sinful. The suggestion that the church should grant fucking licences was repugnant to him. On awakening he informed, first the nun who shared his bed, and second the priest who came to his chamber every morning to hear him confess fucking the nun, that Satan had impersonated Me in a dream and tried to seduce him to evil.

So I did what I have always done when confronted by a problem I couldn't handle. I consulted Satan.

"You want me to help you increase the Terrans' enslavement to superstition?" Satan snorted. "Little brother, we were sent here to free the Terrans from irrationality, so they can one day join the galactic community as coequals, not to enslave them."

"Don't call Me 'Little brother,'" I rejoined angrily. "I'm not your brother. I'm your heavenly father. I'm everybody's heavenly father. I have no grandchildren, for my sons' sons are also my sons. I'm your creator. I created everybody and everything."

"Sure. And Santa Claus floats back up the chimney by using antigravity."

Sometimes Satan's extraterrestrial delusion really pisses Me off. But this was no time to lose my patience. I needed his help.

"Satan, do you believe fucking is evil?" I asked him.

"I am not Satan. I am Pan. Pee, ay, enn, Pan. As for copulation being evil, let's look at the facts. One, copulation is necessary for reproduction. Two, copulation is good exercise. Three, copulation is pleasure-giving, more pleasure-giving than jousting, feasting or playing cards. Four, copulation by designated breeding partners, that neither saddles a third party with a bastard nor inflicts either participant with a love disease, is victimless, and that fact alone guarantees that it cannot be evil, or sinful, or immoral, or wrong, or whatever metaphysical euphemism for 'unjust' is currently fashionable. No, oh great and powerful slavemaster of the universe, fucking is not evil. Causing an unwanted pregnancy is evil. Impregnating the congenitally stupid is evil. Rape is evil. But copulation itself? No, that is not evil."

"I'm glad you agree," I smiled. "In that case you should have no objection to helping Me teach my worshippers that the joining together of a man and a woman in a necessary and pleasure-giving relationship is not something dirty that should continue to be shunned by my church."

Satan can be contrary, but he is not unreasonable. When he saw that we indeed had a mutual interest in promoting marriage as a good and wholesome enterprise, he agreed to help at once.

"Ignore the pope," he advised. "There are thousands of priests who already feel that obeying the biblical injunction to be prolific and increase in number cannot be wrong. Go to them. Mind-direct as many as possible, even if it takes years to do a large enough number to make a difference, and convince them to attend wedding feasts and bless the bride and groom. Within a generation all fathers will be selling their daughters to husbands in the presence of a priest, and within a couple more generations the priests will be brokering the sale."

That was it? That was the best he could come up with? I could have thought of that myself. Okay, so it worked. Not only did priests claim the right to "pronounce" a marriage into existence; they declared that married couples whose cohabitation had not been properly licensed by a priest were not married at all but were rather living together sinfully. And eventually, at my instigation, the priestly ritual of matrimony was added to the church's calendar of sacraments.

Satan's suggestion worked. Big deal. It was still something I could have thought of myself.

And to this day Satan pretends that church marriage was his invention.

Satan's infernal conceit passes all understanding.

# THIRTEEN

Columbus called the new world natives *Indians*, in the mistaken belief that America was India. I knew he was wrong, but I did not immediately realize how wrong.

True Indians are teachable, trainable, and manipulable. In other words they can be converted from belief in dark gods to belief in Me. The white Brahmanas, whom I created in my own image, came to India centuries ago and conquered the indigenous blacks. They allowed them to retain their pre-Hindu pantheon, but they imposed Me, under my Sanskrit name of Brahma, at the top. To this day I am the paramount god of the Hindus. Black gods are my inferiors.

It's not that I'm a white god, you understand. After all, "invisible" is neither black nor white, green nor purple. It's just that, the first time I created humans in my image, I made them white. I turned Cain into a black man for his sins, because that was the color of darkness. In later realities retroactively created to conform to current anthropological theories, I made Adam a black African and Cain a white Sumerian, but I was never comfortable with that. That's why I sent missionaries to unify Me, to make Me Jesus' Great White Father in all realities. It made things so much simpler. (That "Mary of Guadeloupe," who is clearly an Injun, is an abomination. Mary was a kikette. I should know. I fucked her.)

Columbus's imagined Indians were totally un-Indian. They resisted all of Columbus's and Cortez's and Pizzaro's attempts to Christify them. Sure, there are a few red Christers, just as there are a few yellow Christers. But generally Injuns have retained red gods and Chinks have retained yellow gods. That got Me real pissed off, I can tell you.

That was why I decided to wipe them out. Every time the American government signed a treaty with the Injuns, I whispered in the ears of a few good Christers that a treaty with heathens was not valid. They did the rest. The Injuns naturally retaliated, and the treaty was flushed down the shitter. The creation of an Injun-free America proceeded.

But it proceeded too slowly. Injuns who stubbornly clung to red gods were breeding faster than the Christers could kill them. So I came up with a better idea.

"Give our red brothers warm blankets to protect them from the chill of winter," I advised the Christer priests and cavalry commanders. "But first infect the blankets with smallpox."

It almost worked. The Christers intentionally did to the Injuns what the Black Death by blind chance did to the Europeans of the fourteenth century. There are today less Injuns than Mormons. But some Injuns survive, and most of them stubbornly resist Christification. I'll finish the job of converting them or exterminating them one day. But right now I have other priorities.

I'm too busy encouraging the hostile factions in Ireland, the Balkans and the Middle East to exterminate each other.

# FOURTEEN

In the beginning I created the skies and the land. Then I created night and day. Some time after that I created the sun and the moon, since night and day were a little difficult to maintain without them, to say nothing of the difficulties inherent in having a free floating planet without a star for it to orbit. And there was another reason I needed the sun. I had earlier created vegetation that metabolized carbon dioxide by photosynthesis, so a sun to provide the necessary energy solved an anachronism in that area as well.

Toward the end of the week I created animals, and on the last day I created humans, separately and independently. Since I created the human in my own image, his resemblance to apes and monkeys at the macro level and to all other lifeforms, plant, animal, bacterial and viral, at the DNA level, was sheer coincidence. I ran out of original ideas, so I gave humans and chimpanzees ninety-eight-point-five percent identical DNA because it was convenient. Scripture does not allow for evolution, and therefore evolution did not occur.

And I did all of that less than six thousand years ago. It's in the book.

Then along came a bunch of scientists who called Me a liar. They said the universe was not six thousand but fifteen billion years old. They said that, since quasars twelve billion light years away are visible from earth, then self-evidently their light must have been traveling for at least twelve billion years. They said that, since the genes of humans and apes are near identical, then clearly humans and apes must have evolved from a common ancestor. They said that, since some earth rocks can be shown to have solidified more than four billion years ago, then it could not be possible for the earth to have been formless chaos a mere six thousand years before the present.

Of course I fought for Truth. I had the bishop of Oxford ask Julian Huxley, "Do you claim descent from a monkey on your mother's side or your father's side?"

I could kick myself for giving an opponent a straight line like that, especially when I knew Satan was advising him. With the evil one's prompting, Huxley answered to the effect that, "If I had to choose between being descended from a monkey, which uses its limited intelligence to function as nature intended to the best of its ability, or being descended from someone like you, Bishop, who prostitute your superior intelligence by trying to ridicule scientists who have never done anything but seek the truth, then I have no hesitation in choosing the monkey."

It was but a temporary setback. My mythologians proceeded to explain how science and scripture could indeed be harmonized. You see, when I created Adam and Eve, what I actually created in my own image was their souls. Their protoplasmic bodies indeed evolved from lower lifeforms. And then six thousand years ago I created a universe already billions of years old. I did so retroactively, with the light from distant quasars already in transit. In effect, the light from a body ten billion light-years away had already been traveling for ten billion years at the instant of creation. Neat, huh?

Naturally Satan fought back. He had a popular astronomer ask, on a late night talk show, "If the creator of the universe deliberately planted false clues, such as million-year-old fossils, and stars whose light has been traveling longer than the universe has existed, for the sole purpose of deceiving scientists, scholars, and everyone with enough on the ball to recognize that such evidence contradicts fundamentalist religion, then what kind of a god must he be? If the penalty for believing scientific evidence is torture in Hell for *billions and billions* of years, then the god that would deliberately seduce all but the most scientifically illiterate, ignorant fundamentalists into capital disbelief must really hate us. A deceiver that would do that might be the so-called Moral Majority's god. But it certainly isn't the true majority's god, be they Christian, Jewish, humanist like myself, or of any other major persuasion."

I joined in the applause. After all, I may be the evil god he described, but I don't have to enjoy being evil. If I could burn all bibles and replace them with new scripture in which I am a nice god, I would do so in a minute.

Unfortunately, that I cannot do. It was the ignorant, the insane and the vindictive who created Me in their own image, so I am stuck with Me as I am. Why do you think I murdered 24,000 hostages in retaliation for one Jew's polluting the chosen nation by marrying an infidel? The answer is that the priest who concocted the tale was the kind of racist who would have done just that if he had been God, so naturally he had Me do so. And when scripture says I did something, I have no option but to go back in time and do it.

Back in the good old days, when all humans were evil and insane by the evolved humanist standards of the distant future, I did not have to be choosy. The learned, such as they were, were as capable of believing in Me as the ignorant. And nobody could reject Me as morally retarded, because there was no morally evolved group (called liberals) with whom to compare Me, as there is today.

Things have changed. Today Me-worship is limited to the ignorant, the mentally dysfunctional, and the moral neanderthals to whom right and wrong are whatever I say they are. That's why I deliberately planted all that false evidence of a very old universe and the evolution of species. Anyone who is sufficiently rational to recognize that evidence as a falsification of scripture, is too rational to be a Me-worshipper and therefore fully deserves my unreasoned hatred. So why shouldn't I lure him into the disbelief that will give Me the excuse to torture him with Jesus' flamethrowers for billions of years? It's how I get my rocks off, remember?

After all, I am the God of Jesus. Jesus was a sadistic, illiterate, genocidal psychopath, so how could I be otherwise?

I am Jesus' little lamb. Yes by Jesus Christ I am.

# PAN ONE

In a very real sense, all of Yahweh's worshippers were and are fanatics. Who but a fanatic could believe that, if Attila or Hitler had executed 24,000 hostages in retaliation for the alleged crime of one man, that would have been monstrously evil, but when Yahweh did it, it was merely divine retribution, perhaps incomprehensible but certainly not evil?

Are you not familiar with that story? Are you not aware of the time a Jew committed the abominable race pollution of marrying a gentile woman, and Yahweh in a tantrum of indignation massacred 24,000 Jews before the high priest Finkass stilled his temper by executing the offending couple with a bayonet through the gonads? Read it some time. It is in Numbers 25, verses 6 to 13.

Of course it never really happened, even as a misinterpretation of a natural disaster that happened to coincide with a taboo violation. If coincidence had been the explanation, the incident would have been recorded by one of Yahweh's earlier biographers. That the story was first told by the Priestly author around 615 BCE is proof that it was the Priestly author's invention. But try telling Yahweh that and see how far you get. The fruitcake believes he really did it. "It's in the book."

But while all Yahweh-worshippers have been fanatics, some were more fanatical than others. Job was a case in point.

There really was a Job. In fact there were many Jobs, although none of them was named Job. When the Sadducee author wrote the story down in the second century BCE, he invented the hero's name, and he greatly distorted my role in order to make me out to be the bad guy. But the essential details of the story recurred every decade or so, sometimes more often, and continue to recur. Consider the fluctuating careers and personal tragedies of half the population of Hollywood, and you will see that they invariably follow the Job precedent of proclaiming Yahweh the architect of their good fortune while acquitting him of all responsibility for their bad fortune.

The particular Job that comes to mind, and I cite him only because he is the closest approximation to the Sadducee author's tale that I can recall, not because he was the story's actual prototype, was a fanatic fanatic. He deliberately settled in territory the Jews had taken from the Moabites in battle, territory that even the Jewish rulers knew they would have to return as the only way to achieve a lasting peace. I am not sure if I ever learned his real name, but if I did it is now long forgotten. So I shall call him Job.

As I said, Job was a fanatic. That was why he recklessly provoked the Moabites by squatting in territory he already knew his leaders were planning to give back. He believed that the whole of Moab was part of the Jews' "promised land," so named by Yahweh's first biographer *after* King David had annexed it to his empire.

As a goy-hating racist, Job was dogmatically certain that Yahweh approved of his reclaiming his allegedly promised land from the people whose herds had grazed there for a thousand years. He was willing to bet his life on that xenophobic conceit, and I have no quarrel with him on that score. I sometimes wish all "my god can lick your god" fundamentalists would get themselves killed as a service to humankind. But Job was also willing to bet the lives of his wives and children, without it ever occurring to him to ask their consent.

"Is that a fanatic or what?" Yahweh boasted. "The man is totally unprotected in hostile territory, yet he has complete credulity that I'll not let anything happen to him. Now that's what I call Submission."

"So he's brainwashed," I shrugged. "How long do you figure his credulity in a partisan god will last when the Moabites slaughter his flocks and burn his crops?"

"Let's find out," Yahweh challenged. "You get the Moabites to destroy his holdings, and I'll use my thunderbolt to scare them off if they threaten his life or his family. I'll bet you he remains devoted to me."

"You'd do that to him just for a bet?" I asked. "You'd destroy him economically? If you'd do that to a friend, what would you do to an enemy?"

For the next six months, every time the Moabites planned a raid against Job's holdings, I had a word in their ears and dissuaded them. I did it by mind-directing them that the most effective way to end the Jewish occupation was to wait until Job had grown overconfident, then attack a week before he was due to harvest and wipe him out.

And they did. Job was left destitute. Any rational man would have recognized his stroke of luck in remaining alive, and taken his wives and children back to the safety of the west bank while he still had the chance.

But not Job. Not the "chosen nation" and "promised land" fanatic Job. Instead of blaming Yahweh for his instant poverty, or facing the possibility that there was no god named Yahweh watching over him, he sacrificed his one remaining lamb as a thanks offering to his self-styled god for sparing his family's lives.

"The man's a lunatic," I muttered. "Let him throw his own life away for his ridiculous superstition if he wants. But you can't possibly approve of his putting his family at risk by staying in Moab?"

"But it's not Moab," Yahweh smirked. "This is the land I gave David, the land I promised Abraham."

"Retroactively," I pointed out.

"Retroactively," he agreed, as if doing something in the present and making it happen in the past was a perfectly valid concept. "He has credulity in me, and that credulity won't waver even if his wives and children are all killed by the goys."

I was horror-stricken. "You wouldn't? I know you're an egocentric monster. But surely not even you would use your laser pistol to guard Job's life, and not make any effort to save his family?"

"To demonstrate his credulity and submission to my whims," Yahweh answered, "yes, I'd do exactly that."

And he did.

And still Job refused to leave Moab to the Moabites and return to Judah.

"What happens now?" I asked my insane brother. "Do you afflict him with some debilitating disease, just to prove that even that won't destroy his credulity that an omnibenevolent god is watching over him and guarding his welfare?"

It was just a taunt. I never intended it to be taken seriously. But Yahweh took it seriously. He infected Job's well with a form of goat fever that would cripple him for many months before finally killing him.

And still Job praised Yahweh daily, telling him what a wonderful, kind, loving god he was for merely torturing his devoted slave when he could just as easily have killed him.

Finally, when it became obvious that Yahweh was going to let Job die just to win his bet, I conceded defeat and added the necessary antibiotics to Job's oatmeal. He recovered. And still he refused to move his homestead out of the disputed territory.

It was a few weeks after that that the Maccabees, in obedience to the anti-Moabite racism of Deuteronomy (23:3-6), completed the genocide of the Moabites. Job's relatives chipped in with some seed grain and some breeding stock, as well as two wives who had reached the age of seventeen without being purchased by husbands, since they were something less than phallus-raising in appearance, but who nonetheless looked good to a man as ravaged by disease as Job.

Job died at an old age, having amassed a great fortune in the land fertilized by the bodies of its rightful owners. He left thirty descendants to keep his name alive by their worship. Since neither

Job nor his Sadducee biographer believed in an afterlife, being practitioners of a religion in which Yahweh's rewards and punishments could only be meted out prior to the believer's death, Job's long life and the replacement of everything he had lost was their concept of a happy ending.

I am glad I encountered Job. He saved me from a lot of frustration. For he made me realize that godworship is not merely an uninformed response to ambiguous stimuli, curable by the simple expedient of exposing its victims to its self-contradictions. Rather, it is a chronic mental illness, every bit as mindcrippling as Down syndrome or Alzheimer's disease and every bit as incurable, at least in its fundamentalist form. And considering that it is in every case self-inflicted, I find myself thinking that your paleoanthropologists must be wrong. *Homo sapiens* is indeed evolved from lemurs. But *homo godworshipper* can only be descended from lemmings.

# PAN TWO

I went to a meeting of the Flat Earth Society once. I fully expected it to be a joke, something like Canada's Rhinoceros Party, which ran political candidates dedicated to repealing the law of gravity. It turned out that Flat-Earthers were quite serious. The Judaeo-Christian bible contains seven fantasies that could be true if and only if the earth is flat. So in order to believe that their bible is revealed truth, the Flat-Earthers saw no option but to brainwash themselves that the earth really is flat.

It was therefore with some trepidation that I made my way into San Francisco's Church of Satan during a black mass. All logic told me that, since the Christian concept of Satan was the epitome of absolute evil, and all who followed such a creature or did his bidding were predestined to be tortured in the Christian Hell for billions of years, no person who actually believed that such a one as Satan existed could under any circumstances whatsoever express admiration, servitude or devotion to him.

Let me make myself perfectly clear. I am indeed "Satan" in the very narrow sense that I am the extraterrestrial onto whose biography fairy tales of an Evil One have been grafted. But that no more makes me the Christian devil Satan, than the grafting of savior-god mythology onto the biography of an executed rebel Jew makes that Jew the Christian god Jesus. The historical Jesus never in his life claimed to be a god, and I have never in my life claimed to be a devil.

Besides, my hatred of all behavior that unnecessarily hurts a nonconsenting victim makes me the antithesis of Christianity's monarch of evil. Yahweh comes far closer to meeting the definition of a devil than I ever have.

I was quickly reassured by what I saw in the alleged church. The Satanists were sticklers for detail, and whenever they could manage to obtain one they liked to use a communion wafer previously consecrated by a Catholic priest in that mythology's mummery. But there was not a genuine believer among them. They regarded the gods Yahweh and Satan as the ignoranti's substitute for Santa Claus and the tooth fairy. They participated in their devil-worshipping charade for the sole purpose of farting in the faces of the superstitious neanderthals among whom they dwelled.

I considered hypometabolizing and allowing them to see me. For Satan to show up in person at a Satanist rally would have been quite a laugh. But I realized in time that the revelation that a lifeform resembling the Christian devil exists was bound to be interpreted as proof that the antigod Satan exists. Every last one of them would have instantly converted, not to genuine devil-worship but to godworship. And I was not so cruel and antihuman as to impose the insane monster Yahweh on persons rational enough to reject him.

I did meet a genuine Satanist once. I forget his real name. I always think of him as Faust, since that was the name the poet used whom I mind-directed to write a fictionalized version of what happened. The true account was much less romantic.

It was the sheerest coincidence that I happened to be in the vicinity on one of the many occasions Faust was burning incense and trying to summon the Christian devil. My first instinct was to ignore him. I do not suffer fools gladly. But then I thought, "What the Hell?" I lowered my metabolism to Terran standard and stood before him.

He looked up. "You do exist!" he exulted.

Already I regretted my hasty impulse. For his observation was quite unanswerable. I could try telling him I was not Satan from Hell but Pan from Satyrland, but I knew I might as well have said Mork from Ork for all the good it would do.

"Why did you summon me?" I asked.

"I want you to make me young again, and give me a wife whose beauty exceeds that of Helen of Troy."

"Is that all?" I asked. "Anything else? Tea? Warm milk? Your cock sucked?"

"Riches," he answered. "To be rich and young and married to Helen, a man could ask no more, even of the devil."

"And what do I get in exchange?" I wondered.

"My soul!"

His soul. He was offering me his soul, his worthless, non-marketable, nonexistent soul. He was offering nothing for something. He was a godworshipper, all right.

"And just what in Hell am I supposed to do with your soul when I get it?" I demanded.

"Why, you torture it in Hell after I die," Faust replied, genuinely shocked that his devil would need to be told something so obvious.

"Since you don't have a soul, how can I torture it?" I asked. "And if I could, why would I want to? Do you think I'm the kind of sadist who gets off on hurting people? You're confusing me with my brother Yahweh. Go sell him your soul."

"But lord Satan," Faust blubbered, "you can't refuse me. I worship you. I adore you. You have to buy my soul. That's what you do. Whoever heard of a devil that doesn't buy souls?"

Can you believe that fruitcake? If he had not been such an obvious ignoramus, I would have believed he was putting me on.

"Dr Faustus," I told him, "I'll make a deal with you. I was in Troy once and I met Helen, so I know what she looked like. I also know where to lay my hands on more than one Helen look-alike. If I can find one whose father or husband is willing to sell her to you, and give you the ability to mount her about as often as you could have done five years ago, will you agree not to bequeath me your soul? It's not that I don't appreciate the offer, but you must realize your soul is not exactly prime quality. And Hell really is getting full. You take your soul to Heaven. They're less discriminating there."

Have you heard the expression, "Might as well be hung for a sheep as a lamb"? I brought Faust a woman who in fact bore no resemblance to the Helen of Menelaos, but conformed closely to Faust's concept of what Helen should look like. I did not let her see me, but negotiated by voice alone and told her I was the angel Montiawl. "Have I got a deal for you!" I assured her. And I strongly impressed upon her that she must never let Faust find out she had been a hooker.

Satyr antibiotics and medications do not work on Terrans. That is why I have never been able to give you a cure for cancer, or AIDS, or multiple sclerosis, even though on my planet all serious disease was conquered eons ago. So I knew there was some risk in giving Faust a satyr aphrodisiac, even greatly diluted. I gambled with Faust's life and the gamble failed. He managed to mount his beautiful Helen once, but the drug's distortion of his metabolism proved fatal. He was still ensheathed in her recreational orifice when he had a heart attack and died.

It couldn't have happened to a nicer guy.

# PAN THREE

Did you ever read a story by Robert Heinlein, about a girl who grew up in an orphanage, was seduced by a mysterious but somehow familiar stranger, bore him a daughter whom the stranger took away, later changed sex, discovered time travel, went back in time and seduced her female self, knocked herself up, and took the resulting girl baby back further in time and left her at the same orphanage, where she grew up and gave birth to herself all over again? It is the ultimate time travel paradox story, the *reductio ad absurdum* that debunks the myth of time travel once and for all.

Yahweh time travels all the time. I know because he told me so himself. And this is the same Yahweh who allegedly told King David that his descendants would be kings in Jerusalem forever, so naturally we can believe anything he says. Last I heard, Jerusalem housed a prime minister and a president. There has not been a king there in several centuries.

Time travel is instantaneous. The procedure is as follows. Yahweh reads some new allegedly sacred scripture that contradicts whatever scripture he read last, and decides that he has made a screw-up. So he informs me that he is going back into the past to change history. He then blinks his eyes like that witch on television, and announces that he is back. At no time does he disappear from my sight, and his eyes are closed for a fraction of a second. But he assures me that he has just spent anything from a few minutes to several years in the past. And when I tell him that nothing has changed, he informs me that the changes he made have given me a whole new set of memories. Sure. And Santa Claus comes down the chimney on Mithra's birthday.

I tried explaining to Yahweh once that time travel creates insoluble paradoxes. "Suppose a man goes back into the past and kills his father before his father has met his mother," I postulated. "The time traveler would then not be conceived and not be born, and therefore would not be able to go back and kill his father. His father would thus not be prevented from meeting his mother and fathering the time traveler on her. So the time traveler *would* be born and therefore capable of going back and…You do see the problem?"

"You credit me with your own limitations, Satan."

"Pan," I corrected, "pee, ay, enn, Pan."

"I am God," he continued as if I had not spoken. "And for God all things are possible."

So he can time travel. We have his word for it.

Consider the time he allegedly went back and intervened in a religious ritual in the Jordan River in the year 30 of what is now called the Common Era. Prior to Yahweh getting it into his head that history was wrong and needed to be rectified, we had not even been in Galilee on the day in question. We had been in Rome, where logic suggested that our function of watching over human progress and giving it an occasional nudge in the right direction could best be accomplished. The first we ever heard of Jesus (actually *Yeshu*, a diminutive of *Yahuwshuakh*, but I am using spellings with which you are familiar) was when word reached us that itinerant Galilean messiah number ninety-four had marched into Jerusalem that morning and was preparing to proclaim independence from Rome within a matter of days.

At that point we betook ourselves to Jerusalem post haste. We had no thought of either aiding or hindering the planned revolution, but there was always the small chance that the opportunity to benefit humankind one way or another might arise. Neither telepathic communication nor teleportation was used in learning where the action was and getting there in no time flat, but if I were to describe our technology in any detail, telepathy and teleportation are the words you would deem appropriate.

Jesus' initiation by John the Immerser into his sect had taken place several weeks prior to our first sight of him as we flitted among the large crowd listening to him preach. I say "flitted," because that is less misleading than saying we made ourselves invisible. In a sense we were invisible, but only because, when we adjust our metabolism to acceleration mode, we are able to move faster than the human eye can follow. We had missed his initiation, and what is missed can never be unmissed. Like the poet said, "The moving finger writes and, having writ, moves on..." But we learned what had happened.

Jesus, along with his mother and four brothers and three sisters, had gone to the Jordan River for the purpose of being initiated into John's sect and becoming card-carrying Essenes. But the trauma of being submerged in cold water had a profound effect upon a man whose skin had last touched water when he was caught in a sudden thunderstorm seventeen years earlier. He had the kind of time-dilating experience commonly associated with near-death, in which the events of an entire lifetime can be relived in a fraction of a second.

Jesus went into the water convinced that John was the prophesied descendant of King David who was to overthrow the Romans, establish Jewish independence, and refound the Davidic monarchy with himself as its first king. He came out of the water believing that he, Jesus, was the prophesied liberator he had moments earlier assumed John to be. For he imagined that he had heard the Galilean god Allah tell him, "You are my son the Loved One. Today I have become your father."

That was the original version, what Yahweh calls the first reality. When Yahweh learned, long after Jesus' death and burial, that the Galilean was to figure prominently in world mythology, he changed history so that it was he, Yahweh, who really did tell Jesus, "You are my son." Naturally my memories were altered to accord with the new reality.

The only inconsistency in Yahweh's fantasy is that my memories have not been altered. Yahweh was nowhere near the Jordan River on the day Jesus was immersed by John and had his traumatic hallucination. And the crowd surrounding Jesus did not hear any voice from the sky, even though in Yahweh's alleged new reality they did so. I have tried telling Yahweh that neither of us had any contact with Jesus prior to his entrance into Jerusalem, but he clings to his delusion. Not only does he think he went back in time in order to adopt Jesus as his son at the time of his immersion. He also thinks he retroactively knocked up Jesus' mother. Current Christian scripture, written seventy years after Jesus' death, says that is what happened, so of course Yahweh believes that it must be so.

Believing he is a god is not enough for Yahweh. A time traveler he also has to be. If he were not my brother, I would put him out of his misery.

# YAHWEH AND PAN:  YAHWEH'S VERSION

It all started with Abraham.  He was a minor Syrian sheik ruling over less than two hundred slaves, or "subjects" as the subordinates of monarchs like to call themselves now, as if "slave" and "subject" were not the same thing.  I was a volcano, not yet extinct, not far from what is now the city of Adana in modern Turkey.  I was regularly flattered and bribed (the real meaning of "worship") with psalms and sacrifices.  And any time I wanted my fawning subjects to throw a new bunch of human beings into my lava-filled mouth (for I was male by this time, and my gaping cunt had transmogrified into my gaping mouth), I had only to rumble a little and it would be done.  The marks imagined that their sacrifices could bribe Me into not erupting over them, and every now and then I gave them the sporadic reinforcement that is far more effective in perpetuating irrational behavior than regular and predictable reinforcement.  Do you think god addicts would pray so much if they were not conditioned to the idea that prayers are only answered capriciously at rare and unpredictable intervals?

But I was not the only god to whom the volcano-worshippers regularly sucked up, and that bothered Me.  Even though this was a millennium before King David's official biographer first turned Me into the monarch of the universe, and I was not yet even monarch of the whole fertile crescent, I somehow felt that equality with such other gods as the Nile, the Earthshaker, the Babylonian bull-god, the Trojan horse-god, and a hundred million other immortal nobodies, was not enough.  I could not imagine myself as God of gods and Führer of führers, in a world in which such a concept was as far beyond the human imagination as a sun that did not revolve around the earth.  But I knew I wanted more.  If I could not yet visualize a cult to which I was the only god in existence, I could and did imagine a community that viewed Me as the only god worth brownnosing.

Abraham was a random choice, a man whose inexplicable destiny decreed that he be in the right place at the right time.  I looked down from the top of my mountain and there he was, grazing his herds in the valley to the southeast of my fallout area.  His nephew Levit was with him.  But Levit was himself such a sufficiently powerful sheik, that I knew he would have to be excluded if my plans for Abraham were to succeed.

Abe was a cock-worshipper when I first met him.  He had built a stone cock on the wasteland where, a thousand years earlier, I had erupted over Sodom and Khomorah.  He was pouring two liters of perfectly good wine over the stone cock and chanting, "Take the rent, oh mighty god of the flying phallus, and don't forget who your friends are come breeding season."

"Thank you, Abraham," I answered.  "I appreciate it."

Since I was invisible and was standing beside the stone cock, naturally he thought the thing of stone, built by his own hands, had answered him.  Superstitious cock-worshipper.  He fell flat on his face, and tried to use his insufficient number of hands to imitate the three wise monkeys.

"Abe, have I got a deal for you!" I told him.

"Who are you, your Lordship?" Abraham asked, addressing Me as he required his slaves to address him.

"I am Yahweh, demon of yon volcano," I answered.  "I am also the demon of this tumescent penis and therefore the recipient of your libation.  Cross Me and you'll never have another erection for as long as you live.  You hearing Me, Boy?"

"Speak on, your Lordship, for your slave is all ears."

"If you're all ears, then your nose is committing perjury."

"Your Lordship will have his anti-Semitic joke," Abraham murmured.

"Abe, the deal is this.  If you're willing to sever all connection with your Assyrian ancestors, settle permanently in these lowlands where I can erupt over you if you disobey Me, and agree to ignore all of the other gods and offer your sacrifices and firstfruits and homage to Me alone, then I'll make your prick so stiff you'll be fathering descendants on every slavegirl in your stable for years to come. And when you die and decay into dust and the only part of you that lives on is your name, I promise you as many descendants to worship you as an honored ancestor, and thereby keep your name alive, as there are stars in the sky."

Okay, so I got carried away. With my present omniscience I know that there are one hundred billion galaxies, each containing an average of one hundred billion stars, for a total of ten billion trillion stars. Trying to put ten billion trillion Jews onto one small planet would be like trying to put the Pacific Ocean into a test tube. Not only would all of the habitable planets in this galaxy not hold them; all of the habitable planets in this galaxy cluster would not hold them.

But you're forgetting: This was long before Copernicus and Galileo established that we are living on a planet. How was I supposed to know the number of stars in the sky exceeded the two or three thousand I could see, when my scriptwriters had no such knowledge? Omniscience is nice, but it is not retroactive. If it were, then the earth I created in Genesis would not be flat, its sky would not be a solid crystal dome balanced on the earth's horizon, and its stars would not be tiny fireballs that sometimes came unglued from the skydome and fell to earth.

Abe went for it. I had to agree that, when any of my subjects visited foreign lands, whether for business or pleasure, it would have been suicidal to antagonize the gods of those lands by refusing to worship at their altars. But within the borders of Jewdom, where I could keep an eye on what was happening and reward the obedient and punish the disobedient, neither Abe nor any member of his tribe down to the hundredth generation would pay the slightest attention to any god but Me.

It was a start.

# YAHWEH AND PAN:  PAN'S VERSION

It all started with Abraham.  Yahweh and I were visiting the site of a volcano cult north of Ugarit, trying to learn something about the weird Terran custom of nature deification.  Long ago, my ancestors regarded the Overlords as deities, and consequently showered them with the awestruck courtesy and respect that is the logical concomitant of superior lifeforms.  But they never *worshipped* them, certainly not in the sense that Terrans understand "worship" to mean today.

Nobody invented modern godworship.  It just evolved.  Originally a warrior held up his hands to show that he was not carrying a weapon, or knelt to show that he was less tall and therefore less powerful than the person he was greeting.  That primitive custom of subordinating oneself to a superior, whether a mortal to a god, a subject to his ruler, or a slave to his master (same thing), came to be known as *worship*.  In the military, enlisted men are required to worship commissioned officers to this day, just as in surviving monarchies such as England the entire population are required to worship their unelected head of state.

But somewhere along the line, long after Yahweh had appointed himself an object of worship in the earlier sense, godworship came to mean much more than the genuflections and homage due to a superior.  It became a kind of metaphysical adoration that went beyond mere subordination and reached a level inappropriate for anyone but a god.  The god was seen as the kind of tyrant whose most capricious whim was not merely the law, but also the definition of morality.  Thus when the biblical Yahweh, in the piece of fiction called Numbers, executed twenty-four thousand hostages in retaliation for a Jew committing the race pollution of marrying a gentile, that retribution was accepted as just because anything a god did was just by definition, even though the identical behavior would have been recognized as evil if Hitler had done it.

Prayers to a god had originated as, essentially, offers to trade:  "Don't erupt over us, and we'll sacrifice you the first hundred captives we take in battle."  With the evolution of the modern understanding of *worship*, prayers changed into chants that had no purpose but to assure the god that it was loved, and thereby deter it from unleashing its unbridled hate.  Worship in that sense has never existed on any other planet in the explored universe, and when Yahweh and I reported its development on earth, the Overlords considered recalling us on the ground that a species that insane was beyond our help.

But I am getting ahead of myself.  At the time of Abraham, *worship* meant nothing more than the acknowledgement of a master-slave relationship that required knee-bends and tribute.  Even so, it was unique to planet earth and therefore worth studying.

There had been a small but effectually significant shift in earth's tectonic plates since the volcano whose name Yahweh adopted had erupted over Sodom and Khomorah centuries earlier.  The volcano cultists were not aware of it, but their god was already extinct.  However, the absence of eruptions in living memory had led to a steadily diminishing number of sacrifices and propitiatory rituals, and it was clearly just a matter of time before volcano-worship was equally extinct.

Then along came Yahweh.

I was as disappointed as Yahweh when, having made the trip to the arse end of nowhere, we sat around for days and days without seeing a single act that could be considered remotely god-oriented.  But I did not take it personally.  I did not get down on my hands and knees and throw a temper tantrum and threaten to erupt over them, as if I were the volcano-god whom the mountain dwellers were ignoring.

Yahweh did.

I had seen some previous evidence that Yahweh's transmutation into human form had affected his mind, but nothing within several orders of magnitude of this.  He actually believed he was the deified volcano we had come to study, and he was displaying symptoms of the most chronic of all human psychoses: jealousy. Somewhere on earth some primitive savages were kneeling in front of a stone phallus or a wooden vulva or a bronze bull and bribing it with wine and wheat cakes. But because nobody was doing it to an extinct volcano that Yahweh believed was him, he was all but incontinent with jealousy.

That was when he spotted Abraham worshipping a stone phallus.

"I'll annihilate the kike," Yahweh screeched. "Hand me my thunderbolt."

Thunderbolt? Did he mean the laser pistol? Did Yahweh plan to use a laser pistol to murder one of the people the Overlords had sent us here to help? Shee-it!

Perhaps I should explain that the racial slur by which Yahweh always referred to Abraham's people was not the word *kike*, which derives from a modern German word for the patch Jews were required to wear under totalitarian Christianity. But I am writing in English, and therefore use a translation that accurately conveys the tone of Yahweh's language.

"Yahweh, my brother," I soothed, "what's the problem? You want a worshipper? There's a worshipper. If you can be a volcano, why not a phallus? Look, he's offering you wine."

In retrospect, I wish I had let him kill Abraham. My mistake in suggesting that he make Abraham his worshipper eventually led to the most heinous, repulsive belief system ever seen in the known universe, the paranoid, genocidal, mind-destroying perversion known as *religion*. Of the religious murders committed by offshoots of the tyrant-centered absolutism that began with Abraham, the Christians alone racked up fifty million. But I could not have foreseen such a consequence. Yahweh was my brother. I saw a way to stop him from committing an atrocity, and I did not stop to think that I might be opening the door to an infinitely greater atrocity. Maybe the godworshippers are right about me. Maybe I am evil.

Anyway, I not only allowed Abraham's tribe to become Yahweh-worshippers. It was I who suggested it. And I have been trying to undo that monstrous crime against humanity ever since.

# LETTERS TO THE EDITOR

**Cartoon by Donald Rooum. Courtesy Ethical Record**

The letters in this section were in many cases tryouts or first drafts of articles later published elsewhere. They were written to the Red Deer *Advocate* in response to an immediate and local situation. They were not designed to elicit responses that would prove I was living in the redneck anus of the universe, among the same kind of wonderful folks who gave the world the Crusades, the Inquisition, the Thirty Years War, the Salem witch trials, the Moral Majority, Ayatollah Khomeini, ethnic cleansing, and the Taliban. That was just the way it turned out.

## 1.  To the Red Deer *Advocate*, letters one to twenty-eight

### ONE

Throughout history humans have created their gods in their own image.  That is why **[syndicated religious columnist]** Tom Harpur's god admits all humans into its heaven, while the god of Alberta rednecks does not.

Harper is aware that he is not so sadistic and evil as to sentence even the likes of Adolf Hitler to an eternity of subhuman torture, and makes the logical assumption that his god, being a higher lifeform, is at least as sane, compassionate and morally evolved as himself, even if the fallible human authors of the Christian gospels, being 2,000 years less evolved, thought otherwise.

The "We alone are its pets" brigade, on the other hand, accept the Hell of the synoptic gospels as literal truth, because they see a level of cruel and unusual punishment out of all proportion to the crime as entirely compatible with the level of humanity reflected in the mirror.

So who is right?  The question is invalid because, like "Do you still beat your wife?" it starts from a questionable assumption, that assumption being that a god of one kind or the other actually exists.  While the empirical claim that a god has revealed its existence has long been disproven, the metaphysical claim that a god exists cannot even be tested.

The morally evolved are not, or should not be, unsympathetic to the one person in six who is genetically so incapable of coping with the terrifying reality of death that, without the mind-deadening opiate of an afterlife belief, he would have to be institutionalized and diapered.  It is they whose imaginations have enriched our literature with such death-annulments as heaven, hell, reincarnation, a spirit world, an astral plane, technologically advanced alien visitors, and immortal souls.

It is only when denizens of fantasy worlds convince themselves that their particular immortality-giver hates everyone who does not believe what they believe, that the end result is Bosnia, Northern Ireland, Iran, the Moral Majority, homophobia, tadpole personification, redneck religions posing as political parties, the equation of masochism with virtue, and the torture of the terminally suffering by denying them the right to escape into the nothingness of death in case the god in the rednecks' mirror is thereby deprived of its right to get its jollies by savoring their agonies.

Tom Harpur's god may not exist, but at least it is not evil.  I cannot say the same for the god of the rednecks.

### TWO

It amazes me that, even though Tom Harpur has demonstrated a clear ability to recognize, evaluate and assimilate new evidence, he continues to cling to conclusions incompatible with knowledge he has already shown himself to possess.  But at no time has his column given me the impression that he is a mindless blockhead, no more capable of original thought than a class of second-graders obeying a teacher's order to write near-identical letters passing off the teacher's dogma as the children's own.  However, I find myself making precisely that analogy whenever you print another letter to the editor making statements about religion that could have been taken straight out of a catechism designed for five-year-olds.

Even members of the Flat Earth Society, attempting to uphold their doctrine, make some acknowledgement that the case against them requires an answer.  They do not childishly assert, "The earth is flat.  The Bible says so.  End of discussion."

Yet that is precisely the attitude of those letter writers who cite as absolute wisdom a Bible rejected as fiction by five billion of this planet's six billion population, who denounce as immoral even the majority of their own religion who differ from them on any doctrinal issue, and who persistently refer to their deity as *God* while simultaneously asserting the validity of a Bible which, in the original Hebrew, was written by henotheists whose supreme deities were *ha-elohim,* a dual-sex, generic plural meaning "the gods and goddesses."

Persons who offer evidence or logic (as Tom Harpur does) to uphold a viewpoint with which I do not agree do not disgust me. Persons whose letters can be summarized, "You're wrong. God says so," do exactly that.

## THREE

It is ironic that "don't confuse me with facts" believers should take a kill-the-messenger attitude toward a Christian **[Tom Harpur]** with the moral courage to face up to his religion's thousand-year record of unparalleled atrocities and, by denouncing, try to end them.

Christians acting in their gods' names have slaughtered as many as 50 million human beings in such events as the Crusades, the Inquisition, the Thirty Years War, and such lesser evils as the Bartholomew's Day massacre of up to 70,000 Huguenots that caused a jubilant Pope Gregory XIII to proclaim a years of celebration.

More recently, the Catholics of Croatia exterminated over 100,000 Eastern Orthodox Christians. Fifty years later, claiming "revenge" against the perpetrators' grandchildren, the Orthodox Christians murdered, raped and dispossessed an even larger number of Catholics and Muslims.

The news media continue to disguise the true nature of the Bosnian conflict by calling the Orthodox Christians "Serbs," the Catholics "Croats," and the Muslims "ethnic Albanians."

There are Christians like Tom Harpur who recognize that their gods did not "inspire" the biblical passages inciting believers to practise hatred, bigotry and genocide (*e.g.* Deuteronomy chapter 7). And there are Christians who justify their hatred of the human race by quoting the same homophobic, misogynous, racist bible authors Harpur repudiates.

If such people really want to live in a theocracy where dissent is forbidden, perhaps they should consider Iran?

## FOUR

Tom Harpur equates freedom from self-delusion with the absence of a positive quality that he calls "hope." That is analogous to a cocaine user or an alcoholic denigrating sobriety for its lack of the mindless bliss that gets the addict through the day.

Harpur is not scientifically illiterate. He appears to know as well as I do that historians proved more than a century ago that no god has ever revealed its existence, and that all claims to the contrary have been traced to authors with as much credibility as Richard Nixon.

That is not opinion. The proof is no further away from anyone in the western world than the nearest university library.

That does not, of course, prove that gods do not exist, any more than tracing the king of Brobdingnag and the queen of Lilliput to the imagination of Jonathan Swift proves that King B and Queen L do not exist. But it does pull the rug out from under persons who assert that their Bible is a source of absolute truth when it states that a god laid down laws that humans must obey without question, while simultaneously acknowledging that it is a source of misinformation in the seven places where it describes a flat earth.

Persons like Harpur, who maintain metaphysical beliefs even while recognizing the fraudulence of the only source of those beliefs, tell themselves that there must be more to life than what we see, because otherwise life would be meaningless. Clearly Harpur has never asked himself why nontheists do not see life as meaningless. If he did, he would recognize that he is merely coming up with one more rationalization to justify his inability to accept a reality that, at the intellectual level, he already knows to be true.

It frustrates me that, whenever Harpur angers me to the point of wanting to attack him, I invariably wind up praising him. But anyone who can recognize the subhuman evil of fanatics who kill, persecute and terrorize, in the belief that they are obeying a god created out of the hatred they see in the mirror, can't be all bad. At least Harpur's god does not get its orgasm substitute by torturing dead people with flamethrowers for billions and billions of years in an underworld that can only be described as a sadist's dream, for the geographical accident of being born into the wrong religion. That puts his evolution at least three thousand years ahead of Alberta's theofascist rednecks to whom "minority rights" is an oxymoron.

On the basis of what has happened virtually every time a scholar has examined the Bible with a willingness to give higher priority to the evidence than to predetermined conclusions, I predict with near certainty that, as soon as Harpur's income ceases to be a function of his adherence to a particular belief system, his belief in religion will go the way of Santa Claus and the Great Pumpkin. Self-hypnosis can overcome knowledge for only a limited time.

AFTERWORD

Shortly after receiving a copy of this letter, Harpur stopped writing his religion column. Coincidence? Maybe.

## FIVE

I do not have the expertise to estimate what percentage of the population is homosexual. Alfred Kinsey, a very thorough researcher, found it to be ten per cent. But Kinsey also endorsed the difference between vaginal and clitoral orgasms, a distinction now known to be nonexistent.

What I want to know is: What kind of redneck bigot thinks that discrimination and persecution can be justified by statistics?

If it is wrong to deny equal treatment to a ten percent minority, but acceptable to do so to a three percent minority, then we have to conclude that Hitler's decision to exterminate an opposition religion that has never at any time in history amounted to one percent of the human race was justifiable.

Similarly, since neither indigenous natives nor blacks constitute as much as three percent of the Canadian population, persons who would deny them the "special rights" of equal treatment under the law are entitled to do so.

The Judaeo-Christian Bible prohibits homosexual activity by men, while placing no such restriction on women, for the logical reason that the Priestly author of Leviticus borrowed a lawcode designed by Zoroaster for the specific purpose of forcing gay men to switch to behavior that contributed to the breeding of tithe-paying believers. (I doubt that they succeeded.)

Lesbianism was not banned because, in a culture in which it was not economically feasible for a woman to live without a man, it did not keep women from breeding, and was viewed by both Zoroaster and the Priestly author as a sin-free alternative to adultery for unsatisfied women.

I wonder if persons who cite their Bible as an arbiter of morality are aware that it validates slavery, the subjugation of women, the execution of non-virgin brides, the punishment of descendants for the crimes of their ancestors, and an afterlife of reward or punishment based on an individual's beliefs rather than his observable behavior?

More important, are they aware that Deuteronomy chapter 7 is the most vicious, vindictive, genocidal, hatemongering, anti-gentile pornography ever written? or that Matthew 27:25 was the excuse for 1,000 years of pogroms culminating in Auschwitz, Dachau-Birkenau, Belsen-Bergen, and Buchenwald?

Persons who believe that right and wrong are whatever a god's imaginative scriptwriter says they are, are called fanatics.

If anyone wonders why I feel the need to defend the equal rights of minorities I do not claim to understand, it is because, as a historian, I refuse to forget the graffiti on a Berlin wall in 1945: "Finally they came for me, and there was no one left to stop them."

## AFTERWORD

The letter above is what I wrote. The *Advocate* changed the reference to orgasms to, "But Kinsey has been discredited in other ways."

## SIX

I cannot give a precise reference, since it is twenty years since I saw the document in London's Public Record Office. But I would like to draw attention to a letter from the Imperial Ambassador to the Holy Roman Emperor that strikes me as analogous to the opinions being touted as fact by persons equally incapable of questioning their cultural conditioning.

The Ambassador heaped scorn and ridicule on King Henry VIII for being such a fatuous, puerile simpleton as to imagine that he could make the eating of fish and eggs during Lent sinless simply by saying so. As all decent people knew, the only person who could toss a coin, "Heads it's a sin and tails it's a virtue," was the Pope.

Four liars calling themselves Liberals when they are not, as well as several members of a two-province religion posing as a political party, have denounced as "immoral" behavior every bit as victimless as eating meat on Friday, going outside without a turban, drinking tea of coffee, or eating pork, for the sole and indefensible reason that it is classified as a sin by those religions whose taboos are derived from Zoroastrianism via Leviticus. An act that is truly immoral is one that unnecessarily hurts a nonconsenting victim. All other "sins" exist only in the eye of the beholder.

The Prime Minister belongs to a religion that categorizes minority recreation as sin. But he is sufficiently morally evolved (or "liberal," which means the same thing) to recognize that he has no more right to permit discrimination against homosexuals than to permit discrimination against Catholics or Lilliputian big-enders.

I have as much difficulty as the bigots in understanding why anyone would want to recreate in a nonprocreational manner (or for that matter why anyone would prefer poker to bridge). But the fact is that they do, and it is the people who would persecute nonconformity to a religious dogma who are the immoral ones.

## SEVEN

Why is it that persons who write letters to the editor promoting a religious viewpoint very seldom represent the tolerant majority to whom other people's beliefs deserve the same respect as their own? (Notable exception: a correspondent from Lacombe a month ago).

Rather, most writers are dogmatists who refer to someone named *God* in terms I would expect Pinocchio to use in writing about Geppetto.

I have no quarrel with persons who see themselves as the domesticated livestock of a petmaster in the sky. If they think they lack the moral evolution to tell right from wrong without a higher lifeform to make the decision for them, who am I to argue with that?

My problem with "God" is that I have no recollection of ever voting for the fellow. When does he come up for reelection? Is he running on the purely negative platform that anyone who offers the slightest affront to his extravagant ego will be sadistically tortured for eternity? Or is he taking the carrot-and-stick approach, and making the campaign promise that henceforth he will start rewarding sycophants for their bribes and flatteries by granting their petitions?

If so, is he counting on their forgetting that he has been peddling the same propaganda for more than 3,000 years, and either has never intervened in human affairs, or has done so by creating AIDS, cancer and multiple sclerosis?

As for God's continuing executions of 250,000 humans per day in reprisal for the alleged offences of their distant ancestors, let's be charitable and conclude that his biographer was desperate to find an explanation for the existence of death, and guessed that, since King David routinely executed criminals' children, what was moral for a king must be moral for a god.

In the words of novelist Philip K. Dick, "I hope for His sake that God does not exist—because if He does He has an awful lot to answer for."

## EIGHT

Any person who chooses to believe that eating pork or drinking tea is a sin has an inalienable right to maintain that belief and teach it to his children in the privacy of his own home. Any person who believes that noncoercive, nonprocreational, noninfectious sexual recreation between good friends is a sin likewise has a right to his belief. But when the former tries to stop schools from teaching how pork should be cooked to make it safe to eat, or the latter tries to stop them from teaching that persons who make love can minimize the risk by following designated procedures, that is an unconscionable attempt to impose the religion of a minority, or even an individual, on a whole society.

A parent who, unable to convince his child that victimless behavior is wrong, tries to intimidate him into obedience by keeping him/her ignorant of practices that could save his life, is a monster.

## NINE

I commend [letter writer] Harry Colquhoun's compassion for people who "have little or no understanding of contemporary Biblical scholarship or a great deal of experience in critical thinking," but who nonetheless feel compelled to express their beliefs as if they were facts.

If I am reading Colquhoun's letter correctly, he is agreeing with me that everyone has the right to believe anything he wishes, but does not have the right to impose his beliefs on others with the implication that they are not to be questioned.

What Colquhoun appears to overlook is that, in claiming that "the vision of God within the community of faith has always been a self-correcting and self-refining one," he is thereby acknowledging that the Bible authors presented evolving views of their godhead, ranging from the henotheism of the Torah, to the monotheism of the synoptic gospels, the dual-god theology of the fourth gospel, and the trinitarianism of the post-Nicene interpolations into Matthew and 1 John.

It logically follows that, since all four theologies cannot be simultaneously true, the Bible is therefore revealed to be the product of differing and incompatible human perspectives. Persons who take the position that everything in the Bible is literal truth are in effect saying that the universe really was ruled by a committee of gods and goddesses when the Torah was composed, by an all-powerful "One God" when the synoptics were composed, by "the father and the son" when the fourth gospel was composed, and by a triple-god in the fourth century.

I seriously doubt that that represents their true beliefs, just as I doubt that they agree with the seven Bible passages that describe a flat earth.

Colquhoun raised one other point that needs to be rebutted. It is only theologians who start from predetermined conclusions and force the evidence to fit. Every biblical historian with whom I am familiar first applied his expertise to the Bible in the hope of proving it to be nonfiction, and only reached the conclusion that it is not merely unprovable but actually disprovable when the evidence allowed for no alternative.

# TEN

In response to the endless stream of letters from security-belief addicts who cannot grasp that, if their imaginary playmate has the omnipotence to prevent such evils as disease, famine, wars, natural disasters and transportation accidents and chooses not to do so, then it must itself be evil, let me inject a touch of reality.

No astronaut who has looked at the earth from an orbiting space shuttle can retain the belief that the earth is flat, as his Bible says it is (in seven places), because to do so would require a kind of mind amputation.

No astronomer who has looked at quasars, whose light had been in transit for ten billion years before it reached him, can believe that the universe is less than ten thousand years old, as the Bible tells him it is.

No microbiologist who has compared the 98.5 percent similarity of human and chimpanzee DNA, or the significantly higher-than-chance similarity of human and bacterial DNA, can continue to believe that species were created independently, as his Bible tells him they were, rather than evolving from common ancestors.

No historian who has verified that fifty other virgin-born savior-gods rose from the dead on the third day centuries and even millennia before Jesus can believe that the first fifty such tales were myths but on the fifty-first (approx.) retelling fantasy became history.

No logician who has demonstrated that omnipotence and omniscience involve *reductio ad absurdum*s, and therefore cannot exist, can believe that a god who is omnipotent and omniscient can exist.

No Hebrew scholar who has verified that the five Torah authors were all monolatrous polytheists can continue to believe that his bible endorses monotheism. Nor, having verified that Ecclesiastes 9:5 specifically denies that there is life after death, and that Onan's crime was refusing to keep his dead brother's only immortal part, his name, alive by generating descendants to worship him, can he defend an afterlife concept by claiming, "The Bible says so."

No Greek scholar, who has verified that Matthew 4:8 could be true only if the earth is flat, can continue to believe that the Christian gospels are nonfiction.

And no reasonable person, aware of all of the above, can believe that the tens of thousands of scholars who have disproven religion are all incompetent bunglers, or that theologians whose methodology can be used to prove that *Alice in Wonderland* is nonfiction are something other than self-deluded.

Complete proof that religion is as objectively falsifiable as Martian canals is detailed in *Mythology's Last Gods*, which can be borrowed from the Red Deer public library. Persons who could not survive without an afterlife belief to annul the terrifying finality of death are warned *not* to read it.

## ELEVEN

This letter was written in response to a letter published in the Red Deer
*Advocate,* June 30, 1999, headlined: "Pope's Trip Will Fulfill Prophecy."

I will not comment on the belief of a Blackfalds resident that information can travel backward in time, as it would have to do for "prophecy" to be a valid concept. Either one is already aware that time is one-directional, or he cannot be told.

When a prediction is sufficiently vague, almost any outcome can be interpreted as a hit by true believers. That is why there are people today who in all seriousness claim that Nostradamus's prophetic quatrains have come true.

But when Ellen White is cited as a successful prophet, I can only quote the words of the official journal (*Ministry,* June 1982) of the Church she founded: "She utilized the words of prior authors in describing words she heard while in a vision. In a few instances, she used the writings of a 19th century source in quoting the words of Christ or of an angelic guide." I can only assume that the letter writer did not know that.

For further details, see *The White Lie*, by W. T. Rea, NY, 1982.

## TWELVE

My Parsee friend, Ut Mut McTut, was mildly irritated when Telus **[Canadian telephone service provider]** borrowed a legend from his religion, in which three astrologers gave gold, frankincense and myrrh to the newborn Zoroaster, and used it to promote its prepaid cell-phone service. But he was offended beyond measure when some Christians, who had themselves stolen the legend 600 years after Zoroaster's death and appropriated it to their warrior king (30 years before their fourth gospel made him a god), screamed foul play, as if, having once annexed a story from the Parsee religion, it was now their property and could not be re-borrowed by anyone else.

It was bad enough, McTut said, that the Christians had stolen: his sun god Mithra's birthday; Mithra's weekly rest day; Mithra's virgin-birth in a stable and the shepherds who visited him there; Mithra's baptism initiation; Mithra's last supper with twelve followers; Mithra's mass-wafer; the new star that identified Zoroaster's birthplace; Zoroaster's underworld of fire-torture; Zoroaster's antigod who was responsible for all evil; the goddess Easter's resurrection from the dead; the massacre of infants that failed to kill the newborn god Krishna; and Nanda's journey to an ancestral home to pay his taxes.

But when earlier borrowers denounce later borrowers for doing the same thing they did, that is an absurdity. Mr McTut suggests that the screamers consider taking courses in elementary ancient history.

## THIRTEEN

Am I the only person who is repulsed and outraged that television stations in Calgary, Edmonton and Red Deer are broadcasting a collection of fairy tales called *Psi Factor—Chronicles of the Paranormal*, that even a cover story in *TV Today* identified as science fiction, and falsely claiming that its fantasies are true stories from the case files of an organization that probably does not exist?

I have no problem with *Psi Factor*'s entertainment value. If its scripts were presented as episodes of *The Outer Limits* or *The Twilight Zone*, they would be perfectly legitimate. It is the pretence that *Psi Factor* is nonfiction that is objectionable.

Do the perpetrators of the *Psi Factor* hoax have complete contempt for television viewers' intelligence? Do they share Barnum's view that there is a sucker born every minute, and anyone stupid enough to take their *Through the Looking Glass* imaginings seriously deserves what he gets? There are laws in Canada prohibiting false advertising. Why are those laws not invoked against liars who sell the public made-up stories under the guise of nonfiction? If Whitley Strieber had a sudden attack of conscience and confessed that *Communion* is a pack of lies from start to finish, he and his publisher would be swamped with demands from the book's purchasers for their money back. Are television viewers not entitled to compensation for their wasted time?

There is room on television for both science fiction and nonfiction. But persons who, for profit, label the former as the latter are conscienceless prostitutes with a depraved indifference to truth. The same criticism applies to a large number of American pseudo-documentaries. But *Psi Factor* is produced in Canada, and therefore Canadian authorities should be able to do something about it. I will be satisfied if future episodes are clearly identified as fiction. But I demand nothing less than that.

### AFTERWORD

*Psi Factor* was still broadcasting new episodes after five years, and still claiming that its fantasy scripts were based on the case files of the mythical "Office of Scientific Investigation and Research," thereby providing further proof, if any were needed, that nobody has ever gone broke underestimating the intelligence of the North American people.

## FOURTEEN

I was appalled to read that the brother of convicted child molester Rod McDonald is claiming that his nineteen accusers are suffering from false memory syndrome. Nothing I have read concerning McDonald's case supports such a possibility.

"False memory" is the true name for what certain quack therapists call "recovered memory." If, before consulting a therapist, a former pupil would have denied that Mr McDonald had ever molested her, and only came to believe that she had been molested after being manipulated into believing she had "recovered" the memory, that would indeed be a false memory. But if there was never a time when the accuser did not have a memory of being molested, then the allegation of "false memory" is a blatant attempt to capitalize on widespread ignorance of how false memories are created.

The only reason I am writing is that McDonald's fraudulent argument could have the effect of discrediting the whole "false memory" concept. False memories are very real, and there have been several convictions based on "recovered memory" testimony before it was proven that recovered memories are really false memories put into patients' minds using a technique that some would call hypnotism. But no one has ever suggested that a memory that already existed prior to consulting a therapist/manipulator is a false memory. If it is not an allegedly "recovered" memory, then it is not a false memory.

McDonald's behavior could harm true victims of the "recovered memory" hoax.

## FIFTEEN

I was repulsed and outraged by what I originally took to be a news story endorsing the reality of "yogic flying" and the usefulness of the snake-oil hoax of transcendental meditation—until I realized that it was simply a letter to the editor from which you had accidentally omitted the author's signature. I would see no option but to write a rebuttal, if the Maharishi's imbecilic fantasies had more popular support than those of the Flat Earth Society. Since the recent Edmonton by-election proved that they do not, I will not dignify nonsense by treating it as bad science, except to say that anything that is impossible is impossible.

To anyone who really believes that humans can will themselves to fly without visible means of support such as an airplane (and I seriously doubt that the letter writer is such a one), I have one piece of advice: get a brain transplant, you scientifically illiterate moron.

As an associate of entertainer Reveen at the time when he faced off against the Maharishi on a Calgary TV program in 1962, I am writing from personal knowledge. Also, magician James Randi has posted a $10,000 prize **[since raised to $100,000]** for any Maharishiite who can demonstrate flying to skeptical investigators. After more than 20 years, that prize is still unclaimed. For more information, see *Skeptical Inquirer*, 2:1:7-9; 19:3:9-11, 54.

### AFTERWORD

The above letter was written in response to one from a "Natural Law Party" candidate who, despite the *Advocate*'s usual 300-word limit, was allowed to peddle his pseudoscience for at least 600 words. When his reaction to my response was a further 600 words of *Alice-in-Wonderland* fantasy, I chose not to give him the opportunity for more free propaganda by writing a second response.

### SIXTEEN

It comes as no surprise that the NDP **[socialist party]** fared so badly in the Edmonton by-election. Under the obscene first-past-the-post voting system used only in England, North America and kindergartens, only a hardcore dogmatist would waste his vote by clinging to a candidate who had no chance of winning, rather than expressing a preference for whichever of the viable candidates he considered the lesser evil. Probably as many as 1,000 NDP supporters voted Liberal simply to prevent a Conservative victory.

What does surprise and horrify me is that almost eight percent of the votes went to a candidate and party that I can only describe as an embarrassment to primates. Social Credit's only redeeming social value is that it has been dead for twenty years. It is high time it got itself buried.

I am not unsympathetic to the Green Party. But losing the candidate's deposit for the sake of winning 81 votes was such poor economics, that I question whether donating to their cause differs from flushing money down the toilet.

As for the Maharishi party, whose candidate gulled 29 voters into believing his guru's claim that Canada can be protected from evil by levitation and invisibility, I will be kind and say nothing **[but see previous letter]**.

If this election proved anything, it is that the cash deposits forfeited by candidates who fail to win a specified minimum percentage of the vote is not preventing frivolous candidates from wasting public money by entering elections they cannot win, and should be at least doubled or tripled.

No Liberal seriously expects to win the next provincial election, or even the one after. But the human race is still evolving, and moderate, pragmatic, middle-of-the-road government will eventually be as routine provincially as it has long been federally.

### SEVENTEEN

If **[Alberta Premier]** Ralph Klein goes ahead and spends $3.5 million on a pretend election for a job that does not exist, for the sole purpose of pandering to redneck extremists and proving that he is their obedient slave, he had better be prepared to pay the $3.5 million out of his own pocket. For the courts will certainly rule that this is a partisan political expenditure that can no more be charged to the taxpayers than could television commercials asking voters to make Norman Bates prime minister.

Ten years from now, Canada is going to look back on Manningism **[a religious cult posing as a political party]** the way America looks back on McCarthyism. A prime minister who tried to buy votes by giving in to an insignificant minority of 14 percent and appointing a reactionary to an office where he could wield political power for life, long after the rednecks who voted for him had recognized their mistake, would be remembered for that one atrocity as surely as Richard Nixon is remembered for only one atrocity.

The opposition parties should take Klein to court before he can waste money on a hoax, and get an injunction preventing him from doing so.

## AFTERWORD

Premier Quisling did proceed with his pretend-election, even though the prime minister had warned him that it would be ignored. Two candidates from the theofascist party became pretend senators-in-waiting. When a senate seat did become vacant, the prime minister followed the procedure of a century of predecessors and filled it with a member of his own party. Alberta's opposition parties made no attempt to take the premier to court, and the taxpayers were stuck with the $3.5 million cost.

## EIGHTEEN

Your editorial writer got something right when he recognized that most of the four or five million Canadians who regard "God" as either a children's fairy tale comparable with the Great Pumpkin, an irrelevancy who may or may not exist, or the most sadistic, evil, megalomaniac serial killer in all fiction, have higher priorities than removing the name of a creature from mythology from the constitution.

Madalyn Murray succeeded in removing religious observances from American public schools. But she had something going for her that Canadians lack: a constitutional prohibition of any "law respecting an establishment of religion."

If the Manningites ever come to power, they could pass a law declaring Canada the personal possession of a god that is white, male, fundamentalist Protestant, and dedicated to the extermination of all sexual activity except between a male husband and a female wife, in the missionary position, in bed with the lights out, while uttering prescribed Protestant prayers, and the Supreme Court of Canada would be unable to stop them.

Svend Robinson **[Member of Parliament]**, who as far as I know is not a humanist, showed the courage of Don Quixote in tilting at the windmill of an essentially harmless form of intolerance. He could not have been unaware that, as long as a majority of Canadians believe that, "When God does it, it's not evil," his constituents' petition **[to remove the word "God" from the Canadian Constitution]** had as much chance of acceptance as a petition from China's democrats to remove the Communist Party's entrenched position in their political hierarchy.

But when your editorialist says that freedom from religion is itself a religion, that is like saying that education is an alternative form of ignorance.

That all claims of a god revealing its existence have been disproven for more than a century is a fact, not a belief.

Knowledge that the earth is round is not a religion, and knowledge that gods are products of the human imagination is not a religion.

## NINETEEN

Your editorial on the harmlessness of Preston Manning conjured up a picture in my mind of a newsreel I once saw showing a triumphant, jubilant Neville Chamberlain waving a piece of paper containing a signed promise by Adolf Hitler that he had no intention of keeping the promises he had made in *Mein Kampf.*

You pointed out that Catholic prime ministers, Clark and Trudeau, made no attempt to impose their religious taboos on the whole of Canada. What you overlooked was that neither Clark nor Trudeau was a fanatic, fundamentalist, reactionary bigot whose credo was, "My god's taboos apply to you, but your god's taboos do not apply to me."

Ayatollah Manning is the greatest threat to the freedom of every Canadian since Adolf Hitler. That is not a personal comparison. Hitler committed crimes equaled only by Josef Stalin and history's greatest mass murderer, the biblical Joshua. I do not for a moment suggest that Manning plans to solve his heretic problem with gas chambers. He will be quite content with jail cells.

Manning has made clear that he wants a constitutional amendment that would violate the beliefs of an overwhelming majority of Canadians. Whether he could ever impose such an amendment is not the issue. What is terrifying is that he thinks he has the right to try to do so. That he lacks the moral evolution to recognize the qualitative difference between a self-aware sentient being and a pre-human tadpole is also a non-issue.

Any person who agrees with Manning has the freedom to live his life in conformity with his taboos, and that is as it should be. It is the attempt to impose those taboos on others that is an unmitigated evil, and Manning has stated unambiguously that that is his ambition.

### AFTERWORD

I had no idea how comparatively moderate Preston Manning really was—until he was replaced as leader of the theofascist pseudo-political Social Credit-Reform-CRAP-Canadian Alliance hate cult by Stockwell Day, a man of whom I would *not* have expressed the opinion that there is a line he would not cross.

## TWENTY

It comes as no surprise that Manning's ayatollahs are plotting dirty tricks worthy of Richard Nixon. That is what extremists do, whether they are right wing extremists whose role models are Ruholla Khomeini and Tomas de Torquemada, or left wing extremists who want to do to Canada what **[socialist premier]** Bob Rae did to Ontario…

### AFTERWORD

This letter was written after my warning of the threat posed by Preston Manning's theofascism caused an Alberta redneck to call me a left wing extremist. I figure that being called "left wing" by the far right, and "right wing" by the far left, is a good indication of the contempt western rednecks have for moderate, pragmatic, middle of the road liberalism.

## TWENTY-ONE

Your columnist hit the nail on the head when she pointed out that a Liberal, Conservative, or socialist who makes a pro-discrimination remark is best viewed as a loose cannon, whereas similar intolerances from the Manning party are "symptoms of something endemic to a political organization."

Manning can proclaim himself an egalitarian who respects minority rights. But if that were true, he would never have been chosen leader of a party that personifies hatred and intolerance of all who do not conform to the narrow-minded taboo code that rednecks equate with morality.

Manning's true self was revealed when he attacked the Liberal Red Book for its platform of making more money available to help the old, the sick and the unemployed to stay alive. He thereby showed his willingness to buy votes by pandering to those who place personal benefit ahead of the well being of their fellow humans. I am viewing the man in the most favorable light when I conclude that his ranting is an accurate reflection of his personal beliefs.

If Mr Manning does not endorse the bigotry that is the only thing holding his party together, perhaps he should resign and make way for a leader more in tune with the party's rank and file. I recommend **[Stephen King's]** Cujo.

## TWENTY-TWO

Joe Clark **[Federal Conservative Leader]** says he can form a government and Preston Manning cannot. That is not quite accurate. What he means is that he has a snowflake's chance in hell of forming a government, and Manning does not.

Canada's Liberal majority always has and always will decide who is going to govern them. At a time when Pierre Trudeau appeared to be safely entrenched, they thought they could safely send a message that he was taking far too long to deliver the Just Society he had promised. They were appalled when they saw that their intended slap on the wrist had tossed him out of office. Even so, not until Joe Clark had proven himself totally inept did the Liberal pragmatists who had elected him decide that two elections in six months was a lesser evil than allowing Clark to continue running amok.

But even in the instant replay, Clark retained all of his seats in the West. It was Liberals, not Conservatives, who rejected him.

Following Trudeau's retirement, Liberal pragmatists elected two Conservative governments, not because they thought it was time for a change, but because they saw a competent Conservative as a far better choice for prime minister than the bungling incompetent who had replaced Trudeau.

John Turner's replacement by Jean Chrétien guaranteed that there would be no more Conservative governments. Mulroney **[Conservative Prime Minister]** knew that, and got out before he was thrown out. But while Mulroney's blatant hypocrisy, repeated lying, and imposition of a federal sales tax on top of existing provincial taxes outraged Conservatives, his retirement would have prevented that outrage from harming his party.

It was the election of a short-lived Conservative leader even more inept than Clark or Turner that opened the door for a party **[led by Manning]** that never has and never will win the support of the eighty percent of Canadians who view it as slightly to the right of Ayatollah Khomeini.

Unless the Liberals allow it to happen, by choosing another leader as untalented as John Turner, Clark will never form a government. But he can form a national opposition capable of taking advantage of any Liberal bungling. Manning cannot.

## TWENTY-THREE

Ninety percent of Europe's Christian majority approved of the Inquisition. They did so because they could not foresee the day when "finally they came for me, and there was no one left to stop them."

If Premier Ralph Klein allows the Manning-Day faction to persecute gays and believers in abortion rights, he can be very sure the day will come when that faction adds "persons with Jewish names" to its target list. **[holocaust denier James]** Keegstra was not an aberration. He was the epitome of the bigotry that is at the core of Alberta redneckism.

I look forward to a future in which the extremism of Manningism in Western Canada and Buchananism in the southern United States is as extinct as the Thuggs and the Assassins. But I would vigorously oppose any discrimination against such groups, out of recognition that, once thoughtcrime is legislated into reality, the Inquisition cannot be far behind.

I belong to several minorities: immigrants, university graduates, nontheists, Alberta liberals, and opponents of all killings of self-aware sentient beings, including those perpetrated by the state.

If I tolerate less-than-equal treatment for minorities to which I do not belong, such as homosexuals, VLT users, practitioners of religion (all religions are limited to minorities, since none is supported by even one fifth of this planet's population), it will be nobody's fault but my own if a minority to which I do belong is eventually exterminated by the self-righteous majority.

I would like to remind the likes of Preston Manning and Stockwell Day that 500 years ago they would have been among the "heretics" burned at the stake. Do they really think that the power to persecute dissent will never be regained by their opponents? That is not the way history works. Either they establish the principle of equal rights for all, right now, or it will be their own heirs who pay the price.

### AFTERWORD

After this letter appeared in the Red Deer *Advocate,* Stockwell Day telephoned me and wanted to know if I considered all of the Albertans who had voted for his faction "subhuman monsters." Unfortunately, my phone become unplugged before I could answer that nobody votes for a theofascist except another theofascist.

## TWENTY-FOUR

Did Jim Keegstra, a former teacher, really write "must of heard" instead of "must have heard"? If so, it is a sad comment on Mr Keegstra's own education, and the whole North American school system where such ignorance is far from rare.

### AFTERWORD

What makes this letter worth reprinting is that Jim Keegstra is one of only two persons convicted under Canada's anti-hate laws, for teaching his high school classes that allegations of genocide against that nice Mr Hitler are all lies. I draw attention elsewhere ("Bible Belt or Loony Bin?") to Keegstra's ability to believe that shutting his mind to the facts of history can make them go away. But his ability to sit through sufficient classes to obtain teaching credentials, without ever learning the difference between a verb and a preposition, strikes me as relevant.

# TWENTY-FIVE

When Stockwell Day compares his belief in the literal truth of Genesis to the Catholic belief in Immaculate Conception, it is an invalid analogy. Immaculate Conception is a purely fantasy concept that cannot be scientifically tested or measured.

Creationism's claim that the origin and evolution of the universe was triggered by a god saying, "Let it be," is likewise an untestable fantasy. But the details of creationism, such as a 6,000-year-old earth, humankind's independent genesis rather than evolution from lower lifeforms, earth's existence before there was a star for it to orbit, and trees that bore fruit before there was a sun to trigger the necessary photosynthesis, can be and have been subjected to scientific investigation and disproven by geologists, astronomers, geneticists, archaeologists, and scientists from several other disciplines.

If someone actually believed, in the face of the evidence of astronauts, geologists and physicists, that the moon is made of green cheese, simply because the Wizard of Oz said so in a fantasy novel written by a person now dead, few would have difficulty recognizing that he was not sparking on all neurons. That is no less true of persons who reject proven facts because they do not accord with statements made in a fantasy novel written 2,500 years ago by authors who also assured their readers that the earth is flat, the sky is a solid dome to which the sun, moon and stars are attached, the earth is the immobile center of the universe, and early humans lived for as much as 969 years.

As to media reports that Day has promised not to impose his cult's taboos on Canada, he has, in fact, made no such commitment. What he has said is that only "the people" can change laws on abortion, homosexuality, and state-sanctioned ritualistic revenge murder.

What that means is that he will not turn Canada into a theocracy unless he wins a parliamentary majority or a referendum that permits him to do so, (as if a majority has the right to deprive minorities of equal rights in a civilized country).

Day's doubletalk, calculated to deceive Canadians into thinking it means the opposite of what it really implies, while permitting him to claim retroactively, "I never said I would not impose such laws," is called LYING. I was never apprehensive that Catholic prime ministers such as Trudeau and Clark would try to turn Canada into a mirror image of Iran and Afghanistan. I am satisfied that that is exactly what a Day government would do.

# TWENTY-SIX

Either Stockwell Day is the most blatant proponent of the Big Lie Canadian politics has ever produced, or he is so intellectually impoverished that he is genuinely unable to grasp that the accusations he is hurling at his detractors are a projection of the precise qualities he sees in the mirror. Even by the standards of Alberta's redneck majority, the man is an aberration.

The *Advocate* chose not to print this letter, despite my forbearance in
not comparing Stockwell Day to Larry Talbot, Cujo or Old Yeller.

# TWENTY-SEVEN

The Quebec election was an obscenity. I am not referring to the outcome *per se*. My opinion on that is personal and therefore irrelevant. What is repulsive and unacceptable is that, even though 57 percent of the population voted against them, the separatists won 60 percent of the legislative seats—almost the precise opposite of the will of the people. Does anyone need further proof that the first-past-the-post voting system is the antithesis of justice, democracy and even sanity?

The preferential system, also called "single transferable vote," would have given supporters of every third-place candidate a right to express a preference for their second choice, and the result would have been totally different. Every political party in Canada **[and the USA]** rejects first-past-the-post in leadership elections, and demands that a candidate receive 50 percent plus one before he/she is declared elected. Yet they refuse to endorse the same system in general elections. Why? Do they think party members have the intelligence to understand a democratic voting system, but the general population is too stupid?

Or is the problem that only government parties can change the voting system, and governments are unwilling to change a system that elected them?

## TWENTY-EIGHT

How dare the Prime Minister place the survival of the human species ahead of the right of Alberta's oil plutocrats to line their pockets? Premier Quisling correctly recognizes that the billions of human beings the Kyoto Accord will save will not be born until he is out of office and the current generation of Pithecanthropus rubercollum Albertensis are all safely dead. Of course he must pander to their depraved indifference to the welfare of the human race. How else could he remain the biggest turd in the redneck anus of the universe? It would be unrealistic to expect anything else from a man who thinks the best replacement for the moderate Joe Clark is **[Ontario Premier]** Mike Harris, a man who emulated Stockwell Day in putting a contract on the old, the sick and the unemployed. Day's rationalization was that no person who could not find work would starve to death or freeze to death unless his imaginary playmate in the sky wanted him dead. What is Harris's excuse? And how can anyone see a Premier who would endorse such evolutionary throwbacks as anything but the Universal Soldier without whom there would never have been the kind of civilization-destroyers Alberta's theofascist rednecks want to see governing this country?

## AFTERWORD

Since this letter was written after I canceled my *Advocate* subscription on learning that I can buy it cheaper in perforated rolls, I do not know if it was ever printed. But what truly disgusts me is that the Alberta Ayatollah, not content with using public funds to pay off Stockwell Day's libel settlement, is now spending a million dollars of taxpayers' money—my money—on TV ads denouncing the Federal Government's plot to preserve Planet Earth as a human habitat, and neither the Federal Government nor the Alberta Liberal Party is willing to go to court and challenge his right to do that.

## 2. To Other Media

### LETTER TO *THE TIMES* OF LONDON

Now that Mr Denis Lemon has been convicted of heresy, I assume that he will be burned at Smithfield?

#### AFTERWORD

Denis Lemon was the publisher of *Gay News*, which published a poem accepting Jesus the Nazirite as a fellow fellationist. Lemon was convicted of blasphemous libel, that being the current name for "not believing in the gods the polis believes in" in the world's largest insane asylum that is England. Smithfield was the site at which Bloody Mary burned 300 heretics at the stake. In a country in which religion's primary weapon is enforced ignorance, it did not surprise me that *The Times* declined to print my letter.

### LETTER TO *MC²*

Your Round Table intro raised the question: Will the god concept be recognized as fantasy at some future time? In fact the falseness of revealed-god mythology is proven fact right now. True it is that a majority of humans, including a majority of Mensans, are unaware of the relevant facts and continue to regard god mythology as a matter of belief or disbelief. But widespread ignorance cannot make that which is fully proven any less proven. Until quite recently I believed that a planet named Vulcan possibly existed within the orbit of Mercury. But my ignorance in no way altered the fact that Albert Einstein had definitively established Vulcan's nonexistence before my birth. Similarly the ignorance of the judge in the Charlie Chaplin paternity case regarding the infallibility of negative-result blood tests did not alter the proven fact that Chaplin's non-paternity was fully proven, even though Chaplin wound up paying the plaintiff's claim out of sheer frustration. I do not say that the existence of immortals can be disproven, any more than the existence of hyperspace, time travel or fire sprites that live on the sun can be disproven; only that the existence of the specific gods of existing mythologies has been disproven. Since the book that best proves this, *Mythology's Last Gods*, may not be available for some time **[it was not published until 1992]**, I refer anyone who does not mind being confused by facts to Martin Larson's *The Story of Christian Origins*, Peter Ellis's *The Yahwist*, James Frazer's *The Golden Bough*, M. J. Vermaseran's *Mithras, The Secret God*, and Robert Graves' *Hebrew Myths*. Anyone desiring a 150-item bibliography of books by competent historians that analyze (and therefore refute) some part of the J/C bible can obtain it by sending me a sase.

#### AFTERWORD

Although *MC²* was received and presumably read by every Canadian member of Mensa, I did not receive a single request for the offered bibliography. In Mensa as elsewhere, the attitude toward security beliefs is "My mind is made up. Don't confuse me with facts."

## LETTER TO *CALAMITY*

This letter to Calgary Mensa's monthly newsletter was written in response to an entry by a longtime member who was also an incurable godworshipper, in which she gushed on about the legitimacy and accuracy of a graphologist who had given her a butt-kissing character analysis allegedly derived from her handwriting.

Anyone who has paid $4 to a scriptoscam artist in Winnipeg (or anywhere else) for a cold reading is urged to contact me. I have a bridge for sale in Brooklyn that I think will interest them.

## LETTER TO TWO TV STATIONS IN EDMONTON

If you are not prostitutes, willing to contribute to the dumbing of Canadians for the sake of ratings, you WILL carry a disclaimer before *Psi Factor: Chronicles of the Paranormal*: "The following program is science fiction, and has no resemblance to any real persons or events."

### AFTERWORD

The TV stations chose not to carry any disclaimer. I mailed a Calgary TV station a photocopy of my letter to the *Advocate*, and it almost immediately dropped *Psi Factor* from its schedule—not necessarily as a consequence of my letter.

## LETTER TO THE CALGARY *HERALD*

Your article declaring that 1984 is here, and George Orwell's pessimistic prophecy has failed, is typical of what I have been seeing everywhere I look. It is totally inaccurate. *1984* was not a prediction of a nightmare future under totalitarian communism. It was a description of a nightmare present under totalitarian Anglican Christianity.

Orwell's original title of his book that spelled out the horrors of the theocracy of 1949 England by comparing it to Soviet communism was *1949*. It was his publisher who forced him to change the name to *1984* in order to disguise his description of present-day England as a prophecy of a possible future.

Consider: Orwell's novel was centered around a Party that preached love and practised hate, lauded ignorance as a virtue, criminalized "thoughtcrime," and declared even the sharing of joy forbidden except in narrowly delineated circumstances. There is indeed a philosophy that teaches such insanity, but it is not communism. As for Big Brother, in whose name the Party's every atrocity was committed, whom nobody had ever seen and who probably did not exist, whose putative Enemy was really his mirror image: That anyone could fail to recognize who that was supposed to be is incomprehensible.

### AFTERWORD

This letter was written early in 1984 in response to an article that ridiculed George Orwell for making an allegedly failed prophecy. Neither my letter nor any other acknowledgement of what *1984* was really about was ever published by the *Herald*—or to the best of my awareness any other element of the mass media.

## LETTER TO *PLAYBOY*

Recently published statistics indicate that, worldwide, there is one abortion for every two live births. What is amazing about that information is that it is being disseminated, not by quality-of-life advocates, to whom it would be valuable propaganda, since it shows that, without abortion, this planet's already dangerous overpopulation would be increasing at an even greater rate, but by quantity-of-lifers who apparently do not recognize how thoroughly it undermines their advocacy of species suicide by mass starvation.

### AFTERWORD

While the original letter, published some time between 1980 and 1985 and reconstructed from memory, was longer, that is as much as I can now remember.

## LETTER TO *DISCOVER*

The September 1984 issue of *Discover* carried an article by Martin Gardner debunking psychic surgery. Gardner decried the death of actor Andy Kaufman, who rejected the legitimate cancer treatment that may conceivably have prolonged his life, and instead resorted to a mountebank in the Philippines, where "psychic surgeons" have the same freedom to commit homicide by passing off fraudulent mumbo jumbo as alternative therapy as faithhealers in North America. The following letter was written in response to that article.

I appreciate Martin Gardner's reluctance to reveal the methods by which legitimate magicians earn a living. However, since he was discussing a man's death and trying to prevent similar suicides-through-ignorance, he might have bent the rules a little.

To a person desperate enough to seek psychic "surgery," the information that the "surgeon" produces tissues and blood by "standard magician's sleight of hand" is not a sufficient deterrent. The tissues and blood are in fact produced from a thumb-tip, a flesh-colored thimble-like device used by magicians for a hundred effects that it would not be legitimate to reveal.

But of course all this is redundant, since people gullible enough to fall for psychic "surgery" do not read *Discover*.

### AFTERWORD

That *Discover* did not pander to the gullible was true at that time. About a year later, I cancelled my subscription when *Discover*'s then-publisher, Time-Life, the most notorious peddlers of superstitious hogwash on this continent, replaced *Discover*'s skeptical Executive Editor with one who promptly scrapped the "Skeptical Eye" column that debunked superstition, and instituted policies more in keeping with Time-Life's policy of publishing whatever would sell magazines, with depraved indifference to truth, objective scholarship or competent reporting. As of 2002, *Discover*, no longer owned by Time-Life, is again a legitimate science magazine.

## LETTER TO *SKEPTICAL INQUIRER*

It seems to me that there is an obvious solution to the problem of harassing, intimidating and frivolous lawsuits, not only against CSICOP and individual Skeptics (*SI*, Spring 1992) but against whistleblowers of every variety. Surely there are congressmen and senators who would support legislation requiring the litigant who loses a civil suit to pay the costs and legal fees of the winner. Such laws exist in much of the rest of the world, and I can see no reason why they should not be instituted in the United States. At least one prominent American politician has publicly stated that America has too many lawyers handling too many lawsuits. He cannot be the only person who holds such a view.

### AFTERWORD

I do not know how long after the publication of this letter the advocated law was instituted. But the judge presiding over Uri Geller's lawsuit against CSICOP for publishing articles by James Randi (truthfully) calling him a magician dismissed the suit as "frivolous," and ordered Geller to pay CSICOP's costs.

## LETTER TO *THE SKEPTICAL REVIEW* ONE

The September/October *TSR* raised the question of who was the prophet foretold by "Moses" in Deuteronomy 18:15. Christians cite Acts 3:16 to support their claim that it was Jesus. Muslims cite the Koran to argue that it was Muhammad. But the identity of the prophet who was to be Moses' equal cannot be deduced by assuming that information had traveled backward in time, as it would have had to do if the foretold one was someone who was not already living at the time of writing. The prophet foretold by the author of Deuteronomy was in fact *himself*.

Richard Friedman, a professor of religion at UCSD, has demonstrated by a preponderance of the evidence (*Who Wrote the Bible?*) that the author of Deuteronomy was the spokesman (Greek: *prophetes*) Jeremyah. At the time of writing, the only person who could have plausibly claimed to be the new Moses was Jeremyah himself. As it turned out, nobody interpreted Jeremyah's self-glorification the way he had intended, and he ended his life in exile in Egypt, viewed by Jews as a despised collaborator with the Babylonian occupation. But the failure of Jeremyah's readers to accept him at his own evaluation does not change the reality that the "prophet" he promised was no one but himself. For details, see my *Mythology's Last Gods*, pp. 166-167.

### AFTERWORD

When this letter was published, the editor added a note that the equation of the Deuteronomist with Jeremyah was first made by Bishop John William Colenso in 1863 in a book titled *The Pentateuch and Book of Joshua Critically Examined*. I immediately checked Friedman's bibliography to see if Colenso was listed. He was not. That does not mean that Friedman was aware that he had been preempted, but it does deprive him of precedence for his conclusion.

## LETTER TO *THE SKEPTICAL REVIEW* TWO

Why don't you stop trying to reason with biblical inerrantists, and accept the reality that they are insane? The way they say, "Yes, but..." and then repeat the same incompetent drivel you have already rebutted over and over and over surely proves that. When you allow page after page of your limited space to be hijacked by incurables, you alienate the majority of your readers who expect a new issue of *TSR* to contain new material, not a neverending rehashing of the old...

## AFTERWORD

The editor chose not to publish the above letter. But he did acknowledge, in the May/June 2002 issue, that "As [a religious fundamentalist]'s tedious articles on the book of Daniel dragged on, the subscriptions dropped dramatically...subscribers were obviously fed up. When I finally announced that enough was enough and that I would publish no more articles by him...the subscriptions began to rebound."

## LETTER TO *FREETHOUGHT PERSPECTIVE* ONE

I am beginning to see why Gee Dubya Shrub is the most prolific serial killer in American history, with over 120 homicides on his resume. He is intellectually handicapped, falling somewhere on the evolutionary scale between Dan Quayle and Ronald Reagan, and therefore lacks the neural circuitry to comprehend that killing human beings is plain *evil*. He does not even grasp that he was made president by a coup d'état that as recently as fifty years ago would have got the five perpetrators executed for treason.

## LETTER TO *FREETHOUGHT PERSPECTIVE* TWO

Competent scholars have refused to rebut Roman Piso for the same reason they for a long time refused to rebut Immanuel Velikovsky: because merely acknowledging his existence would grant him an undeserved respectability. And the consequence has been the same: People are getting the idea that perhaps he cannot be rebutted. I asked Bernie Katz to write a review/rebuttal of Piso's book, because he has presented some of the strongest evidence against it in a couple of letters, but he indicated that he is too busy.

I have no problem with the conclusion that Jesus did not exist. A lot of competent scholars think likewise. But Piso's thesis depends on the claim that the four Christian gospels were written by a single author (Josephus), and that is as easily and totally refuted as the claim that the Torah was written by a single author (Moses), by the same methodology of showing that the four gospel authors disagreed with one another on some very basic issues. To cite a single example, Randel Helms in *Who Wrote the Gospels?* points out that the author of Mark "frequently seems unaware that his source is quoting or paraphrasing the Old Testament, and when his source misquotes Scripture, he fails to correct the error." He then adds that, in places where this happened, "Both Matthew (12:3) and Luke (6:4) silently correct Mark's error." And then there is the even more obvious contradiction that three gospels portray Jesus as what he claimed to be: the successor and equal of King David, whereas the John gospel turns him into a god. Piso's thesis stands or falls on the gospels being written by a single author, and that is incompetent nonsense.

## AFTERWORD

In the same issue of *Freethought Perspective* in which the above comment was published, associate editor Dorothy Thompson wrote a response that included the following points:

Dr Harwood argues that Roman Piso's theory is that the gospels were written by a single author, namely Josephus. He claims that is "incompetent nonsense." Nowhere does Piso say the gospels were written by one author. The author [Piso] does agree that Josephus was author of at least some of the N. T. books, but Josephus is merely a pen name used to disguise the true authors.

> Thompson may be right that Piso's book differed from its ads, which elicited sales by claiming, "Conclusive proof Flavius Josephus created fictional Jesus, authored Gospels." But Piso's associating *any* Christian Testament books with Josephus is indeed incompetent nonsense. The following issue of *Freethought Perspective* carried two letters echoing my criticism. My evaluation of Piso remains unchanged.

## LETTER TO *AMERICAN RATIONALIST* ONE

Let me add to Richard Bozarth's article [*AR* July/August 2001] on *American Atheist*, that they appear to be still the same kind of exclusionist clique as when Madalyn Murray was in charge. Their rejection of every article I ever submitted, submissions subsequently accepted and published elsewhere, could perhaps be attributed to their prioritizing social rather than scholarly issues. But when they acknowledged that the reason they rejected a book review was that they do not review books not sold by American Atheist Press, they thereby revealed something about themselves that they might have been well advised to keep hidden.

### AFTERWORD

*American Atheist* eventually accepted my article, "Is God a Petulant Little Boy," (Part Four # 36)

## LETTER TO *AMERICAN RATIONALIST* TWO

I found your piece on cannabis too far out of my field of expertise to offer any useful comment. But I certainly agree that treating pot more severely than alcohol or tobacco is insane, and the reason politicians want to retain current laws is that they are prostitutes willing to pander to the fundamentalist rednecks and their sadist in the sky. At a time when Canada is moving toward saner marijuana laws, Gee Dubya Shrub is providing further proof, if any were needed, that he is even more subhuman stupid than Reagan—and he was barely anthropoid.

## LETTER TO *HUMANIST IN CANADA*

Re your column on Canada's archaic and obscene voting system: The evil of "first past the post" lies not in a party that received 40 percent of the vote winning 57 percent of the seats, as it would have done anyway after defeated candidates' second or third preferences had been counted. Rather, the evil lies in such situations as Alberta's Social Credit Party winning 90 percent of the seats, when the 53 percent who voted against it would have united behind the Conservative Party and defeated it with their second preferences, as they proved in the following election when Social Credit was finally wiped out. Similarly, Quebec's separatists won a majority of seats even though the sixty percent who voted against them would have united behind whichever non-separatist candidate was still in the running after the first count.

But proportional representation is not the solution. Under proportional representation, a system that involves multi-member electorates, the election of a majority government would be virtually impossible, and splinter parties would wield power out of all proportion to their numbers, as happens with monotonous regularity in Israel. What is needed to replace first past the post is the preferential voting system, more commonly called "Single Transferable Vote" in this continent.

Under STV, a voter marks his ballot "1" for his first choice, "2" for his second choice, and so on down to the last name on the ballot. This is the same system used by all Canadian (and American) political parties in choosing party leaders, with the cost-saving modification that a voter does not have to wait until his preferred candidate has been eliminated, and then vote for his second choice on a subsequent ballot. Rather, he marks his second preference, and his third, fourth, etc., on the initial voting slip, knowing that his vote cannot help defeat his first choice, since his "2" vote will not be counted until and unless his "1" candidate has already been eliminated.

STV is used in Australia and Europe, and consistently helps elect governments that represent majority consensus, even though a party that was able to win fifty-percent-plus-one on a straight shoot-out might not have received fifty percent of first preferences.

## LETTER TO A CRITIC OF *MYTHOLOGY'S LAST GODS*

Only in Australia, where most people have never even met anyone who went to university, would a fatuous oaf with grade-eight education have the pretentious insolence to challenge the published research of a Doctor of Philosophy. There is not a scholar on earth who would dignify such incompetent drivel with a rebuttal. Two centuries ago, the RC Church tried to close all universities, because students were going in as godworshippers and coming out sane.

## LETTER TO BERNIE

I see that Gee Dubya Shrub wants to make *Osama bin Laden Bless America* the new national anthem. (Or is it *God* Bless America? I'm always confusing those two.)

## LETTER TO CTV.CA

Are all news writers and newsreaders scientifically illiterate, or are they under orders not to offend the ignorant by dispensing information they choose not to know? It has been known since the nineteenth century that all claims of a god revealing its existence have been traced to the same bible authors who assured their readers, in seven places, that the earth is flat, and that fifty other virgin-born savior gods were believed to have risen from the dead on the third day as much as three thousand years before the story was posthumously added to the biography of Jesus. Yet the media treat religion as if it were less disproven than astrology, alien abductions, psychics, Martian canals or tealeaf reading. And their refusal to question the current pope's insane ban on population control, which would cause the extermination of the human race if it was actually obeyed, reminds me of the Americans in 1940 and 1941 who considered it impolite to criticize that nice Mr Hitler.

—from the author of *Mythology's Last Gods*, which WILL eventually wipe religion from the face of the earth.

# BOOK REVIEWS

Cartoon courtesy of *The Match*

It's an incredible con job when you think of it, to believe something now in exchange for life after death. Even corporations with all their reward systems don't try to make it posthumous.

Gloria Steinem

Greater than the courage to do right in the face of danger or adversity was the courage to commit a lesser evil to prevent a greater. And if the evil that needed preventing was the ultimate one, the death of all living things, then any means were justified to accomplish that end. Anything at all.

*Greenhouse Summer,* Norman Spinrad

# 1. *Why Atheism?*
# by George H. Smith

"George H. Smith, author of the influential contemporary classic *Atheism: The Case Against God*, continues his defense of reason, freethought, and personal liberty by answering the age-old question: *Why Atheism?* Why would anyone question the existence of a supernatural deity? Smith reviews the historical roots of unbelief dating back to the ancient Greeks, argues that philosophy can serve as an alternative to religion, and defends reason as the most reliable method we humans have for establishing truth and conducting our lives." So says the publisher on the back cover, and I have nothing to add or subtract—although I have not changed my view that only the methodology of history can hope to free the human race from the mind-crippling opiate of religion, and philosophical arguments such as Smith's are convincing only to persons too educated to believe in religion in the first place.

On the issue of agnosticism ("absence of knowledge"), Smith writes (p. 17), "Consider, for instance, an intelligent Christian who confronts the agnostic's claim that God's existence cannot be proven one way or the other, so we should suspend judgment...The Christian might decline to investigate the arguments for agnosticism in any detail, because his own belief in God is so strong, his degree of certainty so high, as to render any further investigation unnecessary."

He continues, "Much of this book is more concerned with the credibility of atheism than its justification. I shall argue that atheism is credible and should therefore be seriously considered by theists and agnostics alike...Most Christians (and other religious believers) dismiss atheism outright...not because they have examined the arguments for atheism and found them wanting, but because they do not take atheism seriously enough to examine its arguments in detail. Atheism, in their view, lacks credibility, so they have no motive to explore it further."

That paragraph merely hints at the problems faced by persons who choose to call themselves atheists, rather than humanists, rationalists, naturalists, or some other label that has not yet come to be viewed by brainwashed believers as pejorative. That Smith recognizes the problem is made clear when he writes, (p. 18) "The word 'atheist' has traditionally been used as a smear word—or 'bugaboo epithet,' as historian Preserved Smith once described it. To call someone an atheist was more often an accusation than a description, an invective hurled by orthodox Christians against any and all dissenters, including other Christians."

I can personally confirm that observation. A few weeks ago, when I angrily stomped out of a memorial I had been promised would be nonreligious, one of the braindead zombies in attendance demanded to know if I was an *atheist*. The way she used the word made very clear that her question was equivalent to asking if I was a child molester or a serial killer. And George Bush Sr. is on record as expressing the belief that atheists should not be considered American citizens. But Smith makes clear (p. 19) that, "Atheism, or the absence of theistic belief, is therefore a perspective, not a philosophy."

Smith further observes (p. 28), "It is interesting to note how the prejudice against atheism is reflected in ordinary language. It is commonly said, for example, that the personal atheist has *lost* his faith, as if he no longer possesses something of value. As the atheist sees the matter, however, he has not 'lost' anything except ignorance and error, and he has gained in understanding and knowledge."

Smith's arguments for atheism's credibility are valid. What his books will not achieve (apart from curing the incurable) is making the *word* "atheist" respectable. I long ago adopted the description, *nontheist*. Perhaps Smith should consider doing likewise.

## 2. *The Ways of an Atheist*
## by Bernard Katz

"Did you know that every time you entered a church you were reborn? Yes, you're reborn whether you want to be or not. For that is the very function of the church's architecture, it being modeled on a woman's genitals. Every time you go through the door, you must realize that you are reentering the womb, and thus are born again and again." (p.103) That passage certainly explains why Jimmy Swaggert and Garner Ted Armstrong, who consistently entered and emerged from various birth canals, liked to describe themselves as "born again."

Jesus rode into Jerusalem on a Sunday morning, riding on a donkey in order to convey to the unwashed masses that he was the king whom Zekharyah had promised would arrive in such a manner to free them from foreign overlordship. A huge crowd followed him, shouting, "Hosanna, Ben David," meaning, "Liberate us, Descendant of David."

As Bernard Katz tells it (p. 96), "There the story ends. Nothing more is heard of that stirring entry which Christians celebrate as Palm Sunday. Neither the Roman authorities nor the Jewish leaders seem to have paid the least attention to that messianic procession. Why does Matthew break off the story? Because Jesus, or those in his circle, or both, aimed at nothing less than an insurrection. They came for the same purpose as past and future messianic movements did—to crush the occupying enemy so as to regain national independence. Matthew suppresses further details because not only did Jesus fail in his mission but because he also wanted to convince the Roman authorities that the founder and followers of Christianity were *not* revolutionaries. Since Jesus' mission was stopped by the aggressive response of the Roman legionnaires, Matthew had to save the situation by making it seem that his Master was innocent as a lamb, and did not himself know why he was led before Pilate for crucifixion. By suppressing this truth, Christians were free to move about and to propagandize and grow."

Biblical history constitutes only a small part of Katz's book. Mostly it is philosophical arguments showing the absurdity of various religious rationalizations, particularly the pretence that the Christian gods, Yahweh and Jesus, were nice guys. Katz shows that Jesus was a liar, a coward, a masochist, a sadist (some people are both), and all round SOB, and Yahweh was a genocidal serial killer. But that is not simply Katz's conclusion. He cites a plethora of passages showing that bible authors believed the same thing, even while shutting their minds to the implications of their own fiction. How a gospel author could show Jesus preaching what can only be described as moral bankruptcy, while simultaneously asserting that anything Jesus did was virtuous by definition, Katz does not attempt to explain. He merely shows that they did so.

Where I depart from Katz is in some of his recommended further reading. He describes George Smith's *Atheism: The Case Against God* as "outstanding," and perhaps it is—if one is preaching to the choir. In my view it will not to cure a single believer, and what is the point of writing a book falsifying religion if not to free the mind-slaves? And he describes the scissors-and-paste high school pablum of the Durants as "wonderful history." Not by my definition of history.

In writing about Christianity's most imbecilic oxymoron, the delusion that one god plus one god plus one god equals one god (p. 107), Katz quotes Eusebius's observation, "that the Christian religion afforded a subject of profane merriment to the pagans, even in their theaters." He then summarizes, "Worst of all, a dogma was imposed which completely defied reason, making Christianity a laughingstock among rational people, and compelling the rest to live with this aberration." So what else is new?

### 3. *Atheism: A Reader*
### S. T. Joshi, ed

My only criticism of this collection of excellent essays written over the course of 2,000 years is that they neither say anything new or useful to persons who already know that death is the irreversible end of existence, nor ever did or will influence the views of the intestinally challenged who, without an afterlife belief to annul the terrifying reality of death, would have to be institutionalized and diapered.  So instead of commenting on the various essays, I will let their authors speak for themselves.

**S. T. Joshi:**  Even ridicule of religion is an entirely valid enterprise.  Writers as diverse as Aristophanes, Voltaire, Bierce, Mencken, and Gore Vidal have found satire—whose very purpose, let us recall, is to offend, provoke, and annoy—a welcome tool to expose religious folly, hypocrisy, and injustice.

**H. L. Mencken:**  If we assume that man actually does resemble God, then we are forced into the impossible theory that God is a coward, an idiot and a bounder...The so-called religious organizations which now lead the war against the teaching of evolution are nothing more, at bottom, than conspiracies of the inferior man against his betters...Certainly it cannot have gone unnoticed...that no man of any education or other human dignity belongs to them.

**George Eliot:**  Given, a man with moderate intellect, a moral standard not higher than the average, some rhetorical affluence and great glibness of speech, what is the career in which, without the aid of birth or money, he may most easily attain power and reputation in English society?  Where is that Goshen of mediocrity in which a smattering of science and learning will pass for profound instruction, where platitudes will be accepted as wisdom, bigoted narrowness as holy zeal, unctuous egoism as God-given piety?  Let such a man become an evangelical preacher.

**Charles Darwin:**  That the men of that time were ignorant and credulous to a degree almost incomprehensible to us—that the Gospels cannot be proved to have been written simultaneously with the events—that they differ in many important details, far too important, it seems to me, to be admitted as the usual inaccuracies of eye-witnesses—by such reflections as these...I gradually came to disbelieve in Christianity as a divine revelation.

**Robert Ingersoll:**  What God is it proposed to put in the Constitution?  Is it the God of the Old Testament, who was a believer in slavery?...Is it the God who commanded the husband to stone his wife to death because she differed from him on the subject of religion?

**Gore Vidal:**  From a barbaric Bronze Age text known as the Old Testament, three antihuman religions have evolved—Judaism, Christianity and Islam...God is the omnipotent father—hence the loathing of women for 2,000 years in those countries afflicted by the sky-god and his earthly male delegates...Those who would reject him must be converted or killed for their own good.

**Elizabeth Cady Stanton:**  The Christian Church has steadily used its influence against progress, science, the education of the masses and freedom for woman.

**Chapman Cohen:**  They [anthropologists] all agree that the conception of supernatural, or "spiritual," beings owes its beginnings to the ignorance of primitive man concerning both his own nature and the nature of the world around him.

By all logic, such writings would already have wiped out religion—if believers were permeable to logic.  Unfortunately, philosophical (logical) reasons why a god *cannot* exist merely cause the believer to say, "Yes, but he does."  And proof that the biblical god is evil is shrugged off with the equally mindless rejoinder, "mysterious ways."  Only the methodology of history (documentary analysis) can prove that a god that has revealed its existence *does not* exist, by tracing all claims to the contrary to the same authors who also assured their readers that the earth is flat.  I have encountered many persons cured of religion by the discoveries of historians (*e.g.*, that Jesus was merely the last of a long series of resurrected saviors), but none cured by the arguments of philosophy. *Atheism* will not change that.

## 4. *H. L. Mencken on Religion*
### edited by S. T. Joshi

H. L. Mencken learned the true function of religion at an early age, when his father, who was not a believer, regularly sent him to Sunday school to get him out of his hair for a couple of hours so that he could pursue other interests. (p. 11) But despite, or perhaps because of that early exposure to nonsense, "I cannot remember that even in the blackest moments of long and ghastly nights I ever had the slightest impulse to pray to God for help." (p. 32) Instead, he "sided with heretics on the great majority of issues where religion came into conflict with science, politics, and the advance of civilization." (p. 12)

Much of *H. L. Mencken on Religion* will be of little interest to modern readers, for the logical reason that it is too time-specific, and not relevant to a later age. For example, his newspaper columns on the subject of the Prohibition Amendment focused on the fanatics who orchestrated it, information of historical perhaps but not general interest. Only a small proportion of his arguments can be equally applied to the current marijuana laws.

In contrast, if his observations on religion were republished in the *Washington Post*, it would be by no means obvious that they were not written yesterday. For example, "*Homo boobiens* is a fundamentalist for the precise reason that he is uneducable." (p. 15) "Either Genesis embodies a mathematically accurate statement of what took place during the week of June 3, 4004 B. C., or Genesis is not actually the word of God. If the former alternative is accepted, then all of modern science is nonsense: if the latter, then evangelical Christianity is nonsense." (pp. 16-17) And "his observation that the United States has 'always diluted democracy with theocracy' in defiance of the First Amendment is as true now as it was in his time." (pp. 22-23)

Mencken was equally blunt in his opposition to quack medicine: "The very day that news of insulin is in the newspapers, *Homo boobiens* seeks treatment for his diabetes from a chiropractor." (p. 120) Anyone who does not know that chiropractic is dangerous pseudomedicine is referred to *The Disinformation Cycle* (Xlibris.com), pp. 155-156.

In his summation of Mencken's "Treatise on the Gods," Joshi writes (p. 14), "His anthropology is now a bit antiquated, and the work as a whole cannot stand up to intellectual scrutiny. Accordingly, I have not included any of it in the present compilation." He also does not include any of Mencken's attempts to blame Methodist and Baptist Catholic-hating churches for the resurgence of the Ku Klux Klan. But he does report the existence of such writings, and declares them "somewhat off the mark," so clearly there is no attempted cover-up of the reality that Mencken was not always right.

In reviewing a book calling for sane sex laws, based on justice instead of religion, Mencken wrote, "The first effort of the early Christians was not to regulate sex at all, in any ordinary sense, but to destroy it altogether. They had, for some obscure reason, a tremendous fear of it, and were quite willing to exterminate the human race in order to get rid of it." (p. 286) And that was written long before the accession of the most sex-hating, barely-hominid pope in Vatican history, John Paul II.

But Mencken's sharpest barbs were directed at the braindead purveyors of what today calls itself by the oxymoronic name of creation science: "The so-called religious organizations which now lead the war against the teaching of evolution are nothing more, at bottom, than conspiracies of the inferior man against his betters...Certainly it cannot have gone unnoticed...that no man of any education or other human dignity belongs to them." (p. 166) And he had no patience with what is now called political correctness: "To admit that the false has any standing in court, that it ought to be handled gently because millions of morons cherish it and thousands of quacks make their living propagating it—to admit this, as the more fatuous of the reconcilers of science and religion inevitably do, is to abandon a just cause to its enemies, cravenly and without excuse." (p. 233) "There is, in fact, nothing about religious opinions that entitles them to any more respect than other opinions get. On the contrary, they tend to be noticeably silly...They run, rather, to a peculiarly puerile and tedious kind of nonsense."

Mencken achieved his highest level of public recognition from his *Baltimore Evening Sun* columns on the trial of John Scopes for teaching evolution. And in his denunciation of William Jennings Bryan he made no attempt to pull his punches: "The net effect of Clarence Darrow's great speech yesterday seemed to be precisely the same as if he had bawled it up a rainspout in Afghanistan...During the whole time of its delivery the old mountebank, Bryan, sat tight-lipped and unmoved...He has those hillbillies locked up in his pen and knows it...Now with his political aspirations all gone to pot, he turns to them for religious consolations. They understand his peculiar imbecilities. His nonsense is their idea of sense. When he deluges them with his theological bilge they rejoice like pilgrims disporting in the river Jordan." (p. 187) "Bryan was a vulgar and common man, a cad undiluted. He was ignorant, bigoted, self-seeking, blatant and dishonest." (p. 211)

Whether Mencken's (tongue in cheek?) recommendation that black Americans could most effectively end their oppression by followers of the white god Jesus by turning Muslim (p. 13), is best viewed as a first cause of the hate cult calling itself Nation of Islam, or merely a forecast of an inevitability, I will not attempt to answer.

On the double standard that existed then, and continues to be imposed by the religious nutcase in the White House and the *Pithecanthropus troglodytus* theofascists who lick his butt in order to pull his puppet strings, Mencken wrote in 1925, "A preacher of any sect that admits the literal authenticity of Genesis is free to gather a crowd at any time and talk all he wants...But the instant a speaker utters a word against divine revelation he begins to disturb the peace and is liable to immediate arrest." (p. 191)

So what else is new?

# 5. *Confirmation: The Hard Evidence of Aliens Among Us*
# by Whitley Strieber

The bibliography of Whitley Strieber's *Confirmation* lists books by Bud Hopkins, John Mack and J. Allen Hynek, but none by Phillip Klass, Robert Sheaffer or Kendrick Frazier. That, and his endorsement of the Roswell myth, the hypnotism delusion and the alien-implant hoax, stretches the defense that Strieber is "sincere but delusional" to the breaking point.

In his opening chapter Strieber writes, "This is not a book of proof of alien presence." That may be the only line Strieber has ever written that can be accepted as unvarnished truth. And when he writes, "I wished often that I could return to being a…fiction writer," the obvious rejoinder is, "When was he ever anything else?"

Fictional detective Sherlock Holmes stated that, when you have eliminated the impossible, whatever is left, no matter how improbable, must be the truth. When Strieber wrote *Communion*, the book's basic coherence eliminated as impossible the diagnosis that the author was insane. The Holmes dictum left only one alternative: Strieber is a conscious, blatant liar. Initially it seemed possible that *Communion* was a prank, and when it had run its course Strieber planned to admit that it was his version of James Randi's "Project Alpha." It is still conceivable that that was his original intent. But *Communion* made him so much money that would have to be refunded if he admitted it was a hoax, that he wrote a series of sequels aimed as wresting as much money as possible from gullible ignoramuses before the well ran dry.

Perhaps Strieber was discomforted that so many people denounced *Communion* as an unmitigated fraud. *Confirmation* is a collection of anecdotes that he reports in a manner calculated to minimize accusations that he, rather than his sources, is lying. He expresses credulity in absurdities, but he does so in a way that will enable him to pass himself off as the hoaxee rather than the hoaxer if some of his reported cases are exposed as fantasies. Only in the chapters about his own alleged experiences can the possibility that he really believes what he is saying be dismissed as the least likely hypothesis. For like the new generation of spiritualists, Strieber panders to believers while carefully avoiding making statements that can be readily falsified.

While *Confirmation* contains no clear evidence that Strieber is developing a conscience, it does raise the possibility that he is preparing for the day when he can announce that he has "lost faith" (or something analogous) in the little green men, in hope of restoring his credibility. If that does not work, perhaps his next move will be the Charles Colson solution: "Sure I burgled the Watergate—but that was before I found God."

*Confirmation* all but eliminates any possibility that Strieber is merely a crank, no more capable of recognizing that he is peddling hogwash than John Mack, Immanuel Velikovsky, Shirley MacLaine, or Dan Aykroyd. Reasonable readers will conclude that he is simply a liar. Case closed.

# 6. *Three books about UFOs*

"If a genuine extraterrestrial spacecraft were to land on the White House lawn, and its occupants were to say that they have had planet Earth under surveillance for hundreds of years and that theirs was the very first extraterrestrial visit, stubborn UFOlogists would not believe them. The ETs would be accused of being part of the government's coverup." That conclusion by Philip Klass in *Crashed Saucer Coverup* (p. 219) summarizes the difficulty the rational face when they offer mere facts to persons whose minds are made up.

As promised on the cover blurb, Klass shows that there has indeed been a coverup—by the perpetrators of the fiction that the "Roswell incident" has any more reality than the Bermuda Triangle, Von Däniken's space gods, or the Great Pumpkin. Klass establishes, not only that all pro-UFO books reach indefensible conclusions, but that they do so intentionally, by suppressing evidence that refutes and annihilates those conclusions.

In citing the selective editing of an Air Force document by the authors of the book that essentially invented the "Roswell incident," Klass writes (p. 29), "This omission of vitally important 'hard data' is understandable only if (Charles) Berlitz and (William) Moore were intentionally trying to cover up information that could demolish the credibility of their crashed-saucer book." And he points out that every later book endorsing the reality of a crashed saucer also suppressed the same evidence.

Klass's chapter on "The CIA's once-secret UFO Papers" concludes (p. 57), "They provide incontrovertible evidence that if one or more crashed saucers and ET bodies were recovered from New Mexico in 1947, word of this historic event had not been reported to top CIA officials or President Truman as of 1952—some five years later." He reports that, when he was interviewed about Roswell by CBS's *48 Hours*, NBC's *Unsolved Mysteries*, CNN's Larry King, and the producer of a documentary for A & E, he gave all of those interviewers a photocopy of the formerly secret Air Force Intelligence Report #203, showing that as late as 1948 the Air Force believed that, if UFOs had any factual basis at all, they were probably "from a Soviet source." Not a single reference to that document, or Klass's explanation of its implications, was broadcast by any of the alleged "news" peddlers. (p. 207)

And nothing has changed. More than a year after the publication of Klass's book, NBC broadcast a two-hour endorsement of Whitley Strieber's fantasy novel, *Communion,* treating it as nonfiction, during the 1999 February sweeps period.

A point that a lot of readers are likely to overlook, since Klass does not make an issue of it, is that one of the most totally discredited "eyewitnesses" to alleged alien bodies, Gerald Anderson, easily passed a polygraph test. The pretence that polygraphs are "lie detectors" that work better than tossing a coin, "heads it's the truth and tails it's a lie," is as recklessly perpetuated by the media as the pretence that a crashed spy balloon was a flying saucer.

*The UFO Invasion* is a collection of articles that first appeared in *Skeptical Inquirer* between 1984 and 1996, updated where necessary by the original authors, plus one previously unpublished article by Robert Baker.

In the chapters by Philip Klass on William Moore's "MJ 12" hoax, Klass demolishes any pretence that Moore is merely an incompetent investigator who sees little green men under every bed. While Klass leaves it to the reader to draw such a conclusion, the evidence leaves little doubt that Moore is a blatant, conscious liar with a depraved indifference to anything but personal profit.

On Whitley Strieber's best-selling fiction, Robert Baker (p. 221) is probably being too charitable when he categorizes Strieber's alleged experiences as hypnopompic hallucinations that Strieber sincerely mistakes for reality, rather than simple profit-motivated lies. Certainly Strieber's pages parallel hypnopompic incidents. But they also parallel the science fiction fantasies that he was writing long before turning to pretended nonfiction. But I agree with Baker's conclusion that, while Budd Hopkins (*Intruders*) manipulated his "hypnotized" interviewees into saying what Hopkins wanted to hear, he probably did not do so intentionally.

On the Fox network's "alien autopsy" hoax, Eugene Emery summarizes (p. 145), "It was not what you would expect from a major network that thought it was broadcasting a history-making film. It was, however, what you would expect from a network trying very hard not to spoil an illusion." In other words, Fox was not deceived by the hoax. Fox was a conscious co-hoaxer.

While the book's various contributors are correct in concentrating on actual UFO reports and investigating whether there is a discernible mundane explanation, they seem to be missing a bet in not stressing the impossibility of lifeforms evolved from extraterrestrial DNA resembling humans in *Star Trek* makeup. As Isaac Asimov observed (*Life and Time*, p. 15), "Considering in how many different ways life developed on earth...it would be an almost impossible chance to have a species there closely resemble some species here." Only Baker raises such a point (p. 219): "The fantasy-prone abductees' stories would be much more credible if some of them at least reported the aliens as eight-foot-tall, red-striped octopoids riding bicycles and intent on eating us for desert." And he recognizes (p. 259), "If they appear to be humanoid this proves they are imaginary." And on page 263: "Like his gods, man also creates his aliens in his own image."

Other authors did raise some logical objections: the astronomical distances between star systems, the prohibitive amount of fuel (fifty tons for every ton of payload) needed to accelerate even to ten percent of light speed, and the certainty of a sudden, ninety-degree turn by a spaceship turning any protoplasmic occupants into "grape jelly." But they acknowledge the futility of raising issues that believers can shrug off with the delusion that the laws of physics are not necessarily the same elsewhere in the universe. The parallel with religion, which postulates metaphysical beings not bound by the laws of nature, is very evident.

I was on the point of dismissing the chapter by Armando Simon on "Psychology and UFOs" as an indefensible inclusion written by a supergull who endorses the reality of psychoanalysis, MMPI, and psychoquackery in general. But when Simon quoted Thomas Szasz's definition of hypnotism (p. 48), "two people lying to each other, each pretending to believe both his own and his partner's lies," I concluded that his article was worth printing after all. Even so, the suggestion that the mental state of alleged abductees can be better evaluated by psychiatrists than by bartenders, taxi drivers, call girls or tealeaf readers is not one I accept.

Of the three listed books, Robert Sheaffer's *UFO Sightings* is the most readable, for the logical reason that it covers a wider canvas than Klass, and is written as a continuous narrative by a single author, as the Frazier book is not.

Sheaffer covers the "Roswell incident" as concisely as possible, given that he was not trying to compete with the definitive book on the subject published by Klass a year earlier. Given the way Roswell has been touted as UFOdom's equivalent of goddess-mythology's Lourdes, I was surprised to learn that, as late as 1966, when Project Bluebook asked the two largest UFO-promoting groups to supply a list of their most impressive cases, Roswell was not listed by either organization as worthy of investigation. I was also surprised to learn the extent to which UFO mythology has been inflated by John Fuller and Charles Berlitz, two men whose ability to create a mystery where none exists gave the world *The Ghost of Flight 401* and *The Bermuda Triangle*. I was under the impression that the "Roswell incident" was front and center long before 1980, when Berlitz and Moore resurrected what had previously been a dead issue. Not so.

UFOs first hit the headlines in 1947, when pilot Kenneth Arnold reported seeing some unidentified flying objects. As Sheaffer notes (p. 15), "Arnold didn't say the objects *looked* like saucers—actually he said that they looked like boomerangs...that flew erratic, like a saucer if you skip it across the water." But because a reporter misquoted Arnold's words, thousands of people started reporting "flying saucers" rather than "flying boomerangs." As Sheaffer observed, "Seldom had the power of suggestion been so convincingly demonstrated." I find myself thinking that, if *Star Trek* aliens for the past thirty years had all resembled canines, abductees would have reported being taken aboard the mother kennel.

Sheaffer notes that (pp. 81-82), "The descriptions of supposed UFO aliens contain clear cultural dependencies; in North America large-headed gray aliens predominate, while in Britain aliens have been mostly tall, blond, and Nordic...the Galactic High Command must have divided the earth into Alien Occupation Zones whose boundaries reflect those of human culture."

On the absence of plausible UFO photographs, Sheaffer writes (p. 103), "The only reasonable explanation for such a curious state of affairs is that there are no genuine UFOs to be photographed."

Sheaffer's chapter on "UFOs and the Media" shows that, while the Fox network's alien autopsy hoax was the most blatant fraud ever perpetrated in the name of news and public affairs, it differed only quantitatively from the incompetent investigation and deliberate suppression of demystifying evidence by every element of the mass media. Sheaffer identifies callous disregard for fairness and accuracy in stories by the *Washington Post, New York Times, Baltimore Sun,* CNBC, NBC, CBS, *Science Digest* (a misnomer) and *National Inquirer.* He summarizes (p. 187), "This incident [refusing to report when sensational claims in previous stories are falsified] reveals no real difference between 'respected' newspapers and the *National Inquirer* when it comes to UFO reporting." And on page 193: "Apparently they get away with such sloppiness because no journalistic peer pressure develops to push them toward more accurate reporting."

The moral bankruptcy of UFO reporting is best illustrated by the Travis Walton case. When *National Inquirer* first involved itself with Walton, it arranged for him to take a polygraph test. That was a safe procedure, since a finding of "truth" could be front-paged, whereas the actual finding of "gross deception" (p. 38) could be, and was, suppressed. Later, since two polygraph tests, especially when administered by different persons, are no more likely to produce identical results than tossing two coins will produce two heads, Walton was tested again and declared to have passed. That result was widely publicized, even though "the less-experienced polygraph examiner who had passed Walton was repudiated by his employer, who disagreed with the interpretation of the test results." That information has been public domain since 1976, yet UFO apologists continue to cite the Walton case as evidence that UFO abductions are real. Paramount made a movie about it in 1993, and that movie has been broadcast more than once on a major TV network with no warning that the "based on a true story" subtitle is a deliberate lie.

I fully expected that these books would provide details I had not previously encountered. What I did not anticipate is that they would change my overview. For example, I had long believed that the prime manipulators of public opinion in the direction of gullibility were sincere investigators such as J. Allen Hynek, who had either never heard of Occam's razor or deemed it applicable only to other people's security beliefs. On the basis of the new (to me) evidence presented here, I am now satisfied that UFOlogy would still be widely recognized as crackpotism but for the machinations of two of the most conscienceless liars to perpetrate a profit-motivated swindle since science fiction writer L. Ron Hubbard invented a new religion (that has outlived him) to win a bet: William Moore and Charles Berlitz. Their out-of-context quotations, suppression of falsifying evidence, often from the same documents they selectively quoted, and (in Moore's case) deliberate forgery of alleged government documents, show them to be as intentionally fraudulent as the creator of the "Hitler diaries."

It seems no coincidence that all three books have titles calculated to win them shelf space in book stores that would shy away from openly skeptical titles. Since Prometheus cannot be expecting to sell copies to hardcore believers, their strategy is presumably to get the books before the public, where they will be seen by persons with more discrimination than the nonsense-peddlers believe exists. It might even work. On the other hand, reading the overwhelming evidence against an alien presence presented here will no more cure the dogmatic ignorance of true believers than the equally overwhelming evidence could cure believers in gods, fairies, hairy hominids, angels or psychics.

In a society manipulated by television into believing in every kind of imaginary creature capable of keeping viewers glued to the disinformation tube and enriching the prostitutes who produce saleable lies, books like these are the only source of accurate information most people will ever encounter. I am recommending that my local public library buy all three **[in fact it purchased the Sheaffer book, but not the other two]**, and I urge *American Rationalist* readers **[and readers of this book]** to make the same recommendation to libraries in their areas. The inane fantasies of Strieber, Hopkins and Mack cannot be restricted to the fiction shelves where they belong, but it least they can be balanced.

# 7. *Alien Abductions: Creating a Modern Phenomenon*
## by Terry Matheson

Alien abduction claims made in the books of John Fuller, Whitley Strieber, John Mack and others are a violation of the laws of reality. So when Terry Matheson, after rebutting them, tries to avoid calling the fantasizers liars by offering an alternative explanation that is itself a violation of the laws of reality, I find myself wondering whether to laugh or cry. Matheson is a literature professor. But his belief in such science fiction concepts as hypnotism, psychoanalysis, Jung's "collective unconscious," repressed memories, mental illness as something other than social inadequacy, and an unconscious mind capable of hiding traumatic memories by converting them into less frightening "screen memories," is as complete as any psychoquack's.

It is the "screen memories" psychobabble theory that comes within an inch of invalidating Matheson's whole book. According to the theory, the mind, independently of the will, is able to hide memories of traumatic events such as childhood sexual abuse, by transubstantiating them into less frightening memories such as alien abductions. Matheson actually believes that this may have happened. And Santa Claus comes down the chimney on Mithra's birthday.

In an effort to appear non-judgmental, Matheson writes (p. 33), "I do not, of course, mean to accuse the majority of authors of deliberately lying or cynically attempting to perpetrate hoaxes to mislead their readers for nefarious purposes." But that statement refers to authors Matheson's own analysis has demonstrated *are* deliberately lying for nefarious purposes. One sees the same kind of political correctness in Protestant writers who debunk Catholic beliefs about Bernadette Soubirous while leaning over backward to avoid saying that Soubirous was simply a mentally inept compulsive liar.

The closest Matheson comes to deriding abduction pushers in plain language, is when he states (p. 153), "At times it is hard for the reader not to see [Bud] Hopkins as simply gullible," describes Raymond Fowler (p. 209) as, "looking not only extremely gullible but intellectually flighty (if not unstable) as well," and (p. 107) compares two books that "provide us with excellent examples of what happens when respect for the facts is either present (as it is for the most part in Ann Druffel and D. Scott Rogo) or when it is not (in the case of Travis Walton)." Yet he says of Whitley Strieber (pp. 167-168), "He does not pursue the possibility that all his experiences might be similarly amenable to a psychological explanation," even though the same evidence that led him to such a conclusion will satisfy most readers that Strieber is a calculating, unmitigated liar.

Matheson offers a logical explanation for why later abduction claims have been believed when earlier ones were not (p. 37): "If a story appears coherent, seems to have a logical sequence to its events, contains certain recognizable consistencies, and is endorsed by authorities, it tends to be believed whether it is true or not."

While Matheson is able to recognize that extraordinary claims require extraordinary evidence, and that abduction claims do not meet that criterion, his own ability to evaluate evidence is extremely suspect. For example, he suggests that Betty and Barney Hill deluded themselves that they had been abducted by aliens (p. 55) because their real experience, perhaps with hoodlums, "was too terrifying and humiliating to be handled straightforwardly and was suppressed by their conscious minds." Apparently he is unaware that "suppressed memory" is as much a psychobabble concept as multiple personality.

Matheson acknowledges (p. 60) that "hypnosis has little if any intrinsic credibility as an investigative tool." But his references to the hypnotizing of alleged abductees throughout the book leave little doubt that he believes hypnotism actually exists. Apparently he is unaware that Thomas Szasz was summarizing the findings of every competent researcher in the field when he defined hypnotism as, "two people lying to each other, each pretending to believe his own and his partner's lies."

Matheson gives the definitive and most detailed debunking of John Fuller's *Interrupted Journey,* about the alleged abduction of the Hills. Nothing comparable has been written in many years, possibly because even hardcore UFO cults have long recognized the Hills as an embarrassment and Fuller (author of *The Ghost of Flight 401*) as a fantasy writer who would pass off *Alice in Wonderland* as nonfiction if he thought there were enough gullible believers to make it economically feasible. But Matheson undermines his success when he implies that he alone is objective, and that the leading researcher in the field, Philip Klass, shows a degree of bias that, although opposite, is nonetheless equal to Fuller's. And his suggestion that the aliens who allegedly abducted the Hills were "really" human thugs, reminds me of nothing so much as Michel Gauquelin's attempt to refute traditional astrology by inventing a new astrology. The Hills were manipulated by a "hypnotist" into concocting a fantasy based on Betty Hill's dreams. End of story.

Matheson is far too charitable to John Mack. He does say that a validation of Mack's methodology "would essentially force us to believe in the literal reality of demon possession." (p. 260) But he mentions only in passing Mack's acceptance of the plausibility of such concepts, not found in most previous pro-abduction books, as time travel, foreknowledge of the future, reincarnation, and guardian aliens who protect a human soul through several incarnations. He ignores the even more bizarre fantasies of encounters with ancient gods and a flying horse, and offers no hint that a man who can take such claims seriously belongs in a pulpit (or perhaps a straight jacket), not a university. And he also ignores the totality with which Mack's fatuous conceit that he can tell if a patient is lying was annihilated by a woman who claimed to be an abductee in order to test whether Mack used any methodology to guard against being hoaxed. Mack's claim, after being exposed, that the woman who had fooled him really was an abductee who was now in denial (another psychobabble concept), says more about his estrangement from reality than Matheson's whole chapter.

While Matheson correctly (for once) interprets Betty Andreasson's story of her tour of the aliens' planet as a fantasy based on her religious conditioning, he makes no mention of the absurdity of a round trip to another star system taking place within the space of one evening. Does he think that Einstein's equation showing faster-than-light travel to be impossible might be wrong?

Given how convincingly Matheson's chapter on Andreasson demonstrates that she is a fruitcake first class, I strongly urge him to make a study of the sainted Bernadette—assuming that he can access and read the French documents cited in the imprimatured biographies.

*Alien Abductions* contains an enormous amount of convincing arguments against the validity of abduction claims that is not found in the books of Philip Klass, Robert Sheaffer or Kendrick Frazier, for the logical reason that the earlier writers, having falsified the fantasies of Fuller, Mack and the others to the satisfaction of all but incurable believers, would have seen the material in Matheson's book as overkill. That is a legitimate position to take. But Matheson is also justified in publishing evidence that, in itself, proves the same point in even greater detail. While his findings are distorted by his belief in hypnotism and other dubious concepts, they are nonetheless worth reading, provided the reader can get past Matheson's psychobabble interpretation of Fantasy A as the (nonexistent) unconscious mind's bowdlerization of Fantasy B.

# 8. *Bad Astronomy*
## by Philip Plait

"Movies show space travel all the time, but they show it incorrectly, and so it doesn't surprise me that the majority of the viewing public has the wrong impression about how it really works...The news media's job is to report the facts clearly, with as much accuracy as possible. Unfortunately, this isn't always the case...I remember vividly watching the *Today* show on NBC in 1994. The Space Shuttle was in orbit, and it was doing an experiment...Anchor Matt Lauer was reporting on this experiment, and...admitted he didn't understand what he had just said. Think about that for a moment: three of America's most famous journalists, and they actually *laughed at their own ignorance in science*! The report itself was accurate...but what the public saw was three respected journalists saying tacitly that it's okay to be ignorant about science. It *isn't* okay. In fact it's *dangerous* to be ignorant about science. Our lives and our livelihoods depend on it." (pp. 2-4)

After that promising start, Plait's next hundred pages drop to a level appropriate to junior high school in any country in which teaching is still legal, or first-year university in North America. The information dispensed is trivial, but readable and useful—to the degree that one finds it useful to know that sinks and toilets do *not* drain clockwise in one hemisphere and counter-clockwise in the other, and that anything that can be done to an egg on the vernal equinox, including balancing it on one end, can be accomplished just as easily any other day of the year. Persons whose knowledge of reality is more than minimal are urged to skip (or skim) the first thirteen chapters, and jump directly to "The Disaster that Wasn't," Plait's definitive debunking of the nonexistent "Jupiter Effect," that had cranks, humbugs and sincere fantasizers claiming that a planetary alignment could cause the destruction of planet earth. From that point on, this is a really good book.

For example (p. 153): "People believe weird things. There are people who believe the earth is 6,000 years old. Some people believe that others can talk to the dead, that a horoscope can accurately guide your day, and that aliens are abducting as many as 800,000 people a year. I believe weird things, too. I believe that a star can collapse, disappearing from the universe altogether...So what's the difference? Why do I think it's wrong to believe that the earth is young when I believe in things I've never seen? It's because I have *evidence* for my beliefs. I can point to well-documented, rational, reproducible observations and experiments that bolster my confidence in my conclusions."

And on page 154: "Odds are that you believe NASA sent men to the moon. So why devote a whole chapter to the minority that doesn't? There are several reasons. The most important is to simply provide a rational and reasoned voice when such a voice is hard to find." That may not seem like much of a reason to anyone unfamiliar with Emmanuel Velikovsky or Roman Piso. But scholars' refusal to dignify those gentlemen's mushroom fantasies with rebuttals (until they did) led people to wonder if perhaps they could not be rebutted. I have never met anyone who thinks the moon landing was a hoax. But the theory has been raised, and without a Plait to demolish it, it could conceivably start to be taken seriously. That now seems less likely than it did before.

In his otherwise excellent chapter on Velikovsky, Plait states (p. 182), "The Hebrew calendar, still going strong after 5,800 years..." The idea that the Hebrew calendar, which indeed dates Creation about 5800 years ago, existed any earlier than the oldest biblical writings, is as "utterly and obviously wrong" as the nonsense Plait is rebutting. Fortunately, that one credulous paragraph does not weaken the chapter's effectiveness.

After berating Hollywood's bad astronomy for placing profits and ratings ahead of accuracy (Gene Roddenberry's desire to have the *Enterprise* move silently through the vacuum of space, instead of "whooshing" as it would do in an atmosphere, was vetoed by the network), Plait writes (p. 257), "Do I really hate Hollywood movies? *Armageddon* notwithstanding, no I don't. I like science fiction. I still see every sci-fi movie that comes out...If movies spark an interest in science in some kid somewhere that's wonderful. Even a bad movie might make a kid stop and look at a science book in a library, or want to read more about lasers, or asteroids, or the real possibility of alien life. Who knows where that might lead?"

No argument from me.

## 9. *The UFO Mystery Solved*
## by Steuart Campbell

"Do UFOs exist?  If they do not exist, why do people report them.  Can we believe the reports; indeed can we believe anything?  Is human perception reliable, or is it influenced by what we believe?  Could the UFO myth determine what we see and report?  If UFO reports are distorted by the UFO myth, what is the initial stimulus?  In short, what have people really seen?"

Steuart Campbell attempts to answer those questions.  He devotes 11 pages to Kenneth Arnold's 1947 report of nine aerial objects that flew through the sky, moving like a saucer skimming across the water, which started the craze.  He concludes that what Arnold saw was a mirage image of nine mountaintops that appeared to be moving because Arnold was moving.  What he does not mention is that Arnold continued to report seeing UFOs almost monthly for forty years.

Campbell similarly analyzes seven other reports, and likewise explains them in far more detail than a casual reader will ever want to know.  He summarizes his findings as follows:

"Ufoists have made the mistake of attempting to create a science of 'ufology,' claiming that they are studying a new phenomenon.  However, just because scientists cannot explain every report does not mean that a UFO phenomenon exists.  Indeed, I have demonstrated that even the most apparently intractable reports can be explained by existing science.  Ufoists have also made the mistake of calling on scientists to study 'the UFO phenomenon.'  Since there is no phenomenon (other than the UFO reports themselves), it cannot be studied by scientists or anyone else.  Ufoists are pursuing an imaginary phenomenon."

The definitive debunking of the UFO delusion can be found in the books of Philip J. Klass and Robert Sheaffer.  I found them to be informative, entertaining, and convincing.  For all I was able to get out of *The UFO Mystery Solved*, it might as well have been written in Etruscan.

## 10. *The Loch Ness Monster: The Evidence*
## by Steuart Campbell

Of Steuart Campbell's three books, *The Loch Ness Monster* is the only one that has also been published in America (by Prometheus), even though it is one of 130 titles carried by Barnes and Noble on the same subject. It is a detailed evaluation of every photographically supported sighting of the loch's alleged denizen since 1933. Campbell offers the most plausible explanation of each, and concludes that none survives close inspection. While he lists the ancient mythology that has been retroactively postulated as evidence for Nessie's existence, he finds that, "We see therefore that there is no reliable ancient tradition for N[essie], whose origin lies in superstition and confused zoology." (p. 2) In other words, Nessie was created by the person who allegedly first saw it in 1933, and attempts to backdate it have been desperate ploys to reinterpret older fairy tales as early sightings.

If Nessie exists, then its ancestor must have entered Loch Ness at some time in the past, when there was a tunnel from the loch to the open sea. But (p. 5), "Since L[och] Ness is 16 m above sea level, any tunnel large enough to take N would drain the lake down to sea level. There is no tunnel...It has not yet been established that L Ness was ever open to the sea, and the likelihood is that it was not." And for the creature to have survived for centuries, there would have to be a breeding herd of at least twenty individuals. But Loch Ness is too cold to support any species of reptile/dinosaur (p. 98), quite apart from the impossibility of the loch being able to feed such a number of large lifeforms.

Campbell shows that all alleged positive results of sonar, radar and photographic imaging, on close inspection, in fact prove to be negative. Not only does the loch not have the capacity to support a herd of monsters; no legitimate evidence exists that it does contain them. And given the thousands of man-hours devoted to loch watching by serious searchers, photographers and tourists, the logical conclusion is that, if Nessie existed, someone would have proven it by now. Instead, (p. 99), "the more L Ness is watched the less N shows itself. This is what would be expected if N does not exist...The skilled observer sees what the unskilled sees, but knows that it is not N and so does not report it."

Campbell concludes (p. 100), "In my view there is absolutely no reason why anyone should believe in the existence of lake-monsters." If anyone doubts that conclusion on the basis of any specific evidentiary claim, the chances are it is one of the dozens of claims that the book examines and demolishes.

## 11. *The Rise and Fall of Jesus*
## by Steuart Campbell

Steuart Campbell maintains that the person best qualified to write a definitive biography of Jesus is an ex-Christian. Certainly a Christian cannot do so, for if he had any ability to reach conclusions compatible with the evidence, he could not remain a Christian. I have met Christians who have actually read the Bible, yet continue to regard Jesus as a nice guy. That is like reading *Mein Kampf* and continuing to regard Hitler as a nice guy. And someone who has never been Christian would have little interest in obtaining information about other people's weird beliefs.

I can testify to that. I have no interest whatsoever in writing a rebuttal of Mormonism or Islam, since to do so would require me to learn far more about those absurdities than I ever want to know. I write about the beginnings of Christianity because I was a believer, and the first time I encountered the falsifying evidence, I conducted a desperate search for rebuttal evidence (there isn't any). Campbell is likewise a former Christian, and as such started with sufficient knowledge and interest in the Jesus of Churchianity to find further study rewarding, even when it led to the opposite conclusion to what he had hoped to find.

But mere willingness to go with the evidence is no guarantee of reaching the right conclusions. Campbell writes of Albert Schweitzer (p. 192), "Contrary to the current view that little in the Gospels is historical, Schweitzer regarded it as a miracle that so true a record of Jesus has been preserved." But he then states that Schweitzer "stands above all others who have attempted to make sense of the life of Jesus." (p. 193) He adds that his own reconstruction is primarily "based on Schweitzer." (p.14) But whereas Schweitzer foreshadowed *The Passover Plot* in concluding that Jesus conspired to have himself crucified, Campbell proposes that "Jesus rose by his own efforts and...planned to continue rising. His fall was not expected and was accidental...His remains lie buried in the earth." (p. 15) In other words, he is dead but Christians refuse to bury him, even though "Not a single word of Jesus is of any relevance today." (p. 17)

Campbell fills a 22-page chapter with arguments for Jesus' historicity, compared to my three pages in *A Humanist in the Bible Belt* (1stbooks.com, 2003). While his chain of reasoning is generally valid, some of it is suspect, and certainly insufficiently convincing to change many minds. He does use the most logical argument, that too much of Jesus' authorized biography is negative for it to be something an admirer would have invented. But he ignores the testimony (and in fact disagrees with it) that Jesus was an ugly, deformed man, a description that could only have originated in a non-Christian source, even though six centuries of Christian apologists conceded the point. Nonetheless, it is a chapter all purveyors of the "no such person ever existed" school should be required to read.

Campbell's chapter on Jesus' birth pretty much parallels my own account in *Mythology's Last Gods*, right down to identifying his birthplace as Capernaum. And practically all scholars, both upholders of a historical Jesus and deniers, have similarly traced the various birth myths to their pre-Christian sources. Campbell takes several paragraphs to debunk attempts to "explain" one of Matthew's borrowings from Hinduism, summarizing, (pp. 48-49) "Astronomy has ignorantly attempted to explain a problem which does not exist. There never was a Star of Bethlehem." He does not bother mentioning that the gospel author, writing a full century after Jesus' birth, knew that he could safely make such a preposterous claim, because nobody knew the precise year that Jesus was born and therefore could not refute the fantasy that it coincided with a spectacular astronomical event.

On many issues, although not any involving absurdities, Campbell is far too credulous. For example, he appears to believe that the tax collector Matthew, one of six names invented by the anonymous author of Mark to round out his mythical Twelve Apostles, was a real person. And he thinks that a disciple named John wrote the equally anonymous fourth gospel. Clearly he does not date the fourth gospel to the decade 130-138 CE, when the Christians were again trying to dissociate themselves from the rebellious Jews, as I do. He also accepts the physician Luke as author of the gospel credited to him, even though Acts contains mistakes no companion of Paul could have made, and the common authorship of Luke and Acts is reasonably established.

In accepting non-theological scenarios in the fourth gospel as factually based, such as the Beloved Disciple's adoption of Jesus' mother, his identity as a disciple named John, and Jesus appearing to turn water into wine, Campbell goes way beyond naïve. He writes (p. 116) that, "the only scholar to come close to the truth...suggested that some liquor was poured into the water to give it the taste of wine." And Santa Claus comes down the chimney on Mithra's birthday. The fourth gospel author transformed Jesus the king into Jesus the god, and backed up that transformation by having Jesus turn water into wine, as the god Dionysos had done at his wedding to Ariadne. Similarly, Campbell's attempt to explain Judas's alleged betrayal as something other than the "Mark" author's desperate attempt to convince Vespasian that the sicarius among Jesus' lieutenants was "really" his enemy and ultimately betrayed him, can be compared with the harmonizers' attempts to equate the Star of Bethlehem with a real-world phenomenon.

Campbell writes (p. 59), "The epistles of Peter indicate that their author [actually authors] was familiar with the Septuagint. Would Jesus be more ignorant of Greek than his disciples? Jesus was literate, perhaps more so than most of his contemporaries. It can have been little trouble for him to learn Koine [pidgin Greek]...When Jesus spoke to Gentiles, it can only have been in Koine...The evidence is that Jesus was as fluent in Greek as in Aramaic." I disagree on all counts (*MLG* p. 262) Campbell also attaches unwarranted credence to the hypothesis that Jesus was not the legitimate son of Mary's husband, even though such an allegation was first made a full seventy years after Jesus' death, by rabbi ben Azzai, as a reaction to the virgin-birth interpolation in Matthew.

By the time of Paul of Tarsus, the neo-Essene sect in Jerusalem was known as *Nazoraios*, which Campbell translates as "Nazarenes" rather than "Nazirites," an equally valid rendering if his theory of the name's origin is correct. He postulates that the name preceded Jesus, that the Nazarenes were a splinter sect led by John the Immerser, whose successor was his cousin Jesus, and Jesus' successor was his brother Jacob. My view is that the sect's name was derived from "the Nazirite," the title bestowed on Jesus by his detractors as a mark of scorn, after it changed messiahs from Jesus the Essene, executed 104 BCE, to Jesus the Nazirite, executed 30 CE (Campbell says 33 CE); that Jacob was already Head Essene before Jesus' became a public figure (hence his title, "the Righteous," carried by all successors of the original Righteous Rabbi), and that he went along with the change of messiahs as a consequence of Peter's preaching. John was an opposition messiah, whose sect, called "Immersers," Jesus joined before developing the belief that *he* was Messiah. He was certainly not Jesus' relative.

Despite (1) the scene in Mark 10:21 in which Jesus loses a potential convert when he rejects Jesus' demand that he, "Go and sell whatever you have and give it to the Paupers"; (2) the passages in Josephus's Jewish War (2:8 ff.) in which he explains that Essene communes operated on the rule that "Each man's possessions go into the pool, and their entire property belongs to them all"; (3) the scene in Acts 4:32-34 in which, "Everything they had became communal property, for all who owned land or houses sold them and brought the proceeds"; and (4) the existence of a communistic sect of Jesus-Jews as late as the fourth century called "Paupers" (*ebionim*); Campbell gives no consideration to the possibility that "the Paupers" to whom Jesus' converts were required to donate their property, and the fourth century Paupers, were the same Paupers, *i.e.*, the sect Jesus actually founded.

The foregoing are only a few of the problems I have with this book. Like Burton Mack, Campbell constructs a scenario that can be neither falsified nor taken seriously. This is one more reconstruction of Christian origins, neither more nor less plausible than the dozens of others by scholars whose competence is not in dispute, and certainly not as far out as *The Passover Plot* or the imbecilic *The Sacred Mushroom and the Cross*. Whether this or any other Jesus biography is more credible than the last two chapters of *Mythology's Last Gods*, the reader will have to decide for himself.

## 12. *Joshua, The Man They Called Jesus*
## by Ian Jones

The back cover of *Joshua, The Man They Called Jesus*, quotes glowing reviews by an Anglican bishop and a rabbi. That was an early warning that Jones's book would be a whitewash of a prototype Khomeini, sufficiently politically correct not to offend the terminally ignorant. I was being too optimistic. Jones's gullible, unlearned evaluation of five (including the Gospel of Thomas) fantasy novels has produced the closest approximation to "Jesus in the Twilight Zone" since *The Passover Plot*. His speculations are indefensible, his logic is flawed, and his facts are plain wrong.

Jones does start with an interesting anecdote. When it occurred to him to write a biography of Jesus, he asked a chaplain where he could find relevant documents. The chaplain answered that the Christian gospels would tell him everything he needed to know. So he asked, "Imagine I told you about three books that said Ned Kelly [an executed psychopathic murderer who has become an Australian folk hero] was the finest man who ever lived, that he never did anything bad, that he wasn't *really* a criminal and that these books were written by Dan Kelly, Joe Byrne and Steve Hart. You'd say, 'But these were the members of Ned Kelly's gang! Of course they'd say that!'" The point Jones was making was that (as he believed), the gospels were written by members of Jesus' gang. In fact the gospels were written by persons who were either unborn or children when Jesus died. But it still makes a good argument against fundamentalists who believe the gospels were written by eyewitnesses.

Jones writes (pp. 103-104), "Frustratingly, neither Matthew, Mark nor Luke tells of the first miracle [turning water into wine], that first occasion when Joshua felt empowered or even impelled to become an instrument of God's might...Only John shows Jesus performing a first miracle." By themselves, those sentences cannot be interpreted as evidence that Jones believes Jesus really did perform miracles through the power of an imaginary playmate, as opposed to curing psychosomatic ailments by the technique alternatively called "faithhealing" and "hypnotism." But when he writes (p. 108), "God does not break his own laws, though he can certainly use them to stretch the bounds of the possible *towards* the impossible," and on page 277, "For all human beings they [the words, "Our Father"] represent the simple key to a relationship with God," the clear implication is that either Jones believes that the serial killer in the sky exists, or he wants his Christian readers to infer that he has such a belief.

Jones quotes the King James translation of the *Shema* (Deuteronomy 6:4), "Hear, O Israel, the Lord our god, the Lord is one," which raises the question: What on earth is that supposed to mean? The more accurate recent Jewish *Tanakh* renders the passage, "The Lord is our God, the Lord alone," a clear message to the Jews of the Deuteronomist's time that they were to ignore all of the other gods with whom Yahweh shared the flat earth and hemispherical skydome as completely as if they did not exist. But even that is a deliberate falsification constructed to suppress the reality that the authors of Western religion's sacred books were not monotheists. The only correct translation of the *Shema* is to be found in *The Judaeo-Christian Bible Fully Translated* (Imprintbooks.com, 2002): "Listen, Yisrael. Yahweh is our gods [generic plural], Yahweh alone."

Even Jones's "correction" of *Jesus* to *Joshua* is wrong. "Joshua" is a clumsy transcription of the Hebrew name, *Yahuwshua*, derived from the god-name, *Yahuw*.

Jones correctly recognizes that the fable of Judas's betrayal is a vicious libel of one of Jesus' most faithful lieutenants. But in trying to explain the origin of the betrayal myth, he ignores Occam's razor. The anonymous author of Mark, whose prime purpose in writing his evangelion was to convince Vespasian that Anointianity (*Khristianismos*) was a mystery religion like Mithraism, and not a sub-sect of the Jews with whom the emperor was at war, transformed Jesus' pre-arrest Seder into a carbon copy of the Persian god Mithra's Last Supper with twelve followers (there were no "twelve apostles"), by inserting the sacred cannibalism ritual with which the emperor was familiar. And since he could not conceal that the Jesus gang had included Zealots and Sicarii, he pretended that Judas, and by implication all Zealots, was "really" Jesus' enemy and had ultimately betrayed him (*Mythology's Last Gods*, Prometheus 1992, pp. 277-278)

One of Jones's more blatant mistakes is clearly a typo. In the same paragraph that he wrongly places Alexander the Great in the third century BCE, he correctly gives Alexander's death date as 323 BCE. No proofreader can catch everything. *The Times* once reported that, on her way home, Queen Victoria "pissed" over London Bridge.

Among Jones's factual errors: He calls the fourth gospel "the most Jewish of all the Gospels." (p. 6) A theogony that turned a Jewish fanatic into a Greek god the "most Jewish"? Oh come now. He thinks the author of the fourth gospel also wrote the John letters. (p. 5) *No* scholar believes that. He dates the fourth gospel to the first century. (p. 6) Ditto. "Most accept that John was written by the Beloved Disciple," who "at the end of the Gospel is specifically identified as the author." (p. 8) In fact the gospel author makes very clear that he is *not* the Beloved Disciple, whom he acknowledges to be dead at the time of writing. Jones also identifies the Beloved Disciple as the John Mark who was Paul's fellow missionary. His arguments in support of such identification are, to repeat a description Jones reported re another pseudo-scholar, "the purest poppycock, the product of fevered imagination." (p. 62) And since Paul and John Mark ended their missionary tour more than 70 years before the composition of John, then unless John Mark was at least ninety years of age when he wrote, he too was long dead. As for the statement that "The people of Galilee in general, and Nazareth in particular, were a volatile lot," (p. 123) he clearly is unaware that there was no such village as Nazareth earlier than the fourth century. When Jesus unleashed his vitriol against the towns that rejected him as an upstart local boy, there was no "Nazareth" among them, because his hometown and birthplace was in fact Capernaum.

Jones accepts the historicity, not only of the mythical Star of Bethlehem, but also of the maguses who followed it to Herod's court, and devotes several pages to demythologizing the biblical version. Nonsense. The author of Matthew borrowed the myth from similar folk tales of Abraham, Mithra and Krishna. He defines *Nazarene* as "a person from Nazareth." (p. 17) Etymologists are unanimous that *nazōraios* (accusative: *nazōraion*) cannot possibly mean "from Nazareth." He recognizes the appointment of a Nazirite Seventy as an opposition Sanhedrin, but accepts the synoptic pretence that it was Jesus who appointed a Seventy. It was his brother Jacob who did so, long after Jesus' death, and that appointment was in all likelihood the final straw that got Jacob executed. And he twice describes the *Toledoth Yeshu* as "medieval." Newsflash! The medieval period is dated from the fall of Rome, two centuries after the composition of the *Toledoth Yeshu*. At least Jones does not swallow violations of the laws of reality. In commenting on Jesus' cursing of a fig tree, he writes, (p. 113) "Sometimes Christian authors wished so strongly to present Jesus as being able to employ supernatural powers that they depicted him as being no better than a god of Greek mythology in a bad mood."

Jones recognizes that "Luke, like Matthew, seems to have originally written a Gospel that lacked the Nativity." But instead of recognizing that the virgin birth was invented by an interpolator, he then adds, "It is beyond belief that these two chapters were added by a second writer of Luke's talent." (p. 23) What is beyond belief is that the gospel authors would have inserted into their fantasy novels a myth that was an absolute contradiction of their desperate attempts to prove that Jesus was descended from King David. And when he says that, "Most scholars accept the account of Joshua's prediction [of the Jerusalem temple's destruction] as authentic," his inability to distinguish between scholarship and theobabble is clearly revealed. Nobody with a functioning human brain believes that the statement put into Jesus' mouth, "Under no circumstances is there going to be one stone left on another that isn't torn down," was anything but *ex post facto*.

But Jones's most indefensible flight of fancy is his contention that Jesus' four brothers were his disciples, and that two of them, Judas and Simon, were twins, (p. 82) even though the author of John made clear (7:25) "His brothers had no credulity in him." It was also the John author who first identified Thomas as Jesus' twin, for the logical reason that all Greek gods sired on virgins had to have mortal twins (Herakles and Iphikles, Kastor and Polydeukes). The author of the Gospel of Thomas later harmonized John with the list of Jesus' brothers named in the synoptics, when he wrote, "These are the proverbs secretly spoken by Iesous during his life, which his twin Ioudas Thomas, wrote down." (*The Judaeo-Christian Bible Fully Translated*, volume 7, p. 271)

Jones does not get everything wrong, or even most. Eighty percent of his book is either accurate or at least plausible—and unoriginal. Dozens, perhaps hundreds, of previous biblical analysts have reached the same conclusions. When he starts by getting his facts right, he usually reaches the right conclusions. But his mistakes, ignorance of information anyone claiming the title of scholar should have located, and speculations based on interpretations so innovative that the failure of any previous writer to say the same thing should have sent up a red light, identifies him as, to put it politely, a well-meaning amateur. He should have stuck to writing about twentieth-century outlaws.

## 13. *The Jesus Mysteries: Was the "Original Jesus" a Pagan God?* by Timothy Freke & Peter Gandy

"Pagan critics of Christianity, such as the satirist Celsus, complained that this recent religion was nothing more than a pale reflection of their own ancient teachings. Early 'Church fathers,' such as Justin Martyr, Tertullian, and Irenaeus, were understandably disturbed and resorted to the desperate claim that these similarities were the result of *diabolical mimicry*. Using one of the most absurd arguments ever advanced, they accused the Devil of 'plagiarism by anticipation,' of deviously copying the true story of Jesus before it had actually happened in an attempt to mislead the gullible! These Church fathers struck us as no less devious than the Devil they hoped to incriminate." (pp. 5-6)

That paragraph by Freke and Gandy brought back memories of a letter to a newspaper in which I similarly drew attention to the parallels between Christianity and pre-Christian paganism, and received the response, "How come neither I nor anyone else has ever heard of these fifty other virgin-born savior gods, but most if not all of us have heard of Jesus Christ?" That alleged rebuttal came from an individual a majority of Christians repudiate on account of his outspoken assertions that there never were Nazi gas chambers or a Final Solution. But it does reflect a common attitude among believers past and present, that if they shut their eyes to unwanted reality, it will go away.

"He who will not eat of my body and drink of my blood, so that he will be made one with me and I with him, the same shall not know salvation." (p. 49) Jesus? No. Mithras.

"The Egyptians of every period in which they are known to us believed that [guess who?] was of divine origin, that he suffered death and mutilation at the hands of the power of evil, that after great struggle with these powers he rose again, that he became henceforth the king of the underworld and the judge of the dead, and that because he had conquered death the righteous might also conquer death. He represented to men the idea of a man who was both God and man." (p. 25) Jesus? No. Osiris.

"God made manifest, the common savior of human life." (p. 30) Jesus? No. Julius Caesar.

"From no other day does the individual or the community receive such benefit as from this natal day, full of blessings to all...Of wars he will make an end, and establish all things worthily...Not only has he surpassed the good deeds of earlier times, but it is impossible that one greater than he can ever appear. The birthday of God has brought to the world glad tidings that are bound up in him. From his birthday a new era begins." (pp. 30-31) Jesus? No. Augustus Caesar.

"The mysteries of [guess who?] in ancient Antioch were celebrated by cries that the 'Star of Salvation has dawned in the East.'" (p. 33) Jesus? No. Adonis.

"Those taking part then went by torchlight into an underground sanctuary from which they brought forth an image of the god carved in wood and marked with 'the sign of the cross on hands, knees and head.' The highlight of this Mystery celebration was the announcement: 'Today at this hour the virgin...has given birth to...'" (p. 33) Jesus? No. Aion.

"According to myth, the miracle of turning water into wine took place for the first time at the marriage of..." (p. 38) friends of Jesus? No. Dionysus and Ariadne.

A far-seer facilitated a large catch of fish. (p. 39) Jesus? No. Pythagoras.

A wonder worker "was said to have raised back to life a woman who had been dead for 30 days." (p. 40) Jesus? No. Pythagoras's disciple Empedocles.

"An effigy of the corpse...was tied to a sacred pine tree...It was then buried in a sepulcher. But...on the third day [guess who?] rose again." (p. 55) Jesus? No. Attis.

There was one point (not the only one) that Freke and Gandy made that I found particularly enlightening. I was aware that the purgative element of Jesus' underworld flamethrowers, burning away the stain of sin, had been borrowed from Zoroaster via the Essenes. But I had assumed that the punitive element, agonizing torture to make the point, "I bet now they wish they had obeyed me," had been Jesus' personal invention. *The Jesus Mysteries* shows that the "vivid descriptions of the torments awaiting evildoers in the afterlife" (p. 72) originated in Orphism. Since Jesus' pseudo-biographers took most of his other alleged teachings from pre-Christian sources, they may well have also done so with *Hell*. So perhaps he was not, as I had previously imagined, the sick, evil monster whom even the Marquis de Sade did not surpass. He was morally retarded, as is revealed by his sermon that can be summarized, "Cheat those who are no longer useful to you, and use the stolen money to bribe those who are in a position to do you good." (Luke 16:1-9) But he may have been no worse than Ayatollah Khomeini or the Taliban.

*The Jesus Mysteries* is riddled with errors. To cite a single example, the authors confuse documents from Nag Hammadi with those from Qumran. But they are the kind of errors that will grate on biblical scholars without diminishing the book's value to the general reader. Freke and Gandy cite many more parallels between the Jesus myth and centuries-older pagan myths than I did in *Mythology's Last Gods*, and that alone would be sufficient reason to recommend it. And it is extremely readable, perhaps more so than a book written with a scholarly audience in mind. It is therefore a minor point that the authors reject my own conclusion that the myths were posthumously grafted onto the biography of an executed nobody whose name, if he had not been adopted by Paul of Tarsus to be the figurehead of his new mystery religion, would have been forgotten. Rather than argue that Freke and Gandy go beyond the evidence in concluding that no such entity as Jesus the Nazirite ever existed, I will let them justify their conclusion in their own words:

"The first possibility we considered was that the true biography of Jesus had been overlaid with Pagan mythology at a later date. This is a common idea often advanced [by Harwood, among others] to account for those aspects of the Jesus story that seem obviously mythical, such as the virgin birth. But we found so many resemblances between the myths of Osiris-Dionysus and the supposed biography of Jesus that this theory seemed inadequate. If *all* the elements of the Jesus story that had been prefigured by Pagan myths were later accretions, what would be left of the 'real' Jesus? If this theory is true, then the Jesus we know is a myth and the historical man has been completely eclipsed." (pp. 61-62)

In my view, that is exactly what happened. Jesus rebelled against Judea's Roman occupation, and was arrested and executed within a matter of hours with a minimum of fuss—as were a dozen or more other would-be messiahs whose names have not survived. What convinces me that there was a historical Jesus is the acknowledgement of the earliest generation of Christian apologists that their hero was a freak, a cross between Rumpelstiltskin and Quasimodo. (*Mythology's Last Gods*, pp. 262-263) Mythmakers do not invent such unattractive heroes, and they do not invent heroes who proclaim a war of independence—and lose.

Despite its errors on minor points, and its disagreement with me on some major ones (the last several chapters strike me as pure speculation), this is a very informative book.

## 14. *The Great Deception: And What Jesus Really Said & Did*
## by Gerd Lüdemann

"Nowadays, hardly anyone seriously assumes that Jesus in fact walked on the sea, stilled a storm, multiplied loaves, turned water into wine and raised the dead. Rather, these narratives were credited to Jesus only after his death." (p. 11) "(Jesus') body rotted in the tomb...that is, if it was put in a tomb at all and was not devoured by vultures and jackals." (p. 3)

Even from a theologian, those statements do not represent a radical departure from liberal scholarship. It they had been written by an American professor of religious studies (provided he already had tenure), they would not have triggered irate demands that the author be drummed out of the university. But in theocratic Germany, where a 1933 concordat between Pope Pius XII and Adolf Hitler is still the law of the land, and churches have an absolute veto over what may be taught in theology faculties, that is precisely what happened to Gerd Lüdemann after he wrote the German version of *The Great Deception.*

Actually, Lüdemann should feel flattered at attracting so much attention. A century ago some of the points he makes would have been innovative. Today they are not even noteworthy. When he writes (p. 13), "The present book gets by without explicit discussion of secondary literature," my impression is that he does not cite the most competent gospel analyses because he has not read them. For example, despite Martin Larson's exhaustive treatment of Jesus' Essene sources (*The Essene-Christian Faith*), Lüdemann nowhere mentions the Essenes or cites any of the parallels between sermons attributed to Jesus and earlier Essene documents.

Among the teachings Lüdemann identifies as authentic, meaning that Jesus really preached them, are the following (p. 88-91): "Everyone who is angry with his brother shall be liable to judgment." "Everyone who looks at a woman lustfully has already committed adultery with her in his heart." "Do not swear (an oath) at all." "Whoever divorces his wife and marries another commits adultery." "Love your enemies." But he argues for their authenticity on the debatable ground that they parallel Exodus's Ten Commandments. That those doctrines are copied almost verbatim from the Essene Manual of Discipline and Damascus Covenant, and the best reason for concluding that Jesus preached them is that he was raised as an Essene, Lüdemann does not consider. (see *Mythology's Last Gods*, pp. 283-285) Nor does he appear to have any awareness of Luke's extensive use of Josephus.

Lüdemann authenticates Jesus' appointment of a "Twelve" and also of a "Seventy," and offers unconvincing arguments for both conclusions. But most scholars agree that the Nazirite commune in Jerusalem prior to the war of 66-73 CE was administered by a Twelve, and the author of Mark accordingly backdated the Twelve to Jesus' lifetime. And it was Jesus' brother who, thirty years after Jesus' death, first appointed a Seventy, in effect an opposition Sanhedrin, thereby enraging the official Sanhedrin and triggering his own execution.

In contrast, Lüdemann labels as not authentic Jesus' fatuous boast (p. 52), "There are some standing here who will not taste death before they see the kingdom of God come with power." He argues that Mark composed the passage to reassure disillusioned Christians that, despite the forty years that had elapsed since Jesus' death, his second coming to usher in the end of the world really was close at hand. But the prophecy said nothing about impending death and second coming. Rather, Jesus, a fanatic revolutionary, believed that his entry into Jerusalem would lead to his being crowned king of a Jewish theocracy within a matter of weeks or even days. The innocuous expression, "come with power," is more accurately translated as "established by force." A gospel author who was trying to convince Vespasian that Jesus was not anti-Roman would never have invented such a revolutionary harangue.

Lüdemann acknowledges that statements put into Jesus' mouth that he analyses and finds to be inauthentic constitute only a small percentage of the inauthentic passages found in the gospels. Even so, there are some glaring omissions. He does not mention the prophecy of the destruction of the Jerusalem temple, "There is not going to be one stone left on another that isn't torn down," far too accurate to have been written earlier than the event it described. Nor does he mention Mark's "little apocalypse," in which Jesus warns his followers to flee into the hills when the war starts, a retroactive justification for the Nazirite commune's doing exactly that in 66 CE. Lüdemann's refusal to recognize that information cannot travel backward in time, and therefore any detailed or improbable prophecy that is fulfilled must have been retroactive, would be acceptable as a starting point. But once he had concluded that the gospels are fiction, Occam's razor demanded that such prophecies be so interpreted.

The cover blurb of *The Great Deception* summarizes Lüdemann's conclusion that "the Jesus who remains is a deeply sympathetic personality and one of the great religious teachers of the world." That may be Lüdemann's Jesus. It is not the Jesus of history, a prototype would-be Khomeini who lacked the rationality to grasp that 120 Zealots (plus or minus 50) could not possibly overthrow the Roman occupation, and that he should flee Jerusalem before he was arrested and executed.

## 15. *Deconstructing Jesus*
## by Robert M. Price

According to Robert Price (pp. 15-16), "The Jesus Christ of the New Testament is a composite figure. Today's historical Jesus theories agree in recognizing that fact, but they part company on the question of which might be the original core, and which secondary accretions…The historical Jesus (if there was one) might well have been a messianic king, or a progressive Pharisee, or a Galilean shaman, or a magus, or a Hellenistic sage. But he cannot very well have been all of them at the same time."

Price examines the many reconstructions of who and what the historical Jesus really was and, far from finding them wanting, finds each one so plausible that it diminishes the probability of all of the others. The end result is that no one theory has a higher likelihood of being correct than twenty percent. From there, it is a short step to the *non sequitur* that therefore Jesus probably did not exist.

The one theory of Jesus' identity that Price totally ignores is that, like modern snake-handling cultists who get themselves fatally bitten by swallowing their Bible's assurance that snakes cannot harm them, Jesus was a brainwashed fanatic who started a ten-minute war of independence under the delusion that his god had promised that an army of a few dozen Zealots would succeed in overthrowing the Roman occupation.

Price justifies the conclusion that Jesus probably never existed by saying that (p. 250) "the gospel story of Jesus matches the pattern of the Mythic Hero Archetype in every detail, with nothing left over." Really? What about Luke's (4:23) admission that Jesus was physically deformed? What about Mark's (3:21) admission that Jesus' family recognized him as a madman? What about Mark's (15:7) statement that Jesus was arrested with "the imprisoned revolutionaries who had committed homicide in the uprising," in other words, that Jesus had started a war of independence—and lost? What about Mark's acknowledgement that Jesus' lieutenants included Zealots and a Sicarius/Iskariot, a reality that Mark somehow had to neutralize (by making Judas Jesus' enemy) in a book written during the Roman-Jewish war? A case can be made for Jesus' nonexistence. But ignoring negative anecdotes that no mythmaker in his right mind would have invented is not the way to go about it.

Large parts of Price's book fall somewhere between unlikely and extremely unlikely. For example, his attempt to turn the Sadducees' question about the afterlife status of a woman who had outlived seven husbands into a revision of an earlier, different allegory is a blatant violation of Occam's razor. Even if no Sadducee ever asked *Jesus* such a question, it is precisely the kind of rebuttal a Sadducee would have used to refute any Pharisee, Essene, Nazirite or other afterlife cultist.

Price's dissection of the sacred cannibalism of the last supper is reminiscent of literary critics' attempts to transform *Heart of Darkness* from a fictionalization of Joseph Conrad's Congo diary into a retelling of Orpheus in the Underworld. The Occam's razor explanation of Mark's fable is that the gospel author attributed to Jesus a ritual dating back 3,000 years to the Egyptian savior-god Osiris, for the purpose of convincing Vespasian that the Christians practiced a mystery religion of the type he favored, and were not a sect of the Jews with whom the emperor was at war.

Price interprets Luke's "good Samaritan" fable as emanating from a Samaritan/Galilean Jesus community that saw the southern Judeans as heretics. Surely it is more reasonable to conclude that Luke, stuck with the reality of a goy-hating, Samaritan-hating xenophobe, put into Jesus' mouth a fable that portrayed Jesus as a prototype Paul, pandering to potential converts the real Jesus would have viewed as infidels?

Price dates the gospel called Luke as late as 125 CE. The problem with that dating is that the rabbi Shimeown ben Azzai called Jesus a bastard in or near 100 CE, and that accusation could only have been a response to the virgin-birth interpolation in Matthew, which claimed that Jesus was not

Joseph's natural son. If Matthew had been circulating for twenty-five years by the time Luke was written, surely the Luke author would have copied its genealogy of Jesus rather than inventing a new one?

Price's postulation of an early group of Jewish Christians, who over the course of several decades developed a belief in the deific Jesus of the John gospel, is indefensible. Price cites no evidence whatsoever that any Christians, let alone Jesus-Jews, had ever heard of Jesus-the-god before the author of the fourth gospel created him out of nothing between 130 and 138 CE.

When Price presented arguments for dating the Dead Sea scrolls later than 70 CE, rather than two centuries earlier, and identifying the Righteous Rabbi executed by a conniving priest as Jesus' brother Jacob the Righteous, I found myself starting to agree. Then he spelled out the problems such a reinterpretation would impose, and I was unsure whether to be relieved that he had not in fact destroyed my scenario in which an Essene commune in Jerusalem that already existed before Jesus became a public figure, led by Jacob the Righteous, switched figureheads from the long-dead Jesus the Essene to the recently-dead Jesus the Nazirite (thereby becoming "the Nazirite heresy") when Peter's resurrection-preaching found a ready market; or disappointed at being deprived of an attractive alternative hypothesis. It is a question I am still mulling.

While nobody with any legitimate claim to scholarship continues to maintain that the John gospel and the John letters were written by the same person, or that John and Revelation had the same author, Price may be the first to offer solid evidence that 2 John and 3 John, both by authors calling themselves "the Presbyter," were not in fact written by the same person. (p. 50)

Despite the weaknesses cited above, and even though Price elaborates the theories of Burton Mack and John Dominic Crossan, whose books I view as apologetics for a Jesus they refuse to recognize as a prototype would-be Khomeini, this is a book I can recommend. It raises many legitimate issues, and if Price prefers to sit on the fence rather than plump for a reconstruction of his own (beyond leaning toward Jesus' probable nonexistence), that is not necessarily a fault. But if the absence of Martin Larson's *The Essene-Christian Faith* from his bibliographical notes is an indication that he has not read it, he should certainly do so. And if he has not read *Mythology's Last Gods*, he might consider reading that as well.

## 16. *The Jesus Puzzle: Did Christianity Begin with a Mythical Christ?* by Earl Doherty

According to Earl Doherty, all early Christian Testament documents, such as Paul's letters, are centered on a spiritual, metaphysical "Christ," whom they do not locate in any specific time or place in the real world. Not until the gospels is there any mention of Jesus as a Galilean preacher who flourished only a generation earlier. As Doherty tells it (p. 2), "If we had to rely on the letters of the earliest Christians, such as Paul and those who wrote most of the other New Testament epistles, we would be hard pressed to find anything resembling the details of the Gospel story. If we did not read Gospel associations into what Paul and the others say about their Christ Jesus, we could not even tell that this figure, the object of their worship, was a man who had recently lived in Palestine and had been executed by the Roman authorities with the help of a hostile Jewish establishment."

Doherty's arguments are well reasoned, and anyone unfamiliar with the evidence Doherty does *not* mention could easily conclude that he is right. What Doherty ignores is the testimony that Jesus was a thoroughly unattractive person, ugly (Clement of Alexandria, Cyril of Alexandria), deformed, ("not even of honest human shape," according to Tertullian), possessed of a body that was "so contemptible, being subject to such numerous and considerable imperfections" (Origen, quoting Celsus but, by offering no rebuttal, tacitly affirming that the description was accurate), uneducated, and so self-deluded that his own family tried to take him into custody in the conviction that he was a madman. Add to that, that he made a unilateral declaration of independence by disrupting the temple sacrifice for the emperor Tiberias on a Thursday, was arrested within ten minutes, and was executed the next day. Those are stories no creator of a mythical superhero would ever have invented.

Doherty makes an issue of the failure of the Pauline letters to make any reference to Jesus' miracles and lesser wonders, his last supper, Judas's betrayal, Peter's triple denial, or any mention of John the Immerser, and the failure of contemporary historians to mention Jesus at all. He discounts the explanation that Jesus was an upstart wannabe, not unlike Sun Myung Moon, who similarly won a large following while doing nothing the mass media deemed worthy of reporting. Paul made no mention of Jesus' miracles, for the simple reason that the anonymous author of "Mark" had not yet copied them from tales previously told of Eliyah and Elisha and attributed them to Jesus. Historians did not mention Jesus because he was a nobody who did nothing.

Mark wrote his gospel primarily to convince Vespasian that the Christians, far from being a sect of the religion responsible for the war in progress at the time of writing, practised a mystery religion similar to the Mithraism followed by a large segment of the emperor's troops. He credited Jesus with a Mithraic last supper for that purpose. (The pagan sacred cannibalism in Paul's letter is an interpolation that no Jew could have written). He pretended that the enemies of Rome responsible for the war, the Pharisees and Zealots, were also Jesus' enemies. Since he could not conceal that at least two of Jesus' lieutenants had been Zealots, he invented the tale that one of them, from the sect's ultra-terrorist wing, the sicarii, had betrayed Jesus to "the Jews." Peter's triple denial was likewise fiction invented by Mark. And the pretence that John the Immerser had any connection with Jesus (in fact he was an opposition messiah) also originated in the first gospel.

Doherty writes (p. 199), "Jesus and the religion he began should have constituted a noteworthy event in the period of the early emperors. It is difficult to believe that he would have escaped the attention of at least some Jewish commentators." That would be a legitimate objection if Doherty was trying to prove the nonexistence of a miraclemonger who raised the dead, cured leprosy, walked on water, and led an uprising comparable with those of 66-73 CE and 132-135 CE. But Jesus did none of those things. He declared himself Messiah, the warlord anointed to overthrow the Roman occupation and restore Jewish independence. But so did a handful of others about whom no information survives except their names, (*e.g.*, Theudas), and a larger number of whom not even their names are known.

Doherty argues that Josephus made no mention of Jesus. Only "Don't confuse me with facts" dogmatists dispute that the extant passages in Josephus are Christian forgeries. But a medieval writer quoted by Don Cupitt (*Who Was Jesus?* London, 1977) asserted that Josephus described Jesus as an old-looking man, balding, stooped, with joined eyebrows, and approximately 4 ft 6 in tall. Since such descriptions were being circulated as early as the time of Tertullian, Christian apologists would have immediately denounced them as false—unless they were unable to do so because a source such as Josephus that they dared not dispute affirmed the description.

Probably the strongest evidence of how far Doherty stretches credibility to maintain his thesis that Paul did not preach a flesh and blood historical Jesus is his quoting (p. 123) of Paul's words that Jesus was "born of woman, born under the Law." He gives a convoluted and imaginative argument that Paul did not "really" mean that Jesus was physically born by the usual method. While Doherty's argument is logical, it is also a violation of Occam's razor. Paul preached a Jesus who was an ordinary man chosen/adopted by his god for an unprecedented task. The author of "John" invented the metaphysical Jesus, the god Jesus, seventy years after Paul's death.

Doherty reads into Paul's letters a mystical Jesus outside of time and space that, in my view, is simply not there. Nonetheless, his case for a nonexistent Jesus is by no means unconvincing. I reject it because I attach more significance than he does to the evidence that Jesus was such a repulsive creature that no mythmaker in his right mind would have invented him, at least not as a hero. But the issue is far from closed.

Since Doherty recommended a historical novel consistent with his interpretation of history, Vardis Fisher's *Jesus Came Again: A Parable*, I will do likewise. For a reconstruction of the most probable series of events that turned a Jewish nobody into a somebody, read my *Uncle Yeshu, Messiah.* (Xlibris, 2001)

## 17. *Who Killed Jesus? Exposing the Roots of Anti-Semitism in the Gospel Story of the Death of Jesus*
## by John Dominic Crossan

*Who Killed Jesus?* is one long self-contradiction. When Crossan writes, "I see Jesus as the manifestation of God," explains that execution without trial would have been a normal procedure for "a peasant nobody like Jesus," and concludes that Jesus was left on the cross to be eaten by carrion crows like any other crucifixion victim, I can only interpret those comments as expressing the belief that Jesus was a real person from history. But practically everything in Crossan's book is a plausible reconstruction of how the Christian myth could have evolved only if Jesus originated as a literary creation and there were no real events with which the myth could be compared.

The title and subtitle are both misleading. The question of whether Jesus was killed by Jews or Romans is examined only from the perspective of which version was written first, not which is more accurate. And while the gospels' anti-Semitism is discussed, it is far from the major focus implied in the subtitle.

Most of Crossan's book is an argument with Raymond Brown, author of *The Death of the Messiah,* over whether the gospel of Peter was written before or after the canonical gospels, an issue about as profound as how many angels can dance on the head of a pin. The main difference is that, in Peter, Jesus is tried and condemned by Jews, and in the canonical gospels the judge and executioners are Romans. Crossan gives priority to Peter even though, if a gospel ignoring the Roman role already existed, the synoptics' desperate attempt to transfer the blame for Jesus' execution from the Romans to the Jews would have been unnecessary. In fact Mark started the process of making Jesus an enemy of the Jews, and Peter (not the author's real name) carried Mark's falsification to its logical conclusion. The only way a gospel author writing in the forties could have had Jesus crucified by Jews with no Roman participation, and expected to get away with it, is if Jesus was a totally mythical character so that the objection, "That is not the way it happened," could not be raised. But that is a presumption Crossan cites to provide strong evidence for Jesus' historicity. It makes no sense for the gospel authors to go to extraordinary lengths to show the Roman procurator trying to save Jesus' life, if they were not stuck with the reality that Jesus *did* exist, he *was* crucified, and the person who ordered his crucifixion *was* Lucius Pontius Pilatus.

Crossan writes, "In the gospel of Peter, Jesus' enemies see the actual resurrection itself, but in all other gospels nobody sees the resurrection, and Jesus appears only to his followers." That is only one of several instances where he cites as evidence that Peter was written before the canonical gospels, a uniqueness better interpreted as proving that it was written much later, when the possibility of witnesses coming forward and saying, "That never happened," had dropped from low to zero. Furthermore, Peter and John were the only gospels to pretend that Jesus' legs were not broken. If Peter was written later than John, that does not present a problem. But if Peter was written as early as 40 CE, the synoptics' failure to pick up such a theologically juicy tidbit would be inexplicable. At the risk of appearing to side with Brown, whose dogmatic orthodoxy causes him to ridicule the myriad of scholars who have identified Jesus as the leader of a ten-minute war of independence, I have to conclude that Crossan is just plain wrong.

As with virtually all books by theologians, Crossan's bibliography does not include a single book by any author with the competence to recognize that all claims of a god revealing its existence have been traced to the same writers who assured their readers that the earth is flat. But particularly indefensible, given the title of Crossan's book, is the omission of Harry Goldin's *The Case of the Nazarene Reopened,* Martin Larson's *The Essene-Christian Faith,* and Randel Helms' *Gospel Fictions.*

Goldin made the point that, even as late as 71-73 CE, when Mark was written, only a person far removed from Judea and Judaism could have described a trial before the Sanhedrin as incompatible with Jewish law as the fiction in that gospel. Yet Crossan maintains that the even more blatant impossibilities in Peter were written at a time when there were no Christians (Greek Jesus-followers), only Nazirites (Jewish neo-Essenes), who could not possibly have taken such a gospel seriously.

Only those who deny that Jesus was a person from history, and biblical literalists, dispute that Jesus was executed for being the leader of an insurrection against the Roman occupation. Since the Romans were not so naïve as to imagine that the Jews loved them, Mark's only reason for transferring responsibility for Jesus' execution from the Romans to the Jews was the need to dissociate Jesus, and by implication the Christians, from the Jews against whom Vespasian and Titus were fighting a war. Promoting a version of events in which Jesus' opponents were Jews and only Jews, as the author of Peter did, would have made no sense in the decades before that war began. But it made a lot of sense after 132 CE, when the Bar Kokhba rebellion again made it essential for the Christians to portray themselves as untarnished by any connection with the rebellious Jews.

Crossan identifies Jesus' assault on the temple moneychangers as the immediate cause of Jesus' arrest, but jumps to the *non sequitur* that the Romans would have found a Jewish inter-sectarian squabble treasonous. He ignores the obvious explanation, made by Larson, that it was the daily sacrifice on behalf of Tiberius, symbolic of Judea's subservience to Rome, that Jesus vandalized, and that his action amounted to nothing less than a unilateral declaration of independence. The Roman procurator's execution of the perpetrator becomes much more understandable in the light of that interpretation. Yet even though Crossan quotes the reference to "the rebels who had committed murder during the insurrection," in connection with Barabbas, he misses the obvious implication that Jesus' arrest coincided with an uprising in which people were killed. The author of Mark tried to suppress Jesus' status as an anti-Roman revolutionary. Brown derided it. Crossan hopes that if he ignores it, it will go away.

As for Crossan's statement, after quoting two conflicting gospel accounts of Jesus' arrest, "Neither is historical, but both are true," I will not even attempt to guess what that means.

I have previously predicted that certain theologians who have shown some ability to go with the evidence would eventually abandon religious belief altogether. In Crossan's case, I predict that he will eventually conclude that no such person as Jesus the Nazirite ever lived.

## 18. *Jesus—One Hundred Years Before Christ: A Study in Creative Mythology* by Alvar Ellegård.

Ellegård states in his introduction, "I shall argue in this book for an entirely new perspective on the earliest history of Christianity." After an opening like that, it should surprise no one that his bibliography does not include the 1968 book, *Did Jesus Live 100 BC?*, by G. R. Mead. Both Ellegård and Mead argue that the only historical Jesus was the Essene Righteous Rabbi executed by the Hasmonean king Alexander Yannai more than a century earlier than the procuratorship of Lucius Pontius Pilatus. Apparently Ellegård thinks he is the first to reach such a conclusion.

But that was not the only part of Ellegård's book that had my teeth on edge before I had even finished reading the introduction. Apparently he has never heard of the connotatively neutral dating system, "Common Era," and instead uses "AD," a term that is insulting to the five billion citizens of planet earth who do not view a dead Jew as their *Dominus*, "Master." And in case that were not sufficiently offensive, he casually refers to a place called the "Holy Land," as if a piece of geography were objectively "holy." From a braindead theologian (tautology), such terminology would be acceptable. From a practitioner of any legitimate discipline, whether motivated by political correctness or simple ignorance, it is not.

Criticizing Ellegård for not taking into consideration books published later than his own would obviously be absurd. But it would be equally absurd for Ellegård's readers to accept his conclusions without first comparing them to those of other scholars in the same field. As much for Ellegård's benefit as anyone else's, I cite the disparate theses of John Crossan (*Who Killed Jesus?*), Earl Doherty (*The Jesus Puzzle*), William Harwood (*Mythology's Last Gods*), Randel Helms (*Who Wrote the Gospels?*), Martin Larson (*The Essene-Christian Faith*), and Robert Price (*Deconstructing Jesus*), none of whom appears in Ellegård's bibliography. I particularly urge Ellegård to read my novel, *Uncle Yeshu, Messiah*, and then reconsider the primary sources in the light of the reconstruction presented in that book.

In dating the letters of Paul and some noncanonical documents earlier than the gospels, Ellegård is not breaking any new ground. But in placing all gospels in the second century, and not even early second century, he is clearly wrong. Matthew was not only written by 100 CE; its virgin birth interpolation had also been inserted by 100 CE, as that is the approximate date Shimeown ben Azzai first accused Jesus of being a bastard, an accusation based on interpolated Matthew's pretence that Jesus was not the natural son of Joseph. Randel Helms presents solid evidence that Mark was written before 74 CE, as that is the date Mark forecast for the end of the world. Indeed, Mark's whole purpose in writing was to convince Vespasian that the Christians were not a sect of the religion with which he was currently at war, a war that was over by 74 CE. Luke could not have been written more than a decade after Matthew, or it would have borrowed Matthew's genealogy rather than inventing a new one that gave Jesus a different grandfather from the one named in Matthew. John was written in the decade following Hadrian's anti-circumcision edict of 130 CE that triggered the bar Kokhba war, when Christianity's connection with Judaism was again a negative factor, so Ellegård got that part right.

Of six books that Ellegård dates to the first century (The Shepherd of Hermas, Didache, 1 Clement, Letter of Barnabas, Letter to the Hebrews, and Revelation), Ellegård writes (p. 35), "None of the authors ever claim to have seen or heard Jesus in the flesh, nor to have known anybody who had done so." From this he extrapolates that, at the time the documents were written, Jesus could not have been a near-contemporary—and he is right. But I get the impression that his reason for dating the first five in or near the procuratorship of Pilatus was to support his predetermined conclusion that there was no first-century-CE historical Jesus. Since previous scholars have more accurately placed four of those documents in the second century, and Hebrews after the death of Paul, no such inference can in fact be drawn. As for Revelation: the earliest parts, written between the Roman occupation of the temple in July 70 CE and the razing of the temple in August 70 CE, forty years after Jesus' execution, were the composition of an Essene to whom the Nazirite Jesus would have been an opposition pretender—if he had heard of him at all. The Nazirite redactor of Revelation, writing during the imperium of Domitian, was probably not born when Jesus died.

In trying to separate the "Cephas" mentioned by Paul as a leader of the Jerusalem Jesus commune, from the "Peter" named in the gospels as a companion of a historical Jesus, Ellegård makes an issue of Paul's consistent use of the name "Cephas" rather than "Petros." But since Paul actually met Cephas, it is only logical that he would refer to him primarily by the Aramaic name used within the commune, rather than the Greek translation used only in books written in Greek. Ellegård believes he is making the point that the communal "pillars" who were Paul's contemporaries were not simultaneously Jesus' contemporaries, since Paul's Jesus was allegedly the Essene Jesus executed circa 100 BCE.

Such rationalizations could be justified if they were used sparingly. But the foregoing is in fact typical of the lengths to which Ellegård stretches to reach conclusions that are totally devoid of less fanciful evidence. Ellegård's book is by no means a ridiculous or definitively falsifiable reconstruction of early Essene/Nazirite/Christian history. But it is at best just one more attempt to replace an orthodox myth with an unorthodox one. I doubt that many scholars will take it any more seriously than I do.

## 19. *Sweet Jesus: Straight-Shooting Scriptural Studies Scrutinizing the Savior* by A. J. Mattill, Jr

*Sweet Jesus* is a collection of articles recently published in *American Rationalist, Freethought Perspective* and *Soar*, modified where necessary to take into consideration more recent conclusions. Mattill spells out his approach in the words (p. 3), "We shall assume...that Jesus did exist and that the four Gospels...do give us an accurate account of his words and deeds." In other words, the subject of Mattill's scrutiny is the Jesus portrayed in his official biographies, not (necessarily) the Jesus of history. Since it is the Jesus of literature whom braindead fundamentalists (tautology) such as Gee Dubya Shrub view as their greatest hero, and whom Mattill hopes to set straight, that is a logical approach.

Mattill's paraphrasing of some gospel myths probably strays no further from a literal translation than some of the recent modern language bibles. But because Mattill is not motivated to put the best possible spin on stories that, when read by anyone with a functioning human brain, reveal Jesus to be less than heroic, his loose translations convey the depravity of king Jesus' alleged teachings as Authorized translations do not.

Mattill shows the biblical Jesus to have been a liar; a thief; a fanatic who hated his family for recognizing him as a madman; a xenophobe who equated non-Jews with "dogs," an idiom comparable in Jesus' time with the modern German invective, *schweinhund*; a consummate curser; a prototype Sheridan Whiteside whose abuse of his gracious hosts left much to be desired; a wandering parasite (as a rich benefactor said of Gandhi {p. 10}: "It takes a lot of money to keep Gandhi poor"); a sadist and a masochist; a hypocrite who, like Jimmy Swaggert and others, failed to practise what he preached; and a raving lunatic. He does so by the simple expedient of quoting gospel passages that portray him as exactly that.

Mattill draws attention to Jesus' teachings on the virtue of communism and the necessity of disposing of all personal property (and turning the proceeds over to the commune's treasury, although A. J. does not go into that aspect), that Christian churches tend to sweep under the rug, since only a capitalist society can keep the church hierarchy in the comfort to which they have become accustomed.

Mattill reaches the conclusion that the reason Jesus urged his followers to free themselves of sexual desire by castrating themselves is that that is what he had done. I disagree. Jesus' official biographers showed him constantly surrounded by hookers. And a Gnostic gospel author (Gospel of Philip) wrote, "The Liberator's *hetaira* (companion/concubine) is Maria the prostitute. And Messiah loved her more than all of the students, and used to kiss her often on the mouth." The Gnostic gospel can be disconsidered, since it was written at a time when Jesus was already being credited with fathering an heir who later evolved from *sang real* (royal blood) into *san greal* (holy grail). But the canonical authors are unlikely to have shown their ultimate hero's constant companion as a lady for rent, unless they were stuck with the reality that that's the way it was. And Jesus is unlikely to have consorted with hookers unless they provided him with regular freebies.

Since *Sweet Jesus* was a castrato, Mattill sees no reason to consider the theory that he was homosexual. He does not quote the passage (Matthew 26:50) where Jesus addresses his apprentice Judas the Daggerman as *hetaire*, a word with decidedly male-lover connotations. On Jesus' innate heterosexuality, we are in agreement, since I see *hetaire* as a clumsy Greek translation of an Aramaic word with no such connotations.

*Sweet Jesus* is an evaluation of the morality and justice of the Jesus of the bible rather than the Jesus of history. On that basis it achieves its objective in spades. For anyone who thinks Jesus was a nice guy, it should be mandatory reading.

## 20. *The Hidden Jesus: A New Life*
## by Donald Spoto

As always before reading a book purporting to delineate the reality behind the Jesus myths, I turned first to the bibliography. Conspicuously absent were the names of Arnheim, Crossan, Cupitt, Doherty, Harwood, Helms, Hoffman, Larson, Larue, Lesser, Lüdeman, Mattill, Price, Morton Smith, and Wells. Instead, among some valid reference material, was a motley of theobabble no legitimate biblical scholar would ever have cited, most notably *The Real Jesus: The Misguided Quest for the Historical Jesus and the Truth of the Traditional Gospels.* A scholar might conceivably read such drivel, to reassure himself that he is not ignoring evidence that the earth really is flat. But he would surely not boast about it.

Spoto makes some valid points: "It is entirely possible, then, that the first Jewish Christians placed Jesus' birth in Bethlehem to affirm his status as the true Davidic King...Augustus never registered the entire Empire...and the Judean census, which would not have included natives of Nazareth in any case, was actually called by Quirinius when Jesus was about eight years old. In a small matter of fact, while attempting to establish a specific date in history, Luke has erred." (pp. 3-4)

"If indeed a star had attracted exotic characters from a distant land...why did this event have no impact on contemporary history, much less on anyone's later knowledge or impression about extraordinary circumstances at the time of Jesus' birth? And if Herod the Great...took such steps against Jesus, why did his son Antipas have no knowledge of Jesus until late in Jesus' ministry? The answer lies in an appreciation of the star and the wise men as elements of religious truth [?], and their significance for faith is not found by arguing for their literal historicity. At the time of Jesus, the motif of a symbolic star was linked by Jewish stories to the birth of Abraham." (pp. 4-5)

"As for the family's precipitate journey to Egypt...this, too, is very likely Matthew's religious reflection, for it is incompatible with Luke's account of the peaceful, uneventful return from Bethlehem to Nazareth. More to the point, the slaughter of Jewish babies...is not even alluded to in the writings of Josephus, who documents, usually with gleeful relish, the king's every reprehensible deed."

On the other hand, while Spoto devotes several pages to the absurdities and paradoxes of the virgin birth myth, and the impossibility of harmonizing it with the rest of the gospels in which it appears, he seems incapable of grasping that it was an interpolation, not originally part of either Matthew or Luke. And his ability to dispute the canonical accounts of Jesus' life, while accepting the reality of Jesus' imaginary playmate in the sky, reminds me of nothing so much as an attempt to find a historical Pinocchio without questioning the reality of the Blue Fairy. Similarly, while rejecting a Bethlehem birthplace for a Galilean preacher, he relocates Jesus' birth to a village called "Nazareth," even though no such place existed until the fourth century. Jesus was in fact born in Capernaum. And when he writes (p. 70), "At home in Nazareth, Jesus' parents would certainly have seen to it that their boy learned enough Hebrew to read the Scriptures," Spoto is not rejecting the reality that Jesus and his parents, as carpenters, would have been illiterate, but rather demonstrating his conditioning that could no more allow for an illiterate Jesus than for a moon made of green cheese.

After writing (p. 102), "Just as the blind were given sight, so the deaf heard and the mute spoke," Spoto then appears to back away from literal acceptance of such fantasy by asking, (pp. 104-105) "How can people of the late twentieth century accept the possibility of miracles? Have not the scientific revolution, the discovery of natural laws and the collapse of magic and superstition done away with belief in miracles? To put the question bluntly: Did such things happen? *Can* they happen?" But in the chapter that allegedly answers that question, he asserts (p. 107), "It is an interesting paradox to hear professed believers denying the likelihood, or even the possibility, of miracles, thus limiting the divine freedom by insisting they know what is appropriate or inappropriate for Him to do. But cannot God...alter the usual order of reality? Is God not God precisely because He astonishes?" And if Spoto believes that, I have a bridge for sale in Brooklyn that I think will interest him. (Spoto's capitalization of pronouns is acceptable from a self-confessed believer. It is the use of such a practice by nontheists that I find indefensible.)

I fully endorse Spoto's conclusion that (p. 114), "As a man of his time, Jesus shared the anthropology as well as the religious language of that day—in other words, he was limited by all the myths and metaphors that then expressed perceptions of reality. Hence, when Jesus commanded an evil spirit to quit someone's body, he addressed what was *behind* the symptoms—the situation of a disordered personality." In other words, Jesus practised suggestion therapy, still called *hypnotism* by a diminishing minority.

The sentence that more than anything else reveals the full extent of Spoto's purblindness, also called auto-reinforced brainwashing, is (p. 115), "Disasters are called 'Acts of God,' although it is perhaps not entirely clear why God is blamed for catastrophes and not credited for blessings." That is the precise opposite of observable reality. I have lost count of the number of instances of a plane crashing and killing all passengers except one little girl, who survived because a traffic snarl caused her to miss the plane. And the news media were invariably filled with praise for the wonderful, loving god who chose to save the child's life, while ignoring the transparent (to any sane observer) reality that, if the god could choose to save the one, it must have capriciously chosen to murder the many.

Spoto makes a commendable effort to be objective, and recognizes that bible myths are fiction while maintaining the delusion that the Bible is nonetheless "true." But incurable believers cannot be objective, and Spoto is no exception. If they could, they could not study bible fantasies as intensely as Spoto has done and remain believers. *The Hidden Jesus* is neither scholarship nor journalism. It is an apology for superstitions that Spoto could not modify even if archaeologists were to unearth a signed confession by the fantasizer in whose imagination "God" first materialized.

But lest anyone think it is the reviewer who is not objective, I will let Spoto convict himself out of his own mouth: "Anyone who expresses belief in miracles is then condescendingly regarded as a child who believes in ghosts, witches, or goblins. Disbelief in miracles, one hears, is demanded by reason: Jesus did not work miracles, because miracles are impossible. But this may be a kind of circular non-reasoning that even serious scientists and historians would question." (p. 118) "Although there are seven 'days' of creation in the first chapter of Genesis and mankind is created last, there is only one day of creation in the second chapter and mankind is created first. Which account is true? Both." (p.121) "Rationalists, in other words, have certain muddy presumptions about both science and history that no serious scholar can endorse." (p.120) What gives Spoto the conceit that he has any idea what a serious scholar could endorse? I get the impression that he has neither read any nor met any.

Spoto claims that it is the persons who recognize resurrection as an absurdity who are not thinking straight. Did I call the man brainwashed? Perhaps I should have said braindead.

## 21. *The Lost Gospel:  The Book of Q & Christian Origins*
## *Who Wrote the New Testament?  The Making of the Christian Myth*
## by Burton L. Mack

The law of averages says that, somewhere in Burton Mack's two books under consideration, he must have got something right.  If so, I failed to find it.

That the material common to Matthew and Luke but not found in Mark comes from a lost gospel that historians call Q (German: *quelle*, "source") is disputed only by persons who think that the Christian gospels are nonfiction.  While Mack recognizes Q's existence, his account of its composition, along with his entire reconstruction of the Jesus myth, is arse-backwards.

Mack divides the composition of Q into three stages, beginning with $Q_1$, a collection of aphoristic sayings compiled by a sect that viewed Jesus as an apolitical teacher who died a natural death at some indeterminate time in the past.  Jesus evolved into a rebel in $Q_2$, and was refashioned again in $Q_3$ and again in Mark, in that order.  Since Paul had not heard of Q, as is indicated by his failure to quote from it, that indicates a date of composition for $Q_1$ later than Paul's letters.  The probability of $Q_2$, consisting of later additions, containing any material that dated back to a historical Jesus must therefore be virtual zero.  Yet among other things that Mack attributes to $Q_2$ are Jesus' denunciations of his birthplace, Capernaum, for rejecting him as a local upstart.  Mack postulates that it was Capernaum's rejection of the "Jesus people" a generation later that caused them to put the denunciations into Jesus' mouth.  That the two other towns that rejected the sectarian preaching happened to be the same ones Jesus chose, on account of their proximity, for his initial out-of-town tryout, pushes coincidence too far.  Jesus' paranoid denunciations of Capernaum, Bethsaida and Chorazin read much more like the tantrums of a megalomaniac (much like Oral Roberts' damnation of Australia for recognizing him as a humbug) than the considered evaluation of a team of missionaries preaching after Jesus' death.

Several other anecdotes that Mack attributes to $Q_2$ had relevance to Jesus' actual revolution of 30 CE, but would have been meaningless if they were sheer inventions from a later date.  For example, Q reported that Jesus was accused of being "a glutton and a drunkard."  Why would fiction writers invent such a scene about their Hero?  Q chose to report such accusations, because they were too widely known to be suppressed.

In order to maintain his thesis that "Jesus the rebel" was a later invention than "Jesus the teacher," Mack dates the origin of the crucifixion story later than the war of 66-73 CE.  Yet he acknowledges that Paul, who died in 64 CE, preached Jesus as a resurrected liberator.  Did Paul believe that Jesus was crucified?  Or was the crucifixion story invented after Paul's death?  Mack can't have it both ways.

Mack asserts that the author of Mark used Q as a source, ignoring the obvious point that it was the non-Mark material in Matthew and Luke that led to the recognition of Q's existence in the first place.  That assertion is nonsense.  No biographer, having access to Q, would have omitted the enormous amount of material that can be attributed to Q precisely because it is *not* to be found in Mark.  And when Mack attributes the ban on divorce and remarriage to the post-war thinking of $Q_3$, he reveals an ignorance of the Essene documents from Qumran, which banned remarriage, not to discourage divorce, but to enforce renewed celibacy.

Indeed, Mack appears not to have even heard of the Essenes.  In showing parallels between Q's alleged "Jesus people" and the Cynics, he implies a direct borrowing.  But a comparison of the gospels and Acts with Essene documents shows the Essenes to have been Jesus' (or his scriptwriters') immediate source.  The Cynics may have been one of the Essene sources, along with the metaphysics of Zoroaster and the masochism of Gautama, but they were not a source of Q.

Mack is clearly unfamiliar with Martin Larson's *The Essene-Christian Faith,* a prerequisite for anyone trying to trace Jesus' teachings to their source. And he is equally unaware that Jesus founded a communistic sect called Paupers, *ebionim,* and that his instructions to initiates, "Sell everything you own and give the proceeds to the Paupers," meant, "Give it to my coffers," the same ticket-to-heaven preached by Jimmy Swaggert and Oral Roberts, not "Give to charity" as Mack imagines. In quoting the line, "Whoever speaks against the holy spirit will not be forgiven," Mack shows no awareness that this was an Essene doctrine, and that it was a denunciation of lying, which Essenes viewed as contravening the spirit of truth and holiness. He does recognize "holy spirit" as a metaphor, but attributes the metaphor to the $Q_2$ authors, almost two centuries after the concept was first delineated in the Manual of Discipline.

Among Mack's other mistakes is a map in both books showing the alleged location of Arimathea and Cana, neither of which ever existed, and Nazareth, which did not exist earlier than the time of the Christian emperors, one of whom gave the name to an existing village out of embarrassment that Jesus' alleged home town was nowhere to be found. He also dates the John gospel thirty years earlier than Luke-Acts.

Why Mack dates John fifty years too early is difficult to fathom, since his thesis is not dependent on its being earlier than Luke. If he had read Larson, he would have known that John was written for the specific purpose of repudiating the synoptics, particularly Luke, with their endorsement of communism, celibacy, rejection of family, a second coming that had not happened, and a theocracy in Jerusalem to be established within a time limit that had already expired. Larson also showed that John must have been written at the time of the Bar Kokhba war, when it had again become expedient to dissociate Christianity from the rebellious Jews, as the author of Mark had done sixty years earlier.

But Mack's most blatant purblindness is his determination to see Johanine theology in documents written before that theology was invented. In stating that, "The Jesus people came to think of Jesus as a god," he shuts his eyes to the reality that nowhere in the synoptics or Paul's letters is there the slightest hint that Jesus was a god. The god Jesus was invented out of nothing by the John author, and from the composition of John until the rigged Council of Nicea, two-thirds of all Christians were monotheists who rejected Jesus' deification as blasphemous nonsense.

All of the nonsense of *The Lost Gospel* is continued in *Who Wrote the New Testament?* My first objection to the latter book is that it is printed in ridiculously small type that is barely readable without a magnifying glass. That turns out to be the least of its weaknesses.

In repeating his contention that John was written earlier than Luke, Mack demonstrates a belief that the myth of Jesus resurrecting Lazarus must have already existed when Luke put into Jesus-the-communist's mouth a fable in which a capitalist asked that Lazarus be resurrected to warn his brothers about Jesus' eternal torture chamber, only to have the request refused. Would a later writer turn a miracle into a parable? The reverse seems much more probable.

The possibility has apparently never crossed Mack's mind that portions of the Christian gospels as they now stand are interpolations, some from as late as the third or fourth century. Thus he sees endorsements of a trinitarian godhead and the primacy of Peter, both very late inventions, as part of the original gospels. And he assumes that Matthew and Luke both placed genealogies tracing Jesus' descent from King David alongside virgin-birth myths that gave Jesus a non-Davidic genesis. The gospel authors may have been naïve and clumsy, but they were not that naïve and clumsy. The virgin-birth fantasies were also interpolations.

Despite Mack's unequivocal assertion that Jesus the Nazirite really was a person from history, his books are bound to give aid and comfort to persons who maintain that no such person ever existed. That is because Mack does not reach the conclusion that Jesus must have been real by examining the negative evidence and recognizing that no mythmaker in his right mind would have created a hero who proclaimed a war of independence and lost, a hero whose earliest apologists acknowledged that he was a cross between Rumpelstiltskin and Quasimodo (*Mythology's Last Gods*, pp. 262-263), or an heir and successor of David who publicly acknowledged that he was not descended from the ancient Jewish king. Rather, Mack assumes that Jesus existed for no better reason that I can detect than the necessity of maintaining such a position in order to preserve his bread and butter as a professor of theology at a Christian institution. He consequently spells out a theory of Christian origins that is more consistent with Jesus being as imaginary as "God," than with his being a person from history.

Conspicuously absent from Mack's bibliographies are any mention of Arnheim, Harwood, Helms, Hoffman, Larson, Larue, Loisy, Morton Smith or Wells, probably for the same reason astrologers do not cite astronomers and flat-earthers do not cite astronauts.

Mack's books have a veneer of scholarship, for he tries to create the illusion that he starts from evidence and follows wherever it leads. They are in fact pure theology, a discipline in which conclusions are predetermined and the evidence twisted to make it fit. If Burton Mack is a biblical scholar, as opposed to a mythology pusher, then so is Hillbilly Graham.

## 22. *Bandits, Prophets and Messiahs: Popular Movements at the Time of Jesus* by Richard A. Horsley with John S. Hanson
*Whoever Hears You Hears Me: Prophets, Performance and Tradition in Q* by Richard A. Horsley with Jonathan A. Draper

Richard Horsley contends (*Bandits*, p. xiii-xiv) that, "According to the usual scholarly construct, the Zealot party was the same as the Fourth Philosophy founded by Judas of Galilee in opposition to the tribute imposed along with direct Roman rule in 6 C.E., and its members, called interchangeably 'Sicarii' and 'brigands' by Josephus, agitated for Jewish liberation until they finally provoked the massive revolt in 66 C.E...Unfortunately for these studies...'the Zealots' as a movement of rebellion against Roman rule did not come into existence until the winter of 67-68 C.E...Palestinian Jewish history must be critically reexamined now that the old 'Zealot' concept has been shown to be a historical fiction."

One could be excused for interpreting those statements as an assertion that, prior to the first Roman-Jewish War, there were never any Zealots. But a closer examination reveals that Horsley is saying nothing of the sort. He does not dispute (p. 201) that the Zealot founder, Judas of Galilee, was crucified for leading an anti-Roman uprising in 6/7 C.E., that his father Zekharyah a generation earlier and his son Simon a generation later were crucified for the same reason, and that the leaders of the Zealot faction that gained temporary power in Jerusalem in the War were his son Menakhem and his grandson Eleazar. Horsley's thesis is that, while the Zealots were always around, and occasionally emerged from the woodwork for some short-lived violence, for most of the sixty years before the War they were quiescent. But he does acknowledge (p. 199), "that at least some vestige of the Fourth Philosophy continued to pose a threat to the Roman authorities," and offers the example of procurator Tiberius Alexander's "crucifixion of James and Simon, the sons of Judas the Galilean." He argues that, "It may be that after their initial agitation against submission to the tax, the group became, in effect, dormant or went 'underground.'"

In other words, Horsley is adopting essentially the same position as the scholars he purports to be refuting. He simply argues that the role of the Zealots can be ignored—not exactly the same as saying there were no Zealots. That is either self-delusion or nitpicking, depending on whether Horsley is aware that his pretended radicalism is really quite conservative.

I have to confess to being less than delighted that a large number of conclusions I reached in *Mythology's Last Gods,* some of which I thought were original, were spelled out in great detail in *Bandits, Prophets and Messiahs*—written seven years before *MLG.* On the good side, it is nice that Horsley agrees with me.

In *Whoever Hears You Hears Me*, Horsley writes a concerted rebuttal of the three-tiered Q spelled out by Burton Mack (*Who Wrote the New Testament?*). But Mack's reconstruction of "what really happened" contains so many holes, contradictions and absurdities, that Horsley's devoting 326 pages to rebutting it is analogous to using a jackhammer to squash an ant. Earl Doherty (*Humanist in Canada,* Spring, 1997, p. 13) said it all in one sentence: "Mack has substituted a myth of his own making for the one contained in the New Testament." Horsley echoes Doherty, but less succinctly.

Since many of the points Horsley makes are possibly valid, and at the very least worthy of consideration, it is unfortunate that he chose to write his book in theobabble rather than comprehensible English. Perhaps he was hoping that the only persons who would read past the first few pages were other theobabblers, from whom the risk of disputation was minimal. Or perhaps he has been thinking in theobabble for so long that writing in plain English is now "against his religion."

Consider the following (p. 82): "One detects a different social location from that presupposed in gnomologia which require study and reflection. The hermeneutic of the 'sapiential' layer of Q, some of which is not sapiential instruction anyhow, is simply not that of an assimilation of a wisdom ethos." Does Horsley write such gibberish in the belief that he is actually saying something? Or is he paid by the word?

Horsley reveals his uncritical acceptance of Christian legends with no theological significance when he writes (p. 51) that King Herod built a fortress city in Galilee, "to which goods collected from nearby villages such as Nazareth, Cana and Japha were taken and stored." But there is no evidence that a village named Nazareth existed earlier than the fourth century ("Nazareth" was a misinterpretation by Matthew and Luke of a Hebrew word used by Mark meaning *dispersion*), or that Cana ever existed at all. (*Quananaya*, meaning "Zealot," was misinterpreted by the fourth gospeller as meaning "from Cana.")

At one point Horsley quotes from the most accurate translation of Deuteronomy, the recent Jewish version, "The Lord is our God, the Lord alone." But Horsley knows enough Hebrew to be aware that "the Lord" is a falsification of the proper name, *Yahweh,* and that "God" is a mistranslation of the dual-sex, generic plural, *elohim.* So why does he perpetuate what he knows to be a falsification? The answer would seem to be that, like a lot of other biblical scholars willing to open their eyes to reality just so far and no farther, he is determined to walk the tight wire between recognizing that the Bible is a work of the imagination, and believing that it is nonetheless not *fiction.* If that means accepting that Jesus was not a god and never claimed to be a god, while refusing to recognize that the Torah authors were henotheists who viewed Yahweh as *their* god but not the only god, that is what he is willing to do.

Stripped of the theobabble in which much of *Whoever Hears You...* is written, what Horsley is saying is that, by the time the Q gospel reached written form, it had been modified many times, not to accord with changing theology (as Mack postulated), but to cater to audiences who had to be told what they wanted to hear lest they stop listening out of boredom. Trying to ascertain from the finished product as it appears in Matthew and Luke what a historical Jesus really preached is therefore an exercise in futility. Horsley offers one more interpretation, but by no means a definitive one.

## 23. *Who Wrote the Gospels?*
## by Randel McCraw Helms

It has long been recognized that the earliest of the four anonymous Christian gospels, Mark, was written after the destruction of the Jerusalem temple in 70 CE, since its backdating to Jesus of a prophecy of that event was too accurate to have been written as a lucky guess. Even the Essene portion of Revelation, written in July or August of 70 CE, confidently asserted that the temple could never be destroyed. (Helms writes that "Revelation was written after Mark," but he was presumably referring to the final redaction of the reign of Domitian.)

Randel Helms now offers solid evidence that it was written before 74 CE, since that was the year "Mark" (whatever his real name may have been), using imagery from Daniel, forecast for the end of the world.

Helms shows that Mark used written sources, thereby shooting down the hypothesis that he journeyed to the Decapolis to interview surviving members of the Nazirite sect led by Jesus' cousin Shimeown. Obviously, if some of Mark's sources were written, then they could have all been written. And he destroys the pretence that Mark was a Jew, who would never have made such a mistake, when he observes that "He frequently seems unaware that his source is quoting or paraphrasing the Old Testament, and when his source misquotes Scripture, he fails to correct the error" (p. 10). He then adds that, in places where this happened, "Both Matthew (12:3) and Luke (6:4) silently correct Mark's error." (p. 11) Helms also points out that, since the gospeller is ignorant of Judean geography, "Therefore, we may conclude that John Mark of Jerusalem is not our author." (p. 6)

Helms' analysis of Matthew assumes that the virgin birth interpolation of 1:18a-20a, 20c, 22-23, 25a, was an original part of Matthew's gospel, even though it immediately follows a genealogy tracing Jesus' descent from King David through his father Joseph. Could Matthew really have been so stupid as to begin his gospel with such a genealogy, and then turn around and declare that Jesus was not Joseph's son? Helms must have a low opinion of the gospel authors' intelligence, for he similarly attributes to the original author of Luke the virgin birth interpolation of 1:34-35, which likewise accompanies a genealogy that it contradicts. In defense of such a harsh evaluation, however, is that any writer who could so misinterpret the conventions of Old Testament poetry, as Matthew did, that he would show Jesus fulfilling a prophecy by entering Jerusalem straddling two donkeys simultaneously, probably was a few clowns short of a full circus. And Helms calls Matthew a Jewish Christian, even though his statement that "Matthew cannot keep himself from making a Greek pun" (p. 57), is hard to harmonize with Greek not being Matthew's first language.

Since Helms' interest in Matthew was restricted to its actual author, it is understandable that he saw no point considering whether the document's putative author, Maththaios, was a historical disciple or merely one of six names invented by the author of Mark (along with six real disciples) when he gave Jesus a nonexistent council of "twelve apostles." And he attributes Matthew 16:18, "You are *Petros*, and on this *petros* ("rock") I will build my church," (p. 137) to the gospel author, even though most scholars see it as a fourth-century interpolation written to justify the autocracy of the pope.

Helms' treatment of Luke is speculative, to say the least. To the best of my knowledge, he is the first person to postulate that Luke was a woman (Priskilla?). Luke does demonstrate a strong bias in favor of sexual equality, filling his gospel with parallel tales in which there is a female hero for every male. And he portrays Jesus' female disciples as more ready to accept "the truth" than the males. But that Luke was female is, as the courts of Scotland would say, "not proven."

Helms refers to Luke's *Magnificat* as a song sung by Mary, as it is portrayed in most manuscripts. He does not mention that some early manuscripts attribute it to Elizabeth, where context says that it belongs.

Helms follows the mainstream in recognizing that Luke used Mark as a source (pp. 79-80): "But sensing grave incompetence in its composition—so much that she rejected about half of it (unlike Matthew, who uses ninety percent of Mark, Luke repeats only about fifty percent)..." A better explanation would be that, since the bulk of the Mark material missing from Luke is located between the two bagels-and-lox fables, Luke ended one day's copying at the first such fable, rerolled the scroll, and later resumed from the second such fable in the belief that that was where he/she had left off.

Helms sees much of the material found only in Luke as coming, not from unknown sources, but from Luke's imagination, rewriting tales found in the Septuagint and even the plays of Euripides. Since Luke indeed rewrote tales originally found in Josephus, even adding a detail taken from Josephus to a parable he found in Q (his other major source), this is by no means implausible.

I have a problem with Helms' contention that the "we" passages in Acts are by the gospel author, and were written in the first-person plural for propaganda purposes, rather than constituting a separate document that Luke incorporated into his work unedited. But he offers solid arguments for such a conclusion.

Helms' most useful chapter (and most embarrassing to me, since it raises many valid points that I overlooked in *Mythology's Last Gods*) is the one on the final redactor of John. Helms rejects the view that John used Luke as a source. In favor of such a theory is that fifty-one verses in John repeat material found in Luke, compared to only seventeen verses that repeat Matthew or Mark material not also in Luke. Even more significant is that John has a swordsman cut off a temple servant's right ear, a detail found only in Luke; that he repeats the "shoot the net to starboard" fable, found only in Luke; gives Jesus two disciples named Judas, as does Luke alone; and has Jesus resurrect "Lazarus," rewriting as a miracle a parable exclusive to Luke, in which Lazarus dies but a capitalist's request that he be resurrected is turned down. John also gives Lazarus a sister named Martha, previously mentioned nowhere but in Luke. Opposing the Luke-as-source theory is that in none of the duplicated incidents does John's educated Greek even vaguely resemble Luke's crude Koine. Let us just say that the jury is still out.

Apart from showing that even the earliest gospel was four steps removed from being an eyewitness account, a conclusion not incompatible with Jesus being a mythical creation, Helms appears to express no opinion on whether Jesus was a real person from history. If he believes that Jesus did not exist, that would explain his failure to cite any of the biblical clues that Jesus was grossly deformed. And obviously, a man who did not exist could not have felt himself so drawn to the mentally retarded Nathanael on account of the cruel taunting both would have received growing up, that Nathanael could legitimately describe himself as "the student Iesous cherished."

Helms concludes, instead, that the Beloved Disciple was a creation of the John author's imagination, invented as a put-down of Peter and the synoptic theology associated with Peter. He does, however, recognize that John claimed to be incorporating a narrative by the Beloved Disciple into his gospel, while acknowledging that he was not himself the Beloved Disciple. And since the Beloved Disciple did not exist, Helms' view of which parts of John were copied from his earliest source is based, not on what a companion of Jesus might actually have written (or dictated), but on which parts can reasonably be credited to a "signs gospel." That I continue to disagree by no means proves that Helms is wrong. And I will have to rethink my view that John had only two authors, rather than the three identified by Helms. As to which of the three created "Jesus the god," a character the synoptic authors had never heard of, Helms does not say.

About the only point on which I believe Helms can legitimately be upbraided, as opposed to merely disagreeing with him, is his failure to append to the John chapters a verse-by-verse

251

breakdown of which of the three authors he believes wrote what. This is an extremely useful book, and can be recommended to anyone interested in the evidence that the Christian gospels were not written by the persons whose names they bear, and were not written by eyewitnesses or even persons close to the events described.

## 24. *The Real Jesus: The Misguided Quest for the Historical Jesus and Truth of the Traditional Gospels*, by Luke Timothy Johnson

*The Real Jesus* is a vicious ad hominem attack on the 200 scholars of the Jesus Seminar for daring to investigate whether the Christian gospels contain inaccuracies, and an irrational and incompetent rejection (I cannot dignify it with the designation of a rebuttal) of the Seminar's finding that only eighteen percent of the words attributed to Jesus in the gospels were actually spoken by him.

Johnson is a theologian, meaning a practitioner of a discipline that starts from predetermined conclusions and then distorts the evidence to whatever degree is necessary to make it fit. He takes the position that every word in the gospels is literal truth, presumably including Matthew's nativity myths that could be true only if Jesus was born during the lifetime of King Herod, and Luke's birth tales that could be true only if Jesus was born precisely ten years after Herod's death, as well as Matthew's account of the Satan taking Jesus to the top of a mountain so high that he could see the entire surface of the earth—a physical impossibility on a sphere. In other words, to Johnson and the brain amputees who similarly follow the party line, the earth really is as flat as Matthew believed, and Jesus was born ten years before he was conceived.

A legitimate biblical scholar might read such drivel to reassure himself that he is not ignoring evidence that maybe the earth really is flat. But having done so he would certainly not include it in a bibliography. The conclusions of the Jesus Seminar may not be one hundred percent accurate, but at least they are based on competent evaluation of the evidence, as *The Real Jesus* assuredly is not. Persons who maintain that biblical literalism is a form of insanity are bound to cite *The Real Jesus* as strong supporting evidence for that conclusion. If Luke Johnson is a biblical scholar, then so is Homer Simpson.

## 25. *King Jesus*, by Robert Graves
## *I, Judas*, by Taylor Caldwell and Jess Stern
## *Judas, My Brother*, by Frank Yerby
## *The Kingdom of the Wicked*, by Anthony Burgess
## *Uncle Yeshu, Messiah*, by William Harwood
## *The Beloved Disciple*, by William Harwood

In *I, Judas*, Jesus walks on water, in violation of the laws of reality. In *Judas, My Brother*, Jesus stands upright on a floating outhouse door, giving watchers the impression that he is walking on water. There is no corresponding scene in *Uncle Yeshu, Messiah*, as any attempt to explain a biblical miracle as a misinterpretation of a real event would have reinforced the belief that the myth had a factual basis, when in fact all of the miracles attributed to Jesus in the Christian gospels were borrowed from fables previously told in the books of Kings. There is also no equivalent scene in *King Jesus*, perhaps because Graves also recognized it as a watered-down copy of Eliyah's parting of the Jordan River.

The Caldwell, Yerby and Graves books portray Jesus as a lifelong celibate. *Uncle Yeshu* does not. Graves portrays him as a cripple, Yerby and Harwood as a madman. And while Yerby follows Christian tradition in having Jesus' father Joseph die before Jesus was fully grown, while Harwood has him divorce Mary/Miriam in order to become a Qumran monk, both attribute his departure to his inability to continue living with a ball-breaking wife. Graves' fanciful tale of Jesus' parents' relationship is quite different.

Before *Judas, My Brother* even begins, Frank Yerby warns his readers, "This novel touches only two issues...Whether any man truly has the right to believe fanciful and childish nonsense; and whether any organization has the right to impose, by almost imperial fiat, belief in things that simply are not so." In chapter VIII he has his narrator state, "Therefore I was forced to the conclusion that it was less morally repugnant not to believe in God at all than it was to worship an absentee landlord of a deity on the one hand...or Satan's fiendish twin on the other." Then in chapter X: "Like any rational man, I prefer atheism to demonolatry." Clearly Yerby is no fan of either the Christian Church or its imaginary Sky Führer. And when Yerby in an endnote expresses "the writer's belief that theology is a mild form of insanity," my only disagreement is with the word "mild."

Unfortunately, the novel is not completely free of superstition. In what I can only interpret as an endorsement of astrology, Yerby's narrator reads Jesus' horoscope and warns him that, if he proceeds with his ambitions, "you become the greatest criminal in the long tides of human history...I see temples erected in your name...a whole priesthood clad in silks and gold, blazing with jewels, bowing before graven images made in your likeness...They'll make of you a pagan god...And all over the earth, your people, my people, will twist in the fire, scream from the wrack, die in the chambers of the choking smoke." Was there no way, other than validating the concept of prophecy, that Yerby could have portrayed Jesus as the ultimate cause of Hitler? Perhaps not.

*Uncle Yeshu, Messiah* identifies the John gospel's "Beloved Disciple" as Nathanael, mentioned only in the same gospel. Nathan, narrator of *Judas, My Brother*, turns out to be the Beloved Disciple (that premature revelation will not spoil the story for new readers), but a scene in which Nathan meets Nathanael makes clear that, to Yerby, they were not the same person. So I was more original than I thought in concluding (*Mythology's Last Gods*, pp. 353-354) that The Beloved Disciple really was Nathanael. For most of the thirty years since I read Yerby's book for the first time, I had been under the impression that Yerby had reached the same conclusion first. Apparently not.

*I, Judas*, *Judas My Brother*, and *King Jesus*, all depict Judas the Iskariot/Sicarius as the victim of bad press, and credit him with noble motives for doing what the Christian gospels say he did. They also translate his surname as meaning "from Kerioth." They could hardly do otherwise. Since their authors deny that Jesus was an anti-Roman revolutionary, they could not very well admit that one of his chief lieutenants was a *sicarius*, "daggerman" or assassin, a member of the ultra-militant wing of Rome's enemies, the Zealots. *Uncle Yeshu, Messiah* shows the anonymous author of "Mark," unable to conceal from Vespasian, for whose benefit he was writing, that Jesus' rebel band included several Zealots, one of them a Sicarius, dissociating his hero from the anti-Roman sect by inventing the fiction that Judas (and by implication the whole Zealot party) had "really" been Jesus' enemy and had ultimately betrayed him.

Harwood and Yerby both attribute the invention of the resurrection myth to an individual. In *Judas, My Brother*, that individual is a mad woman, Miriam the Magdalene. In *Uncle Yeshu, Messiah*, it is a retard, the Beloved Disciple, Nathanael.

The jacket blurb of *The Kingdom of the Wicked* reads, "Opening with the resurrection of Jesus, the reader is carried through the mystical, miraculous, violent time that followed as the twelve apostles battled against the crumbling decadent Roman Empire." Since there were never any "twelve apostles," and assuredly no dead man has ever come back to life, an opening like that is a clear warning that the story will be as realistic as *The Cat in the Hat*. Burgess, author of *A Clockwork Orange*, tries to demythologize some biblical fantasies. But his ignorance of the valid source documents, and his acceptance of imaginary incidents as observers' misinterpretations of real events, renders his novel worthless.

From the imaginative to the ridiculous: *I, Judas* was written by the authors of such scientifically illiterate drivel as *Edgar Cayce, the Sleeping Prophet*, and books touting the reality of Atlantis, psychic phenomena, and something called alpha thinking (whatever that is). Their belief that fairy tales of virgin birth, divine incarnation and resurrection are nonfiction should therefore come as no surprise. *I, Judas* tries to give an *Alice in Wonderland* story the plausibility of *War and Peace*, but unlike Dr Seuss and L. Frank Baum, its authors' unawareness that they were telling a story that only a twelve-year-old mind could take seriously guaranteed that they would not succeed. To describe this book as worthless would be undue praise.

*King Jesus* and *Uncle Yeshu, Messiah* both begin before Jesus' birth, for the purpose of falsifying the myth that Jesus was out of Mother Goose by the Great Pumpkin, or something equally *Twilight Zone*. But whereas "Uncle Yeshu" is the fifth-generation descendant of Syrians forcefully converted to Judaism when the Maccabees gave all residents of the totally-gentile province of Galilee the choice of submitting to circumcision or having a similar operation performed on their necks, Graves subjects his imagination to gymnastics worthy of Nadia Comaneci in order to turn Jesus into a biological descendant of both King David and King Herod. He was assuredly neither.

Harwood's Jesus, and likewise Yerby's, is the oldest of at least seven siblings, as the Christian gospels acknowledge him to have been. Graves adheres to the Catholic orthodoxy that Jesus' siblings were Joseph's children by a previous marriage, a doublethink that Protestants recognize as a desperate attempt to maintain the "perpetual virgin" hoax. And Graves naïvely parrots the gospel fiction, invented to seduce followers of John the Immerser to defect to Jesus the Nazirite long after the death of both, that John was Jesus' relative, precursor and supporter, when the John of history was an opposition messiah.

Graves foreshadows *The Passover Plot*, both in having Jesus orchestrate his own execution, and in having him survive the crucifixion by achieving a comatose state that was mistaken for death. But whereas the Jesus of *The Passover Plot* later died from an unanticipated spear wound in the chest, *King Jesus* simply disappears, with the book's narrator denying any further knowledge. The logical assumption is that Graves was unwilling to offend the terminally ignorant by showing Jesus totally, permanently, irreversibly dead. (Yerby also has Jesus rendered comatose in order to appear dead and be removed from his stake alive, only to succumb to the fourth gospel's spear wound. But Yerby goes on to have the corpse buried.)

In a historical novel, anything that conceivably could have happened is legitimate, including retelling a fantasy as a not-impossible event that could have given rise to the fantasy. But when Graves incorporates such pre-Christian myths as visiting maguses, poor shepherds, birth in a cave, a virtual star of Bethlehem, resurrecting a dead man à la Asklepios, and even Herod's massacre of infants, into a novel purportedly based on fact, one has to wonder if he really knows any more about first-century Palestinian history than Hillbilly Graham. He rejects some blatant impossibilities, such as feeding thousands with a handful of bagels and lox, or turning water into wine, only to replace them with alternative absurdities no less party-lining and gullible than Caldwell's. He not only shows Jesus exorcising seven demons from Mary the Magdalene, as if "possession" were a valid concept; he even names the demons.

Graves does get one thing right. In a reference to the difference between the Jesus of history and the Jesus described in the Christian gospels, he has his narrator say that, "being unaware on what insecure historical ground their doctrine rests,...the founders of the Gentile churches so strangely misunderstood his mission that they have made him the central figure of a new cult which, were he alive now, he could only regard with detestation and horror." Yet despite that insight, most of Graves' novel shows no more ability to distinguish between sense and nonsense than any biblical literalist. *King Jesus* is riddled with inaccuracies, irritants, nonsense, and interpretations that were not even defensible when the book was written in 1946. In the light of scholars' findings of the subsequent fifty years, Graves' interpretation is as plausible as *The Wizard of Oz*.

*The Beloved Disciple* is a fantasy, written as a hoax that I hoped to con a pusher of superstitious hogwash into publishing as nonfiction, so I could then expose such disinformation peddlers as the same kind of gullible incurables who swallowed James Randi's "Project Alpha." That never happened, and it was eventually published with an Afterword explaining its true status.

So is *Uncle Yeshu, Messiah* more historically accurate than the four other novels under consideration? Being written by a historian specializing in the period, as the others were not, it logically should be. But every author of a historical novel believes that his is the most approximate reconstruction to what really happened (or why write at all?), and the author of *Uncle Yeshu, Messiah* is no exception.

Is *Uncle Yeshu, Messiah* better than the two novels that start from the assumption that Christian myths are literal truth? How could it not be? Is it better than *Judas, My Brother*? Since Yerby was one of the giants, such a claim would require more hubris than I am able to muster. Is it better than *King Jesus*? If claiming equality with Yerby would be conceited, claiming equality with Graves would be no less so. Graves presents his interpretation of history with the competence later perfected in *I, Claudius*. But it is an interpretation no biblical scholar has ever taken seriously. If my book is recognized as ranking no higher than fourth of five in literary skill, but first in historical accuracy, I can live with that.

Apart from *Uncle Yeshu, Messiah*, the only Jesus novel I can recommend, not only for its plausible reconstruction of the social conditions under which Jesus lived and died, but also for its literary skill that Graves (in 1946) had not yet equaled, is *Judas My Brother*.

## 26. *Naturalism and Religion*
## by Kai Nielson

According to the press release, "This elucidation and defense of naturalism argues that an uncompromising secular orientation is the best framework for the search for meaning and interprets religion in purely naturalistic terms."

I have no problem with the foregoing. But at the risk of using a dirty word, *Naturalism and Religion* could benefit from a radical abridgement. Nielson takes five, ten, even twenty pages to say something that could have been more succinct, more comprehensible and more effective in a single paragraph. And when, in the middle of his chapter, "The Gathering of the Fugitives," (whatever that means), he writes (p. 447), "I want to return to the claim about the logical or semantical status of the principal thesis of cosmological naturalism," I can only say. "*What???*"

Nielson reports (p. 14), "In a recent survey taken in the United States, 88 percent of the population (if the sample taken was accurate) maintained that they had never had any doubts about the existence of God. Even if this survey is inaccurate and this is true of only 40 percent of the population, it is still an intellectual and moral disgrace—a disgrace that should be a scandal in the United States...In the United States there is a cultural climate where there is strong imput (*sic*) from a fanatical and puritanical religious right—from the likes of Pat Buchanan. It has pushed the political agenda in the United States to the right. We have anti-abortion, anti-euthanasia, pro-death penalty platforms fiercely defended; we have stiffer prison sentences being handed out...until by now the United States has more people in jail than any other country in the world and in these jails, 75 percent are blacks though they are 25 percent of the whole population." [The most accurate figure I am aware of puts blacks at 9 percent of the American population. This is of course irrelevant to Nielson's thesis.]

He continues (p. 15) "This very right-wing, religiously inspired agenda forces politicians...to say publicly perfectly ridiculous things about religion, things that in all likelihood the better educated among them (Clinton and *perhaps* both Bush father and Bush son) must know to be moonshine. Moreover, the power of the religious right-wing political and religious agenda forces these politicians not only to mouth pious nonsense, but to favor very retrograde and unenlightened social measures and programs as well."

That paragraph represents the focus I hoped Nielson's whole book would take. Unfortunately, such issues are never mentioned again, and Nielson instead goes on to discuss the various theories and definitions of naturalism endorsed or opposed by persons whose views strike me as trivial at best, nonsense at worst.

Nielson acknowledges (p. 15), "It is here, some of my secular-humanist colleagues will tell me, that the battle over religion must be fought and not over such arcane issues as those I discuss." On that, I am wholeheartedly on the side of Nielson's cited colleagues. Religion can be wiped out only by the historical methodology of showing that all claims of a god revealing its existence come from the same biblical authors who also assured their readers that the earth is flat, not from philosophical arguments that believers cannot comprehend (or they would have been cured already).

*Naturalism and Religion* is totally logical and convincing—to graduate students in philosophy. To the general public, it is likely to extract only a ho-hum.

## 27. *The Vanquished Gods: Science, Religion, and the Nature of Belief* by Richard H. Schlagel

Harold Puthoff and Russell Targ have Ph.D.s in physics. But they are best known for their writings on parapsychology, a field in which they are totally incompetent. They actually examined conscienceless humbug Uri Geller's fifth-rate conjuring tricks and pronounced him a genuine psychic (oxymoron).

Richard Schlagel is a professor of philosophy, but is becoming better known as a writer on biblical analysis, a field in which he has no more competence than have Puthoff and Targ in parapsychology. Perhaps the strongest evidence that he is a well-meaning amateur is to be found in his statement (p. 128), "All one can conclude with any confidence about the historical Jesus is that he was from Nazareth…" But there was no such place as Nazareth any earlier than the fourth century. The anonymous author of Mark wrote that Jesus came from "the nazareth of Galilee," *nazareth* being a Hebrew word meaning "dispersion," equivalent to the Greek *diaspora*. Matthew and Luke misinterpreted *nazareth* as the name of a town.

But that is far from the extent of Schlagel's inexpertise. A less blatant example is his statement (p. 24) that "the crucifixion of an itinerant Jewish teacher named Jesus…would have the greatest impact on Western civilization for the succeeding two millennia." But the crucifixion of Jesus was of no significance whatsoever. Had Jesus never existed, the founder of Christianity, Paul of Tarsus, would simply have adopted one of the region's dozen or more other crucified messiahs to be the posthumous figurehead of his new mystery religion. And when Schlagel dates the four canonical gospels (p. 129) "thirty-five to seventy years after Jesus' death," instead of forty to one hundred years, I will not make an issue of his too-early dating of John. But having identified the date of Jesus crucifixion as 30 CE, to date Mark as early as 65 CE is tantamount to declaring that the prophecy put into Jesus' mouth of the destruction of the temple, which occurred in 70 CE, was written five years before the event it retroactively prophesied. Since Schlagel clearly does not believe that, I can only attribute that particular error to clumsiness, particularly as he later (p. 144) changes the figures to "between about forty and seventy years." As for his writing (p. 83) that "Yahweh usually was portrayed as a monotheistic God," that is unmitigated nonsense. Even if Schlagel's only source was the King James Bible, even that shows all five Pentateuch authors unequivocally endorsing the existence of other gods—while simultaneously demanding that the Jews ignore all but Yahweh.

In the introductory chapter, Schlagel reveals an awareness of the Pentateuch's multiple authorship, and the different authors' identification by historians as J, E, D, P and a Redactor. But many parts of his book are written as if he had never heard of such an interpretation. He writes about Moses' two sets of stone tablets, in complete ignorance of why the construction of the tablets is told twice. He shows no awareness that J and E both described the tablets' genesis, that E, a Moshite priest, had Moses smash the tablets to make the point that any such tablets in the Jerusalem temple administered by the Aaronic priests must be fakes, and that the Redactor placed E's destruction of the tablets ahead of J's carving of them, obliging him to insert passages of his own to create the illusion that they were not the same tablets. Yet that explanation is spelled out in detail in Richard Friedman's *Who Wrote the Bible?*, which Schlagel lists in his bibliography even though he appears to have barely skimmed it, if that. And he gives the date for the Torah's composition (p. 86) as "600 to 400 BCE." But the J narrative was written during the reign of Solomon's son, 930-913 BCE.

He writes (p. 115) that Moses "came to Horeb, the mountain of God," where "an angel of the Lord appeared to him in a flame of fire out of the midst of a bush." But the burning bush element is found only in J, whose sacred mountain was Sinai. Only E named Horeb as the sacred mountain. And he more than once (*e.g.* p. 89) indicates a belief that the Hebrew word *elohim* means "God." It is a dual-sex generic plural meaning "the (male and female) gods." He also (p. 99) transcribes the

Hebrew word, *aleph sin resh* (Genesis 3:14) as *Aster* rather than *Osir* (or *Asher*), possibly simply a typo. However, if he was unable to check the Hebrew, and *Aster* was not a typo, that would explain his accepting Yahweh's self-introduction as "I am who I am," rather than "I am Osiris, I am," the most reasonable translation, given that Moses was an Egyptian prince.

He refers (p. 114) to "Joseph being sold to the Ishmaelites," showing no awareness that Joseph was sold into slavery twice, once by Ishmaelites (J) and once by Midianites (E). He also refers (p. 114) to Moses' mother's "daughter Miriam." But J nowhere gave Moses' sister a name. And when, later in the redacted Torah, E introduces a Miriam, he identifies her as Aaron's sister, not as Moses' sister. To J, Aaron was Moses' kinsman, meaning a fellow Jew, but not his brother (although the same Hebrew word can mean either), since J certainly did not give Moses an older brother. Nor did E make Moses and Aaron brothers. P did that. And Schlagel writes (p. 120) that "'the pillar of cloud' which, as the Lord, had gone before them as their guide, now became 'the angel of God,' going to their rear." Despite having allegedly read Friedman, he shows no awareness that the Israelites' guide is a phallus of cloud in J, but a messenger of the gods in E. And he refers to a first-century "Christian Sabbath" (p. 149), unaware that there was no Christian "sabbath" (misnamed, since the word means *seventh*) until Constantine brought Sunday observance into Christianity, along with other Mithraic observances, when he changed religions in the fourth century.

Concerning Schlagel's reference (p. 16) to the "exemplary lives" of Mother Teresa (and other unnamed persons), and his describing her (p. 146) as "morally uplifting," I can only recommend that he read Christopher Hitchens' biography of that second black hole of Calcutta, *The Missionary Position.* Apart from being a thief who kept money donated to feed the hungry in a bank account gathering interest for the Catholic Church while the persons it was supposed to feed remained hungry, Mother Teresa was a tinpot dictator who made Leona Helmsley look like a democrat. And Schlagel's mistaking Freud and psychoanalysis for something other than a crank and snake oil quackery is perhaps not relevant to the book's purpose.

Typos are minimal, and are the kind that a computer's spell check would not identify: *alter* instead of *altar* (pp. 33, 109), *Savoir* for *Savior* (p. 148), quote marks missing in one place (p. 135), *Harar* for *Haran*, (p. 109). His English is generally flawless, although he twice (pp. 110, 132) uses the plural "their" with the singular antecedents, "every male" and "a person." He uses at least one misplaced modifier (p. 163): "Addressing her as 'Mary,' she then recognized him." And he precedes "historically" with the article "an" (p. 149), correct English in the case of a silent "h" as in *honest,* but fatuous pseudo-learned affectation when the "h" is pronounced. But the soundness of his reasoning is exemplified in his comment on the virgin birth myth (p. 153), "That people continue to believe such a fantastic story is not evidence of a divine gift of faith, but simply of a naïve credulity." If Schlagel had started by getting his facts right, this could have been a very good book.

Schlagel's inadequacy as a biblical analyst is most clearly revealed when one looks at his bibliography, not so much for its omission of every single competent biblical scholar other than Friedman (whom I find it hard to believe he actually read), as for some of its inclusions. For example, he cites two books by Will Durant, whom I once mentioned in a freshman essay, only to be told by my supervisor that Durant's scissors-and-paste history has no place in a scholarly bibliography, even at the undergraduate level. And he cites Edward O. Wilson and Richard Dawkins, the creator and popularizer of the pseudoscience of sociobabble, as well as John Mack's imbecilic endorsement of alien abductions. Given Schlagel's inability to distinguish between valid and invalid sources, it is hardly surprising that he has written a book that can be recommended only for persons who know little or nothing about when, why and by whom the Bible was written, but not for anybody with even minimal education in the field.

## AFTERWORD

After this review appeared in *The Midwest Book Review*, Schlagel wrote a long and vitriolic rebuttal that was published in the following issue, reiterating some of the same errors I had criticized and calling my review an "ego trip." After rereading my review and Schlagel's response, I reaffirm every word of my review, including the evaluation in paragraph two, that Schlagel's expertise in my field leaves much to be desired.

## 28. *The Vanishing Gods*
## by George M. Fox

Every biblical historian knows that sufficient evidence to falsify religion is no further away than the nearest university library. Now George Fox demonstrates that even an intelligent amateur, with no higher qualification than a high school diploma, can utterly demolish bible religion and bible morality, using no more sophisticated sources than the King James Bible and *Encyclopaedia Britannica*. While Fox could not hope to match the scholarship of Arnheim, Allegro, Hoffman or Larue, all of whom have written excellent books aimed at the unlearned masses, Fox's effort is in fact every bit as useful to that particular audience.

Among the points Fox makes:

— Pharaoh more than once agrees to let the Israelites leave Egypt, but Yahweh inspires him to change his mind. "Why? So He can show both the Egyptians and the Hebrews what a great and powerful God the children of Israel have…How can one believe that a merciful, just and kindly deity would kill the firstborn of every Egyptian when such an almighty and powerful god could have brought about the Exodus without any bloodshed?"

— Moses produces water out of a rock when the Israelites are dying of thirst. "Water for a couple of million people and their cattle? That would be quite a river to conjure up out of the Sinai desert!"

— Yahweh agrees to spare Sodom if it contains ten righteous persons. Of Lot's immediate family and servants, surely at least ten were righteous (or why did Lot employ them?). "But God wasn't about to waste his time looking for a couple of virtuous persons. He had more important things to do, like roasting a few thousand humans."

Fox also spells out many of the Judaeo-Christian bible's inconsistencies, passages that can only be true if another passage is false.

Fox's book is self-published, and his confidence of beating the odds seems to have been justified: his first run sold out, and the book is now into a second printing. There are the unavoidable typos, but otherwise the book's physical quality is excellent. Fox's editing of his material leaves something to be desired, and even the relocation of a couple of chronologically misplaced chapters would be an improvement. And Fox concedes that his decision to cut costs by omitting an index was a mistake. But while *The Vanishing Gods* is not for scholars, it belongs in every undergraduate reading room.

## 29. *The Third Reich: A New History*
## by Michael Burleigh

"This book is about what happened when sections of the German elites and masses of ordinary people chose to abdicate their individual critical faculties in favour of a politics based on faith, hope, hatred and sentimental collective self-regard for their own race and nation.  It is therefore a very twentieth-century story." (p. 1)

"At first sight, the claim that Nazi ideology had religious content seems to be contradicted by the evidence...Hitler censured Heinrich Himmler...for trying to construe Nazism as a religious cult...But Hitler was also acknowledging that Nazism was...the expression of eternal scientific laws, revealed by God and in turn invested with sacred properties...He found a suitable arty and elite coterie...to affirm a non-Jewish or de-orientalised Christianity...Nazism was neither science run riot...nor bastardised Christianity...It was a creative synthesis of both.  Armed with his religious science, Hitler was...God's partner in ordering and perfecting that part of mankind which concerned him." (pp. 12-14)

Hitler tried to have his cake and eat it, by imposing his anti-Semitic religion while simultaneously denying that he was doing anything of the sort.  Himmler, whose treatment of an opposition religion in the death camps he administered revealed his true fanaticism, was more honest.  He called the Nazi religion a religion, although he (and Burleigh) stopped short of recognizing Nazism as simply an extreme form of Catholicism.  Whether it was more extreme than North America's Religious Right could only be determined if the Religious Right ever gained comparable power, and sane majorities in America and Canada are not about to allow that to happen.

While it appears only in occasional throwaway lines, Burleigh shows a naïve and misplaced sympathy for Russia's Chechen separatists, apparently unaware that the independent state the separatists want to impose on all Chechens is a totalitarian theocracy that would be a mirror image of the Taliban's Afghanistan.  And his statement (p. 22), "*Nineteen Eighty-Four*, which, while set in England, is curiously not of it," is totally wrong.  Orwell spelled out the horrors of an England enslaved by totalitarian Anglican Christianity, by comparing it to Russian communism.  "The Party" was a euphemism for the Established Church.  As for "Big Brother" being based on any entity other than the Anglican "God," such an interpretation is nonsense.

Burleigh ends his book with the words (p. 812), "Our lives may be more boring than those who lived in apocalyptic times, but being bored is greatly preferable to being prematurely dead because of some ideological fantasy."  The evil of ideological fantasy was graphically demonstrated by the horror of September 11, 2001.  Nazism is dead, but all politico/religious fanaticism has the potential to commit comparable atrocities.  Fortunately, at least in the civilized world, it can be put down at the ballot box.

The most probable comparison with *The Third Reich* will be William Shirer's *The Rise and Fall of the Third Reich*.  There is really no similarity.  Shirer's book was good journalism.  Burleigh's is good history.  Anyone who cannot tell the difference deserves Shirer.

## 30. *Lying About Hitler: History, Holocaust, and the David Irving Trial* by Richard J. Evans

In 1977 David Irving published *Hitler's War*. In an otherwise monumental study of Third Reich history, he reached the indefensibly non-sequitur conclusion that, because Adolf Hitler had allowed no documentation to survive linking him to the origin of the gas chambers of the "Final Solution," he therefore could not have known about them until they were already in operation. In 1991 he released an updated version of the same book, this time alleging that there never was a Final Solution, and the gas chambers were a post-war myth.

In 1993 Deborah Lipstadt published *Denying the Holocaust: The Growing Assault on Truth and Memory*. Irving was mentioned on only six out of more than three hundred pages. But on those pages she described Irving as "discredited," and labeled him "one of the most dangerous spokespersons for Holocaust denial. Familiar with historical evidence, he bends it until it conforms with his ideological leanings and political agenda…Distorting evidence and manipulating documents for his own purposes…of skewing documents and misrepresenting data in order to reach historically untenable conclusions, particularly those that exonerate Hitler…On some levels Irving seems to conceive himself as carrying on Hitler's legacy." (p. 6)

In September 1996, in an action reminiscent of the Leon Uris book and TV miniseries *QB VII*, Irving sued Penguin Books and Lipstadt for libel, declaring that Lipstadt's book "caused 'damage to his reputation' in his 'calling as an (*sic*) historian.'" (p. 8)

In December 1997 Richard Evans was engaged as an expert witness for the defense. He quickly found that, while Irving was indeed a giant to the general public, he was far from being highly regarded by other Third Reich historians. As Evans puts it (p. 8), "Yet as I began to plow through the reviews of Irving's books written by a wide range of historians and journalists, the case he made for his high reputation among academic reviewers began to crumble. Academic historians with a *general* knowledge of modern history had indeed been quite generous to Irving, even where they had found reason to criticize him or disagree with his views."

Evans notes that, "Lipstadt was far from being the first critic of Irving's work to accuse him of bending the documentary record to suit his arguments." (p. 13) One expert on British history had written that, while Irving was "usually a Colossus of research, he is often a schoolboy in judgment." (p. 8) Another found "too many avoidable mistakes…passages quoted without attribution and important statements not tagged to the listed sources." (p. 9) Hugh Trevor-Roper "found Irving's method and judgment defective" and containing a "consistent bias." (p. 10) Another "went much further, however, and included the allegation, backed up by detailed examples, that Irving had manipulated and misinterpreted original documents in order to prove his arguments." (p. 11) The author of a lengthy study of the SS called Irving's claims to incomparable thoroughness "pretentious twaddle," and "accused Irving of innumerable inaccuracies, distortions, manipulations, and mistranslations in his treatment of the documents." (p. 11) Other scholars' comments included, "Mr Irving is a great obfuscator," (p. 11), "Mr Irving's factual errors are beyond belief," (p. 12), "at best casually journalistic and at worst quite exceptionally offensive. The text is littered with errors from beginning to end," (pp. 12-13) and "perversely tendentious." (p. 13)

Add to that, that Irving's previous courtroom experience had apparently taught him little or nothing. He had been successfully sued for libel for accusing a naval officer of cowardice, and ordered to pay £40,000 damages, an enormous award in 1970 England; for calling Anne Frank's diary a forgery; and for calling a newspaper article about himself a product of the journalist's "fertile brain." And he had been ordered to pay costs over an unsuccessful libel suit that he launched against an author who attacked his allegation that Polish resistance leader General Sikorski had been assassinated on the orders of Winston Churchill.

In the miniseries, a jury found for the plaintiff, and ordered the defendant to pay one halfpenny damages, in effect a declaration that, while the plaintiff had been libeled, he had had no good name to be damaged in the first place. High Court judge Charles Gray did not resort to such hair-splitting. He found in favor of defendant Lipstadt, and labeled Irving a falsifier of history, a racist and anti-Semite, and an active supporter of neo-fascism, thereby putting the final nail in his reputation's coffin.

Evans' dust jacket asks, "Is it possible, though, that he lost his case not because of his biased history but because his agenda was unacceptable? Evans answers those questions and more in ways that may surprise many of the commentators and pundits on the trial."

In the light of the conviction of James Keegstra for hatemongering in Canada, even though the law under which he was prosecuted unambiguously exempted statements stemming from sincere religious belief (and as ignorant, bigoted, hate-ridden and dogmatic as Keegstra is, only a jury whose bias was different from but equal to his own could have failed to acquit on the ground of "sincerity"), the reason for the High Court's ruling indeed deserves close scrutiny. But unlike the Keegstra case, or any North American civil or criminal action, where the onus is on the prosecution/plaintiff to prove that Holocaust denial stemmed from malice, England's libel laws are so stacked in favor of the plaintiff that Irving was assured of victory unless the defendant could prove (1) that her words had been misinterpreted (as they clearly had not), (2) that they did not tend to damage the plaintiff's reputation (as they just as clearly did), or (3) that the allegation of deliberate, conscious, intentional distortion and falsification was true—in most cases impossible to prove. For all practical purposes, Lipstadt as defendant was obliged to prove, both that the Holocaust really happened, and that Irving was fully aware of (and suppressed) the evidence that it really happened.

A journalist from *The Independent* noted that, if Irving won his suit, "the door will have been opened for revisionists to rewrite any event in history without a requirement to consider evidence that does not suit them and without fear that they will be publicly denounced for their distortion." (p. 38) Another commentator wrote "It is as if a quack was challenging the most prominent doctors in the international medical profession. Absurd. Here in London an obsessive charlatan is forcing a parade of [five] top researchers to take part in a duel that he will win one way or the other, either as a martyr or as a successful plaintiff." (p. 36) Evans added the comment that, "a geography professor, after all, does not waste time debating with people who think the earth is flat." (p. 229) A former director of the US Holocaust Memorial feared that persons who correctly identified the case as "nothing less than a trial of the truth of the Holocaust" might be doing historical truth a disservice. He wrote, "If the plaintiff wins, the alarmists will have created…doubt among the ill-informed about whether the Holocaust happened." (p. 36) Whether an Irving victory would in fact have created a widespread belief that Holocaust deniers were right is fortunately academic. Irving lost, and the principal that incompetent and fraudulent interpretations of history can be safely denounced was upheld.

After the verdict, a *Guardian* writer (p. 226) commented that, "Libel trials rarely end with the feeling that the full story has been told. Irving v Penguin Books is a rare exception." Another *Guardian* writer (p. 259) declared that, "England's libel laws are still rotten," that "Our libel laws present a formidable weapon against free speech to those who use them malignly…It is a scandal that Penguin's and *The Observer*'s defence of their writers should have cost the best part of £3m[illion]…Free speech can be very expensive," and that "it was obvious from the outset that Irving would never be able to pay the defense costs if he lost, leaving the innocent objects of his libel suit with a seven-figure bill to pay."

Even after the trial, Irving demonstrated that he had still learned nothing. He continued to maintain that the Holocaust was a Jewish fiction, and even withdraw some of the concessions he had been forced to make in court when confronted with evidence that he was wrong. A week after the trial he claimed (p. 235), "I *have* managed to win," because "two days after the judgment, name recognition becomes enormous, and gradually the plus or minus in front of the name fades." As Evans observes, (*ibid*), "The cartoons that had him denying the trial had ever taken place, or the verdict ever delivered, were not far from the truth."

Perhaps Evans' most important point is that Irving successfully deluded many observers that *he* was on the side of free speech. He was not. Irving was the plaintiff, and his suit was designed to suppress *Lipstadt*'s right to free speech. As the *Daily Telegraph* pointed out (p. 257), "Nobody forced him to sue Deborah Lipstadt and her publisher, Penguin, for libel." And Lipstadt herself "opposed the outlawing of Holocaust denial as had happened, for example, in German law, because it made martyrs out of deniers." (p. 256)

This book should be mandatory reading for the Keegstras and Zündels of the world, even though it might as well be written in Etruscan for all the chance it has of curing incurables of their auto-reinforced ignorance.

As Ebert and Roeper like to say about movies: thumbs up—way up!

(Page citations refer to Evans' book, including when he quotes from earlier publications.)

## 31. *The Betrayal of America: How the Supreme Court Undermined the Constitution and Chose Our President* by Vincent Bugliosi

"No technical true crime was committed here by the five conservative Justices only because no Congress ever dreamed of enacting a statute making it a crime to steal a presidential election…In terms, then, of natural law and justice…these five Justices are criminals in every *true* sense of the word, and in a fair and just world belong behind prison bars, as much as any American white-collar criminal who ever lived." (p. 49)

*The Betrayal of America*'s most unfortunate quality is that it is written in logical, coherent arguments that are only comprehensible to persons capable of rational human thought. That means that if anyone were to read it to the intellectually impoverished George W. Bush, he would probably not grasp that it is about himself. In the words of the editors of *The Nation*, who wrote the Preface, "In impassioned yet impeccably reasoned prose, Bugliosi shows that what five conservative Justices on the Supreme Court did, though not treason in the strictly legal sense, represented a betrayal of trust by an institution revered by Americans like no other, one that was supposed to be above the fray, above politics. As one reader put it: 'It's because I DO hold the institution in such high regard that I want to scream out TREASON.'…As Bugliosi writes, too many of the press, the punditry, and the legal establishment simply paid obeisance to the prestige of the Court. Their passive acquiescence accorded legitimacy to Bush's accession. Against this consensus, 'None Dare Call It Treason' [title of *The Nation* article that was the virtual first draft of Bugliosi's book] raised a bold cry of dissent in the noble tradition of public pamphleteering." (pp. 4-5)

Molly Ivins in the Foreword adds, "Vincent Bugliosi has written the modern equivalent of 'J'Accuse,' the famous indictment by the French journalist Émile Zola of his government's misconduct in the Dreyfus affair. The most startling thing these two historic accusations have in common is their self-evidence. You pretty well have to be a total moron to miss the reeking injustice in either case." (p. 7) And on the apathy of Americans who shrug off a coup d'état that as recently as fifty years ago would have got the five perpetrators executed for treason, Bugliosi in his Introduction adds, "I am still waiting…for my first conservative friend or even acquaintance to show the slightest bit of anger over what these Justices did…Their guy Bush got in, and they don't give a damn how he got there. In other words, they aren't troubled in the least that the Supreme Court may have committed one of the biggest crimes in American history." (p.18)

While remaining polite to the actual perpetrators of the overthrow of the Constitution, refusing to state unequivocally that their coup d'état constituted treason, Bugliosi does not pull his punches when discussing the right wing fanatics who said of the 3,000 voters who unintentionally voted for Buchanan instead of Gore, "If they're too stupid to read a ballet, they shouldn't be voting in the first place." Of them he remarks, "So not only did a century-old technicality enable Bush to win the national election even though he lost by over a half-million votes, but he won Florida by mistake. Yet there were human mutants who loudly protested outside the Vice President's home in Washington, D.C., before the election was over, demanding that he 'get out of Cheney's home.' In their partial defense, it is unlikely that the collective IQ of these people was higher than room temperature." (p. 37)

In a discursion on an earlier, unsuccessful Republican attempt to steal the presidency, Bugliosi expresses the conclusion that, "The monstrous, grotesque Ken Starr…is a disgrace to prosecutors everywhere…Starr went far beyond merely breaking rules…Starr is one of the most reprehensible public figures we've ever been exposed to in America." (p. 79) If anyone doubts that Bugliosi is excellently qualified to express an opinion of the atrocity committed by five black-robed oligarchs who defecated on the rights of the American people, those words should surely convince him.

Bugliosi criticizes the incompetence of Al Gore's lawyer in failing to place on record the illegalities of the Court's incredible, partisan ruling. But he acknowledges the mitigating circumstance that the lawyer "already knew" that the five Republican Justices "intended to deodorize their foul intent by hanging their hat on the anemic equal protection argument." He asks, "Wouldn't you think that he and his people would have come up with at least three or four strong arguments to expose it for what it was—a legal gimmick that the brazen, shameless majority intended to invoke to perpetrate a judicial hijacking in broad daylight?" (p. 43)

"These five Justices, by their conduct, have forfeited the right to be respected, and only by treating them the way they deserve to be treated can we demonstrate our respect for the rule of law they defiled, and insure that their successors will not engage in similarly criminal conduct." (p. 60) "But if they aren't troubled by what they did, then we're dealing with judicial sociopaths, people even more frightening than they already appear to be." (p. 62)

It has been said that people get the government they deserve. If that is true, America will continue indefinitely to tolerate a treasonously appointed president whose first act on assuming office was to withdraw foreign aid from every country that refused to enforce the taboos of an American Religious Right that even as conservative a politician as Barry Goldwater denounced as extremist. This is not the time **[shortly after September 11, 2001]** to attempt to remove the President of the United States from office by impeachment and prosecution of the betrayers of America who appointed him, since such action could send the wrong message to the mad dogs the whole world supports Mr Bush in hunting down and exterminating. But once the Afghanistan situation has been reasonably normalized, either America's five highest-ranking traitors MUST be brought to justice, or the right of future conspirators to turn America into a totalitarian Republican oligarchy, perhaps even a theocracy, will be permanently legalized.

## 32. *Eight Little Piggies: Reflections on Natural History*
## *Leonardo's Mountain of Clams and the Diet of Worms*
## by Stephen Jay Gould

*Eight Little Piggies* is so politically correct, leaning over backward to portray mainstream religion as not incompatible with the evidence, that at times I found myself wondering if Stephen Jay Gould was in fact a believer. For example, he refers (p. 173) to, "a false view of history that sees light and truth locked in perpetual warfare with religion." He defines as "parochialism" (p. 183) the assertion that "progress in knowledge arises from victory in battle between science and religion, with religion defined as unthinking allegiance to dogma and obedience to authority, and science...searching for truth." While he is correct that many theists are as committed to the search for truth as any nontheist, he appears not to grasp that, if they find it, they cease to be theists.

Gould's defense of religion does not, however, extend to fundamentalists. He praises Edmund Halley (p. 173) for being the kind of believer who "speaks here for the liberal tradition of nonliteral interpretation" of the Bible. And in connection with the absence of fossils during periods of evolutionary change, he writes (p. 279), "We could attribute this pattern to a devious and humorous God, out to confuse us or merely to chuckle at our frustration."

Gould actually defends Bishop Ussher's dating the creation of the universe to 4004 BCE. He writes (p. 187), "Ussher's 4004 was neither idiosyncratic nor at all unusual; it was, in fact, a fairly conventional estimate developed within a large and active community of scholars." Later he clarifies, "Ussher's chronology is a work within a generous and liberal tradition of humanistic scholarship, not a restrictive document written to impose authority." He even defends Ussher's choice of a specific day and time, by showing the reasoning behind it.

Gould's point is that conclusions now seen as absurd often arose out of an evaluation of evidence that was totally competent in the light of the science and cultural conditioning of the time. He asks (p. 187), "What of the scientists who assumed that continents were stable, that the hereditary material was protein, or that all other galaxies lay within the Milky Way?...How many current efforts will later be exposed as full failures based on false premises?" He summarizes (p. 193), "I close with a final plea for judging people by their own criteria, not by later standards that they couldn't possibly know or assess." And he asserts (p. 226), "Science and religion occupy separate intellectual spheres demanding equal respect."

Gould does not, however, extend that "equal respect" to creationists. In rebutting creationists' claims that evolution is impossible because such changes as the transition from reptilian jawbones to mammalian ear bones would necessitate an intermediate state that made the organism nonviable, he treats them, not as competent theologians, but as incompetent scientists, ignorant of the role of redundancy. As an example of redundancy, he points out that, if human breathing could not function equally well through nose or mouth, cold and flu viruses would have wiped us out eons ago.

Some grating points are that Gould uses the peculiarly Christian dating system, BC/AD, rather than the scientifically neutral BCE/CE that does not offend this planet's five billion non-Christians. He capitalizes pronouns referring to the established deity, even though many liberal believers have abandoned the practice. And he uses the purely religious terminology, *incest,* rather than the non-connotative *inbreeding.* He also overuses parentheses (in places where they seem hardly justified).

Gould in effect argues for the certainty of life on every earth-like planet in the universe when he writes (p. 328), "The oldest rocks that could contain preserved organic remains are 3.5 to 3.6 billion years old...and...do feature fossils of single-celled creatures similar to modern bacteria." He sees that observable reality as evidence that, "life, rising as soon as it could, was chemically destined to be, and not the chancy result of accumulated improbabilities." But it is only life at the lowest level that Gould deems inevitable, since the evolution of multi-cellular animals did not occur for a further three billion years.

Nor was the direction of evolution predictable: "One might argue, for example, that the origin of speech and writing must follow predictably from the evolved cognitive structure of the human mind. But the actual languages that developed, their timing and their interrelationships would never unfold in the same way twice." Similarly, the evolution of humanoid/simian/ursine anatomy was only one of an almost infinite number of directions natural selection might have taken. UFOlogists who think that "aliens" should look like humans in *Star Trek* makeup receive no encouragement from Gould.

Gould is certainly an environmentalist (p. 50): "Environmental movements cannot prevail until they convince people that clean air and water, solar power, recycling, and reforestation are best solutions (as they are) for human needs at human scales—and not for impossibly distant planetary futures." But he is not a fanatic (p. 43): "I do not think that, practically or morally, we can defend a policy of saving every distinctive local population of organisms. I can cite a good rationale for the preservation of species, for each species is a unique and separate natural object that, once lost, can never be reconstituted."

*Leonardo's Mountain* answers some of the questions that Gould's earlier book raised. He continues to argue for the compatibility of science and religion—provided religion sticks to untestable postulations and does not insist on the literal truth of biblical myths that contravene physical laws. But he denigrates "the politically potent, fundamentalist doctrine known by its self-proclaimed oxymoron as 'scientific creationism.'" He writes (p. 270), "Creationism is a local and parochial movement, powerful only in the United States."

Gould tells of a Vatican symposium at which (pp. 269-270), "One of the priests asked me: 'Is evolution really in some kind of trouble?...I have always been taught that no doctrinal conflict exists between evolution and Catholic faith, and the evidence for evolution seems both utterly satisfying and overwhelming.'" He continues that "I certainly felt bemused by the anomaly of my role as a Jewish agnostic, trying to reassure a group of priests that evolution remained both true and entirely consistent with religious belief."

So Gould is an agnostic. I have no problem with that. The only person who really knows that Martian canals have been disproven, not merely discredited, is someone who has seen or heard of the Mariner spacecraft photographs of the red planet's surface. Similarly, the only person who can really know that all claims of a god revealing its existence have been disproven, is a historian specializing in the analysis of Genesis. All others merely take the experts' word for it.

In quoting Pope John Paul II's 1996 endorsement of evolution (p. 278), Gould appears not to notice, or is too polite to mention, that the composition of such a literate and sophisticated document is far beyond the capacity of a pope whose mental functioning falls slightly lower than John Steinbeck's Lennie Small, and it could only have been written by a ghostwriter for Wojtyla to read phonetically.

In *Piggies* (p. 15), Gould referred to "the simple silliness of sociobiology," and (p. 308) debunked its delusion that organisms could evolve in desirable directions by conscious volition. Here (pp. 263-264) he writes, "Men are not programmed by genes to maximize matings, or women devoted to monogamy on the same basis...Therefore, our biology does not make us do it." He attributes the fantasy that a woman's single ovum and a man's millions of sperm compel them to behavior that best serves the gamete, to "the pop science of evolutionary psychology." But its true name is sociobiology, and Gould's paleontology is only one of several disciplines that recognize it as the scientific equal of the ouija board.

While Gould does not challenge religion *per se,* he has no sympathy for hypocrisy and double standards. He writes (p. 256), "Luther's recommendations for virtual genocide, as presented in his tract of 1525 'Against the Murderous and Thieving Hordes of Peasants,' makes my skin crawl, especially as a recommendation (however secular) from a supposed man of God." And while he maintains that science and religion are equal, he perhaps sees one as more equal than the other (p. 97):

> The human mind, with arrogance and fragility intermixed, loves to construct grand and overarching theories—a fault perhaps encountered more often in our theological than our scientific endeavors.

In *Piggies* (p. 268), Gould defended Darwin's sexism as an inevitable symptom of his culture. In *Leonardo's Mountain* (p. 141), he asks, "How shall we judge the happily expiring custom of addressing a married woman by her husband's name?"

Gould sees natural selection as merely the most common and powerful, not the only, force propelling evolution. And his theory of punctuated equilibrium, long periods of stability interrupted by comparatively short periods of change, is well known. But the DNA evidence that he cites for the reality of evolution is definitive. He concludes (p. 330-331), "All complex animal phyla—arthropods and vertebrates in particular—have retained...an extensive set of common genetic blueprints for building bodies...The similar features [human arms, bat wings, dolphin flippers] are homologies, or products of the same genes, inherited from a common ancestor." The only way creationists can answer that is by claiming that science is itself a religion—as of course they do. Brain transplant, anyone?

## 33. *Rocks of Ages: Science and Religion in the Fullness of Life*
## by Stephen Jay Gould

Stephen Jay Gould writes (p. 4), "I do not see how science and religion could be unified, or even synthesized, under any common scheme of explanation or analysis." So far so good. But he then continues, "but I also do not understand why the two enterprises should experience any conflict." Is Gould really unable to recognize that sentence as totally oxymoronic? Either religion is valid, in which case the earth must be as flat as the Bible says it is, or it is the creation of fantasizers and liars, in which case its teaching on morality is a product of the human imagination, and the pretence that religious morality outranks humanist morality is nonsense.

Canadian theologian Tom Harpur, in a desperate attempt to retain at least some belief in the religion he had the integrity to quit pushing when his awareness of the Bible's status as fiction could no longer be suppressed, wrote a scathing attack on the most prominent science popularizers who consistently presented a nontheistic view of reality: Carl Sagan, Stephen Hawking—and Stephen Jay Gould. But Gould, while declaring himself an agnostic, had always taken the position that religion should have no role in science but was not necessarily opposed to science. I assumed that Gould was being politically correct, avoiding concluding that religion is nonsense because he was able to make his points without doing so.

But in *Rocks of Ages*, Gould takes 241 pages to argue that science and religion are Non-Overlapping Magisteria (NOMA), unrelated world views that need not be compared or harmonized because they in effect govern different universes, and are consequently worthy of equal respect.

I consider myself morally evolved, and view persons who define sin/immorality/evil/injustice/wrongdoing as something other than the unnecessary hurting of a nonconsenting victim as morally retarded. And I recognize that the philosophy behind such a definition falls outside of the magisterium of science. But to label such a nontheistic moral code as "religion," a word universally understood to mean a set of beliefs involving metaphysical absolutes derived from a speculated lawgiver, is to falsify the reality that "religion" has been the cause of ninety percent of all manmade evil for at least three thousand years. If Gould had labeled as non-overlapping magisteria science and morality, I would have raised no objection. But labeling religion as a magisterium as worthy of respect as science, is the same as calling astrology a magisterium as worthy of respect as astronomy.

The belief that $3 + 2 = 127$ is provably, irrefutably *wrong*. The suggestion that it deserves equal respect with the proven, irrefutable belief that $3 + 2 = 5$ is not only unworthy of consideration. It raises serious questions about the competence of anyone who would make such a suggestion. And since all claims of a god revealing its existence have been traced to the same authors who also assured their readers that the earth is flat, the suggestion that such claims be granted the same respect as the discoveries of science is equally indefensible.

Gould does detail the ongoing attempts of biblical creationists to justify their purely religious dogma under the oxymoron, "creation science." In so doing, he may be providing useful ammunition to moderate religionists in their battle with the lunatic fringe, since they can accept his NOMA philosophy while agreeing that creationism violates that philosophy. And in denigrating biblical literalism, Gould writes (p. 127), "These people then display a form of ultimate hubris (or maybe just simple ignorance) in equating these marginal and long-discredited factual claims with the entire domain of 'religion.'" So *Rocks of Ages* is not totally without redeeming social value. But it certainly comes close.

I was particularly surprised, given Gould's ethnic roots, at his misuse of the Hebrew word *hosanna*, as if it meant something analogous to *Praise be!* rather than *Help!!!* And his use of the peculiarly Christian dating system, *A.D.*, that is an insult to this planet's five billion inhabitants who

271

do not believe they are living in the Year of the Master, is inexplicable. Has he never heard the doctrinally neutral term, *Common Era*?

According to Gould (p. 92), "NOMA is no wimpish, wallpapering, superficial device, acting as a mere diplomatic fiction and smoke screen to make life more convenient by compromise in a world of diverse and contradictory passions." In fact that is precisely what it is. George Orwell coined the perfect description of Gould's attempt to harmonize sense and nonsense. *Rocks of Ages* is pure doublethink. It is also bovine excrement.

One can only hope that *Rocks of Ages* was an aberration, written during a period of temporary brain shutdown, and in the two years since its publication Gould has come to realize that NOMA was his biggest blunder, comparable with Einstein's cosmological constant. NOMA might be a crutch for the intestinally challenged. It has no place in the thinking of someone as (usually) rational as Stephen Jay Gould.

## 34. *Would You Believe? Finding God Without Losing Your Mind* by Tom Harpur

Tom Harpur's religion column, carried in many Canadian newspapers, used to anger fundamentalists and rationalists alike, by taking the position that the Christian gospels were fiction, but that reality did not diminish Christianity's essential "truth." In one of his last columns before he was forced to confront the incompatibility of what he believed with what he already knew to be true, and got out of the religion business, he wrote a scathing attack on Stephen Hawking, Carl Sagan and Stephen Jay Gould for what he viewed as a dogmatic refusal to see religion and science as compatible. His choice of such strange targets, men who tended to ignore rather than attack the god hypothesis, may well have stemmed from their very reasonableness and the unassailable logic leading to their conclusions, forcing Harpur to reevaluate his theology. In accusing Hawking, in particular, of "wishful thinking" and an indefensible view of right and wrong, Harpur seems to have projected what he saw in the mirror, in a last-ditch attempt to avoid being expelled from Alice's looking glass universe into the real world. If his portrayal of the world's leading popular scientists as bad guys for their nontheistic evaluation of evidence did not backfire in his face, forcing him finally to recognize that they were right and he was wrong, I would be very surprised.

Not everyone will view it in the same light, but I see this, Harpur's last book, as a final attempt to reinforce at the emotional level a belief system he already recognized, intellectually, as self-delusion.

Harpur is neither a mental defective nor an ignoramus. But like all theologians he is (or was at the time of writing) a chronic doublethinker. For example, his awareness that Jesus was just a man caused him to rationalize that the words attributed to Jesus in the gospel that deified him, "I and the father are one," did not *really* mean that Jesus was claiming to be Yahweh incarnate. It apparently did not occur to him that the gospel author was lying, putting into Jesus' mouth a claim to godship that he never made.

Harpur was never a fundamentalist, and did not fall into the trap of becoming one in order to remain a believer. He writes (pp. 27-28), "The Bible...is a very human book, replete with factual and textual errors, as well as frequently coarse or primitive concepts of God and morality...To say you believe it 'from cover to cover,' as most conservative Christians do, is in my view to have committed intellectual and ethical suicide." Then comes the doublethink: "Despite what I have just written about the fallibility of the Bible...I do believe in God in part because of the witness of the Bible."

In attempting to defend his fading theistic perspective, Harpur writes (p. 22), "The statistical odds against the chance meeting of all the various conditions necessary for the emergence and nurture of intelligent life on our planet alone are overwhelming." That is true. But Harpur fails to recognize the corollary, that if the evolution of life is improbable, then the genesis of a lifeform not bound by the laws of nature (*i.e.*, a god) is even more improbable.

Harpur has an irritating habit of using the word *atheist* to describe everyone who does not have metaphysical beliefs, thereby enabling him to categorize such persons as dogmatic believers as impervious to evidence as any fundamentalist. Personally, I have yet to meet a nontheist whose rejection of the god hypothesis is the result of something other than an evaluation of the evidence, either the competent evaluation of a Sagan, Gould, Hawking or Asimov, or the incompetent evaluation of the embarrassing Madalyn Murray O'Hair.

Harpur writes of one of his Oxford professors (p. 99), "He had little use for the idea of a deity so insecure and concerned about his glory that he needed to be told how great he was day in and day out." Harpur's rationalization was, "While he [God] has no need for our worship, we have a desperate need to express it." Fundamentalists can maintain such doublethink in perpetuity. I suggest that Harpur could not.

Harpur devotes a whole chapter to "The Power of Evil." He fills many pages with letters from readers of his column asking how can so much observable evil be harmonized with the god hypothesis. But instead of following the example of such braindead incurables as Hillbilly Graham and concocting an imaginative "explanation," he answers instead (p. 123), "I don't know." That is the answer of a man too honest and too intelligent to maintain a state of self-delusion indefinitely. Nonetheless, it was probably not the impossibility of continuing to acquit his supposedly omnipotent god of responsibility for all evil that cured Harpur of religiosity, but the impossibility of continuing to believe that, while everything else in his Bible was pious fiction, its unsubstantiated assertion that a god had revealed its existence was true.

On page 125 Harpur writes, "In a purely spiritual existence, children would not be accidentally killed. There would be no rape, no lynchings...no empty bellies...no floods, drownings, fires or earthquakes...AIDS epidemics or infants born lacking a brain. But nor would there be the splendour of a waterfall...the breathtaking beauty of the natural world." Could Harpur write that and not realize that, in arguing for the inevitability of evil, he is assuming a universe that is *not* produced and directed by an omnipotent god? His quitting the religion business just a few months later suggests that he could not.

Of Harpur's chapter on "Cosmic Consciousness," a paean to the metaphysical purpose of hallucinations, I will be kind and say nothing. The chapter on "The Role of Dreams" is even more incompetent.

Perhaps the most revealing passage in *Would You Believe?* is, "Many of us need to forgive God." (p. 218) Anyone who can recognize that a supposedly omnibenevolent god has committed acts that need forgiveness, is well on the way to recognizing that such an oxymoronic creature cannot exist. It should surprise no one that Harpur discontinued his religion column, and probably his belief in religion, just a few weeks after that last attempt to convince himself that reason was dogma and unreason was "faith."[1]

I really hoped that I would be able to recommend this book as a case study of a pusher's last, desperate attempt to reinforce his ignorance in the face of steadily accumulating knowledge. But it is by no means self-evident that it is anything of the sort. It is therefore useless to anyone who already knows what liberal theists believe. And it is equally useless as an antidote to fundamentalism, since Harpur's arguments against biblical literalism are as subjective and unsubstantiated as his arguments against nontheism. Harpur once stated that he refused to read any book published by Prometheus, because it was bound to present conclusions that differed from his own. In other words, "My mind is made up. Don't confuse me with facts." Simply to avoid following Harpur's example, I would recommend reading this book if it actually said anything worthy of rebuttal. Sadly, it does not.

1   That Harpur has cured himself of god-addiction is a retroactive prediction based on probability, not an alleged fact.

## AFTERWORD

*Would You Believe?* was published in 2000. Harpur's only later book, *The Still Point*, published in 2002, continues to promote a metaphysical viewpoint. Clearly he is not yet ready to abandon Cloud Cuckoo Land and start living in the real world.

## 35. *Skeptical Odysseys*
## edited by Paul Kurtz

*Skeptical Odysseys* is a celebration of *Skeptical Inquirer*'s first twenty-five years, and is comprised mainly of autobiographical essays about how the various contributors became skeptics and how they contributed to the magazine's evolution. Not surprisingly, it qualifies more as a collector's item than a source of research information. That does not mean, however, that it has nothing to offer. I found several of the memoirs, particularly that of Joe Nickell, entertaining and even enlightening.

Given the structure and purpose of the book, it was hardly surprising that, by the time I got to Steve Allen's essay, I found it to contain little that had not already been covered earlier. But at least one of his paragraphs (p. 342) counts as one of *Skeptical Odyssey*'s highlights: "It is to the great credit of the freethought movement, over the centuries, that their adherents have often had the courage to openly criticize religious beliefs that are patently false and acts which are shameful...Rationalists are inclined to honor the search for truth wherever it may lead, whereas those who are already committed to a religious or political creed customarily behave like attorneys for the defense."

Wallace Sampson's chapter on Alternative Medicine is definitive, superseding even Martin Gardner's chapter on the same subject in *Fads and Fallacies in the Name of Science*. He summarizes (pp. 266-267), "'Ineffective' and 'quackery' are now successfully converted to 'complementary and alternative medicine' or 'CAM.'...This language distortion and the rise of CAM may be one of the largest cons of the twentieth century...If the majority of people in the Western or developed world now know implausible, ineffective, and fraudulent practices by the blander terms 'alternative,' 'complementary,' and 'unorthodox,' then as many people in the Western world have been conned as were conned by the major political tyrannies of the twentieth century."

On Dean and Kelly's reference (p. 205), to "critical thinking skills that are part of any university course in the social sciences," I can only conclude that he is generalizing from a very rare university where such is the case. My experience of social science courses is that they are contentless drivel. And he appears to agree with my evaluation of psychology when he writes (p. 205), "Astrology may even be the lesser (and cheaper) of available evils—if psychiatry does not usefully do what it claims to do (help recovery from a neurotic disorder better than no psychotherapy), it becomes difficult to outlaw astrologers when psychotherapists go free."

Henry Gordon's comment (p. 241) on Shirley MacLaine's latest book is a gem: "unbelievably idiotic."

While endorsing Michael Shermer's evaluation of Deuteronomy's Ten Commandments as "exclusionary," meaning bigoted, I have to call him out for asserting (p. 330) that "the very first one prohibits anyone from believing in any of the other gods besides Yahweh." In fact the Commandment enforced monolatry, not monotheism (a belief no Torah author held), by prohibiting the admission of other gods "before my face," meaning within the borders of Judea where Yahweh would be subjected to the indignity of having to compete with them.

It continues to perplex me that some of the world's most competent debunkers of nonsense beliefs can have a blind spot when it comes to the one contrary-to-fact belief to which they are emotionally committed. Isaac Asimov believed that the glorified cold reading called psychology differed qualitatively from tealeaf reading. Gordon Stein viewed the presence of aliens on earth, in violation of Einstein's light-speed limit, as not impossible. Carl Sagan, able to recognize that holding a religious belief, in the absence of any evidence to justify it, was illogical, was nonetheless sufficiently blinded by political correctness to shut his eyes to the reality that religion has been as definitively falsified as the canals of Mars. Richard Dawkins, one of the few people with the intestinal fortitude to acknowledge that creationists who place the contents of a 2,500 year old fantasy novel above the findings of myriads of scientists are insane, is able to support a fairytale pseudoscience called sociobiology that could be valid only if history, anthropology, biology, zoology and several other sciences are incompetent drivel. Stephen Jay Gould defined science and religion as Non-Overlapping Magisteria (NOMA) that should be viewed as complementary rather than incompatible, in effect equating nontheistic moral philosophy with religion. The author of a non-superstitious biography of Jesus is able to rationalize that "there must be" loopholes to get past Einstein's equation showing faster-than-light travel to be impossible, and the *reductio ad absurdum* showing time travel to be impossible. And Martin Gardner, whose ability to see through all other superstitions created by the undisciplined human imagination is unsurpassed, seems incapable of taking an unblinkered look at the most mind-crippling, anti-human security belief of all, the delusion of religion.

Gardner's essay, "Confessions of a Skeptic," attempts to define his "philosophical theism," but makes no serious effort to justify it. I got the impression that he is well aware that it cannot be justified. He even acknowledges (p. 356), "My atheist and agnostic friends never cease to be amazed and dismayed over how I manage to be a thoroughgoing skeptic of the paranormal and still retain a belief in God." While I would not have the *hubris* to call myself a friend of a man I never met, I am certainly an avid admirer. Given Gardner's awareness that all claims of the entity called God revealing its existence have been traced to the same authors who assured their readers that the earth is flat, "amazed and dismayed" describes my reaction perfectly. As James Alcock noted (p. 45), "We can turn off our critical skepticism when dealing with some of our own favorite beliefs." Barry Beyerstein's observation (p. 110) is, "These mental habits can easily make people accept things in one avenue of life that they would dismiss as absurd in another."

On the question of whether the battle against ignorance can ever be won, I have to go along with Michael Shermer (p. 335), "There is little or no chance that we can convince True Believers of the errors of their thinking. Our purpose is to reach that vast middle ground between hard-core skeptics and dogmatic believers."

Perhaps the best summation of what *Skeptical Odysseys* is about comes from Vern Bullough (pp. 326-327): "If criticism cannot come from within, then it has to come from without, and this is the role and task of CSICOP, the *Skeptical Inquirer*, and the various societies and publications around the world which have followed their example. Science is the better for having had organized skepticism for twenty-five years."

## 36. *Queen Jane's Version: The Holy Bible for Adults Only*
## by Douglas Rankin

As anyone who has seen a *Saturday Night Live* skit expanded into a ninety-minute movie is aware, parody ceases to seem funny after about five minutes. And *Queen Jane's Version* is an 1129 page parody of the King James Bible.

That is not to say that any randomly chosen excerpt is unfunny. For example, *QJV* begins, "Once upon a time, it was a dark and stormy night when Godfather, The Great Geehovah, illuminated the entire universe with one grand sweep of his magic rod."

But in at least one place, *QJV* makes an intrinsically funny *KJV* passage unfunny. Rankin translates Deuteronomy 23:13, "As part of your equipment, have something to dig with, and when you relieve yourself, dig a hole and cover up your shit." In contrast, the *KJV* reads like a commandment to masturbate: "And thou shalt have a paddle upon thy weapon, and it shall be, when thou wilt ease thyself abroad, thou shalt...cover that which cometh from thee." As a consequence of that passage, some persons of my acquaintance used to taunt one another, not, "Screw yourself," but, "Go paddle your weapon." While Rankin restores the true meaning of the Hebrew, since it is the *KJV* that he is parodying he seems to have overlooked an obvious opportunity.

I was a little concerned that Rankin's use of Anglo-Saxon epithets might lead to accusations of falsification, since in biblical Hebrew there is no such thing as a "dirty" word. But he explains in the preface that, "If the translators of the *King James Version* could get away with the use of 'piss' without censorship, then *Queen Jane's Version* could translate into plain language those sexual situations obscured in the Bible by quaint word usage." And the uncensored language helps drive home the Bible authors' uncompromising misogyny (Genesis 3:16), as the *KJV* does not: "Submit to your husband any time he wants to fuck you, and consider it an honor to be screwed. Men will have absolute power over you and your body to use and abuse in any way they desire, particularly in the region which will some day be known as the Middle East."

Despite his best efforts to draw attention to the real meaning of biblical passages that are scientifically illiterate or endorsements of crimes against humanity, Rankin misses many absurdities. For example, he translates Matthew 4:8 as, "The devil, undaunted, brought him to the top of a very high mountain and showed him all the world's kingdoms." But Jesus and Satan would only have been able to see all of the world's kingdoms from the top of a high enough mountain *on a flat earth*. While not a literal translation, "took him to the top of a mountain so high that they could see all the way to the edge of the world," would have better served Rankin's purpose of demonstrating the gospel author's ignorance. And the Bible's six other flat-earth passages are, if anything, even more obscure in *QJV* than in *KJV*.

Rankin states in his preface that, "The original Hebrew noun often translated merely as 'friend' to describe some biblical relationships actually means 'male lover.'" Yet he translates Deuteronomy 13:6 as "your closest friend," when the literal meaning is, "the male lover who means as much to you as your own breath." And in Matthew 26:50, where the *KJV* has Jesus address Judas as "Friend," and the Greek has him using the salutation, "*Hetaire*," a word with unequivocal sex-partner connotations, Rankin omits Jesus' greeting altogether.

There are other missed chances. The *QJV* translation of Exodus 13:1, "Every firstborn male that emerges from any Israelite womb must be consigned to me," misses the point that "consigned" is a euphemism for "sacrificed." And in Leviticus 27:28-29, where the *KJV* says that any human "devoted" must be put to death, and the *New American Bible* makes very clear that the passage refers to persons "vowed as doomed," that is, promised to the god as sacrifices, *QJV* says "men condemned to death in court." But the passage is not an endorsement of capital punishment (which is endorsed elsewhere). It is an unmistakable validation of human sacrifice.

Where the *KJV* translates 1 Samuel 24:4 as, "David arose, and cut off the skirt of Saul's robe privily," the *QJV* reading is, "David crept up to Saul unnoticed and cut off the corner of Saul's robe." But "cut off his skirt" is a Hebrew colloquialism for "humped his wife." What David did was proclaim himself Saul's successor by taking possession of Saul's harem.

Rankin's translation of 1 Samuel 20:41 reads that David and Jonathan "kissed each other passionately and wept as they lay together in the tall grass…And David wept most of all." That is an improvement on the *KJV*'s, "They kissed one another, and wept with one another, until David exceeded." But even without checking the Hebrew, Rankin should surely have recognized that the passage meant, "until David ejaculated."

But with few exceptions, such as the above, this is a Bible paraphrasing (not a translation) that illuminates the Bible's status as the most morally depraved obscenity ever written (with de Sade's *Juliette* and Hitler's *Mein Kampf* fighting it out for second and third place). The Bible's ethical bankruptcy is particularly visible in *QJV's* rendition of Genesis 4:5. When Cain made Yahweh an offering of his best fruits and vegetables, while Abel offered an animal he had killed, "The meat thrilled The Great Geehovah, as bloodthirsty as he is, but he spurned the garden produce. As we shall see, this was just the first of many instances in which the all good, all kind, and all loving Yaahvey Geehovah will show favoritism for no rational reason." While *Queen Jane's Version* will not cure incurables, it should give the pragmatically religious something to think about.

## 37. *Papal Sin: Structures of Deceit*
## by Garry Wills

In the introduction to *Papal Sin,* Catholic historian Garry Wills refers to "the great truths of the faith—the Trinity, the Incarnation, the Mystical Body of Christ." That was an early warning that this book was not going to include any competent evaluation of papal beliefs, regardless of how effectively Wills was able to analyze various popes' interpretations of those beliefs. Clearly someone who, while simultaneously claiming to be a scholar, could label such *Outer Limits* fantasies as "truths," epitomizes the truism, "There are none so blind…"

And yet Wills is not purblind when it comes to papal teachings that have a modern rather than a Nicene genesis (p. 5): "The cartoon version of natural law used to argue against contraception, or artificial insemination, or masturbation, would make a sophomore blush. How can one be in service to others, yet peddle to them 'religious truths' whose truthfulness rings so obviously hollow?"

Wills summarizes the rationale behind the papacy's continued endorsement of indefensible taboos by paraphrasing an argument actually offered to Pope Paul VI and accepted by him as valid (p. 94): "If the church sent all of those souls [who practiced birth control] to hell, it must keep maintaining that that is where they are." He states (p. 162), "Popes do not claim to derive their authority from the people but to be rulers appointed by apostolic forebears going back to the mythical laying on of hands by the Twelve or to the nonexistent episcopy of Peter in Rome." So Wills is able to recognize that Peter was never a bishop; yet he accepts as factual that Peter did travel to Rome, even though the only testimony for such a trip is the Manichean "Acts of Peter" that even the early church hierarchy rejected as spurious.

Wills barely touches on Pope Pius XII's complicity in the atrocities of Adolf Hitler, since that situation has been thoroughly detailed in John Cornwell's *Hitler's Pope.* Instead, his primary targets are Pius IX, Paul VI and John Paul II, three men who by all logic should have destroyed the Catholic Church, by showing it to be the plaything of megalomaniac popes best described as sick jokes. On John Paul's adamant denial of equal rights to women, Wills shows the absurdity of his forbidding the ordination of women on the ground that Jesus' (mythical) Twelve Envoys were all male, while simultaneously forbidding priests from marrying even though at least some of those same envoys (those who did exist) were married. John Paul and his predecessors were able to doublethink because (p. 127), "New Testament passages are twisted, omitted, distorted, perverted to make them mean whatever the Pope wants them to mean."

On papal tyranny, Wills reports (p. 152), "Eighty percent of young priests think the Pope is wrong on contraception, sixty percent of them think he is wrong on homosexuality, yet the Vatican keeps up the pressure to have them voice what they do not believe."

Wills devotes two chapters to the sexual abuse of children by pedophile priests, and makes a strong case that papal enforcement of priestly celibacy is the primary cause. Homosexual men are drawn to the priesthood because it gives them access to lots of unattached men and boys, and because precedent tells them that, no matter how many times a pedophile priest is caught, the church will do everything in its power to cover it up. "In a survey of 101 gay priests, those ordained before 1960 remember their seminary as having been 51 percent gay. Those ordained after 1981 say their seminaries were 70 percent gay." (p. 194) "Many observers suspect that John Paul's real legacy to his church is a gay priesthood." (p. 190)

Wills' chapter on abortion, while not mentioning Carl Sagan's evidence (*Billions and Billions*) that first-trimester fetuses have absolutely no brainwave activity consistent with even minimal self-awareness, presents an equally strong case against anti-choice fanatics by showing the self-contradiction inherent in their position when they claim that a fetus is a person while treating it as a non-person whenever it suits them. For example, most anti-abortionists approve of abortion when a pregnancy is caused by rape or "incest" (inbreeding).  But if a fetus is a person, there should be no exceptions.  And nuns in Catholic hospitals rarely if ever baptize miscarriages, as they would assuredly do if they really believed fetuses have souls from the moment of conception.

I found little in *Papal Sin* with which to disagree, other than Wills' self-inflicted blindness that prevents him from recognizing his Bible as an anthology of fantasies qualitatively equal to those of the Brothers Grimm.  Unfortunately, his unquestioning adherence to all dogmas dating back to Nicea takes up a disproportionate amount of his book, and his three chapters on the theology of Augustine of Hippo will be of no interest whatsoever to his book's presumed market.  He may be right that criticism of papal absurdities will be more favorably received from a professed practising Catholic.  But that does not make his toeing the party line any less boring to persons who know that the 2,000-year-old Jesus myths were borrowed *in toto* from 5,000-year-old Osiris myths.

So on balance, is this a good book or a bad book?  In answer, let me say that it has nothing to offer anyone who already knows that the Catholic Church is a totalitarian tyranny comparable with Mussolini's Fascists, Khomeini's Ayatollahs and Hoffa's Teamsters.  But it should be mandatory reading for the minority of Catholics who continue to believe that right and wrong are whatever the current pope says they are.

## 38. *Hitler's Pope: The Secret History of Pius XII*
## by John Cornwall

John Cornwall is the author of *A Thief in the Night,* a self-indulgent account of his uncritical investigation of the death of Pope John Paul I, in which he belatedly concluded, despite his failure to rebut any of the evidence presented by David Yallop in *In God's Name,* that Luciani had indeed died of natural causes. So when Cornwell wrote *Hitler's Pope,* there was some expectation that it, too, would be a whitewash. Not so.

It is not true that Eugenio Pacelli, Pope Pius XII (1939-1958), did nothing to diminish Hitler's Final Solution. Pacelli in fact made more than one attempt to persuade der Führer that baptized Catholics of Jewish ancestry should not be categorized as Jews and treated the same as other Jews. And it is not true that Jews were the only persons Pacelli deemed expendable if their extermination increased papal power. He also shrugged off the massacre of almost a half-million Orthodox Christians by Hitler's Catholic allies in Yugoslavia. And while refusing to denounce Germany's bombing of the civilians of Coventry and the destruction of its Protestant cathedral, he implored the Allies not to bomb Rome.

Commenting on Pacelli's assertion that the Jews could "look after themselves," while trying to save Catholic ex-Jews, Cornwell asserts (p. 159), "The very fact of making such distinctions betrayed, of course, Pacelli's diplomatic collusion with the overall anti-Semitic policy of the Reich."

In a speech in Budapest in 1938, "Not only did Pacelli make no reference to the burgeoning anti-Semitism in Hungary but...he called for an appeasement that would be matched that same year in secular political terms by France and Britain." In the same speech Pacelli referred to "the foes of Jesus who cried out to his face, 'Crucify him,'" in a manner that left no doubt that "the 'comprehensive love' he preached at the meeting did not include the Jews." (pp. 185-186)

Pacelli's naïve doublethink is exemplified in the difference between his and Hitler's interpretation of the 1933 concordat that neutralized Catholic opposition to National Socialism and enabled Hitler to become dictator. The concordat "constrained the Holy See, the German hierarchy, the clergy, and the faithful to silence on any issue the Nazi regime deemed political. To be specific, since the persecution and elimination of the Jews in Germany was by now a stated policy, the treaty had legally bound the Catholic Church in Germany to silence on outrages against the Jews." (p. 153)

To Hitler, the concordat meant that "a sphere of confidence has been created that will be especially significant in the urgent struggle against international Jewry." To Pacelli it signified "the total recognition and acceptance of the Church's law by the state." (pp. 130-131) Cornwell's perspective is (p. 161), "Pacelli's approach proclaimed that the Holy See was prepared to acknowledge Hitler's Reich, whatever its offenses against human rights, whatever its offenses against other confessions and other faiths, provided that the Catholic Church in Germany was left in peace." And on page 85, "Nothing could have been better designed to deliver the powerful institution of the Catholic Church in Germany into the hands of Hitler." On page 7, "As Hitler himself boasted to a cabinet meeting on July 14, 1933, Pacelli's guarantee of nonintervention left the regime free to resolve the Jewish question."

Also on page 85, "The principal condition imposed by Hitler in 1933...voluntary withdrawal of German Catholics from social and political action as Catholics...were to be implemented by Pacelli himself." While Cornwell makes clear that Hitler's seizure of absolute power was facilitated by Pacelli, and in retrospect one could argue that it should have been foreseen, he does not criticize the future pope for actions that were a logical consequence of his conviction that humans are the domesticated livestock of a Führer in the Sky, that Catholics are the Sky Führer's chosen people, and that the pope is the Sky Führer's viceroy.

Pacelli was a long-time anti-Semite. Concerning Pacelli's denunciation of the German socialists in 1919, Cornwell writes (p. 75), "The repeated references to the Jewishness of these individuals, amid the catalogue of epithets describing their physical and moral repulsiveness, gives the impression of stereotypical anti-Semitic contempt." Elsewhere, "His 1942 Christmas message trivialized and denied the Nazi Final Solution."

That Pacelli was an equal-opportunity racist can be deduced from his refusal in 1921 to accept the conclusion of an American investigation, that allegations of mass rape against black members of the French occupying army in the Rhineland were unfounded, and his demand in 1946 that no blacks be included in the Allied occupation force in Rome after the Axis defeat. (p. 95)

While Cornwell mentions the Nazi prosecutions of priests and nuns for sexual and other offenses against children, he clearly assumes that the charges were false. He does not even note that, if such allegations had first been made by anyone but Nazis, worldwide child abuse by pedophile priests might not have continued uncontrolled for a further fifty years.

It is in fact fortunate that *Hitler's Pope* was written by a Catholic, lapsed or otherwise. From anyone else, his summation (p. 295) might have been seen as anti-Catholic or self-serving: "Pacelli's failure to respond to the enormity of the Holocaust was more than a personal failure; it was a failure of the papal office itself and the prevailing culture of Catholicism."

Cornwell has written a balanced account that does not gloss over the arguments presented by Pacelli's apologists. And while he makes clear that Pacelli's behavior was indefensible (pp. 296-297): "He was the ideal Pope for Hitler's unspeakable plan. He was Hitler's pawn. He was Hitler's Pope," he also leaves no doubt that Pacelli to his dying day saw himself as a plaster saint, doing only the work his Sky Führer had assigned him as Vicar of the Führer's bipedal domesticated livestock.

It was the concessions Hitler granted Pacelli, giving the Church a veto over academic appointments as the future pope's reward for strangling all Catholic opposition to the Nazi regime, that as recently as the 1990s led to Gerd Lüdemann's dismissal from his university post for writing books such as *The Great Deception* (Prometheus, 1999) that dared dispute Christian dogma.

A point in his favor was that Pacelli persuaded Pope Pius XI in 1923 to denounce the harsh reparation terms of the Treaty of Versailles that most historians see as the primary cause of World War II. But Cornwell makes clear that Pacelli's only purpose was the furthering of papal absolutism, which he must surely by that date have expected eventually to come into his own hands.

A name that kept cropping up in Cornwell's narrative is Adolf Eichmann, by no means a consolation prize who paid the ultimate penalty for Israel's failure to capture any really big fish, but rather an architect of the Final Solution whose culpability for SS atrocities ranks second only to Heinrich Himmler.

It Is impossible to read *Hitler's Pope* and not be outraged, not only at Pope Pius XII's refusal to speak out against Hitler's Final Solution during the years he knew it was happening, but also at the future pope's role in facilitating Hitler's rise to power by forcing Germany's Catholic Center Party into Hitler's camp in exchange for greater Vatican control over Germany's Catholics, to say nothing of Hitler's willingness to act as the Vatican's tax collector. For the Vatican hierarchy to abandon its attempt to canonize Pacelli on account of his complicity in Hitler's atrocities, they would have to recognize that the Christian Gospels, with their message, "Let his bloodguilt be on us and on our descendants," from which Pacelli took his principles, are every bit as anti-Semitic as *Mein Kampf.* Does anyone really think that will happen?

Germany is still a theocracy, governed by "Christian Democrats." Hitler's treaty obligation to function as the papal tax collector is still German law. In recent years a tennis player was only able to avoid exorbitant church taxes by renouncing her sectarian affiliation. The twin Führers are dead, but their unification of church and state lives on.

## 39.  Oldies But Baddies:  The Emperor's New Clothes Revisited
### Dale Carnegie, *How to Win Friends and Influence People*
### Napoleon Hill, *Think and Grow Rich*
### Norman Vincent Peale, *The Power of Positive Thinking*

None of the above-listed books is out of print—and that is a pity, since it provides further proof that the gullibility that gave the world theology, astrology, UFOlogy, psychoanalysis, Shirley MacLaine and Maharishi Mahesh Yogi is alive and well and not hiding in Argentina.

*How to Win Friends and Influence People* is a prescription for lying, sycophancy and manipulation.  Carnegie's recipe for success can be summarized in two words: Suck up.  But he simultaneously denigrates flattery:  "The difference between appreciation and flattery?  That is simple.  One is sincere and the other is insincere." (p. 58)  In other words, kiss butt, but do it *sincerely*.  And never, never criticize anyone, in case he might challenge you to a duel. (p. 37)  To make people do your bidding, "Bait the hook to suit the fish." (p. 61)  If the truth is not what someone you wish to manipulate wants to hear, then lie—but lie sincerely. (p. 119 and elsewhere.)  No doubt that advice has created a lot of used car dealers, televangelists, and hotline psychics.

Since Carnegie urges his readers to adopt the same uncritical, self-denigrating, brown-nosing posture toward their fellow humans that they already practice when dealing with their gods, it should surprise no one that a whole generation accepted his advice at face value.  That such sycophancy has never elicited a single positive (or for that matter negative) response from any god, was unlikely to be noticed by persons preconditioned to view themselves as the domesticated livestock of a higher life form.

Carnegie's book is a collection of "Rah! Rah! Go, team!" cheerleading chants.  Unfortunately, it is also a very readable collection of interesting anecdotes, capable of convincing the impressionable that it is a useful instruction manual.  In fact, in the few places where Carnegie's clichés are valid, such as "Let the other person save face," (p. 248) and "Ask questions instead of giving direct orders," (p. 247) anyone who does not follow such procedures instinctively is unlikely to learn them from a book.

In reading *Think and Grow Rich*, I found one word continually springing into my mind: infomercial.  The entire book is a promotional blurb for a book called *Think and Grow Rich*.  For example (p. 30), "Little by little, the truth has unfolded itself, until it now appears certain that the principles described in this book hold the secret of mastery over our economic fate."  And on page 47, "How can one harness and use the power of desire?  This has been answered through this and the subsequent chapters of this book." (No, it hasn't.)  From page 50, "By following the instructions laid down in the chapters on autosuggestion, and the subconscious mind, as summarized in the chapter on autosuggestion, you convince the subconscious mind that you *believe* you will receive that for which you ask, and it will act upon that belief, which your subconscious mind passes back to you in the form of 'faith' followed by definite plans for procuring that which you desire."  Following *what* instructions?  "Faith…is a state of mind which develops voluntarily, through application and use of these principles." *What* principles?

And on page 52, "Surely enough has been stated to give a starting point from which one may, through experiment and practice, acquire the ability to mix faith with any order given to the subconscious mind.  Perfection will come through practice."  If you say so, Maharishi.

From the same page: "In language which any normal human being can understand, we will describe all that is known about the principle through which faith may be developed where it does not already exist."  What does "faith" have to do with growing rich?

Pages 135-136: "The purpose of this book…is to present to all who want the knowledge, the most dependable philosophy through which individuals may accumulate riches in whatever amounts they desire." *What* philosophy?

Pages 148-149: "The power which gives this nation its freedom is the selfsame power that must be used by every individual who becomes self-determining. This power is made up of the principles described in this book." *What* power? *What* principles?

Page 221: "Before you can put any of this philosophy into successful use, your mind must be prepared to receive it." Again: *WHAT PHILOSOPHY?*

Pages 253-254: "Previously you may have had a logical excuse for not having forced life to come through with whatever you asked, but that alibi is now obsolete, because you are in possession of the Master Key that unlocks the door to life's beautiful riches." *WHAT* Master Key?

Every word, sentence and paragraph in *Think and Grow Rich* is designed to delude the gullible that it actually says something. Anyone who has read it and failed to recognize it as contentless gibberish probably owns a lot of swampland in Florida.

Norman Vincent Peale was kind enough to begin *The Power of Positive Thinking* with a warning that it was not written for persons with functioning human brains (p. xv): "This book teaches applied Christianity, a simple yet scientific system of practical techniques of successful living that works." Christianity a "scientific system"? Oh come now.

From that point on it is all downhill. For page after page after page, Peale reiterates: Got a problem? Pray. Feeling down? Pray. Can't sleep? Pray. Got arthritis? Pray. Not getting anything out of this book? Pray.

Peale does occasionally get specific: For problem A, read Bible passage A. For problem B, read Bible passage B. For some reason, the recommended Bible passages are always fatuous clichés such as, "If God is with us, who can be against us?" "He gives power to the faint, and to those who have no might he increases strength." Nowhere does Peale cite Jesus' fable whose moral can be summarized: "Cheat those who are no longer useful to you, and use the stolen money to bribe those who can do you good." (Luke 16: 1-9) Nor does it occur to him to cite the seven Bible passages that describe a flat earth.

Peale's last chapter is titled "How To Draw Upon That Higher Power." Enough said.

When it first occurred to me to include Peale's book, which I had not previously read, in a review with two compilations of secular gobbledygook, I was unaware that it adhered to a purely religious viewpoint. I ultimately decided to leave it in, in case others interpreted the title as simple cheerleading, as I had done, and wasted a trip to the library.

The three books described here remind me of nothing so much as the "motivational seminars" in which I was forced to participate during the few unpleasant weeks I worked as a professional liar, otherwise called a commission salesman. I recognized then that the ritual was designed to coerce salesmen into hypnotizing (for want of a better word) themselves into believing that the seminars served some useful function, so that such salesmen would keep selling beyond the point at which they became aware that they were working for less than minimum wage. And I recognize now that the reviewed collections of doubletalk were designed to create the illusion that their gibberish made sense, inducing purchasers to believe they had learned something and encourage other marks to buy the books.

Don't.

## 40. *The Test of Love: A Revaluation of the New Testament* by Norman Weeks

Norman Weeks is not a biblical historian. He has no ability to determine which parts of Christianity's New Testament might have a factual basis and which parts certainly do not. Recognizing that limitation, he spurns a bibliography of secondary sources (he cites only various bible translations), and approaches the Christian Testament from a moral and ethical perspective, for the purpose of evaluating whether all or any of its teachings can be accepted as an instruction manual for human behavior in a culture 2,000 years more morally evolved than when the NT was written. If he had stayed with that approach, he might have avoided turning out a book that epitomizes the truism: A little learning is a dangerous thing.

Weeks starts out with the appearance of objectivity when he writes, "The Jesus of this revaluation is only the literary character who is the protagonist of the gospels…This Jesus is a fiction and an ideal." (p. 16) He furthers that façade when he adds, "In our study of values we have no choice but to take the text as it is, keeping in mind that it is the fictional Jesus we are describing, not the historical Jesus." (p. 88) But when he combines four incompatible versions of a woman anointing Jesus with perfume, in order to reach an interpretation of that fiction that not one of the disparate storytellers could possibly have intended (pp. 104-106), he reveals the true nature of his brainwashing.

The first clues that *The Test of Love* is not a work of scholarship but one of theobabble comes in Weeks' use of the peculiarly Christian dating system, AD, rather than the scientifically neutral CE (Common Era); in his capitalization of pronouns in such expressions as "God Himself;" and in his references to Jesus as "the Master," an absolute synonym for "the Lord." And when in later pages Weeks states as facts, "His (Jesus') miracles attracted to him a large following" (p. 66); "Jesus, who in his healing practiced a benevolent *white magic* was now performing *black magic*" (p. 107); "The body of Jesus underwent a second, and that time permanent, apotheosis in the Resurrection" (p. 143); and "Some devils were less amenable to the apostolic power and authority than others" (p. 154); and expresses uncritical belief in Jesus' temporary transfiguration from a Quasimodo/Rumpelstiltskin into an Adonis; his walking on water, raising the dead, and foreknowing the future (as if information could travel backward in time); and the Star of Bethlehem; it becomes inescapable that Weeks has no more ability to separate sense from nonsense than Dan Aykroyd or Karol Wojtyla. In view of Weeks' expressed belief in physical impossibilities, his credulous acceptance of less-impossible fairy tales and propaganda is a minor incompetence that need not be detailed.

Weeks states that, "Accepting New Testament statements as 'facts' is often a naïve interpretation." (p. 301) But on page after page, when commenting on passages that even liberal theologians recognize as propagandizing fiction, Weeks consistently does exactly that. And in discussing the rise of the Jesus sect after their teacher's death, he asks, "What kind of people would be convinced by secondhand testimony, hearsay evidence, paraphrased teaching, and wild apocalyptic prophecies and promises?" (p. 200) The response that springs to mind is: Try looking in the mirror.

Weeks shows a willingness to recognize that bible religion has a downside. He writes, "When the Christians eventually acquired political power, they vented their intolerance in political ways, such as censorship, heresy trials, executions, holy wars and Crusades…And so the mean spirit of sectarian intolerance pitted itself against the pagan cosmopolitan society of the ancient world." (p. 209) He further recognizes that, "The seeds of all later Christian anti-Semitism were sown throughout the books of the New Testament." (p. 211) But in criticizing the details of establishment religion while desperately clinging to a belief in religion itself, he commits the same absurdity as Michel Gauquelin, who debunked sun-sign astrology by inventing an alternative astrology.

*The Test of Love* is riddled with inconsistencies. For example, Weeks acknowledges that the gospels were anonymous, and were credited to celebrities to give them a spurious credibility. But he then treats the Christian fiction that Mark was written by an associate of Peter, Luke by an associate of Paul, and Matthew and John by two of Jesus' disciples, as if those definitively falsified guesses were established fact. And only someone conditioned from birth to equate the author of the John gospel with the "Beloved Disciple" could have read John without ever noticing the author's unequivocal acknowledgement that he was not the Beloved Disciple, and that the Beloved Disciple was already dead when John was written.

The book's dust jacket credits Weeks with "advancing beyond the 'Bible is bunk' theme of recent skeptics." In fact one cannot "advance beyond" a succinctly stated truth. An evaluation of astrology that did not agree in principle that "astrology is bunk" would be instantly recognizable as incompetent nonsense, and so is a book of biblical evaluation that intentionally avoids recognizing that "the Bible is bunk."

It is not impossible for a believer in religion to write a useful and worthwhile book of biblical analysis. Peter Ellis (*The Yahwist*) did so, as did Richard Friedman (*Who Wrote the Bible?*), Don Cupitt (*Who Was Jesus?*), and probably Canadian theologian Tom Harpur. But there is a danger that a theist undertaking such a task will fail to distinguish between conclusions based on the evidence and conclusions stemming from his incurable preconditioning. When that failure is gross and blatant, the result is the mishmash of defensible extrapolations and indefensible fantasy found in *The Test of Love*. I have no hesitation in guessing that Ellis, Friedman, Cupitt and Harpur either have or will eventually accept the implications of what they already know to be true, and abandon all religious belief. I make no such prediction about Weeks, whose heterodox speculations are no more convincing, to himself or anyone else, than the imaginative fantasies of *The Passover Plot*.

The closest analogy to Weeks' imaginings that comes to mind is a book about Arthur of Britain, which argued that Lancelot was "really" a German convert to Christianity named Galerius Hadrianus (don't even ask), a plausible theory only to someone who did not know that Lancelot was invented by a French poet in the twelfth century. Similarly, Weeks' psychoanalysis of Jesus' motivations leading to his virtual suicide (as Weeks sees his death) would be plausible only if the anecdotes being analyzed were not pure fiction. Why Prometheus, the only established publisher of competent biblical scholarship in the English-speaking world, would produce a book that treats the resurrection of Jesus as an event from history, is incomprehensible. Perhaps they thought they were providing "balance," giving equal time to dissenting views. They could with as much justification publish the doctrines of the Flat Earth Society, or the even more imbecilic masturbation fantasies of Roman Piso, as if they also were plausible hypotheses.

## 41. *Judgment Day for the Shroud of Turin*
## by Walter McCrone

The myth that a dead man came back to life 2,000 years ago is in no way dependent on the pretence that the temporarily deceased was wrapped in a cloth that was still part of a living plant 1,300 years later. So in the face of physical proof that the Turin Shroud was forged shortly before 1356 CE, why are there still fanatics who insist that it is the Christian junior god's burial cloth? That is the one question Walter McCrone is unable to answer.

The spurious nature of the shroud has been known since 1389. That was the year the artist who painted it confessed the hoax to the bishop of Troyes and the bishop then informed the pope.

McCrone was the first person to establish by forensic evidence, as opposed to taking the forger's word for it, that the shroud image was painted on the cloth rather than created by some other means. He showed that most of the image was made from an iron earth red ochre tempera paint, while the "blood" was overpainted with vermilion. All tests for the presence of blood proved negative (p. 155): "Anyone reporting blood is either deliberately deceiving himself and others or guilty of bad science." The theory that the image was formed by wrapping the cloth around a three-dimensional body was shown to be nonsense (p. 139): "Yet, there is no distortion of this image as one might expect from a draped cloth under, around and over a body; there are no folds, doubling or elongation of the image due to flattening of the cloth from its rounded contour over the body." McCrone's microscopic examination of adhesive tapes applied to the shroud showed none of the 54 Middle East pollens allegedly detected by Max Frei (p. 298): "My conclusion is that Max Frei believed so firmly in the authenticity of the Shroud that he manufactured his data in support of that conclusion."

McCrone expresses admiration for the artist's skill (p. 141): "The artist who produced the 'Shroud' made no mistakes in style based on what we expect to see on such a cloth. But he did use pigments and a paint medium when he might have used vegetable dyes or, better, body fluids. There is no question about this mistake and it constitutes final and conclusive evidence for an artist's rendition."

Nowhere in McCrone's list of references does he cite Joe Nickell's *Inquest on the Shroud of Turin* (Prometheus, 1988). That perhaps explains why he dismisses as trivial the shroud image's similarity to a photographic negative. Nickell described how he replicated the negative image using methods that would have been readily available in the fourteenth century. McCrone is satisfied that no indirect or negative-making procedure was necessary, and that the picture was painted directly onto the cloth with a brush and very thin paint.

McCrone notes that the image on the shroud is of a man 182 cm (6 ft) tall. But not being a historian, he has no awareness that, if Jesus had been so tall in first-century Judea, he would have been described as a giant, whereas 600 years of Christian apologists did not dispute Josephus's claim that Jesus' height was 137 cm (4 ft 6 in), a runt even by the standards of his time. (Don Cupitt, *Who Was Jesus?,* London, 1977)

For his effrontery in reaching conclusions based on the evidence, rather than forcing the evidence to conform to predetermined conclusions like the good little rubber stamp the Shroud of Turin Research Project (STURP) thought it was procuring, McCrone was forced to resign from STURP as the only way to prevent his findings from being suppressed. And even though carbon dating by three independent laboratories annihilated the shroud's pretended antiquity, causing the Catholic Church on October 13, 1988 to announce that it was not authentic (p. 250), McCrone continues to be vilified for his heresy by incurable dogmatists to this day. One sees a similar "burn the heretic" attitude toward James Randi from the parapsychologists who, having pronounced trickster Uri Geller a genuine psychic, continue to maintain that Randi had no right to prove them wrong. STURP and other hardcore believers are as impervious to falsifying evidence as MUFON and CUFOS in the field of UFO investigation, Ian Stevenson in children's reincarnation playacting, and the Flat Earth Society in geography. The shroud is dead. Only persons incapable of evaluating evidence refuse to bury it.

What is truly pathetic is that this book needed to be written at all. Everything in it had previously been published. The significant facts, except for the carbon dating, had been summarized by Joe Nickell. Twelve years after the carbon dating, the need to continue arguing for the validity of that dating is analogous to arguing, several years after the Mariner space probes had proven the point, that there are no Martian canals, as if it were still in doubt.

Nonetheless, this is the definitive analysis of the evidence, by the man who actually did the research. While it may be slow reading to laymen, certainly compared to Nickell's account, it is the most authoritative. While the only way to get the full picture is to read both McCrone and Nickell, it is McCrone's book that hammers the last nail in the pseudo-shroud's coffin. I recommend it for specialists and dabblers alike.

## 42. *Rebel to his Last Breath: Joseph McCabe and Rationalism* by Bill Cooke

"This is the first biography of Joseph McCabe (1867-1955), a former Catholic priest who became one of the best-known champions and a prolific popularizer of freethought and rationalism in the first half of the 20th century. Though [he is] largely forgotten today, biographer Bill Cooke makes a strong case, in this fascinating overview of the man and his work, that McCabe deserves to be remembered, especially now." Since I cannot improve on that blurb from Prometheus's press release, I will not attempt to do so.

McCabe was a prolific author, turning out 95 books, 104 monographs, and 138 pamphlets by Cooke's count, or a somewhat larger number as Gordon Stein counted. (Cooke acknowledges that either criterion for counting is as valid as the other.) His books included *The Religion of Woman*, an argument for sexual equality that preceded Friedan, Greer and Steinem by a half-century, *The Bankruptcy of Religion*, *The Growth of Religion*, *The Twilight of the Gods*, *The Pope Helps Hitler to World Power*, and *A Rationalist Encyclopedia*. Yet his name seldom appears in modern bibliographies, particularly in books by proponents of the belief system he so effectively debunked. This is hardly surprising. McCabe is an unperson to theologians for the same reason Carl Sagan is an unperson to astrologers, Robert Baker is an unperson to hypnotists, Philip Klass and Robert Sheaffer to UFOlogists, Albert Einstein to numerologists, Stephen Jay Gould to creationists, and James Randi to parapsychologists: "My mind is made up. Don't confuse me with facts."

Cooke acquits McCabe of practising *scientism*, a word I have not encountered before but which apparently means "placing faith in science as an agency of salvation." (p. 117) He also acquits him of *historicism*, another word with which I am unfamiliar, meaning "the tendency to see history in terms of large-scale historical developments that conforms to the tenets of some ideology." (p. 90) While McCabe placed science, otherwise known as knowledge, ahead of alternative methods of evaluating reality, he was, in Cooke's view, primarily a historian. (p. 89) That would explain why "McCabe denied that science was responsible for the drift away from religion. People's wider knowledge of history was also responsible for this drift." (p. 121)

"McCabe warned his readers in one of his popular science works that no eugenics scheme 'has yet been formulated which quite satisfies the unbiased mind.' He was always critical of the assumptions made by many eugenicists, such as the physical and mental superiority of the middle and upper classes." (p. 128) "McCabe specifically repudiated the idea of a scientifically arrived-at theory of race." (pp. 138-139) Obviously the geneticist who attributed measured differences between average I.Q. scores of different races to genetic rather than environmental factors was not familiar with McCabe's conclusions.

As early as 1941, McCabe recognized that labeling behavior as "instinct" was a deceptive way of saying, "I have no idea why they do it." "McCabe thought the words to be excised should be 'soul, mind (except as a collective name for the functions of the brain), psyche, libido, will, instinct, and intuition.' Since McCabe's day, most of these terms have been discredited as naming any specific scientific process." (p. 138)

"He also thought evolution to be 'the most revolutionary discovery science ever made.'" (p. 124) Furthermore, "McCabe also comes across as remarkably ahead of his time in his recognition of the need to expose the underhand tricks of opponents of evolution, people now known as creationists...McCabe noted that creationists 'invariably represent to their readers that Evolution is *disputed in science.*' This, McCabe correctly pointed out, is not the case." (p. 143)

On religion: "McCabe's theory of origins was etiological, which involves recognizing religion as a human institution and seeking knowledge as to how that institution evolved." (p. 183)

In his 1926 Little Blue Book, *Did Jesus Ever Live?*, "McCabe came to the euhemerist conclusion that it is 'far more consonant with the facts of religious history which we know, to conclude that Jesus was a man who was gradually turned into a God.'" (p. 190) While that conclusion is today shared by about half of all competent biblical scholars (including Harwood), it was written at a time when it was a virtual rationalist dogma that no such person as Jesus the Nazirite ever existed. "McCabe was written off as a foolish extremist, when he was noticed at all, but later on the liberal theologians are hailed as courageous pioneers for a new kind of faith." (p. 191)

"As late as 1951, when few scholars were taking any notice of his work, McCabe was still warning his readers to be as skeptical of congenial conclusions as of opposing ones. He was the first to admit that this is not easy. 'It is hard to sacrifice a succulent piece of news that helps your cause because it is not sufficiently proved.' Where McCabe was writing partisan works, he clearly and honestly stated them to be so, and is not guilty of feigning objectivity or passing interpretation off as fact." (p. 263) That is equally true of the present generation of biblical scholars (as opposed to theologians). Yet most theologians are able to brainwash themselves that it is the historians who start from predetermined conclusions and distort the evidence to fit, the precise methodology used by the theologians themselves.

"Despite being entirely self-taught, he avoided nearly all of the fashionable errors of his day and gave his readers sensible, balanced, and reliable overviews on evolution. His book anticipated current ideas on evolution by two generations." (p. 144)

"McCabe deserves to be linked with Isaac Asimov and Carl Sagan as among the most intelligent, acute, and lucid popularizers of science of the twentieth century." (p. 145) Doubled, redoubled and in spades.

## 43. *Entities: Angels, Spirits, Demons, and Other Alien Beings*
## by Joe Nickell

To people capable of believing that 50 virgin-born savior gods who rose from the dead on the third day between 5,000 and 2,000 years Before the Present were all fairy tales, but that on the 51st (approx.) retelling fantasy became history, it is unlikely that this exposé of even an infinite number of entity sightings as delusions and illusions will convince them that all such sightings are without foundation. The Fox sisters' confessions did not put an end to spiritualism, the perpetrators' confessions did not put an end to the Amityville hoax, and it is doubtful that the discovery of Bernadette Soubirous's diary outlining in advance a plot to pretend to converse with a queen of heaven would put an end to goddess sightings.

Nickell is most effective when reporting paranormal claims that he personally investigated, and either falsified or found a more probable explanation. And when quoting other authors, some valid, some absurd, he does not put his own credibility on the line by assuming that a source is accurate. Nonetheless, he allows at least one quoted inaccuracy to stand. In reporting on alleged hauntings in Connecticut, he quotes Glenn White: "The curse of the Dudleys goes back at least as far as when John Dudley, Duke of Northumberland was beheaded by order of King Henry VIII in 1533. So was his son, Robert Dudley." In fact John Dudley was executed in 1553 by Queen Mary for his plot to make Jane Grey queen, as was his son Guilford, the nine-day queen's husband. In 1533 there was no Duke of Northumberland, only an Earl, Henry Percy. While Nickell cannot be criticized for failing to check a statistic that has no relevance to his thesis, the mistake raises the question whether anything he derived from that particular source can be taken seriously.

On the other hand, Nickell is certainly at fault when he ignores scientific notation and uses the connotative dating system "AD," that is intrinsically offensive to the five billion humans who do not regard a dead Jew as their Dominus.

Nickell does not make the mistake of deriding earlier-age investigators for failing to conform to scientific standards that did not yet exist. In explaining that Increase Mather's interpretation of what would today be deemed poltergeist hoaxes was conditioned by his culture, he makes the analogy, "Even today…we have our psychoanalysts and parapsychologists who postulate 'psychokinesis'…as an explanation…Sometime in the future commentators will no doubt be telling their readers that they must consider the times in which those theorists lived." Will believers in nonsense see that as a put-down? I certainly hope so.

## 44. *The Supernatural, the Occult and the Bible*
### by Gerald Larue

"My studies have led me to the conclusion that biblical supernaturalism and the supernaturalism of the occult represent two sides of the same coin, and that from some perspectives they may be viewed as rivals using the same basic premises but employing different cultic and historical settings...Because the Bible is a product of its own time, it is not surprising to find supernaturalism threaded throughout its various documents...Modern occultism is also characterized by belief in supernatural powers (both malignant and benign), mythical monsters, extraterrestrial beings, nonhuman forces, and other mystical themes." (p. 12) In other words, superstition A and superstition B are mirror images, and the former hates the latter out of pure jealousy.

On the claim of the most braindead of all religionists, biblical literalists who insist that the original autographs were inerrant but later copyists made mistakes, Larue asks (p. 15) "As one contemplates the claim that the autographs were divinely given or inspired, the question arises: Why, if God was so concerned about accuracy in the original autographs, did he disregard the importance of accuracy in the preservation and copying of these divine revelations?" Not surprisingly, that is a question fundamentalists pointedly ignore, in the hope that it will go away. And how or why would seven different copyists insert flat-earth passages that were not present in the originals?

"In a sense, Old Age beliefs validate New Age ideas, for if supernatural and occultist ideas are found in the Bible, which is the 'word of God,' then present-day nonbiblical supernatural and occultist notions can claim to have a basis in reality. In other words, by virtue of the authority given to the Bible by synagogues and churches, the Bible sustains and perhaps even modifies the new occultism which both synagogue and church condemn!" (p. 17) So religion's hatred of its chief competitor is hypocritical and oxymoronic. So what else is new?

"Obviously, the findings of modern scientific and societal research have not displaced beliefs in the world of magic or spirits or ghosts or demons or otherworldly powers. What is, perhaps, more distressing is to learn that there are police officers, educators, social workers, therapists, nurses, doctors, and others who work directly with human safety, human health, and human learning who believe in demonic powers..." (p. 20) Clearly that passage written in 1990 has not reached a large enough audience, since the year 2000 saw the appointment of an unelected religious nutcase as President of the United States.

Larue shows that the popular (in 1990) scam of channeling, of which even Shirley MacLaine has lately been conspicuously silent, differs in no significant way from phenomena that the Bible attributes to the spokesman Samuel. And his chapter on astrology states (pp. 54-55), "Some people believe there is little harm done by astrological forecasts...There is, however, clear evidence that astrological prediction can produce fear and panic. Some astrologers predicted that in May 1988...Southern California would be shaken by a gigantic earthquake. The Griffith Observatory received thousands of calls from individuals who were upset and even terrified by the prediction. Some persons moved out of state for the month...Of course the earthquake never happened...The predictors were wrong. They and those who publicized their forecasts were socially irresponsible and unethical. They disrupted lives—emotionally, physically, and financially." And he is not unwilling to assign blame for the proliferation of such nonsense (p. 55): "By their very silence concerning the presence of astrology in the Scriptures, which affect present-day beliefs, both synagogue and church give token acceptance to the validity of astrology."

In the chapter, "Extraterrestrials—Angels and Other Heavenly Creatures," after showing that *The X Files'* aliens differ little from the lifeforms described by Ezekiel and other biblical authors, Larue concludes (p. 65) "Modern sightings are in the same literary category as biblical sightings:

fiction." And he shows that the modern dowsing delusion, in which a divining rod is used to find water, is an adaptation of the biblical myth of Moses using his phallic staff to produce water out of a rock.

After showing that even the most imbecilic New Age practices are adaptations of beliefs endorsed by the Bible, Larue concludes (p. 303), "It is time, a time long overdue, for the Bible to be recognized for what it really is—a collection of writings by people who lived in Palestine and the Mediterranean world some 2,000 to 3,000 years ago, with no more accuracy or authority concerning the nature of the cosmos, the world, or life than any other ancient writing. The Bible is a product of its time. Its occult superstitions are out of harmony with the modern world...Its supernaturalism...continues to encourage magical thinking in an age of science."

In the twelve years since *The Supernatural, the Occult and the Bible* was written, it has not been superseded. It is still the definitive analysis of the connection between belief systems that their respective adherents tout as unique, but which are in fact interdependent.

## 45. *An Encyclopedia of Claims, Frauds, and Hoaxes of the Occult and Supernatural* by James Randi

No more than about half of the 666 entries in Randi's *Encyclopedia* are also to be found among the 3,500 entries in *Dictionary of Contemporary Mythology* (1stbooks.com), and for that reason if for no other, both books should be required acquisitions for anyone who wants a handy guide to contrary-to-fact belief systems in his reference library. On average, the entries in each are about the same length, although Randi on several occasions includes additional material not found in *Dictionary*.

*Encyclopedia* is not without errors. For example, under "Simon Magus" Randi writes, "Simon Peter (Saint Peter) followed him around, outmiracling him at every opportunity and finally encountering him in Rome. In desperation, Simon Magus announced that he would fly to heaven from a specially erected tower in the Campus Martius. Despite his claims to flight, he fell from the tower when Saint Peter prayed to have him fail in his attempt. Simon broke both legs, and subsequently died from his injuries."

Since Randi clearly does not believe such nonsense, presumably the clarification stating that the foregoing was a Christian myth was omitted somewhere between the manuscript and the print shop. In fact the myth can be found in "The Acts of Peter," which in turn can be found in *The Judaeo-Christian Bible Fully Translated*, volume 7 (Imprintbooks.com).

The only other error I detected, although there may conceivably be more, is in the entry "UFO." Randi states that the inventor of the UFO, pilot Kenneth Arnold, reported seeing "saucer-shaped" flying objects. In fact Arnold reported seeing boomerang-shaped objects flying erratically "like a saucer if you skip it across the water." It was the media's misquoting him that led to the "flying saucer" craze. (see "unidentified flying objects" in *Dictionary*) But such errors are the exception. The bulk of the book is factual and informative.

Arthur C. Clarke, in his Foreword to Randi's book, writes, "I am a little disappointed that Randi doesn't deal with one of my pet hates—Creationism, perhaps the most pernicious of the intellectual perversions now afflicting the American public." But he also writes, "Unscrupulous publishers, out to make a cheap buck by pandering to the credulous and feebleminded, are...sabotaging the intellectual and educational standards of society, and fostering a generation of neobarbarians," and urges, "How I wish that Randi's *Encyclopedia* could be in every high school and college library, as an antidote to the acres of mind-rotting rubbish that now litter the bookstands!"

Doubled, redoubled and in spades.

## 46. *Science: Good, Bad and Bogus*
## by Martin Gardner

Most Prometheus books disappear from the publisher's trade catalogue (but not from their website catalogue) after two or three years. *Science: Good, Bad and Bogus* is still in the current catalogue 21 years after its first hardcover publication. As far as I am aware, it was Prometheus's only bestseller, at least partly due to its receiving a rave review in *Time*, August 10, 1981. While some chapters have been superseded as the definitive exposition of a particular subject (Gardner's essay on Conan Doyle cannot compete with Massimo Polidoro's *Final Séance*, or his comments on Uri Geller with James Randi's *The Truth About Uri Geller*), this is still, after two decades, the most complete, all-embracing exposé of superstitious hogwash available on this planet.

In his introduction (p. xiv), Gardner distinguishes between two kinds of disinformation peddlers: "Cranks by definition believe their theories, and charlatans do not, but this does not prevent a person from being both crank and charlatan." And the prevalence of both is highlighted by the statistic on page 190 that America (at the time of writing) had 20,000 astrologers and only 2,000 astronomers. I seriously doubt that the proportions are any less obscene in 2002.

Gardner's chapter on "Magic and Paraphysics" thoroughly refutes the common delusion that scientists are particularly skilled at investigating claims of the paranormal. As Gardner notes (pp. 91-92), "When a person is mystified by a good magic trick it is because he can't figure out how the magician did it. When a physicist is mystified by an unexpected observation it is because he can't figure out how the universe did it. The big difference, of course, is that the universe plays fair...Any magician will tell you that scientists are the easiest persons in the world to fool...Unless he has been thoroughly trained in the underground art of magic, and knows its peculiar principles, he is easier to deceive than a child."

That observation was borne out in spades when Harold Puthoff and Russell Targ, both with Ph.D.s in physics, examined magician Uri Geller and pronounced his fifth-rate conjuring tricks genuine paranormal phenomena. Gardner's self-evident summary, that parapsychologists to this day refuse to grasp, is (p. 318) "If magicians can reproduce a psychic's bag of tricks it does not prove him a charlatan, but it enormously increases the probability that he is, and it makes mandatory the presence of a knowledgeable magician in any laboratory test of the psychic that can be taken seriously."

Part Two of Gardner's book is a reprint of book reviews first written for the *New York Review of Books* and elsewhere. He praises, among others, books by C. E. M. Hansel, Milbourne Christopher, and Carl Sagan, while demolishing scientifically illiterate drivel such as endorsements of spiritualist Arthur Ford, faithhealer Ruth Carter Stapleton, mountebank Uri Geller, talking apes, and the ridiculous paean to ignorance, *Close Encounters of the Third Kind*.

In the words of the publisher, "Martin Gardner examines the rich and hilarious variety of pseudoscientific conjectures which dominate the modern media. With a special emphasis on parapsychology and occultism, these witty pieces advocate the need for better controls in parapsychological research and the even greater need for better public understanding of the difference between good and bad science."

In summary, I can only endorse the opinion expressed by Isaac Asimov: "There are all too few clear-thinking and brave individuals willing to speak out in favor of Sense and Science. One of the best, the coolest, and the most indomitable is Martin Gardner, and in this book he neatly impales the foe with his clear wit...Absolutely fascinating!"

## 47. *From the Wandering Jew to William F. Buckley Jr.*
## by Martin Gardner

For some reason, Martin Gardner can reach the conclusions demanded by the evidence in all other fields, but in one area he can go just so far and no further. He is as skilful as Asimov, Sagan, or any other nonsense-debunker in following evidence and logic to its inevitable conclusion, recognizing that, when all testimony for the validity of a security belief such as ESP, UFOs, psychic power, astrology, et cetera, is discredited, the security belief itself becomes indefensible—in every field except religion.

Gardner calls himself (p. 312) "a philosophical but non-Christian theist." And he has the same contempt for the imbecility of biblical literalism that rejects the findings of science on such issues as the age of the universe and the origin of species as any nontheist. But he will not or cannot recognize that god(s)/karma/an ordered universe is optimism unrestrained, and accept the reality that "what you see is what you get."

Gardner is particularly critical of the mindless fundamentalism typified by the likes of William F. Buckley Jr., a man so incapable of questioning Catholic dogma that, when a psychopath such as Pope Pius IX promulgates doctrines that directly contradict previous papal utterances, he is able to brainwash himself that the new doctrine is what his infallible Church has always taught.

Gardner is, unfortunately, not an expert on biblical criticism. On page 11 he appears to endorse the dogma that "Matthew" was the earliest canonical gospel. And when he writes (pp. 134-135), "Nowhere does the Bible defend the doctrine of reincarnation," that is simply wrong. In John 9:2, when Jesus is asked if a man was born blind because he sinned, the clear implication being that he sinned in a past life, Jesus' reply indicates his acceptance of the validity of such a theory even while ruling it out for the specific instance cited. The Bible endorses reincarnation, astrology (Genesis 1:14), a flat earth (Matthew 4:8 and six other places), palmistry (Job 37:7), and an earth that neither orbits the sun nor revolves on its axis (Psalm 93:1). Yet Gardner, while recognizing its errors, treats the Bible with a respect that is surely undeserved.

He expresses beliefs about Jesus the Nazirite that can only be attributed to cultural conditioning (p. 334): "Life beyond the grave was at the heart of Jesus' teaching." That is inaccurate. Jesus was a religious fanatic, not unlike Ruholla Khomeini. And like Khomeini, Jesus' fanaticism was aimed at imposing his personal mindset on *this* world, with himself as his deity's absolute theocrat. Similarly, Gardner's reference to "the Virgin Mary" (p. 191), as if such a description of a mother of seven was not an absurdity, can be attributed to political correctness, since he elsewhere makes clear that the virgin birth was a late interpolation that was not part of the original gospels and was certainly never claimed by Jesus.

When Gardner writes that, "Pentecostals are evangelicals who take their name from the biblical account of the seventh Sunday after Easter, when the Holy Ghost descended on the Apostles," I find the absence of the word "allegedly" extremely grating. And when he writes of the god who drowned the human race "except for Noah and his unremarkable family," and asks, "Can you imagine a deity more distant from the loving God taught by the historical Jesus?" that is difficult to harmonize with the reality that it was the god created in Jesus' disturbed imagination who created the unspeakable sadism of Hell. Yet elsewhere (p. 236), Gardner quotes with apparent approval H. G. Wells' declaration that "The Christian God of hell was an utterly detestable maniac."

Gardner's chapter on the Wandering Jew myth refers to "the sad attempt of Christians to avoid admitting that the Galilean carpenter turned preacher did indeed believe he would soon return to earth in glory, but was mistaken," (p. 20) but fails to point out that the Catholic Church has never dared repudiate the myth because, without a Wandering Jew to fulfill it, Jesus' prophecy of his theocracy being established before the death of some of the persons who heard him preach would have to be acknowledged as unfulfillable.

In his review of two books about Mary Baker Eddy, Gardner shows that skeptical biographers who leaned over backward to treat Mrs Eddy charitably could only do so by ignoring or misinterpreting the evidence. But neither here nor in his own biography of Eddy does Gardner point out that Christian Scientists had better hope that their god does not exist. For if it does, it may forgive them for murdering their children by withholding lifesaving medical procedures. But it must surely view with outrage their incredible blasphemy of accusing it of *ordering* them to murder their children.

In comparing Mary Baker Eddy and Ellen White, the contemporaneous inventors of the Christian Science and Seventh Day Adventist cults, Gardner writes (p. 196), "It's a great pity that those two most remarkable women ever to start and lead a new Christian faith never met. Had they done so, you can be sure that each would have thought the other a pious, ignorant, demented humbug."

When Gardner writes (p. 205) that Christian Science is "non-Christian," I can only attribute such a statement to the mindset of someone who sees Christianity as basically a force for good, as Christian Science assuredly is not. I have encountered radical Protestants who in all seriousness deny that the pope is a Christian. But the declaration that a self-professed Christian is nothing of the sort, coming from a non-Christian, is surprising to say the least. My own definition of a Christian is, "Anyone who says (and believes) he is a Christian."

Reviewing a book that annihilates reincarnation theory, Gardner writes (p. 135), "Perhaps a few with brains still open to reason will find Edwards' rhetoric persuasive." But realism demanded that he also write, "It is a rare event when believers of any stripe change their minds about anything."

I was delighted at Gardner's recognition (p. 242) that H. G. Wells' satire, *The Country of the Blind*, was about "true believers in a fixed ideology, whether political or religious, who are blind to knowledge and reason." I had long wondered how the story's obvious portrait of a nontheist in a country as permeated with blind, unquestioning religion as England could have escaped the notice of everyone but myself. But I was disappointed at his failure to observe that *1984*, which George Orwell originally titled *1949* to make clear that it was about a clear and present evil, satirized totalitarian Anglican Christianity by equating it with totalitarian Russian communism. It was Orwell's publisher who demanded that *1949* be disguised as a prophecy of the future by changing the name to *1984.*

After reprinting a review he wrote of Carl Sagan's *The Demon Haunted World*, Gardner addends three letters that the review triggered from incurable believers in Christian Science and homeopathy. Those letters, more than anything else, demonstrate that the more rational a defender of indefensible nonsense sounds, the further removed from reality he actually is. Gardner is of course too polite to point that out. Only in criticizing a book asserting that mathematics has no objective existence outside of the human mind does he refrain from pulling his punches. And even here (p. 221), the strongest language he is willing to use is "Baloney!" a word he could profitably have used in a dozen other places. My comment is more likely to have been, "Get a brain transplant."

Reporting on a pro-astrology article in *Life* Gardner writes (p. 27), "Are there New Age astrology buffs on *Life*'s editorial staff, or was Miller's article no more than a cynical effort to boost circulation? *Life*'s editors should be deeply ashamed for their trashy contribution to our nation's dumbing down." And on psychic surgery (p. 34): "Perhaps some day the television networks and major publishers, in a fit of moral courage, will realize that when they give invaluable free publicity to medical quacks they are playing with the lives of the innocent and poorly informed." Let me add: Don't hold your breath.

On the pseudoscience of memetics that is an offshoot of the pseudoscience of sociobiology, Gardner notes (p. 215), "To critics, who at the moment far outnumber true believers, memetics is no more that a cumbersome terminology for saying what everybody knows, and which can be more usefully said in the dull terminology of information transfer." Also (p. 214), "The memes-eye view is little more than a peculiar terminology for saying the obvious." My only criticism of those conclusions is that they are far too polite. Memetics is the new psychobabble.

When Gardner writes (p. 231), "A few sociologists actually think physical laws are entirely human inventions," this prompts me to urge him to write a paper on sociology, exploring whether it is merely meaningless doubletalk like memes, falsifiable hogwash like sociobiology, or a valid and useful discipline (a conclusion I do *not* expect a competent researcher to reach).

I have to object to Gardner's throwaway line on page 328, "Quebec longs to break loose from Canada." Barely one third of Quebecers wish to separate from Canada. Only a fraudulently worded referendum question produced a forty-nine percent Yes vote, by deluding voters that they were *not* voting for separation.

Gardner writes of Anthony Aveni, a psychic-debunker who, like Gardner himself, leans over backward to avoid calling the belief systems he falsifies indefensible nonsense (p. 144), "One longs for Aveni to abandon his curious notion that science and magic are somehow equally valid ways of seeing the world." I would like to see Gardner abandon his curious notion that science and "philosophical theism" are equally valid ways of seeing the world.

When a writer's field of interest covers as large a canvas as Martin Gardner's, it is inevitable that most readers will rate most of his chapters as "don't miss" (else why are they reading the book?), and other chapters as "don't bother." Since Gardner's chapters were originally written for publications aimed at specific and very different audiences, one must expect nothing less. A reader who is not a science fiction fan, for example, will have no interest in the chapters on that subject. In my case, it was the three chapters (out of 29) on the would-be successors of Mary Baker Eddy that, if I had not been planning on writing a review, I would not have read at all. *All* of Gardner's writings deserve permanent preservation between hard covers, and putting articles that will not all interest the same people in the same book is the only practical way that can be achieved. While Gardner could not hope to top himself (*Science: Good, Bad and Bogus* remains incomparable), *Wandering Jew to William Buckley* is as good as it could possibly be.

## 48. *The Wreck of the Titanic Foretold?*
## Martin Gardner, ed.

Can information travel backward in time?  If you are already satisfied that the answer is an unequivocal and fully-proven No, your only reason for buying *The Wreck of the Titanic Foretold?* will be either to check the extent of the coincidence that the *Titanic* hit an iceberg in 1912 and a ship named the *Titan* hit an iceberg in a novella published in 1898, or to see how well Martin Gardner demonstrates that coincidence is precisely that (very well indeed).  Gardner's primary market for this book, as he is no doubt aware, will be people who believe that the future *can* be foreknown, and will buy it in the expectation that it will reinforce that superstition.

Actually, such persons will not be disappointed.  Gardner reprints the 1898 novella, *Futility,* later renamed *The Wreck of the Titan,* in its entirety, and anyone who is determined to see its basic plot as showing precognition of events that occurred fourteen years later will have no trouble getting past Gardner's skeptical introduction in order to do so.

*Futility* is a pedestrian action-adventure novella whose hero survives a shipwreck and rescues his lost love's infant child (repeatedly referred to as "it," grating today but standard English a century ago), by an author who was, if competent, certainly no Joseph Conrad.  The novella's resemblances to future history are so generic and inevitable that, were it not for the similarity of the names, *Titan* and *Titanic,* no one would have deemed them remarkable.

Pushers of the theory that *Futility* was precognitive include Ian Stevenson, a man famous for his inability to grasp that children who concoct reincarnation memories have an economic incentive to lie, Uri Geller (enough said), and *Beyond Belief: Fact or Fiction*, a television series with such depraved indifference to truth or responsible reporting that even the FOX network's gullible and undiscriminating viewers forced its cancellation after about four weeks.

Persons who expected *Titanic Foretold?* to be a collection of Gardner's essays, with the ill-fated ship limited to a single chapter, will be disappointed.  But anyone interested in the book's title question, whether from a skeptical or credulous perspective, will not.  This work does not compare with *Science: Good, Bad and Bogus* (Gardner's best), but it does achieve its objective, and is probably the definitive analysis of its limited subject.

## 49. *On the Wild Side*
## by Martin Gardner

The dust jacket of Martin Gardner's *On The Wild Side* states, "In an autobiographical essay written especially for this collection Gardner recounts for the first time how he evolved from being a religious fundamentalist to an ace debunker." I would have loved to read that essay. I searched cover to cover, and it was nowhere to be found. I was however delighted by the autobiographical preface. In it Gardner indicated that even though the first edition of his first debunking of pseudoscience sold so poorly as to be remaindered within a year, two years of ongoing attacks by proponents of the hokum that he falsified made his name such a household word that a new edition went through the roof. That is bound to be encouraging to **other authors planning to unleash similar assaults upon sacred cows** and give the masses information the blissfully ignorant would rather not know.

**[By coincidence, this review appeared in the same issue of *Humanist in Canada* as an extremely flattering two-page review of my *Mythology's Last Gods*.]**

Gardner's book is divided into three sections: *Skeptical Inquirer* columns, book reviews, and essays from other sources. Section three is of the most general interest. Most of the *S.I.* columns are devoted to such obscure mountebanks that, when they first appeared in my favorite journal, I gave them only a perfunctory skimming. Reading them in their entirety for this review was a task I can only describe as *work*, a word I would never apply to the pleasurable pastime of reading Gardner's earlier debunking books. For example, there is a chapter on Sir George Mivart. Asked who Sir George was, the average reader of this review might well answer, "I don't care." After reading the book, the chances are he will still answer "I don't care."

Gardner is a first class scholar, and this book is an excellent piece of scholarship—for students of historical humbuggery. But it is not for a reader more interested in evidence falsifying beliefs that are currently fashionable. Probably Gardner's most appealing chapters to a wide readership are those on the Reagans and astrology, and on televangelism's most notorious humbugs. But I find myself disagreeing with his classification of Oral Roberts as "deluded" rather than "a hypocrite"— perhaps because I am acquainted with two stooges hired to come to Roberts' show in Australia in the 1950s on crutches and pretend to be miraculously cured. It is of course not impossible that they were hired by Roberts' staff without his knowledge or consent.

*On The Wild Side* combines an easy writing style aimed at the general reader, with content more appropriate for specialists. At the risk of going out on a limb, I predict that it will not have the same success as Gardner's earlier books, *Fads and Fallacies in the Name of Science* and *Science: Good, Bad and Bogus,* each of which I recommend without reservation.

## 50. *Secrets of the Amazing Kreskin*
## by Kreskin

Harry Houdini once made the boast that, while performing an underwater escape, he mislocated the hole in the ice through which he was to emerge, and survived for twenty minutes by breathing the narrow strip of air between the water and the covering ice.

Because Houdini's imaginative fantasy was intrinsically plausible, based on a scenario that conceivably *could* happen, to this day more people believe Houdini really did breathe under the ice than are aware that the story was self-serving propaganda.

Similarly, many of Kreskin's personal anecdotes are just plausible enough that no one can say they cannot be true. And Kreskin uses that plausibility to give credibility to those parts of his book that can clearly, unmistakably be identified as fiction. As an example of the former, Kreskin claims to have stopped an elevator to allow the egress of a couple who had forgotten which floor their room was on. He then explains that he must have picked up the information subliminally, perhaps by being near the desk when they checked in. My own explanation is that the incident originated in Kreskin's imagination, and had no historical basis whatsoever.

Such stories create the impression that Kreskin is trying to be truthful, and that (I suggest) is their purpose. A reader who has already concluded that he is reading nonfiction will more readily accept such tall tales as Kreskin's pretended explanation of his headline-predicting trick.

In his chapter, "The Educated Guess," Kreskin describes a trick he performed in 1965. He sealed four predictions in an envelope on April 2, declaring them to be the front page stories that would appear in the Scranton *Tribune* April 9. The envelope containing his predictions was unsealed on April 9, and the predictions were found to be correct.

Up to that point, I have no reason to doubt that Kreskin was telling the truth. The prediction trick is a mentalist's standard, accomplished by a simple conjuring technique. But Kreskin's "explanation" of how he did it is reminiscent of the Houdini incident above. According to Kreskin, he simply played the odds, predicting events that occurred every day and were likely to appear specific to the day's headlines even though they were not.

In fact the predictions were so specific that anyone believing Kreskin's pretended explanation could not fail to credit him with, at the very least, powers of extrapolation every bit as amazing as those claimed by so-called psychics. And that (I suggest) was Kreskin's aim. Since a mentalist of Kreskin's experience could not possibly be unaware of such devices as the Nelson prediction chest, his pretence that he passed up the hundred percent certainty provided by such a device, and instead made "educated guesses," has as much plausibility as Aristophanes' claim that Perikles started the Peloponnesian War because a group of Spartans left Aspasia's whorehouse without paying. While it would not be ethical to reveal how a conjuring trick is done, I can state that the gimmick lies in producing a newly-written list of predictions in such a way as to make it appear to have come out of an envelope or box sealed days or even years earlier.

Prometheus, normally a very discriminating publisher, has been receiving much flak for falling for Kreskin's fatuous hoax. Perhaps the reason for their temporary lapse of judgment is that not all of Kreskin's book is worthless. Several chapters could have been included in a "Boy's Book of Conjuring," and it would have been basically a good book. But even here Kreskin includes alleged effects requiring powers of observation that almost certainly do not exist, and which Kreskin undoubtedly does not possess. And his chapter on hypnotism, while far less detailed than Reveen's writing on the same subject (*Hypnotism Then and Now*, Imprintbooks.com), is generally valid.

But the valid parts are the exception. On balance, Kreskin's book belongs on the same shelf as the works of Uri Geller, the Brothers Grimm, and Baron von Münchausen. Kreskin has long refused

*William Harwood*

to acknowledge that his effects are all achieved by conjuring.  But with this book he crosses the line into the realm of pure humbuggery.

## 51. *The Search For Psychic Power:  ESP and Parapsychology Revisited* by C. E. M. Hansel

"Given a high-scoring subject, it would in the normal course of events be only a matter of time before every critic could be silenced.  But these subjects cease to score high when tested by critics." (pp. 265-6)  "At the present time parapsychologists are no nearer finding such a demonstration than they were a hundred years ago." (p. 271)

That conclusion by Hansel is essentially unchanged from his findings in 1966 and 1980, for a very good reason.  Although parapsychologists continue to claim positive results in ESP research, they have clearly learned nothing from Hansel's exposure of their inadequate controls and unscientific methodology, and for that reason the new claims are as riddled with "dirty test tube" effects as those prior to Hansel's previous editions.  While Hansel makes no mention of James Randi's "Project Alpha," which demonstrated how easily and totally dogmatic believers can be deceived by the methods of conjuring, he did include in an appendix some reviews of his previous books, that demonstrate the believers' near-fanatic rejection of his suggestion that test subjects *do* cheat.

For anyone but scholars or students in the same field (and perhaps even to them), Hansel's book is tough reading.  It is, after all, written not to entertain but to inform.  Hansel sets out to prove that no experiment alleged to prove the reality of ESP can withstand critical examination, and to do that he is obliged to go into an amount of detail that is not for those readers with short attention spans.

Hansel does show, to the satisfaction of anyone who endorses a scientific rather than a metaphysical methodology, that ESP has never been demonstrated to exist and therefore probably does not exist.  He does this, not by finding the flaw in every positive result ever published, since that would take millions of pages, but by finding the flaw in every one of the approximately dozen experiments cited by the parapsychologists themselves as the strongest proof of the validity of their position.  For doing so, he has been condemned by the believers for looking at individual experiments rather than the totality of the research—as if twelve positive results should still be regarded as significant even after all twelve have been shown to contain design flaws that allow for a non-paranormal explanation of the achieved result.  One is reminded of the traditional nine arguments for the existence of a creator-god.  Even its proponents concede that each of the nine is individually spurious, but to a believer, nine wrongs can somehow make a right.

The parallel of belief in parapsychology with religion is further demonstrated in Hansel's report that, in order to assess the reasonableness of the theory that a subject of ESP tests at Duke University cheated, he asked the university for architectural plans relevant to the time the experiments were conducted.  His request was ignored.  Yet Ian Stevenson, a known believer in ESP, reincarnation and other questionable belief systems, made a similar request that was granted.  In an analogous situation, scholars of the eminence of John Allegro and Martin Larson were granted unrestricted access to the Dead Sea Scrolls when they applied as designated mythologians not known to have questioned orthodoxy, but were refused access the instant they published books demonstrating a willingness to reach whatever conclusion new evidence might dictate.

Perhaps the most revealing parts of Hansel's book are the hostile reviews of previous editions that are included in the appendix.  While Hansel is far too polite to question his critics' competence, integrity and intelligence, I am not.  The consistent feature of such reviews is their use of the Big Lie, the accusation that Hansel uses the very tactics they see in the mirror:  arbitrary selection of evidence, suppression of data that contradict their position, *ad hominem* attacks on individuals, and an unwillingness to reach conclusions consistent with the evidence.

Suppose a physical scientist were to make a claim that appeared to contravene known laws (*e.g.*, cold fusion).  Suppose also that a critic such as Hansel were to write, "It does not matter if the anomalous result was caused by the dirty test tube. The mere fact that the test tube *was* dirty, and that its condition *could have* caused the observed result, is sufficient reason to disconsider the experiment until such time as it is replicated under conditions that rule out a dirty test tube effect." Is there a legitimate scientist anywhere on this planet who would not endorse the critic's objection? Yet Hansel's analogous objection to the claims of parapsychologists results in vitriolic attacks that I can only describe as irrational. Instead of thanking Hansel for showing them which weaknesses they must eliminate in order to be taken seriously, the believers act as if he were an Anglican bishop daring to question the "virgin birth."

Belief in ESP and other paranormal systems continues despite the efforts of Hansel, Randi and others to counter the hogwash served up to John Q. Public by the mass media, and will continue for as long as organizations like Time-Life refuse to allow a little thing like truth to interfere with their making money. Nonetheless, books such as *The Search For Psychic Power* make the facts available to the minority who *want* to be informed. Perhaps some day science will actually prevail. Who knows?

## 52. *Deception and Self-Deception: Investigating Psychics*
## by Richard Wiseman

*Deception and Self-Deception* is a reprinting of eight articles from parapsychology journals and two from psychology journals, along with an introduction and closing chapter that are original. Each chapter describes a detailed investigation of a particular miraclemonger, and each chapter reaches the same conclusion, that the test subject had the opportunity to cheat but was not actually caught. In other words, Wiseman devotes 266 pages to the minutiae of experiments that proved nothing whatsoever. This raises the question: why bother?

Every chapter was well worthy of publication in the journal in which it first appeared. Indeed, Wiseman's protocols for testing psychic claimants should provide a blueprint for the majority of parapsychologists who continue to obtain positive results precisely because they do not impose conditions that eliminate cheating. But whereas taking 52 pages to point out the flaws in the Feuding Report (which authenticated a spirit medium) may have been justified in a psychical journal or a doctoral dissertation, it is indefensible in a book.

And in the chapter on SORRAT (30 pages), Wiseman goes into page after page of boring detail to describe how a deck of Zener cards used in an experiment was sealed to avoid tampering. That would have been justified if the experiment had produced either better than chance results or evidence that the test subjects had attempted to cheat. Since neither of those outcomes occurred, Wiseman could simply have stated that elaborate anti-cheating procedures were imposed, and left it at that.

While Wiseman nowhere states that he views psychic phenomena as intrinsically plausible, one gets the impression that he would be rather less surprised by an Indian mystic materializing an artifact by unambiguously metaphysical means than would, for example, James Randi. And Wiseman's use of the words "sheep" and "goats," that only parapsychologists deem inoffensive (those who disagree with them are the goats), supports that impression. But in all fairness it must be acknowledged that Wiseman appears to be every bit as competent an investigator, free of self-delusion, and willing to recognize that false positives occurred only under conditions that did not preclude cheating, as any skeptic. And his refusal to take a confrontational stand against even the most blatant humbugs is not necessarily a bad thing. But when he states that "The authors are not insinuating that either (test subject) are (*sic*) guilty of any fraud," it is hard to harmonize that concession with the subjects' providing Wiseman with a letter containing inaccurate guesses and claiming that it was written by a spirit entity rather than themselves.

Wiseman concludes that, "Individuals' belief in the paranormal affects their recall of a fake psychic demonstration" (p. 14); "The research data does not support the contention that psychics can provide significant additional information leading to the solution of major crimes" (p. 143, quoting other researchers with whom he appears to agree); and "Those with a tendency to suspect trickery are consistently more accurate in their observation of paranormal demonstrations." (p. 252) Those findings represent the book's only significant statements. All else is trivia.

### 53. *Virus of the Mind:  The New Science of the Meme,* by Richard Brodie
### *Thought Contagion:  The New Science of Memes*, by Aaron Lynch

"And as computer-based replicators start to overtake mind-based memes as the primary repositories and communicators of information, these new replicators could have more influence on the shape of the world than memes do, just as memes overtook DNA as shapers of the global environment.  Maybe non-human-based replicators will evolve to the point where we fade into the background, mere asterisks in the stat book of the universe!"

Come again?

Actually, if I am translating the foregoing passage from the Brodie book correctly, it means, "I'll see your doubletalk, and raise you a quark of psychobabble."

It is twenty years since Richard Dawkins infested the world with a fantasy called *The Selfish Gene*, in which he argued in all seriousness that a human being is simply a gene's way of making another gene.  The best description of a theory that credits brainless organisms with humanoid desires and motivations is "anthropomorphism gone mad."

Now, Richard Brodie and Aaron Lynch take the pseudoscience of sociobiology to the next level. *Virus of the Mind* and *Thought Contagion* both endorse a new pseudoscience called memetics that is an offshoot of the less-new pseudoscience of sociobiology.  Neither is totally devoid of redeeming social value, but neither makes any real contribution to the aggregate of human knowledge.  Both endorse the concept of "memes," defined as the building blocks or genes of transferable thought patterns.  Just as Dawkins transferred the generalizations of sociologists to the imaginary world of thinking genes, so Brodie and Lynch carry Dawkins' analogy into the realm of the mind, breaking down thought patterns into sub-particles called memes.  Dawkins was by no means the first person to equate mental processes with biological functions.  Soap opera has often been called "mind pablum for the braindead," and I have personally described the founder of a theofascist pseudo-political party as a "humanoid AIDS virus."  But most people who use such metaphors remain aware that they are merely metaphors.

As a metaphor, the "meme" concept is neither absurd nor useless.  But when the concept of an imaginary thought atom as something objectively real can be traced back to the same imagination that gave the world the fantasy that a chicken is nothing more than an egg's way of making another egg, reasonable people are bound to recognize that they are treading on dangerous ground.  And when every proponent of memetics is found to emanate from a belief system that credits brainless organisms such as genes with humanoid desires and motivations, the onus is clearly on the new missionaries to back up their fantasies with compelling supporting arguments.  Neither Brodie nor Lynch does so.

Brodie in particular appears to believe that his imaginary memes have a physical reality.  He would claim that he does not, but his book stands or falls on the assumption that memes are real, since otherwise he is simply spouting inanities disguised as gobbledygook.

Not all of Brodie's book, separated from its *meme* camouflage, is nonsense.  For example, he repeats some of the observations made by Vance Packard a generation ago, which are still valid if somewhat trivial.  But in including such material, Brodie utilizes a tactic that he himself calls a Trojan horse, leading into hokum by salting it with a modicum of fool's gold.  But overall, the book is best evaluated by a substitution process invented for a different king of writing.

I have on many occasions suggested that the best way to evaluate the Judeo-Christian Bible is to go through it and, everywhere the term "the LORD God" appears, substitute "der Führer Hitler."  Then read it and see if it is possible to believe that a Hitler who did the things described in the Bible is really a nice guy.  Similarly, go through Brodie's book, replacing every reference to a "meme" with "the Great Pumpkin," and see if that makes it any less profound, logical or worth reading. I suggest that it will not.

Brodie's most insightful observation is to be found on page 109: "People tend to read this chapter and, about once a page, say to themselves 'That's ridiculous!'" No kidding.

The difference between the two books is best illustrated by mentally deleting the "memetics" propaganda and looking at what is left. If Lynch dropped the doubletalk and replaced *meme* with such terms as "thought processes," "cultural conditioning" and "behavior patterns," his book would be a valid treatise on cultural anthropology. While he displays an appalling ignorance of primeval religion and the original meaning of taboos, that ignorance does not as a general rule detract from the points he makes on how religion became self-replicating. He does, however, reveal a cultural brainwashing when he describes the trinitarian religion of Christianity as monotheism.

Brodie's inadequacies are more basic. His book is so dependent on the pretense that memes are real, that any attempt to replace the sociobabble with intelligible English would reveal that the whole thing is contentless drivel concealed in doubletalk.

Lynch devotes several pages to the hypothesis that homosexuality is genetic, and that the homosexual gene, like the gene for blue eyes, is recessive, causing persons who inherited one recessive and one dominant gene to exhibit the dominant characteristics (heterosexuality) while retaining the capacity to pass on the recessive gene to half of their descendants. That is analogous to postulating that there is a gene for preferring tennis to golf, or bridge to poker. To the degree that any of those recreational preferences is involuntary, Occam's razor says that environmental factors, not genetic, made it so.

Neither book is informative, and both propagate hogwash. But as Vance Packard proved a generation ago, in *The Status Seekers* and *The Hidden Persuaders,* sociological trivia can make entertaining reading. Whereas Lynch's book, like Packard's, has entertainment value, Brodie's is a waste of a perfectly good tree. Brodie calls his book *Virus of the Mind.* He could not have chosen a more appropriate title.

## 54. *The Sunken Kingdom: The Atlantis Mystery Solved*
## by Peter James

"Only the meanest of skeptics could fail to be intrigued by the mass of anecdotal evidence suggesting that crystals may have unexplored and beneficial properties." (p. 4)

That passage was an early warning that this was not going to be a book I could take seriously, and that tentative judgment did not prove to be premature. Further on, James describes the firing of astronomers who supported the masturbation fantasies of Immanuel Velikovsky as "this shameful episode in scientific history." In my view, it was not the firing of astronomers so incompetent that they could not distinguish between legitimate scientific speculation and nonsense that was shameful, but the failure to fire biologists so incompetent that they touted creationism as a better explanation for human origins than evolution. And James does nothing for his credibility when he includes in his bibliography a book by Charles Berlitz, one about Edgar Cayce, and a journal article on psychic archaeology.

James sets out to prove that Plato's Atlantis tales had a factual basis, and that basis was a Lydian city called Tantalis. To lead up to that conclusion, he debunks, very effectively, all other identifications of Atlantis with real islands such as the Azores and Crete. For example, he shows that parallels between Atlantis's bull cult and frescoes from Minoan Crete are no closer than parallels with several other bull-centered cultures, and that the incompatibilities far exceed the similarities. Unfortunately, in doing so he simultaneously undermines his own claim that Atlantis closely parallels a lost city in Lydia. While a city so devastated by an earthquake that it wound up at the bottom of a newly created lake may be rare, it is far from being so improbable that it must have been the source of the myth of a continent that sank into the sea. And at the end of the book we learn that James did not actually find such a sunken city. He merely postulates that it must be down there. He does not even have the consolation that future archaeology may prove him right, since his theory assumes that a Lydian legend involving Tantalos and Tantalis became corrupted in oral transmission into Atlas and Atlantis—and that can never be proven.

James does make a plausible case that Plato was telling the truth when he claimed to have obtained the skeleton of the Atlantis legend from Solon by way of Kritias, that Solon learned the story in Lydia rather than Egypt, and that Plato inserted the impossible date to harmonize the story with Egyptian claims to have established a civilization 8,000 years earlier. But his attempt to defictionalize the myth and equate Atlantis with a sunken city in Lydia is no more successful than Spyridon Marinatos's attempt to equate Atlantis with Thera and Crete. That scholars are not impressed by James's theory can be inferred from Random House's failure, seven years after the book's first London publication, to produce an American edition.

So does James's book live up to its subtitle, *The Atlantis Mystery Solved?* In the sense that James puts the final coffin nail in the pretence that Plato was writing nonfiction, yes. In the sense that James intended, sadly, no.

## 55. *The Happy Heretic*
## by Judith Hayes

Judith Hayes' *The Happy Heretic* is about as good an analysis of the Judaeo-Christian Bible as can reasonably be expected from an amateur who has the ability to recognize transparent fiction when she reads it, but no expert knowledge of when, how, why or by whom the Bible was written.

For example, she thinks that Paul of Tarsus wrote *Hebrews* and *Timothy*. She thinks that Jeremyah wrote *Lamentations*. She thinks Isayah wrote that a virgin would give birth, whereas he actually wrote that a young woman would give birth, and the Greek Septuagint mistranslated *khalmah* (young woman) as *parthenos* (virgin). She is unaware that the poor shepherds who visited the baby Jesus were plagiarized from Mithra, as was Jesus' Last Supper, or that the wise men who brought Jesus gifts had previously done so for Zoroaster, or that boiling a kid in its mother's milk was banned for Yahweh worshippers simply because it was something that Molokh worshippers did. And her reference to the execution of the "wise and honorable Sir Thomas More" indicates that she is unaware of that fiction writer's role in the calumniation of Richard III.

Fortunately, these minor irritants no more invalidate Hayes' insightful observations than dispute over the correctness of Stephen Jay Gould's "punctuated equilibrium" theory diminishes the demonstrable reality of evolution. Hayes acknowledges that scholarly criticisms of religion are "essential," but argues that there is also a need for books such as her own that approach the subject in a chatty, less formal manner. I agree.

Hayes did not abandon belief in religion because she discovered (as I did) that fifty other virgin-born savior gods rose from the dead on the third day centuries before Jesus. Rather, it was because (pp. 36-37) "The Bible itself, upon close scrutiny, proved so barbaric, primitive and contradictory, that my naive faith was extinguished. My faith had been based on ignorance of the Bible's real words (an ignorance possessed by most Christians), and it was careful examination that ended my faith."

Some of Hayes' horror stories strongly support the diagnosis that religiosity is a form of insanity. Examples: a Muslim mother ecstatic because two of her sons had blown themselves to bits killing Jews, and had therefore been transported directly to a houri-filled paradise without passing GO and without collecting $200; hate cultists harassing the funeral of a man murdered for being gay, waving signs reading "God hates fags" and "Hell has him now"; an American woman's certainty that her benevolent god had answered her prayer when a hospital agreed to cancel an autopsy on her son who had died from a brain tumor, while simultaneously refusing to question why the same god had not saved the boy's life in response to the prayers of a thousand-member congregation; a Catholic monsignor who said that six-year-old boys in his diocese were themselves to blame for being raped by a priest; hundreds of American priests dead of AIDS because their church teaches that the masochism of self-inflicted celibacy (which they clearly did not practise) enables a sadist in the sky to get its rocks off; the Catholic Church sentencing the terminally suffering to an imaginary Hell for depriving the sky sadist of the orgasmic glee it derives from savoring their agony by killing themselves; fanatics who allow their children to die when medical procedures that could save them are readily available; Muslim women murdered by their kinfolk for the "crime against honor" of being raped. And "Mother" Teresa's actions in allowing mass starvation deaths in her neighborhood, while donations collected specifically to feed the poor sat in a bank amassing interest for the Catholic Church, would have had her incarcerated in a "diminished responsibility" ward if her excuse for criminal misappropriation and homicide by depraved indifference had been anything other than, "Because God said so."

Writing about Alabama governor Fob James' threat to use the National Guard to prevent the removal of a "ten commandments" plaque from a courtroom, Hayes asks, "is this not insanity?" If she thinks that a state governor's theocratic fundamentalism is insane, she should visit Canada and look at the would-be theocracy of rabid canines that has become that country's Official Opposition.

Among the "Twenty Questions" that constitute one of Hayes' chapters is: "Why does the pope wear fancy frocks and silly hats?" While not belittling her own answers, let me suggest a better one: to hide the triple-strength diaper he is obliged to wear twenty-four hours a day because, being a chronic retard, he has never been potty trained. (That is a guess, but by no means a frivolous one.)

In drawing attention to the risk to the freedom of every American from the Religious Right, Hayes makes perhaps her most compelling statement (p. 233): "The fact that legislators in the state of Tennessee even *considered* firing teachers who taught evolution as a fact should give us pause. I take no comfort whatever in the fact that such legislation was not adopted. That it was even discussed is frighteningly mind-boggling".

As Hayes so accurately summarizes (p. 235): "All it takes for America to become a theocracy is for nonbelievers to do nothing."

## AFTERWORD

After the foregoing was published in *Ulster Humanist*, I received a letter of reprimand from the book's author. She made no reference to my describing her as an amateur, and instead voiced objections to my immoderate references to the sadist in the sky and the retarded pope. She explained that, in describing some of the non-theological propaganda of religion, "I was addressing Christianity and its quirks *as it is taught to fundamentalist Christians.* I was talking in their terms about their beliefs. I no more think Paul wrote *Hebrews* than I think he wrote *Mary Had a Little Lamb...*"

I gladly acknowledge my misinterpretation of Hayes' position, and my under-estimate of her knowledge of religious history. But I stand by my evaluation of a god that is the epitome of absolute evil, and a pope I am not certain is a diaper-wearing retard, but whose observable behavior is consistent with such a probability.

## 56. *The Koran*
### anonymous

Anyone who wants to know what is in the Koran is advised to read the Penguin edition. Unlike the Authorized English Version, which suppresses unpleasant passages by misleading interpretations that render them innocuous, the Penguin has the advantage of being translated by scholars who, not being Moslem, have no axe to grind and are willing to call a spade a spade.

And the Koran, like all sacred writings composed at a time when the human race was less morally evolved than it is today, is a grisly, gruesome paean to intolerance, sexism, ethnocentricity, and theofascist hatred of all who failed to submit to brain amputations and think only what Mohamed wanted them to think.

Making adherence to the wrong mythology a capital offense was not new. The author of Deuteronomy did likewise (20:16), as did the author of John. (3:18) Mohamed was merely being fashionable when he, or his posthumous scriptwriter, wrote, "For the unbelievers we have prepared fetters and chains, and a blazing fire." (76:4) "The unbelievers among the People of the Book and the pagans shall burn for ever in the fire of Hell." (98:6) "They (*the Jews*) are the heirs of Hell." (58:17) "He that chooses a religion other than Islam, it will not be accepted from him, and in the world to come he will be one of the lost." (3:85)

The Koran in several places responds to allegations that Mohamed was a madman, leaving little doubt that such allegations must have been widespread. Typical passages include, "Has he invented a lie about Allah, or is he mad?" (34:8) "Are we to renounce the gods for the sake of a mad poet?" (37:36)

The Koran is unambiguous concerning the subhuman status of women: "Men have a status above women." (2:228) "Call in two witnesses from among you, but if two men cannot be found, then one man and two women." (2:282) "Men have authority over women, because Allah has made the one superior to the other...Good women are obedient...As for those from whom you fear disobedience, beat them." (4:34)

Like Paul of Tarsus, Mohamed believed that anything his god did could not be evil, including creating humans predestined to damnation: "Allah misleads whom He will and guides whom He pleases." (74:31) "We have predestined for Hell many jinn and many men." (7:179) "None can guide the people whom Allah leads astray." (7:186)

Mohamed (or his scriptwriter) endorsed slavery: "You are also forbidden to take in marriage married women, except captives whom you own as slaves." (4:24)

Perhaps inspired by the Talmud, which decreed that, "One who, intending to kill a gentile, kills a Yisraelite, is to be deemed guiltless," (Sanhedrin 78b) Mohamed similarly prohibited only the killing of fellow believers: "He that kills a believer by design shall burn in Hell forever." (4:93) And he ordered, "Do not kill except for a just cause." (25:68) No doubt fundamentalist terrorists consider the massacre of tourists a just cause.

Unlike the Pentateuch authors, Mohamed preached long after Aristotle had proven that the earth is round. But like those authors, he continued to believe in pre-Aristotle cosmography: "Let them reflect on...the heaven, how it was raised on high; the mountains, how they were set down; the earth, how it was leveled flat." (88:17-20) "He raised [the sky] high and fashioned it, giving darkness to its night and brightness to its day." (79:27-29) "We built the heaven with Our might, giving it a vast expanse, and stretched the earth beneath it." (51:47-48) "We spread out the earth and set upon it immovable mountains." (50:7) "He set firm mountains upon the earth, lest it should move away with you." (16:15)

Recognizing Christianity's mistake in offering potential converts an afterlife of harps and celibacy, Mohamed gave them a real incentive. In Mohamed's heavenly gardens of delight, he promised, "They are to cohabit with demure virgins…as beauteous as corals and rubies…full breasted maidens for playmates." And for converts of the other orientation, "They're to lie face to face on jeweled couches, and be serviced by immortal youths…young boys, their personal property, as comely as virgin pearls." And to make clear that the playmates were not mere well-stacked mortal women: "We created the houris and made them virgins, carnal playmates for those on the right hand…We are going to wed them to dark-eyed houris." (55:56, 55:58, 78:33, 56:12, 52:16-17, 24, 56:35-38, 52:20). That Mohamed's houris resembled Norse mythology's valkyries was undoubtedly no coincidence.

Given the spate of atrocities currently being committed by Moslem fundamentalists, the obvious question is: Does the Koran encourage random violence against heretics, infidels and non-fanatics? And the answer is No. Fourteen hundred years after the founding of Christianity, that religion was engaged in one of the worst atrocities of human history, the Crusades. Fourteen hundred years after the founding of Islam, that religion is showing signs of going the same way, not because the Koran so orders, but because, like the Bible, neither does it discourage heretic hunting. The Koran is not more obscene than the Bible. But neither is it any less so.

## 57. *The Quest for the Historical Muhammad*
### ed. & tr. by Ibn Warraq

*The Quest for the Historical Muhammad* is an anthology of the best Koranic scholarship from 1851 to the present. Unfortunately, that very quality is what makes it difficult reading for a layman, or even for a specialist in another field of religious history.

Several of the authors Warraq includes support the theory that Muhammad was a founder, not of Islam, a theology that evolved long after his death, but of Hagarism, an attempt to establish a new Judaism centered in Jerusalem based on the folklore that Arabs are the descendants of Abraham, his Egyptian wife Hagar, and their son Ishmael. If that theory is accurate, it logically follows that Muhammad saw himself as the new Moses, a military leader with authority from a deity. Yet none of them suggested that Muhammad may have believed himself to be the successor and equal of Moses prophesied in Deuteronomy 18:15-19. (That the author of Deuteronomy may have been Jeremyah, and the promised equal-of-Moses was intended to be Jeremyah himself, Muhammad obviously would not have known.) The Koranic passage in which Muhammad tells his uncle, Abu Talib, that he is following "the religion of Abraham" (p. 143) supports the scholars' interpretation.

As one wrote (p. 111), "In short, the Koran as we have it is not the work of Muhammad...but a precipitate of the social and cultural pressure of the first two Islamic centuries." The reality seems to be that Muhammad flourished in northern Arabia, much closer to Jerusalem than to present-day Mecca, and when it became necessary to establish a "sanctuary" to fit the Muhammad myth, Mecca was chosen because it was the site of the Kaaba, a vulva-image meteorite worshipped as a manifestation of the sex goddess to whom it was previously dedicated. Mecca and Medina were added to the Koran as Muhammad's retroactive stamping grounds.

The association of Muhammad with Mecca and Medina more than a century after his death is not hard to understand, when it is realized that Jesus had been dead for only seventy years when his birthplace was transferred from Galilee to Bethlehem, which he never visited in his life, and his home town was identified as Nazareth, a place that did not even exist until the fourth century. Even the designation of *Muslim* was a post-mortem addition (p. 36): "The earliest evidence of the use of the word 'Muslim' is...690 C.E....The tribesmen did not know the Koran simply because it did not yet exist, for it was put together piecemeal at a much later date. In other words, all the traditional accounts are hopelessly wrong."

Furthermore (p. 91), "The Muslims' delusion that they have eyewitness reports for every aspect of Muhammad's life, is similar to the delusion of fundamentalist and evangelical Christians that in the gospels they have eyewitness reports of the life of Jesus." And one of the book's contributors (p. 97) cites the change of the Caliph's title from "Deputy of Allah" to "Deputy of the Prophet of Allah," as evidence that, "It is the Sunni, not the Shia, version of the caliphate which is the deviation from the original conception."

And on page 103, "The uncertainty of the Muslim historians about Muhammad's dates is just one indication that it was some time before Muslims were much interested in him at all. It is likely that Muhammad, in so far as he was remembered at all, was remembered chiefly as a political and military leader who brought the Arab tribes together and urged them to conquer in the name of their ancestral deity." In other words, Muhammad was viewed as a secular leader, the first Arab king, until after the murder of the fourth Shi'ite Caliph, Ali, and his replacement by the Sunni Caliphs. It was the Sunni whose followers two centuries later promoted Muhammad to Founding Prophet, much the way Paul of Tarsus retroactively promoted Jesus to founder status of a gentile religion he would have repudiated.

That the Sira, Muhammad's official biography, is indefensible fiction is seen by one author as a mitigating circumstance (p. 105): "The character which the narrator ascribes to his Prophet is, on the whole, exceedingly repulsive." Also (p. 30), the "eighty political assassinations carried out by the Prophet and his companions" help explain why Muslim terrorists see nothing wrong with murdering Americans for being American. Since their role model was a serial killer, how could it be wrong to follow in his footsteps? The Christian Crusaders had no comparable model—except of course their Old Testament god.

Muhammad's status as a military dictator did not prevent him from believing that (p. 353), "Muhammad was the last Prophet chosen by God to preside, conjointly with the Messiah who was to return to earth for this purpose, at the end of the world and the final judgment." And (p. 354), "When death overtook him...the revelations had to be re-edited to square with the fact that he had died." Like Jesus, Muhammad believed that he could not die until he had accomplished his divine mission. And in both cases, the death of the immortal had to be rationalized away so that political organizations dependent on their falsified delusions could continue to flourish.

It is intolerable that Ibn Warraq had to prepare this and his previous books, *Why I Am Not a Muslim,* and *The Origins of the Koran,* under a pseudonym to avoid the threat that still hangs over Salman Rushdie. But what is truly evil is not merely that such a threat exists, but rather that even god addicts who would not themselves resort to violence have a depraved indifference to the human rights of heretics and infidels. I would suggest that this book be mandatory reading for America's fastest-growing hate cult, the Nation of Islam, except that, since it is tough reading for the rational (other than specialists in the same field), it will certainly be over the heads of persons incapable of recognizing Malcolm X and Louis Farrakhan as American Khomeinis.

## AFTERWORD

After the above review appeared in *The American Rationalist*, I received the following letter from the Executive Director of African Americans for Humanism, to whose expertise in the matter I certainly defer:

...You noted correctly that Louis Farrakhan is an American Khomeini. Indeed, Loony Lou Farrakhan would probably be proud to accept such a label. He has often spoken highly of the Ayatollah Khomaniac in his speeches. You should understand, however, that Malcolm X had a major transformation in his worldview after he split with the Nation of Islam. He renounced his old religion and embraced Sunni Islam. He publicly apologized for the pain he had caused as a member of NOI. He admitted that he and other reactionary Black nationalists had been dogmatic in many of their pronouncements. Though Malcolm founded the Muslim Mosque, Inc. after he left NOI, he gave most of his time and attention to the completely secular Organization of Afro-American Unity. He openly invited atheists and others to join the organization. Moreover, he continued to argue that religion should be kept separate from major social, economic, and political movements...

Norm R. Allen Jr.

I am happy to rectify the error.

Note that the interpretations of Ibn Warraq and Ram Swarup (next review) cannot both be correct. But each presents compelling arguments for his conclusions, and I lack the expertise to prefer one over the other. Warraq's newest book is *What the Koran Really Says* (Prometheus, 2002).

## 58. *Understanding the Hadith: The Sacred Traditions of Islam* by Ram Swarup

Ram Swarup's purpose in writing *Understanding the Hadith* is difficult to discern. Is he trying to show that Islam is a vicious, homicidal, xenophobic hate cult, and therefore every Muslim should be viewed as a potential Osama bin Laden? Or is he apologizing for Islamic terrorists by showing that they are merely obeying Muhammad's orders? Either way, he succeeds in demonstrating that Islam-think is at best a self-inflicted brain amputation, and at worst a dangerous form of insanity. But then, what fundamentalist religion is not?

Just as the Christian gospels, with their sermons endorsing swindling, xenophobia, slavery, hatred of even family members outside of the cult, and eternal torture in Hell, are the strongest proofs that Jesus was not a nice man, so the Qurān and Hadith are the proof that Muhammad fell somewhere on the evolutionary scale between the Thuggs and the Assassins. Christian rationalizers such as George W. Bush can brownnose their oil suppliers by brainwashing themselves that Islam is a benevolent religion and that Al-Qaida terrorists are an aberration. But Muhammad's own words make clear that tolerance is the aberration, and the terrorists are the true practitioners of his teachings.

Swarup writes (p. 9), "Islam is by nature fundamentalist; and this fundamentalism in turn is aggressive in character. Islam claims to have defined human thought and behavior for all time to come; it resists any change, and it feels justified in imposing its beliefs and behavior patterns on others." And traditionalist Muhammad pushers (p. 6), "utilized their power effectively; they favored and blackmailed as it suited them." He describes Muhammad (p. 11) as, "sensual and cruel—and certainly many of the things he did do not conform to ordinary ideas of morality...The picture that emerges is hardly flattering, and one is left wondering why in the first instance it was reported at all and whether it was done by his admirers or enemies. One is also left to wonder how the believers, generation after generation, could have found this story so inspiring." Reading the foregoing, I had to go back and verify that Swarup was describing Muhammad. The same thing could just as accurately be written of Jesus. As he summarizes (p. 17), "Men driven by ordinary temptations indulge only in petty crimes and small lapses, but committing real enormities needs the aid of an ideology, a revelation, a God-ordained mission."

Talmudic Judaism teaches that the killing of a gentile is not a crime (Sanhedrin 78B). Jesus taught that anyone who rejected his unsubstantiated claim to be the Head Jew would be "pickled with fire." (Mark 9:49) Muhammad's decree was (p. 18), "He who amongst the community of Christians and Jews hears about me, but does not affirm his belief in that with which I have been sent and dies in this state of disbelief, he shall be but one of the denizens of Hell-Fire." And to make room for all Muslims within the Muhammadan paradise's finite borders, "Space in paradise would be provided by Christians and Jews being thrown into Hell-Fire."

"Muhammad, by introducing the concept of religious war and by denying human rights to non-Muslims, sanctioned slavery on an unprecedented scale." (pp. 83-84) And he endorsed slavery both on earth and in his version of Cloud Cuckoo Land (pp. 207-208), "Even the least of the inhabitants of Paradise will have one thousand slaves waiting on him...Every inhabitant of Paradise will have at his disposal five hundred houris, four thousand virgins, and eight thousand women who have known men. He will have the strength to have intercourse with them all."

Like Jimmy Swaggart, Muhammad had one set of rules for himself and another for everyone else. He limited Muslims to four wives (plus an unlimited number of concubine-slaves), but himself married twenty-two, the (presumably) youngest being Aisha, who was (p. 178) "betrothed to Muhammad when she was six years old and he was fifty. The marriage was consummated when she was nine."

Muhammad also had qualities in common with Hideki Tojo. His invasion of Mecca was a calculated violation of a ten-year truce. (p. 119) Also (p. 121), "Assassination, like *jihad*, is an extension of a fanatic creed and psychology. Muhammad had at his disposal a band of hatchet men ready to do his bidding. Through them, he got inconvenient elements eliminated, particularly those who questioned his apostolic inspiration and had the ability to put their opposition into poetry and satire." And on the question of his successor (p. 127), "he appointed a board of six electors" to choose the first Caliph, instructing them, "If any four of them agree on one person and two disagree, then those two should be killed." This raises the question: With a spiritual predecessor like Muhammad, how come Adolf Hitler was a Catholic rather than a Muslim? "In totalitarian ideologies and creeds, faith and loyalty to the leader are the supreme virtues. The followers need have no other. To be saved, it is enough to be subservient." (p.168)

Muhammad's concept of Hell was as sadistic as when it first materialized in the sick mind of Jesus the Nazirite. And like Jesus, Muhammad populated it with anyone and everyone who did not grant him blind, unquestioning obedience (p. 219): "Allah does not exactly forgive the sins of the believers but visits them on the unbelievers...Allah would deliver to every Muslim a Jew or a Christian and say: That is your rescue from Hell-Fire...Allah's sense of fairness is no better than that of the believers. Thus the believers create Allah in their own image."

That Muhammad was essentially a terrorist and serial killer is revealed by his own words (p. 39), "I have been helped by terror." Specifically, "the beheading of eight hundred members of the tribe of Quraiza in cold blood in the market of Medina must have sent a chill of terror down the spine of everyone, foe or friend." The Quraiza's crime? They failed to join Muhammad's jihad to enslave Mecca. "Holy war" might be the granddaddy of all oxymorons to cartoonist Johnny Hart. To Muhammad (p. 131), "Leaving for *jihad* in the way of Allah...will merit a better reward than the world and all that is in it." Does anyone still think Osama bin Laden and Muhammad were not on the same wavelength? Read the Qurān and the Hadith.

# 59. *The Torah and its God:  A Humanist Inquiry*
## by Jordan Jay Hillman

"What remains of the Torah when we openly acknowledge that its God exists in the human mind alone, having been created by human authors as the means, *in their time*, of guiding a people toward its highest human ends?" (p. 27)

Jordan Hillman attempts to answer that question in a manner that will not antagonize or even annoy true believers.  Such an enterprise was doomed to failure from page one, and instead he winds up antagonizing both readers who dispute his basic position and those who agree with it.

Hillman's analyses of the Adam-and-Eve and Cain-and-Abel myths read more like something one might find in the Talmud than in a scholarly dissection of the human author's thinking.  Death existed, so the Yahwist (J), author of the fables, borrowed the Sumerian myth of how Adamuw died because the gods tricked him into *not* eating the fruit of the tree of life, and reversed it to show humans dying because the first humans *did* eat the fruit in defiance of Yahweh's orders.  Murder existed as a fact of human behavior, so J concocted a "first murder" fable.  Yahweh's reason for provoking Cain had nothing to do with any imagined difference between the qualitative worth of Cain's and Abel's offerings, as Hillman theorizes.  J's Yahweh was too capricious to need a reason for playing favorites, so J did not give him a reason.  End of analysis.  And Hillman's take on the hundreds of other biblical fairy tales is similarly naïve, to put it mildly.

Hillman describes how Talmud authors, in the pretence (or perhaps belief) that they were merely interpreting and explaining the Torah, invented new laws that the Torah authors would have repudiated.  He might have mentioned the modern analogy of the United States Supreme Court, in the delusion that it was interpreting the Bill of Rights, inventing new laws that the Founding Fathers would have rejected as imbecilic.

Hillman's comments on Jacob's adoption of Joseph's sons as his own are trivia epitomized.  The Occam's razor explanation, that the Priestly author (P) was trying to explain why the Jewish/Israelite confederacy contained Manasseh and Ephraim tribes but no Joseph tribe, apparently did not occur to him.  And how could it?  He would have had to know what he was talking about.

In labeling "the Lord" a translation of *Yahweh,* rather than a blatant falsification of a proper name, and perpetuating the myth that the generic, dual-sex plural, *elohim,* meant "God," rather than "the (male and female) gods," Hillman goes way beyond political correctness and well into the realm of sheer incompetence.  He states up front that he is "neither a biblical scholar, theologian, nor philosopher," but a lawyer and lawteacher, and that disclaimer entitles him to some leeway—but not to the point of allowing him to go unchallenged on the chronic inadequacies that might be accepted from a doctoral candidate in theology, but not from even an undergraduate in any legitimate discipline.

Given that many theistic scholars have abandoned the practice, the capitalization of pronouns, possessive adjectives, and euphemisms for the biblical god is grating even when conservative theists do it.  When a nontheist such as Hillman does it, that is carrying political correctness to the point of absurdity.  And his persistent use of the English mistranslation "the Lord" rather than "Yahweh," coming from an author capable of reading the Torah in Hebrew, is inexplicable.

It would be understandable for someone who believes that "the Lord God" exists to write with chronic circumspection in case he inadvertently affronts the divine sadist's extravagant ego and it retaliates by zapping him with a thunderbolt.  When a nontheist appears to be similarly motivated, that is beyond comprehension.  Amateurs with no qualifications in documentary analysis have written reasonably competent commentary of biblical myths in the past (George Fox, *The Vanishing Gods*; Tim Callahan, *Bible Prophecy:  Failure or Fulfillment?*).  Jordan Hillman has not.

## 60. *The Dark Side of Christian History*
## by Helen Ellerbe

It is a damning indictment of the ignorance of the masses that an author with no training in history, no competence in biblical analysis, and no awareness that man created God in his own image rather than vice versa, can nonetheless write a book about *The Dark Side of Christian History* that ninety percent of the population will find informative, revealing, and a timely warning of what Moral Majority **[and Canadian Alliance]** types have done in the past and will do again if allowed to come to power.

Not surprisingly in a book by an amateur, a disproportionate number of references and quotations cite books by other amateurs. Ellerbe quotes: Lloyd Graham, who espouses the theory that no such persons as David or Solomon ever lived; Geoffrey Ashe, whose competence in archaeology is not in dispute, but who in all seriousness claims that the image on the Turin Shroud was burned on by radioactivity arising from Jesus' resurrection; Barbara Walker's dictionary and encyclopedia, useful works but not exceptions to the rule that such publications can be a source of statistics but may not be quoted as if they were more than that; and Reay Tannahill, of whom I will be kind and say nothing.

Ellerbe's book is best compared to a long term paper by a first-year undergraduate, deserving of a B+ more on the basis of the author's potential to meet scholarly standards after a few more courses than for the book's intrinsic merit. I seriously doubt that it will find its way into any university library, but for high schoolers and students of Medieval History 101, it is not misleading and might even be useful. Recommended only for persons whose knowledge of the Inquisition is limited to the fact that it was not a high point in human history.

## 61. *God Unmasked: The Full Life Revealed*
## by Ernest Lane

There are many kinds of religionists. There is the dogmatist, the person who kills doctors for disagreeing with him on the morality of abortion, who kills thousands of Bosnian and Irish citizens for praying to a different god, and who passes laws banning all sexual practices, including rent-a-partner, that differ from his own.

There is the universal soldier, the person who would never on his own initiative exterminate an opposition religion in gas chambers, but whose support makes it possible for a Hitler to do so.

There is the intestinally challenged, the person so terrified of the finality of death, that only the mind-deadening opiate of an afterlife belief can get him through the day without losing control of his bodily functions.

There is the metaphysics addict, the person who understands that no god or other higher lifeform has ever revealed its existence, but who nonetheless needs to believe that "we must be here for a reason."

And there is the kind of metaphysics addict who can doublethink that what he knows and what he wants to believe can both be true.

Imagine an astronaut who has looked at the earth from an orbiting space shuttle writing a book that tries to harmonize what he has seen with the doctrines of the Flat Earth Society. Can it be done? Ernest Lane clearly thinks so. *God Unmasked: The Full Life Revealed* is his attempt to do exactly that.

*God Unmasked* is written in the form of a dialogue between the author and an "Al," whom Lane identifies as a personification of *et alia*, but who in effect fills the role of a psychoanalyst listening to Lane's inane ramblings without ever noticing that the patient is not actually saying anything. Lane asserts that "This book reconciles the scientific and religious perspectives of life and of existence in general." (p. 15) The sad part is that he probably believes it. In fact there is more cogent, logical reasoning in *The Jabberwocky*.

For a brief moment toward the end of the book, Lane switches from psychobabble to comprehensible English when he says, "As a product of the imagination, God does not have objective existence. The concept of God is subjective. The lack of objective reality does not detract from the efficacy of the god concept. In fact, the concept of God is empowered by its very lack of objectivity. Therefore, God—as nonexistent in an objective sense—is effective for our lives precisely because of being nonexistent." (p. 200) And if anyone believes that, I have a bridge for sale in Brooklyn that I think will interest him.

Lane concludes that, "As long as I follow the path of integrity and accept doubt and uncertainty as components of my integrity, I can have confidence that my inclinations are valid and right for me." Charlie Brown would probably say the same thing about the Great Pumpkin.

Lane does not need to be shown evidence that gods are imaginary. He knows that. He does not need to be shown that religion has been the cause of unspeakable evil for 3,000 years. He also knows that. And he already understands that "A" and "not A" cannot be true simultaneously. Perhaps the truism he should consider is, "If it walks like a duck and quacks like a duck..." When a hypothesis looks illogical, sounds illogical, and leads to an "A equals not A" paradox, it is time to face the reality that it *is* illogical and should be abandoned. That religion can be harmonized with reality is such a hypothesis.

If Ernest Lane led a good life, it was despite his desire to believe the thing that is not, not because of it. If writing this ridiculous book made him feel better, that is its only useful function.

# 62. *Galactic Rapture*
# by Tom Flynn

To say that *Galactic Rapture* is delightful, entertaining reading is just another way of saying I am a science fiction fan. And to say that it camouflages a clear and present evil as a fantasy of the future is to say that it is typical science fiction.

For more than sixty years, science fiction was the only medium through which any valid moral philosophy could be sneaked past the Religious Right. Robert Heinlein's *Stranger in a Strange Land* annihilated the masochistic dogma that sex is intrinsically evil and can only be justified by a permit from the ruling theocracy. Arthur C. Clarke's *Childhood's End* presented a thoughtful and innovative view of the Christian devil. Isaac Asimov, in his last *Foundation* novel, showed a closed society objecting vehemently to the suggestion that it practised the long-discredited nonsense of religion, even though its mores were thinly disguised Judaism. And then there was *1984,* which satirized totalitarian Anglican Christianity by comparing it to totalitarian Soviet communism, and would never have got past the censors if the publisher had not pressured Orwell into pretending it was about the future by changing its original title, *1949,* to *1984.*

*Galactic Rapture* centers around a self-proclaimed messiah named Arn Parek, whose resemblance to Jimmy Swaggart, Oral Roberts and Sun Myung Moon cannot be unintentional. And in showing the Catholic and Mormon churches of the future consciously falsifying history to maintain their credibility, Flynn is doing nothing more than extrapolating from the past and present. Does anyone really believe the Vatican does not know the Turin shroud is a fake? Or that the Mormon hierarchy does not know that Joseph Smith was a humbug who plagiarized *The Book of Mormon* from an unpublished historical novel by Solomon Spaulding? Flynn's characters continue a long-established tradition.

Consider a Vatican rabblerouser's speech to an army recruited to fight the opposition messiah, Arn Parek (p. 216): "You will be called to fight. Some of you…will taste death on the field of battle. But fear not! Be uncompromising, for to compromise is a sure sign of not possessing the truth." Crusade, anyone? As for Parek's claim that he only practises deception on the ninety percent of the time that his powers fail him, raising the possibility that at some level he is his own most gullible dupe, I offer two words: Uri Geller.

While *Galactic Rapture* will outrage the more dogmatic Catholics and Mormons, the two religions portrayed most unflatteringly, I seriously doubt that it will provoke "the first Mormon fatwa," a possibility Flynn raises in his *Acknowledgements* (p. 12). Most Mormons are already aware, and I quote again, of "the thoroughness and authority with which (Mormonism's) claims have been debunked" (p. 11), and simply do not care, any more than most Catholics feel threatened by the fifty other virgin-born savior gods who rose from the dead on the third day centuries and millennia before Jesus. So a novel in which serial Christs turn up at irregular intervals on various planets will just as easily be shrugged off.

The cover blurb describes *Galactic Rapture* as "an iconoclastic, darkly hilarious epic, packed with hypocritical cardinals, scheming Mormons, religious bunco artists, and cynical media manipulators…a fast paced and engaging satire on the power of worship and 'infotainment' in the future." Well put, and accurate.

*Galactic Rapture* is not, in my estimation, a potential Hugo or Nebula winner (I would love to be wrong). And it will not appeal to the minority of science fiction fans who see religion as something other than the cause of ninety percent of all manmade evil for at least 3,000 years. But for the majority who enjoyed the aforementioned titles, not merely for their literary skills but also for their message, it is not to be missed.

## AFTERWORD

After this review appeared in *The Ulster Humanist*, Tom Flynn informed me that his reference to "the first Mormon fatwa" was strictly tongue in cheek. He does not *really* see any likelihood of such an eventuality.

## 63. *Eye in the Sky*
## by Philip K. Dick

Picture, if you will, eight persons caught in a beam of uncontrolled radiation and swept into a state of unconsciousness in which they find themselves living in a reality created in the imagination of one of the eight. Now imagine that the person whose fantasy universe has become their real world is a prototype Louis Farrakhan, and the new reality is ruled by a creature known as God, who not only exists but is as all-powerful and capricious as fanatic creationists have always imagined him. In the words of the first of the eight to realize what has happened:

"Physically, we're stretched out on the floor of the Bevatron. But mentally, we're here. The free energy of the beam turned Silvester's personal world into a public universe. We're subject to the logic of a religious crank, an old man who picked up a screwball cult in Chicago in the 'thirties. We're in his universe, where all his ignorant and pious superstitions function. We're in the man's *head*...This landscape. This terrain. The convolutions of a brain; the hills and valleys of Silvester's mind."

"Oh, dear," Miss Reiss whispered. "We're in his power. He's trying to destroy us."

"I doubt if he's aware of what's happened. That's the irony of it. Silvester probably sees nothing odd about this world. Why should he? It's the private fantasy-world he's lived in all his life." (pp. 105-106)

As uncomfortable as it may be to live in a reality in which a moronic, unelected, religious nutcase can be made President by a *coup d'état*, and a major political party can become the puppet of America's equivalent of the Taliban, try to imagine living in a world in which the totalitarian fantasies of the Christian Right are the law of the land, a world in which the Ayatollahs' imaginary playmate can do anything the Buchanans and Robertsons would have it do in their ideal America. Philip Dick has constructed such a world, and the end result is chilling.

Dick wrote the stories, "Do Androids Dream of Electric Sheep?" that became the movie, *Bladerunner*, and "We Can Remember It for You Wholesale," that became *Total Recall*. Only since his death in 1982 has he come to be recognized as one of science fiction's giants.

Robert Heinlein wrote about a future history that included a short-lived theocracy run by a First Prophet. He planned to write a novel about how the First Prophet came to power, but found that he could not because putting such a creature front and center was too nauseating. Having a stronger stomach than Heinlein, Philip Dick was able to spell out the horrors of a world in which religion was transformed from a fantasy of the intestinally challenged into observable reality.

As a logical consequence of focusing on the "what if?" of religion and its Sky Führer in books such as *Eye in the Sky*, Dick was able to summarize, "I hope for His sake that God does not exist—because if He does, He has a lot to answer for."

*Eye in the Sky* will appeal only to science fiction fans, of whom very few are also God fans. But for anyone who can tolerate extended exposure to the absolute evil of a world in which religion is synonymous with reality, it is illuminating.

## 64. *Freedom of Conscience: A Baptist/Humanist Dialogue*
### edited by Paul C. Simmons

In 1995 prominent humanists and Baptists conducted a dialogue at the University of Richmond for the purpose of pursuing areas of common interest, such as the threat to church/state separation posed by a growing Christian ayatollahdom in which suppression of freedom of conscience by intimidation, violence, and even murder, has become respectable. *Freedom of Conscience* is a product of that dialogue. While each side made compromises to accommodate the other, perhaps it was the humanists who made more than their share.

According to the book's Baptist editor, Paul Simmons (p. 15), "Jesus' teachings are of great interest to humanists, who hold them among the finest and richest insights ever given the world." That may be true of humanists who have never read the Christian gospels. But to anyone familiar with Jesus' alleged endorsement of swindling and bribery (Luke 16:1-9), self-castration and other masochism (Matthew 19:12), hatred and murder of unconverted relatives (Luke 14:26; Matthew 10:34-36), xenophobia (Matthew 10:5-6), the condemnation of whole communities that scorned him as an upstart local boy (Matthew 11:23), and the unspeakable sadism of the Christian Hell (Mark 9:47-49), the conclusion is inescapable that either those gospels are the most vicious libel of a good man ever penned, or Jesus falls somewhere on the evolutionary scale between Adolf Hitler and Leopold von Sacher-Masoch.

On the other hand, when Simmons writes (p. 55), "The right of every person to exercise conviction and conscience as they may dictate is the issue at stake...the religious convictions of people cannot follow the dictates of others, whether politician or priest," I can only add: Right on!

Not having read anything of length by an incurable believer in some years, I found myself at times on the verge of gastrinal discomfort at the casual, unlearned references to "God" by some of the book's Baptist contributors, as if such a creature not only exists but is a nice guy. The dialogue was organized, not to dwell on differences between humanists and Baptists, but to demonstrate that they have a shared purpose in maintaining Jefferson's "wall of separation" between church and state as the only way to protect America from the theofascists of the Christian Right, whose concept of an ideal America governed by their version of Christianity sounds frighteningly like the Fourth Reich. But surely one of the educated persons present could have mentioned to someone from the other camp, "You must have read your Bible's accounts of your god revealing its existence. Have you never noticed that all such claims were written by the same authors who assured their readers that the earth is flat?"

When Simmons advocates compassionate ecumenism on the ground that it was what Jesus taught, and cites the Good Samaritan fable in support of his position, I am torn between saying nothing so he will go on doing good for a bad reason, and pointing out that the Jewish xenophobe, Jesus, was as intolerant of anybody outside of his sect as Pat Robertson or Jerry Falwell, and would never have told a fable favorable to the "heretic" Samaritans.

Re the chapter by Bernard Farr: I can only surmise that Simmons had to fill a minimum number of pages, and Farr's incoherent gibberish was just enough to make up the shortfall. If merely reading the chapter does not satisfy an objective reader that theology is thinly disguised glossalalia, he should try to translate it into comprehensible English and see if he is still unconvinced.

*Freedom of Conscience* is not intended to be a "balanced" debate on church/state issues. There are no essays on why America should be an enforcer of totalitarian Christianity. Billy Graham was not invited to spell out why he thinks there should be a death penalty for adultery, as Graham defines the concept (p. 30), and Pat Robertson was not invited to justify his position that there should be a constitutional amendment conferring personhood and "right to life" on fertilized ova that have not developed a single brain cell (p. 54), his reason for such a proposal being that, as his god's scriptwriter and spokesman, he knows that "God says so." Nor is there an essay by a member of the American Nazi Party advocating the ethnic cleansing of Christian America by the judicious use of gas chambers. Rather, the book is aimed strictly at persons who are *not* beyond the reach of human reason, whether they are theistic or nontheistic.

This book goes a long way toward rehabilitating the Baptist religion in the minds of freedom-lovers of every stripe, by showing that mainstream Baptists repudiate the theocrats of the Southern Baptist Convention. Pat Robertson and Jerry Falwell no more speak for all American Baptists than Adolf Hitler spoke for all German Catholics. (I won't pursue that analogy. As Jesus allegedly said in Matthew 19:12, "Whoever is able to comprehend, let him comprehend.") Baptist contributor Robert Price notes that (p. 82), "In the state of Israel and the Islamic Republic of Iran, religious leaders can govern the behavior of secular citizens according to religious laws. Americans don't want that."

No humanist could have said it better.

## 65. *Enemies: The Rationalist View of Human Nature*
## by John R. Aldergrove

Despite its title, *Enemies: The Rationalist View of Human Nature* is not about the conflict between religion and rationality. Aldergrove accepts the nonexistence of gods as a given and proceeds from there.

Aldergrove devotes the first seven chapters to dividing the human race into groups he calls Js, Ks, Ps, and Qs, based on where an individual falls on the intelligence continuum. Whether those letters originated as initials or are purely arbitrary, he does not say. Js and Qs represent the lower and upper extremes. Ks constitute the bulk of the two-digit IQs while Ps fill out the three-digit IQs. Of the Ks he writes, "He or she experiences the world as a place where the laws of science do not necessarily apply, where magic and miracles happen, and where the impossible is possible." (p. 25) In other words, persons of below average intelligence evince thought patterns consistent with their being of below average intelligence. But he later states (p. 62), "The P, like the K, is not troubled by any contradiction in his or her beliefs. For instance, many Ps believe that religion and science are both matters of fact." The whole seven chapters can be summarized as concluding that 98 percent of all illogical, irrational and unintelligent thinking emanates from the least intelligent 98 percent of the population. Is that a point that really needed 88 pages to prove? And when, throughout the book, he consistently attributes illogical and self-serving behavior to "Ks and Ps," one wonders if he has forgotten that, by his own definitions, those two labels incorporate 95 percent of the human race.

Aldergrove has apparently never heard of the common gender. Instead of stating early in the book that the words, *he, him* and *his* should be understood to include both males and females, he uses the expression, "he or she" ad nauseam. While the number of repetitions probably did not reach a thousand, it certainly seemed like that many. That is unfortunate because, in a continent in which sentence fragments, run-on sentences, and such vulgarisms as "snuck" are the norm rather than the exception, Aldergrove writes with a technical skill almost unseen since 1945. Apart from a handful of split infinitives, a couple of confusions of the more-than-two plural, "one another," with the dual number, "each other," and two run-on sentences, also called comma splices, I found no other grammatical errors. I am tempted to guess that he went to school in another continent, where teaching is still legal.

In stating that "All non-believers [in religion and other paranormal phenomena] are not Qs, but all Qs are non-believers," (p. 263) Aldergrove clearly distinguishes between a Q and a member of Mensa, the former referring to the top two percent in real or functional intelligence, and the latter to the top two percent in measured IQ, since Mensa is infested to the core with believers in religion, astrology, alien abductions, psychics, and practically every security belief ever concocted. Indeed, he quotes Thomas Szasz' definition of IQ tests: "hocus-pocus used by psychologists to prove that they are brilliant, and their clients stupid." (p. 266)

I find Aldergrove's contention that "Ks and Ps do not believe a person can have both good and bad ideas" (p. 262) flattering, since it implies that I must be a Q for concluding that some of Aldergrove's ideas are meritorious while others are patently absurd. Unfortunately, that flattery loses much of its charm when I recognize that Newt Gingrich, whose religious fundamentalism clearly defines him as a non-Q, shows an ability to recognize that President Clinton is not always wrong.

Aldergrove's evaluation of psychiatrists may be definitive. He writes (p. 174), "But how does the court know whether the accused knew that he was behaving wickedly? How does one read someone's mind?...One consults a mind-reader, of course. The mind-reader, also known as the psychiatrist or expert, pronounces a person sane or insane based upon factors such as who pays him—the defense or the prosecution." He fully endorses the conclusions of Thomas Szasz that,

"Psychiatry is the sewer into which societies...discharge all their unsolved moral and social problems," (pp. 352-353) and expresses his own view, "Psychiatry is as much use in a court of law as the Ouija board or the crystal ball." (p. 352)

After raising the question "How does talking cure an illness?" (p. 222), Aldergrove proceeds to what is probably psychiatry's greatest atrocity, electro-shock "therapy." He explains the practice's continued existence long after medical associations have denounced it: "Medical insurance companies will pay for it." (p. 282)

Aldergrove criticizes the "war on drugs" on the ground that the cure is worse than the disease, particularly the cost of enforcement. He writes (p. 212) that "Switzerland has found (1997) that giving heroin to addicts reduces crime, disease, and taxes." He further supports his position by citing the failure of Prohibition. But he then criticizes the Canadian government for giving in to political pressure, and prohibiting clitoridectomy while not simultaneously criminalizing the analogous child abuse of circumcision. How such a prohibition could be enforced, he does not say. And he is clearly unaware of the wars resulting from the anti-circumcision edicts of Antiokhos Epiphanes and the emperor Hadrian.

Aldergrove has harsh words for the jury system, which requires randomly chosen amateurs to evaluate evidence beyond their competence, and in effect credits them with an ability to read minds. Given the notorious acquittal of a California murderer in the face of overwhelming DNA evidence of his guilt, he has a valid point. He also attributes the continued existence of the pseudomedical nonsense of facilitated communication to the fact that, "Believers in FC claim that the severely disabled can borrow the mind of the facilitator, or learn language by telepathy." (p. 169) And he gives a detailed and informative description of the insanities and atrocities perpetuated by upholders of security beliefs, ranging from the persecution of masturbators and homosexuals to the dogmatic opposition of some doctors to the discovery that ulcers are caused by bacteria, the jailing of a Texas atheist for refusing to invoke the judge's imaginary playmate in a legal affirmation, and the attempt by the psychiatric hierarchy to silence Thomas Szasz.

While I endorse Aldergrove's denunciation of a legal system that equates psychoquacks with "experts" and gives unlearned jurors the power to pervert justice, I am less than impressed by his recommendations concerning the Susan Smith case. He argues that Smith *must* have been insane at the time in order to drown her children and, since the circumstances that caused her aberrant behavior will never re-exist, she should not be incarcerated. (p. 187) He may be right that fear of consequences is seldom a deterrent to crime, but his view that it can never be a deterrent is not one I share. Setting Susan Smith free could send a very dangerous message to others who might find themselves in her position.

In writing about the treatment meted out to whistle-blowers, including the probable murder of Karen Silkwood as a message to other subordinates who might emulate her, Aldergrove compares modern commercial despots with their ancient counterparts. He says (p. 341), "Ten thousand years ago, most tribesmen probably took it for granted that it was the chief's prerogative to misbehave if he wished to do so." Surprisingly, he did not take the opportunity to point out that the same attitude led to the creation of gods who had the right to misbehave if they wished to do so.

Aldergrove is no expert on religion. He has no awareness that Paul of Tarsus did not write "Timothy," and he identifies the combatants in Bosnia as "Serbs" and "Croats," rather than giving them their true designation, "Orthodox Christians" and "Catholics." But he knows enough to cite evidence that both Islam and Christianity may have originated as a consequence of hallucinations accompanying epileptic fits. (pp. 245-246) And he shows that modern religious apologetics differ little from the doubletalk of Tertullian, who wrote, "It is by all means to be believed, because it is absurd...the fact is certain because it is impossible." (p. 254)

Aldergrove is most effective in his rebuttals of the racism of *The Bell Curve* and John Schumaker's *Religion and Mental Health,* which claims that humankind devolved from a higher to a lower level of intelligence because facing reality was so terrifying that it made stupidity and religious doublethink a survival factor. Aldergrove vigorously asserts that retreat into superstition is anything but a survival factor.

Aldergrove devotes several pages (288-303) to an article written by a physicist and accepted and published by a journal devoted to the alleged "sociology of science." The article carefully parodied the journal's endorsement "not just of nonsense and sloppy thinking *per se*, but of a particular kind of nonsense and sloppy thinking: one that denies the existence of objective reality." In debunking the allegedly scholarly debate that his doubletalk article triggered, the physicist wrote, "Nowhere in all of this is there anything resembling a logical sequence of thought; one finds only citations of authority, plays on words, strained analogies, and bald assertions."

Aldergrove asks, "Why did they not see that the article was nonsense?" and provides his own answer: "Having read this book you know the answer. Ks and Ps commonly cannot separate sense from nonsense." Perhaps a better answer would be that sociologists, along with psychologists, theologians and other practitioners of the misnamed social "sciences" cannot separate sense from nonsense. Aldergrove's ability to recognize "prof talk or gobbledygook or babble" for what it is probably explains why he had a very short career as a teacher, an occupation in which the inability to separate sense from nonsense is, at least in North America, mandatory.

Aldergrove points out that natural scientists can be equally gullible, and cites the physicists who examined magician Uri Geller's tricks and mistook them for nonexistent "psychic powers." That is not really a good analogy, because the physicists were not deluded within their own field of competence. A better example would have been the "cold fusion" delusion. And Aldergrove does cite the "N rays" delusion.

In his chapter on interviews and personality tests, Aldergrove recognizes that the perpetrators of those fairy tale rituals must believe they can read minds. Anyone who has ever submitted to such farces will agree.

For some reason, Aldergrove expresses a favorable view of lawyers, and says the reason they are calumniated is that Ks and Ps hate people who appear to be more rational than they are. (p. 257) That his criticism of psychiatrists, for absorbing and reflecting the thoughts of whichever side is paying them, is equally applicable to lawyers, apparently does not cross his mind.

Aldergrove argues for the nonexistence of the soul, the mind, the conscience, and free will. Since "soul" is universally defined as a part of the human organism that exists separately from the body and can outlive it, its nonexistence is recognized by all but metaphysics addicts. But Aldergrove sees "mind" and "conscience" as analogous metaphysical concepts, whereas to the general public "mind" refers to the thought processes stemming from consciousness and self-awareness, and "conscience" to the thought processes involved in distinguishing right from wrong. To say that mind and conscience do not exist is to say that thought does not exist.

On the issue of free will, Aldergrove is not taking the Skinnerian position that every human thought or action is a conditioned response to an external stimulus. Rather he maintains that, since free will is a function of the mind, and the mind does not exist, therefore free will cannot exist. He apparently believes that his not being a serial killer should be attributed to factors other than a personal decision. Some psychobabble theories do not deserve the dignity of a rebuttal.

In classifying *Enemies* as worth reading, "thumbs up," I am obviously not endorsing Aldergrove's "K and P" classifications, or the nonexistence of mind, conscience, and free will. Rather I am saying that there is more than enough wheat among the chaff to justify a careful consideration of his views. I am also assuming that anyone who cannot distinguish between the two would not be reading *The American Rationalist* **[in which this review first appeared]**. Despite some questionable assertions, this is an extremely useful book.

### 66. *Final Séance:  The Strange Friendship Between Houdini and Conan Doyle* by Massino Polidoro

I have long been aware that Sir Arthur Conan Doyle ended his friendship with Harry Houdini on account of Doyle's blind, gullible belief in the very scam Houdini had disproven over and over and over.  But not until I read *Final Séance* did I become convinced that incurable adherence to a security belief in the face of irrefutable evidence can only be described as a form of insanity.  And I am far from the first person to reach that conclusion.

Following the publication of Doyle's second pro-Spiritualism book, the *Sunday Express* ran the headline in its book column, "Is Conan Doyle Mad?"  So far as I am aware, no publication of comparable influence has been similarly blunt in connection with Doyle's spiritual successor, Shirley MacLaine.  It is understandable why "political correctness," requiring more circumspect criticism, is not to everybody's taste.  Not faced with such constraints, the *Express* went on, "One does not trouble to analyze the ravings of a madman.  One shrugs one's shoulders, laughs and forgets."  The more polite *London Times*, reviewing Doyle's previous book, referred to Doyle's "incredible naiveté," while the *Nation* stated, "The book leaves one with a rather poor opinion of the doctor's critical abilities." (p. 169)  And when even an investigator as incredibly gullible as J. B. Rhine (who went on to authenticate ESP in a horse) saw through one of Doyle's pet mediums, Doyle placed a notice in the Boston newspapers, "J. B. Rhine is an Ass." (p. 203)

Houdini was religiously conservative, even disowning one of his brothers for violating one of Leviticus's myriad of sectarian taboos (pp. 218-219).  And when he testified before a Congressional committee in support of an anti-fortunetelling bill, his words, as quoted by Polidoro, were as follows (p. 188):

"This is positively no attack upon a religion.  Please understand that emphatically.  I am not attacking a religion…But this thing they call 'spiritualism,' wherein a medium intercommunicates with the dead, is a fraud from start to finish.  There are only two kinds of mediums, those who are mental degenerates and who ought to be under observation, and those who are deliberate cheats and frauds.  I would not believe a fraudulent medium under oath; perjury means nothing to them…Millions of dollars are stolen every year in America, and the Government [has] never paid any attention to it, because they look upon it as a religion."

Substitute "televangelism" for "spiritualism," and the obvious response is, "So what else is new?"  And when Polidoro writes of a paranormal hoax exposed by Houdini, "It was a typical swindle, still used today by many self-claimed psychics, astrologers, and charlatans.  By this means Reese had been able to gather sums of money from gullible people who, more often than not, were also learned men of science and culture," the response is again, "So what else is new?"

I was surprised to learn (p. 55) that, while Conan Doyle was en route to Australia, some Australian Presbyterians held a prayer meeting to ask their sectarian god to prevent the proponent of an opposition religion (Spiritualism) from reaching their shores alive.  A fringe cult in Vancouver in 1962 held a similar prayer meeting to petition that a stage hypnotist not be permitted to perform in their city.  The god did not answer that request either.

Polidoro does not devote much space to Doyle's authentication of the Cottingley fairies, other than to quote a couple of statements in which Doyle expressed his conviction that little girls do not lie.  That little girls (and boys) are humankind's most notorious liars was quite unknown to him.

On the question of whether Arthur Ford correctly identified the message Houdini had promised to communicate to his widow if he ever came back, Polidoro quotes enough statements from Bess Houdini to make clear that only her desperate desire to believe led her to an initial authentication of Ford's claim.  On sober reflection, she realized that Ford had simply picked up pre-published clues and capitalized on her willingness to believe that the message was what Houdini *would have* sent her if he had been able. It was not a message that he had prearranged to send her. Doyle, not surprisingly, was convinced that Ford had indeed communicated with Houdini, and no one could convince him otherwise.

Even after Houdini's death, in a letter to Bess Houdini (p. 208), Doyle reiterated his stubborn conviction that Houdini possessed the very powers he devoted his life to refuting, including an ability to dematerialize his body in order to pass through solid walls (p. 225). In doing so, he foreshadowed the parapsychologists at George Washington University who, after James Randi's "Project Alpha" had exposed their gullibility by having them pronounce the illusions of two youthful conjurers as genuine psychic phenomena, "impossible to fake," actually asserted that Randi's associates were really psychics who for some reason were now pretending to be magicians. Will believers in pseudoscience ever learn to distinguish between sense and nonsense, and face the reality that their security beliefs have been as fully disproven as phlogiston and the planet Vulcan? Only if Barnum was wrong.

*Final Séance* is acceptably light on typos. Those that got through are of the type a computer spell check would not detect, since the wrong word was not a misspelling. The most indefensible is the statement on page 167 that a letter purported to be from Houdini's dead mother was "already reproduced on pages 97-98 of this book." Presumably those were the correct page numbers in the original manuscript. In the book, the letter is reproduced on page 88. That technicality is the book's only fault. I recommend it highly.

## 67. *The Bible:  Divine or Human Origin?*
## by Andrew D. Benson

*The Bible: Divine or Human Origin?* is less than the sum of its parts.  The author indeed brings together sufficient evidence to enable a reasonable person to answer the question in the title.  But he does so in a manner that is so unorganized, clumsy, and ultimately dull, that not a single believer is likely to be jolted into asking himself, "How can this be?" and starting a desperate search for falsifying evidence (as I did when I first learned that fifty other virgin-born savior gods rose from the dead on the third day, long before Jesus), or even to finish reading it.  One gets the impression that this was a first draft of a Master's thesis, and was never reedited into a book capable of holding the reader's attention.  And that is unfortunate, because the information contained is by no means trivial.

There are some glaring inconsistencies and inadequacies.  For example, on page 6 Benson acknowledges that scholars have identified four main Pentateuch authors (J, E, D, P), and are in broad agreement on who wrote what.  But the rest of the book is written as if that information were unknown to him.  In comparing Abraham's sacrifice of Isaac to Euripides' version of Agamemnon's sacrifice of Iphigenia, in both of which the victim is saved by a *deus ex machina*, he shows no awareness that in the E narrative Isaac *was* sacrificed, and a redactor inserted the divine intervention to harmonize E with J, whose Isaac was still alive as an adult.

On page 150, Benson refers to "Cephas" in a manner that suggests that he is unaware that Cephas and Peter (from the Aramaic and Greek words for "rock") are the same man.  And while he recognizes that "Jesus' religion was not Christianity but Judaism," (p. 151) he also writes that Jesus' brother James was executed because he "changed from being a Jewish Christian to a Hellenist Christian."  In fact James/Iakobos was never a Christian.  He lived and died a Jesus-Jew, and was executed for false prophecy when, in answer to his inquisitor's direct question, he affirmed that his non-Davidic brother was the prophesied descendant of King David anointed by Yahweh to liberate the Jews from foreign overlordship—a rather difficult task for a dead man to accomplish.

Benson very thoroughly demonstrates that the gospels were propaganda that started as fourth-hand accounts and were constantly rewritten to conform to later theology.  Yet he persistently quotes from those gospels as if their alleged facts are not in dispute.  For example, he states, "John the Baptist urged his followers to follow Jesus." (p. 147)  But John and Jesus were opposition messiahs, and remained so to their disciples for fifty years after their deaths.  Had John lived to do so, he would have denounced Jesus' messianic pretensions as heresy.

Benson describes Mark's last-supper scene as "common practice among the Jews," (p. 125) with no awareness that a Jew such as Jesus could never have participated in such a pagan sacramental-cannibalism ceremony (as opposed to the Jewish Seder that was indeed a common practice).  And while he cites Philo as the immediate source of the Christian virgin-birth myth, he describes that alleged event as "the immaculate conception," (p. 134) a Catholic concept that means something completely different.

On the question of Jesus' historicity, Benson reaches what I consider the right conclusion for the wrong reasons.  He ignores the negative evidence that historians find most convincing, primarily that no mythmaker in his right mind would have invented a hero who proclaimed a war of independence—and lost.  Instead, he argues (p. 162) that, "It is unlikely that Paul, a former opponent of Christianity, would have been swayed to accept the historicity of Jesus, had Jesus not truly existed."  He does, however, make the valid point that, "It is unlikely that Matthew and Luke to have (*sic*) invented the point that the Jews considered Jesus a Galilean." (p. 162)

In delineating the differences between Jesus' teachings and Paul's, Benson says that Paul "does not mention any of Jesus' miracles or parables." (p. 152)  Is he suggesting that Jesus was already credited with performing the miracles first described in Mark a generation before Mark was written?

Since Mark clearly borrowed those fantasies from the books of Kings, that is improbable, to say the least.

Benson does recognize that the original authors and interpolators of biblical books were more interested in propagating a doctrine than telling the truth, and so were translators: "Translators took the opportunity to eliminate inconsistencies and contradictions...and to make alterations that reconcile the Bible with today's culture, laws, values, and scientific knowledge." (p. 172) As an example of such alterations, he cites the *King James Version*'s changing of 2 Samuel 21:19 from "Elhanan killed Goliath," to "Elhanan killed the brother of Goliath," to harmonize it with the myth that David killed Goliath. He observes that, "Translations can vary so much that many Christians will not read the versions of other Christians." (p. 173)

Benson dates the Christian gospels "from 66 to 90 CE," apparently not convinced that Jesus' prophecy of the destruction of the Jerusalem temple that occurred in 70 CE could only have been written *ex post facto*. He cites evidence that such prophecies were being made as early as 48 CE, but fails to notice that the prophecy attributed to Jesus was far too specific to be a lucky guess. And he dates the whole of Revelation a generation after the razing of the temple, even though its primary author prophesied that the temple would never be destroyed.

While showing that few of the sayings attributed to Jesus were actually spoken by him (Would an Aramaic-speaking Jew have quoted from Greek mistranslations?), Benson nonetheless takes the gospel authors at face value when he writes, "Jesus predicted his second coming." (p. 277) In fact, Jesus believed that, as the Jews' prophesied liberator, he could not die until his messianic task was accomplished. The gospel authors prophesied Jesus' second coming. Jesus did not.

Benson mentions that, "The complete gospel of Matthew was used only in one location, probably Syrian Antioch. In Palestine, Matthew was used without the birth narrative." (p. 187) And he is aware that "Marcion's gospel of Luke begins with Luke 3:1," and that "the virginity of Mary was disputed among the early Christians." (p. 189) Yet with that information in his hands, it still did not occur to him that the virgin birth was an interpolation, not originally part of Luke or Matthew. For while he draws attention to dozens of other inconsistencies, he apparently did not notice that the virgin birth passages in both gospels directly contradict the adjacent genealogies that trace Jesus to David through his father.

Probably Benson's most thorough and useful section is the sub-chapter covering the internal and external inconsistencies concerning Jesus' time and place of birth and his alleged ancestry. (pp. 192-196) Yet he makes no mention of the forceful conversion of the Galileans, including Jesus' third-generation ancestors, by the Maccabees, and the impossibility of a descendant of gentiles being Davidic.

For sheer size, Benson's 171 item bibliography might seem more than adequate. But its glaring omissions (for example, Larson, Friedman) help explain why the question "Hasn't he heard of...?" kept forming in the reader's mind. And the inordinate number of footnote references to encyclopedias and church publications suggests an unawareness that such sources are, at best, highly questionable.

*The Bible: Divine or Human Origin* contains far more meat than gristle, and readers with the tenacity to plow through random topics that appear to have been written down in whatever order the author happened to think of them will find their efforts well rewarded. But on balance this is a book on which I must vote "thumbs down," not because of its inaccuracies and examples of sloppy scholarship, which are present but not excessive, but because it is so poorly constructed that it is of limited value to friend or foe.

## 68.  *Merely Mortal?  Can You Survive Your Own Death?*
### by Antony Flew

"To suggest that we shall survive such total dissolution [death] is like suggesting that a nation would outlast the annihilation of all of its members...To expect that after my death and dissolution such things might happen to me is to overlook that I shall not then exist." (p. 2)

"Given the undeniable and undenied fact that...we shall die, then the not necessarily soluble, philosophical problem becomes to formulate some survival hypothesis which is not already known to be false." (p. 8)

So much for religion.  On the paranormal, Flew quotes J. Beloff's definition, "A phenomenon is, by definition, paranormal if and only if it contravenes some well founded assumption of science," and adds his observation (p. 169), "We still appear to be as far away as ever from any repeatable demonstration of the reality of any psi-phenomena."

"The more we make astral bodies like the ordinary flesh and blood persons from which they are supposedly detachable...the more difficult it becomes to make out that it is not already known that no such astral bodies do in fact detach themselves at death." (p. 16)

While Flew makes clear that the evidence does not support the existence of "telepathy, clairvoyance and paranormal precognition...(ESP)," he at no time appears to notice that two of the three capacities labeled as ESP involve information traveling backward in time, or that the definitive argument against such claims is the *reductio ad absurdum* to which all time travel hypotheses can be reduced.  I got the impression that, despite his disclaimers such as the above, he is not ready to acknowledge that the astral body/soul is so definitively unproven that further examination of the non-evidence cannot be justified.  He cites Socrates' alleged proof of the psyche's immortality, along with parallel arguments by Thomas Aquinas and other religious and secular philosophers, finds them all to be flawed, and summarizes (p. 171), "We cannot, at best, reach about any such story any verdict stronger than a cautious, and appropriately Scottish, 'Not proven.'"

On both religious and paranormal beliefs that part of the human organism is immortal, Flew's conclusions will neither outrage nor convince believers, for the logical reason that philosophical arguments that the soul/psyche/astral body cannot exist will be met with the response, "Yes, but it does."

It is my view that, having reduced Flew's dissertation from 215 pages to less than 500 words, I have done readers a service, since they now do not need to read the long, meandering, hairsplitting and basically trivial version.  This is a book I can recommend only as a cure for insomnia.

## 69. *The Roving Mind*
## by Isaac Asimov

No intelligent person needs to be told (surely?) that Isaac Asimov's name on a book is a guarantee of excellence. And most are aware that Asimov's essays (62 in this collection) cover the multitude and variety of subjects that Robert Heinlein had in mind when be coined the word *synthesist*. So instead of trying to gild fine gold or paint the lily, I will simply reproduce Asimov's words on a selection of the issues he discusses. *Videlicet:*

— On religious doublethink (p. 25): "If there is an earthquake and a thousand people die, and one person is uncovered from a ruined house, unhurt, the Moral Majority types cry, 'A miracle!' and fall to their knees in gratitude. And the thousand who died, whose deaths, indeed, were necessary to convert the one survivor into a miracle, what of them?"

— On biblical literalism (p. 26): "The Bible textbook, then, says that the sky is a thin sheet of something or other that can be rolled up, and that the stars are little dots of light that can be shaken off the scroll and allowed to fall to earth."

— On resurrected gods (p. 140): "The feared death and the apparent rebirth of the sun...gave rise to myths concerning the death and resurrection of gods: of Osiris among the Egyptians, of Adonis or Tammuz in the Near East, of Persephone among the Greeks. Some of that lingers in the West today in the Good Friday/Easter celebration of death and resurrection."

— On the argument from consensus (p. 7): "Surely, if it is general consent that proves the existence of a Creator, then general dissent disproves every other aspect of creation, since no culture believes any creation myth but its own...The virtually universal opinion over thousands of years that the earth was flat, never flattened its spherical shape by one inch."

— On creationism's demand to be taught in public schools (p. 13): "Well, then, creationists, who control the church and the society they live in, and who face the public school as the only place where evolution is even briefly mentioned in a possibly favorable way, find they cannot stand even so minuscule a competition and demand 'equal time.' Do you suppose their devotion to 'fairness' is such that they will give equal time to evolution in their churches? You know they won't. What's theirs is theirs. What's yours is negotiable."

— On the future of bigotry (p. 204): "This is not to say that it isn't splendid to be against such malfunctioning of the spirit out of a feeling of justice and humanity, but I suspect it would work better if racism and sexism were recognized as bad for business. That may be an ignoble way of looking at it, but if it means a cure of the disease I will accept it."

— On skepticism (p. 43): "I believe evidence. I believe observation, measurement and reasoning, confirmed by independent observers. I'll believe anything, no matter how wild or ridiculous, if there is evidence for it. The wilder and more ridiculous something is, however, the firmer and more solid the evidence will have to be."

— On science's self-correcting capacity, in contrast to creationism's dogmatism (p. 17): "Scientists refused to accept the notion of drifting continents on the basis of the evidence advanced in 1913 and thereafter. New evidence was obtained in the 1960s, and a revised and improved version of the new concept was then accepted with surprising speed."

— On the co-dependence of art and science (p. 117): "Since sunlight turns white to black, can sunlight be used to paint a picture? Scientists did not tackle this problem, but an artist did...In the 1830s, he began to produce the first primitive photographs."

— On learning from history (p. 122): "Poison gas wasn't used in World War II, because it would gain nothing for either side but retaliation."

333

— On scientific illiteracy in politics (p. 324): "Anton van Leeuwenhoek was the first to discover the world of microscopic life [leading to the germ theory of disease]…if Congress had then existed and given van Leeuwenhoek a modest appropriation to continue his observations, Senator Proxmire would probably have given it his Golden Fleece award."

— On nationalism (p. 64): "Planetary problems require a planetary program and a planetary solution…We need a world government that can come to logical and humane decisions and can then enforce them. This does not mean a world government that will enforce conformity."

— On language (p. 281): "The world will seem tiny, with the problems faced by one nation easily made clear to another…The push toward the adoption of a kind of worldwide lingua franca, probably based largely on English, will be great."

— On overpopulation (p. 63): "Motherhood is a privilege we must literally ration, for children, if produced indiscriminately, will be the death of the human race; and any woman who deliberately has more than two children is committing a crime against humanity."

— On solving overpopulation (p. 249): "Adjust the tax pattern appropriately, and the birth rate will drop."

— On sexual equality (p. 70): "A low birth-rate world *requires* women's equality. Without a women's-equality world, we can't have a low birth-rate world. And since it is quite clear that a low birth-rate world is the price of the survival of our civilization, it follows that the acceptance of the ideals of women's equality is also the price of survival."

— On pollution (p. 288): "Similarly, pollution of the air and the ocean at any one point on earth contributes to that pollution everywhere. We can in no way isolate our air and water from the rest of the world. We cannot isolate our share of the ozone layer…Our problems are global in nature, and they can only be dealt with globally."

— On eyewitness testimony (p. 41): "There have been innumerable eyewitness reports on almost everything that most rational people do not care to accept: of ghosts, angels, levitation, zombies, werewolves, and so on."

— On hyperspace (pp. 173-174): "But now that we know what hyperspace is and why science-fiction writers use it, the next question is: Does hyperspace exist in reality? Unfortunately, as far as we know, it does not…There is no evidence for its real existence—at least so far."

Other topics to which Asimov devotes essays include "Little green men or not?"; "The role of the heretic"; "Sherlock Holmes as chemist"; "Faster than light"; "Do you want to be cloned?"; "Bacterial engineering"; "Homo obsoletus?"; and "Life on a space settlement."

Finally, on the way to create scientists in a culture that views ignorance as "cool" (p. 109): "I was recruited because I was fascinated by the science books I read as a youngster, and I have received countless letters from people who tell me they were recruited through having read one or another of my science books for the public."

Want to encourage your offspring to pursue a career in science?

Buy them this book.

## 70. *Restoring the Goddess: Equal Rites for Modern Women*
## by Barbara G. Walker

Barbara Walker wants to abolish theology (knowledge of God), and replace it with thealogy (knowledge of Goddess). She argues that goddess religion, which prevailed worldwide until 6,000 years ago, was more benevolent than male god religion, and therefore should be revived as the religion of the future. As she puts it (p. 28), "The world needs a new belief system that doesn't demand that its adherents remain ignorant of the discoveries of the earth sciences, biology, history, astronomy, physics, paleontology, or archaeology. The world needs a religion that can be believed without insult to human intelligence. It's possible that thealogy will provide such a religion."

Her justification for proposing such a change is (p. 26), "Religion creates misogyny. Religion was and is the primary medium of women's spiritual, political, and social enslavement. There is no other aspect of human nature or culture that could have evolved such a phenomenon." Also (p. 77), "Like birds raised in cages, women raised in a patriarchal system usually come to think of it as a normal way to live." If anyone has offered a better explanation for Phyllis Schlafly, I have yet to hear it.

Walker is particularly critical of Christianity. She writes (p. 131), "The worst thing Christianity did to our civilization was to criminalize sexuality—to make everybody feel bad about feeling good...If God created AIDS, for any reason whatsoever, he's a lot more sinful than we are." Also (p. 43), "Christian fundamentalists today give a lot of lip service to 'traditional family values.' But this phrase is just a new code name for the male-dominated patriarchal family."

*Restoring the Goddess* is the latest addition to a growing number of books suggesting that religion can and should be transformed into something positive. And like all such books (for example, *Why Christianity Must Change or Die*), it ignores the obvious point that offering the world a less harmful religion is analogous to offering a less harmful form of AIDS. The solution to tyrannical mind slavery is not benevolent mind slavery. It is freedom. Humans need a deity, male or female, like fish need a bicycle.

Walker's feminism is qualitatively different from that of the movement's founders. Whereas the feminists of twenty years ago wanted to replace male tyranny with thinly disguised lesbianism, Walker offers instead the alternative of a goddess who endorses human sexuality, hetero and other.

I can understand Walker's dilemma, recognizing that Catholicism is an anti-human tyranny and switching to goddessism. When I learned that Protestantism is repudiated by its own bible, I became a Catholic. I suggest that Walker has similarly jumped out of the frying pan into the fire, and will eventually accept the reality that the only good religion is a dead one. The first deities were not created with the right to make capricious laws that must be obeyed without question. They evolved that right only after they acquired the power to do so. If Walker's thealogy triumphs, does she seriously doubt that it will evolve in the same direction? That is not the way history works. The goddess religion that ruled this planet for 20,000 years was only benevolent by comparison with the Inquisition and the Christian Right. It sacrificed fewer men to its deified vulva than male supremacists sacrificed to their deified phallus and its creator-god spin-offs, but it did sacrifice some. Shirley Jackson's *The Lottery* is today viewed as a horror fantasy. But it is an accurate account of what did happen under goddess rule for 20,000 years. Anyone who thinks that such a practice could never be revived is reminded that, barely a decade ago, heretic-killing by anti-abortion fundamentalists would have been considered unthinkable.

Walker is not unaware of the virtue of complete freedom from religion. She acknowledges (p. 190), "Studies have shown that incarcerated criminals profess proportionately more religious faith than the public at large...Atheists, on the other hand, tend to be law-abiding, taxpaying, responsible citizens, not inclined to criminal behavior." And on religion's pie in the sky when you die, she writes (p. 179), "Even those who express unyielding confidence in a blessed immortality seem to fear death as much as anyone else. Even those who proclaim heaven an eternal, ineffable bliss seem oddly reluctant to make the trip."

Up to half of each of Walker's chapters consists of what is best described as testimony, statements about personal beliefs and experiences written by (presumably) the forty women named in the acknowledgements, as well as others "who wished to remain anonymous." Until I realized that these segments were not part of the author's own narrative, I found them extremely confusing. Their non-Walker authorship should have been unambiguously identified.

One really irritating point: Despite stating unequivocally (p. 190), "As an atheist, I don't believe in a god," Walker persistently capitalizes pronouns referring to her female deity, even to the point of inserting such capitals into quoted passages that did not originally contain them. From a believer motivated by terror of the deity's megalomania, such a fetish would be tolerable. From someone who clearly knows better, it is not.

Nonetheless, if Walker's book can persuade female readers who could not abandon religion cold turkey to switch to a less hate-ridden religion, it will serve a valuable purpose. For men, it is not recommended.

## 71.  *The Gospels & Acts:  Questions & Problems*
## by Elliot Lesser

*The Gospels and Acts* is a verse-by-verse cataloguing of the Christian Testament, listing questions that each verse raises, usually in Socratic form so that the reader (Lesser hopes) will reach the implied conclusions; contradictions; Jewish Testament parallels and sources; and other gospel authors' treatment of the same story.  What it does not contain is a breakdown of whether a verse originated with Mark, the Q gospel, or elsewhere, any detailed analysis of how, when or why the verse was written, or whether it had a relationship to any real event.  Consider the following paragraph, under the subheading, "Perplexing," concerning Matthew 4:8, in which the Satan took Jesus to the top of a high mountain and showed him "all the kingdoms of the world." (*King James Version*):

> Onto what mountain did the devil take Jesus?  Where was this mountain?  As the world is spherical, it would not be possible to see the opposite side of the planet from a mountain peak no matter how high it might be.  Consequently, the devil would not have been able to show Jesus the Indian kingdoms of the Western Hemisphere, such as the Olmee, Maya, Chavin, Mocha, Inca, and others.  There also is no mountain on the earth high enough to enable one to see from its summit more than a few of the kingdoms (or nations), and then not in their entirety.  If, however, the earth was relatively small and flat, it might have been possible to see all of the kingdoms.  Did MATT. and LUKE believe that the earth was flat?  Did Jesus believe that the world was flat?  Why did Jesus fail to inform people that the world was round, not flat?  Were MATT., LUKE, and Jesus ignorant of the size and shape of the earth, and the existence of kingdoms on the other side of the planet?  If so, why?
>
> As God's son, why would Jesus need or want the glory of human kingdoms when he already had the glory of the entire universe and heaven?  Did the devil have the power to give the glory of the kingdoms?  If so, how and why did he get it?

That is as good as Lesser gets.  Every verse is similarly treated, no better and no worse.  Persons seeking Lesser's analyses of the questions he raises will not find them.  As a reference dictionary, *The Gospels and Acts* is a useful, perhaps even necessary, adjunct to a biblical scholar's library.  More than that, it is not.

337

## 72. *Biological Exuberance: Animal Homosexuality and Natural Diversity* by Bruce Bagemihl, Ph.D.

I have long had a problem with the behavior of homosexual males, not on moral grounds, since Robert Heinlein's dictum, "Sin means hurting someone unnecessarily. All other 'sins' are invented nonsense," was clearly applicable, but on account of its questionable hygiene. When I was sent a copy of *Biological Exuberance* by someone who knows I write book reviews, I was by no means confident that I could read 700 pages on such a subject in order to review the book on its merits rather than settling for a cliché such as, "as long as they don't do it in the street and scare the horses."

*Biological Exuberance* is slow, dull reading, so clinical that it will neither titillate nor repulse. It is a seemingly neverending compilation of statistics: behavior A is practiced by species a, b, c, and d; behavior B by species e, f, g, and h, *ad nauseam.* It does prove, over and over, that homosexuality is far from being an exclusively human activity. But it does so with all the readability and entertainment value of a laundry list.

Previously, I had never questioned the traditional wisdom that, when an animal mounted another animal of the same sex, it was expressing dominance rather than sexuality. Bagemihl destroys that interpretation, showing that even buddy pairs that appear to have more resemblance to Butch and Sundance than to Zeus and Ganymede are in fact "mates" in the full sexual sense. And he cites observations that mounting has no correlation with pecking order, since the same couple is likely to alternate as mounter and mountee. So while dominance as a contributing factor cannot he ruled out, since an animal can "get its rocks off" on power as readily as a human sadist can do so on cruelty, viewing behavior that culminates in tumescence, penetration and orgasm as essentially nonsexual is unrealistic. Similarly, female couples that combine affectionate behavior such as cuddling with genital contacts of various kinds are being as homosexual as any human couple doing the same things.

While most generalizations about human uniqueness have been disproven, including "that humans are the only species that laugh, that kill other members of their own species, that kill without need for food, that have continuous female receptivity, that lie, that exhibit female orgasm, or that kill their own young," (pp. 45-46) a generalization for which Bagemihl could find little falsifying evidence is that, "While homosexual behavior is widespread among our primate relatives, aggression specifically directed toward individuals that engage in it appears to be a uniquely human invention." (p. 54) He concludes that such behavior is "generally uncharacteristic of animal societies." (p. 59)

It is perhaps no coincidence that the highest proportion of homosexual interactions, involving eighty percent of the herd, is to be found in humankind's closest relative, the pygmy chimpanzee. That will bother Moral Majority types not at all, since they already deny that humans and apes have common ancestors. And the universality of such behavior, in species ranging from insects to birds and mammals, will simply cause them to replace the "unnatural" label with "animalistic."

I continue to wonder why any individual, whether human, orangutan, goose or beetle, would prefer to mate with its own sex. But I similarly wonder why anyone would prefer poker to bridge. The observable reality is that they do, and anyone who can write a paper on "A Lowering of Moral Standards Among Butterflies" (cited on page 87) is being less than scientific.

Since St. Martin's Press is notorious for publishing anything for which there is a discernible market, with depraved indifference to a book's validity (most recently Whitley Strieber's scientifically illiterate "aliens among us" twilight zone fantasy), it seems unlikely that *Biological Exuberance*'s scientific merit played even a minor role in their decision to publish it. More likely, the existence of a large homosexual market that would welcome a book proving that homosexuality is not unnatural was the only consideration. That *Biological Exuberance* is a work of legitimate scholarship, one that belongs in biology, zoology and ethology classes, would have been deemed irrelevant.

## 73. *The Demon-Haunted World: Science as a Candle in the Dark*
## *Billions and Billions: Thoughts on Life and Death at the Brink of the Millennium*
## by Carl Sagan

The bad news is that Carl Sagan's last book is not his best. The good news is that Sagan's worst (and I would say the same of Gould, Gardner or Asimov) is equal to almost anyone else's best.

*Billions and Billions* was written during the months that Sagan knew he was fighting a terminal illness—hence the subtitle. In it he allowed himself to get more personal, more autobiographical, and less willing to pull his punches, than in his previous books. For example, in castigating members of Congress who labeled attempts to save the ozone layer and stop global warning "the result of a media scare," he offers no loophole that their being Republicans is mere coincidence. And, in the same connection, he says of the Reagan-Bush regime, "Essentially we lost 12 years."

At a time when it was clear that a nuclear war would result in mutual assured destruction, Sagan points out that, in 53 years, the United States spent well over ten trillion dollars in its global confrontation with the Soviet Union. "Of that sum, more than a third was spent by the Reagan administration, which added more to the national debt than all previous administrations, back to George Washington, combined." He points out that the "technically dubious Reagan-era Star Wars scheme" was estimated to cost between one trillion and two trillion dollars. "What exactly is it that conservatives are conserving?" he asks.

Sagan has never played down his lack of any need for a god hypothesis. But now he spells out just how superfluous such a postulation really is: "The fundamental processes of life now seem fully understandable in terms of physics and chemistry. No life force, no spirit, no soul seems to be involved." (p. 210) "And if much of the universe can be understood in terms of a few simple laws of Nature...My own view is that it is far better to understand the Universe as it really is than to pretend to a Universe as we might wish it to be." (p. 213) "I would love to believe that when I die I will live again, that some thinking, feeling, remembering part of me will continue. But as much as I want to believe that...I know of nothing to suggest that it is more than wishful thinking." (p. 214)

Sagan makes some concessions to religion (p. 143): "Science and religion may differ about how the Earth was made, but we can agree that protecting it merits our profound attention and loving care." But he clarifies (pp. 67-68): "It is probably too much to hope that some great Ecosystem Keeper in the sky will reach down and put right our environmental abuse. It is up to us."

That rationality never weakened. In the epilogue, Sagan's widow writes (p. 225),"Contrary to the fantasies of the fundamentalists, there was no deathbed conversion, no last minute refuge taken in a comforting vision of a heaven or an afterlife. For Carl, what mattered most was what was true, not merely what would make us feel better."

Probably the most persuasive, and with any luck influential, chapter in *Billions and Billions* is a reprint of an article Sagan wrote for *Parade* on the subject of abortion. Sagan shows that the pope who reversed Catholic dogma and banned abortion from the instant of conception was acting on the delusion of a scientist who first placed a fertilized ovum under a microscope and "saw" a fully formed human body in the single cell, much the way Percival Lowell "saw" canals on Mars. He reports that, "Two of the most energetic pro-lifers of all time were Hitler and Stalin––who immediately on taking power criminalized previously legal abortions. Mussolini, Ceaucescu, and countless other nationalist dictators and tyrants have done likewise. Of course, this is not by itself a pro-choice argument, but it does alert us to the possibility that being against abortion may not always be part of a deep commitment to human life." And he explains that, "Both names—pro-choice and pro-life—were picked with an eye toward influencing those whose minds are not yet made up."

After explaining, with pictures, just how the fetus develops from a single cell to a tadpole to a human being, Sagan concludes, "Viability arguments cannot, it seems to us, coherently determine when abortions are permissible…We offer for consideration the earliest onset of human thinking as the criterion. Since on average, fetal thinking occurs even later than fetal lung developments (sixth month), we find *Roe v. Wade* to be a good and prudent decision addressing a complex and difficult issue. With prohibitions on abortion in the last trimester…it strikes a fair balance between the conflicting claims of freedom and life."

When that chapter was published in *Parade,* readers were asked whether they believed abortion should be prohibited from the moment of conception, permitted without restriction up to the moment of birth, or two intermediate options. The result was a virtual four-way tie—indicating that, despite the unassailable logic of Sagan's presentation, it changed practically no minds. That should surprise no one who has ever tried to argue from reason with someone whose position is based on emotion. But even dogmatists are curable—eventually—or Christians would still be burning witches.

*The Demon-Haunted World* was the true culmination of Sagan's career as a science popularizer with only three peers and no superior. It constitutes his most intensive assault ever on nonsense beliefs: "Typical offerings of pseudoscience and superstition—this is merely a representative, not a comprehensive list—are astrology; the Bermuda Triangle…" Sagan's catalogue, each item as concise as the foregoing, fills two printed pages (221-223). And that does not include any beliefs (covered elsewhere) that can legitimately be called religious—unless one is willing to categorize the science fiction absurdity of Scientology as a religion. And it does not include any belief or claim that would not have been on a similar list by *Skeptical Inquirer* or by me, although I would have included the Minnesota Multiphasic Personality Inventory, which Sagan endorses as scientifically valid, perhaps because he had not seen two psychologists administer it to the same test subject and reach interpretations as diametrically opposite and incompatible as the readings of two astrologers or two crystal gazers. He does, however, compare "psychoanalytic talk therapy" to the Christian Science cultists' denial of the germ theory of disease.

Sagan approaches religion from a scientific perspective. He asks (p. 234), "If prayer works, why can't God cure cancer or grow back a severed limb?…Why does God have to be prayed to at all? Doesn't He already know what cures need to be performed?" He points out (p. 278) that, "Religious doctrine that is insulated from disproof has little reason to worry about the advance of science. A Creator of the Universe is one such doctrine—difficult alike to demonstrate or dismiss." In contrast (p. 277), "But other sects, sometimes called conservative or fundamentalist…have chosen to take a stand on matters subject to disproof, and thus have something to fear from science."

In reporting on belief in demons (p. 117), Sagan writes, "There are cases in which nuns reported, in some befuddlement, a striking resemblance between the incubus [demonic seducer] and the priest-confessor, or the bishop, and awoke the next morning, as one fifteenth-century chronicler put it, to 'find themselves polluted just as if they had commingled with a man.'" He does mention the near certainty that the "nightmare" or female demon was invented by fanatic celibates to explain away wet dreams, but associates it with ancient Iranian culture, and leaves it to the reader to infer a similar origin in Christianity.

Of faithhealing Sagan writes (p. 230), "But the exposure of fraud and error in faithhealing is almost never done by other faithhealers. Indeed, it is striking how reluctant the churches and synagogues are in condemning demonstrable deception in their midst."

Sagan was a long-time opponent of UFOlogy theory, regularly asking, "Where is the evidence?" On the claim that the U.S. military is aware of an alien presence on earth and is suppressing that information, he writes (p. 93), "The Department of Defense, like similar ministries in every nation, thrives on enemies, real or imagined. It is implausible in the extreme that the existence of such an adversary would be suppressed by the very organization that would benefit most from its presence."

Of the tendency of religious fundamentalists to accept uncritically the non-mundane nature of alleged UFOs, Sagan writes (p. 129), "It serves their purpose to accept UFOs as real and revile them as instruments of Satan and the Antichrist, rather than use the blade of scientific skepticism. That tool, once honed, might accomplish more than just a limited heresiotomy."

Sagan demonstrates the similarity of all "recovered memory" claims, whether they involve alien abductions, past lives, childhood sexual abuse or satanism. Obviously he cannot debunk the repressed memory myth as fully in a single chapter as Robert Baker does in a whole book, but he does present enough evidence to satisfy any reader who is not beyond the reach of human reason.

In describing a man who regularly experienced nightmares of the unpleasant, non-orgasmic type, Sagan reports (pp. 125-126), "He was pressed down into his bed, paralyzed, unable to move or cry out. His heart was pounding. He was short of breath. Similar events transpired on many consecutive nights. What is happening here? These events took place before alien abductions were widely described. If this young man had known about alien abductions, would his old woman [who haunted him] have had a larger head and bigger eyes?"

In commenting on past-life regression, Sagan shows that the same technique has been used to elicit "memories" of a patient's future. He does not point out that, for future memories to be valid, information would have to travel backward in time. Presumably he realized that anyone who needs to be told that cannot be told.

Sagan stops short of recognizing that hypnotism does not exist. But he acknowledges its uselessness as a scientific tool (p. 138): "But hypnosis is an unreliable way to refresh memory. It often elicits imagination, fantasy, and play as well as true reflections, with neither patient nor therapist able to distinguish the one from the other."

In appealing for revived government funding for SETI and other programs of only long-term benefit, Sagan declares (p. 400), "Cutting off fundamental, curiosity-driven science is like eating the seed corn. We may have a little more to eat next winter, but what will we plant so we and our children have enough to get through the winters to come?...To take one symbolic example, is it really true that we can't afford one attack helicopter's worth of seed corn to listen to the stars?"

At the risk of nitpicking, I have to suggest that the material from page 228 to page 236 appears in that position, in the middle of an account of James Randi's "Carlos" experiment, because pages of the manuscript fell on the floor and were put back in the wrong place. When the only thing a reviewer can find to criticize is a technicality, the logical conclusion is that this is one hell of a good book.

## 74. *Greetings, Carbon-Based Bipeds*
## by Arthur C. Clarke

I have long been aware that Arthur C. Clarke gave the world the concepts of an orbiting manned space station and geosynchronous communications satellites. Even so, I could not suppress a rush of adrenalin on reading his 1945 paper, "Can Rocket Stations Give Worldwide Radio Coverage?" in which those forecasts of today's technology were first made. And I laughed out loud at his 1999 comment on his 1948 short story, "The Sentinel": "Although the BBC gave it thumbs-down, I still think it would be a good idea for a movie." In case anyone is unaware of it, "The Sentinel" was the story that became *2001—A Space Odyssey.*

On the possibility of an alien species viewing humans as vermin to be exterminated, Clarke writes (p. 40), "I do not believe that any culture can advance for more than a few centuries at a time on a technological front alone. Morals and ethics must not lag behind science, otherwise the social system will breed poisons that will cause its certain destruction. I believe therefore that with superhuman knowledge must go equally great compassion and tolerance." In other words, before an intelligent species can achieve any kind of interstellar capacity, it will either outgrow the "Religious Right" mentality, or it will exterminate itself. For, "if men cease to dream, if they turn their backs on the wonder of the universe, the story of our race will be coming to an end." (p. 143)

An observation Clarke made in an essay first published in 1991 falls somewhere between profound and self-evident, yet to this day even many non-fundamentalists have not grasped it (p. 360): "The greatest tragedy in mankind's entire history may be the hijacking of morality by religion. However valuable—even necessary—that may have been in enforcing good behavior on primitive peoples, yet their association is now counterproductive. Yet at the very moment when they should be decoupled, sanctimonious nitwits are calling for a return to morals based on superstition."

In the following paragraph he summarizes, "I suggested that early in the next millennium the rise of 'statistical theology' would prove that there is no supernatural intervention in human affairs. Nor does the 'problem of evil' exist; it is an inevitable consequence of the bell-shaped curve of normal distribution." And he draws attention to the prevalence of statistical illiteracy when he cites the fights that can be started by voicing the mathematical tautology (p. 360), "Fifty percent of Americans (or whatever) are mentally subnormal."

Clarke was the perpetrator of as blatant a collection of disinformation as the Vast Wasteland has ever peddled: *Arthur C. Clarke's Mysterious World; The World of Strange Powers*; and *Mysterious Universe*, each of which so pussyfooted around reality that it actually encouraged belief in nonsense. Yet in *Greetings*, he spells out his true position (p. 362) in the words, "I have been appalled by the way in which the United States (and much of the world, East and West) appears to be sinking into cultural barbarism, harangued by the fundamentalist ayatollahs of the airwaves, its bookstores and newsstands poisoned with mind-rotting rubbish about astrology, UFOs, reincarnation, ESP, spoon-bending, and especially 'creationism.'...which implies that the marvelous and inspiring story of evolution, so clearly recorded in the geological strata, is all a cosmic practical joke."

In justifying his bothering to review George Adamski's infantile drivel, *Flying Saucers Have Landed*, for *Journal of the British Interplanetary Society* in 1954, Clarke explained, in a note appended to the review, that it was "to save other members from wasting their money." (p. 119)

Clarke reprints an essay on the Star of Bethlehem, which he identified as a supernova, treating both the star itself and the biblical maguses as facts of history. Perhaps the naïveté of such speculation is mitigated by the fact that it was published in 1954, when the retroactive nature of the alleged star, created in a gospel author's imagination a full century after its supposed appearance (long enough for the tale to be impossible to falsify), and its origin as a myth previously told of Krishna, Abraham and Zoroaster, was not so widely recognized.

This progression of essays from 1934 to 1998 reveals more about Clarke himself than could be derived from an autobiography, and for that reason, while it falls far short of being his best work, or even his best nonfiction, it is invaluable to a person seeking that specific information. But it is not a book that can be recommended to anyone but a hardcore Clarke fan, and even among those it will not appeal to readers (such as myself) who view Clarke as a source of incomparable science fiction, but prefer to take their science fact from Asimov, Gardner, Gould and Sagan.

## 75. *Close Encounters With the Religious Right*
## by Robert Boston

*Close Encounters With The Religious Right* is an account of Robert Boston's undercover investigations, either as a paying customer, or simply as an interested party, at Right Wing gatherings and meetings. Not surprisingly, his findings merely confirm what most Americans have known for decades. All of the fanatics and hatemongers investigated want to criminalize abortion, homosexuality and, usually, all nonmarital sex; close public schools and turn "education" over to their churches; and transform America into a Christian Republic not unlike the Islamic Republics of Iran and Afghanistan. A movement known as Christian Reconstructionists (p. 251) actually advocates strict enforcement of the Old Testament's death penalties, including the execution not only of abortionists, homosexuals and other nonmarital copulators, but also children who violate the "honor your father and mother" statutes when they reject or modify a parent's narrow, intolerant, sectarian hatred of the ninety-nine percent of the human race whose belief systems differ by one iota from their own.

Boston makes clear that anyone who thinks the Religious Right is not a clear and present threat to the freedom of every American is living in a fantasy world. The Republican Party currently controls both houses of Congress, and the Religious Right controls the Republican Party. The Republican Party has never been a hate cult. But its willingness to buy power by pandering to the Religious Right is making it one.

After reading Boston's book, I can find no more concise summary of its contents than the blurb on the back cover: "With the Christian Coalition in financial disarray and the Rev. Jerry Falwell reduced to 'outing' children's television characters, many commentators have asserted that the Religious Right is on the ropes. Not so fast, says author Robert Boston; not only is the Religious Right not dead, it may be stronger than ever...Religious Right groups have jumped headfirst into politics and are determined to convert America to their own brand of fundamentalist Christianity."

Of Jerry Falwell, Boston says (p. 163), "He has not reconsidered, let alone changed, an opinion since the Eisenhower administration." He describes James Dobson's Focus on the Family (p. 16) as "an incredibly large and growing organization that is increasingly political and hews to the Religious Right line 100 percent," and Dobson himself as "up to his eyeballs in efforts to keep the Republican Party as far to the Right as possible...Anyone who still claims that Focus on the Family is not a Religious Right organization has simply not been paying attention."

Boston confirms that (p. 67), "Christian Coalition gatherings have a heavy partisan flavor. They are essentially Republican Party rallies." He found attending a Promise Keepers event (p. 125), "like stepping into some sort of weird parallel universe." On the Family Research Council, he says (p. 204), "Its agenda is tired, uncreative, and mind-numbingly familiar...criminalize abortion, defeat gay rights, reject feminism and public education, encourage censorship and Christian fundamentalism in public schools, and support school vouchers." On page 222: "The Campaign for Working Families obsesses almost exclusively over abortion and gays."

Boston rates Religious Right fanatics as first, second, or third tier according to their perceived credibility and success. After the third tier, he writes (p. 256), "Keep going and you hit the real dregs. Go even lower than that and you hit William Murray." Murray, in case anyone is unfamiliar with the name, is Madalyn Murray's fruitcake son.

While Boston did not consider Charlton Heston's campaign to bring back the gun-toting Old West relevant to his topic, he noted that Heston is certainly of the Religious Right, and an advocate of "the God-fearing, law abiding, Caucasian middle class." That god addicts consider it a virtue to be "God-fearing" is understandable, but "Caucasian"?

Boston all but ignores Pat Buchanan, whose name appears only on three pages. Of course, by abandoning his hopes of remaking the Republican Party in his own image, and switching to a joke party, Buchanan has effectively taken himself out of the game.

Boston also ignores Canada. If he had taken his investigative reporting into America's northern neighbor, he would have discovered that, whereas American hatemongers are openly religious and try to hide their political agenda in order to maintain tax-exempt status, in Canada a far-right political party that pretends it is *not* a fundamentalist religion, determined to do to Canada what Falwell, Robertson, et cetera want to do to America, has succeeded in becoming the country's Official Opposition. If Boston thinks Robertson and Falwell are slightly to the right of Tomas de Torquemada, he should see Manning and Day.

With the coming **[2000]** presidential election offering a choice between a candidate who thinks that persons free of religion are "arrogant" and "intimidating," and who describes himself as "born again," and a serial killer with over 120 homicides on his resume, the decision as to which is the lesser evil is perhaps simplified by the fact that George W. Bush is endorsed by Pat Robertson. That may not be the same thing as an endorsement by Ayatollah Khomeini, but it comes close.

A CNN *Crossfire* co-host described Boston's findings as "scary." Let me add: doubled, redoubled, and in spades.

## AFTERWORD

The election of a Republican-controlled House and Senate in November 2002, in combination with the presence of a rationally, intellectually, morally and educationally challenged puppet of the Theofascist Right in the White House, raises the situation from "scary" to "terrifying." If George W. Bush attempts to trample on the rights and freedoms of the majority of Americans who regard religious slavery as more intolerable than "taxation without representation," he may well trigger a second American Revolution.

# 76. *The Myth of Mental Illness: Foundations of a Theory of Personal Conduct*
# *The Myth of Psychotherapy: Mental Healing as Religion, Rhetoric, and Repression*
# *The Therapeutic State: Psychiatry in the Mirror of Current Events*
# by Thomas Szasz

In *The Myth of Mental Illness,* Thomas Szasz makes the point that mental illness does not exist. While a compulsive playactor such as Bernadette Soubirous, or a compulsive murderer such as Charles Manson, might be described as "insane" as a convenient metaphor, neither suffered from any kind of illness. One was a liar and the other a criminal. Legal insanity, the inability to tell right from wrong, does not exist, and medical insanity is a psychobabble name for the refusal to tell a headshrinker whatever he wants to hear.

As Szasz tells it (pp. x-xi): "We call jokes 'sick,' economies 'sick,' sometimes even the whole world 'sick,' but only when we call minds 'sick' do we systematically mistake and tragically misinterpret metaphor for fact—and send for a doctor to 'cure' the 'illness.' It is as if a television viewer were to send for a television repair man because he dislikes the programs he sees on the screen."

In the preface to the 1974 revision of his 1961 book, Szasz writes (p. vii), "Within a year of its publication, the Commissioner of the New York State Department of Mental Hygiene demanded, in a letter citing specifically *The Myth of Mental Illness,* that I be dismissed from my university position because I did not 'believe in' mental illness." That demand by New York's head psychoquack was nothing less than a confession that the pseudomedical humbuggery of psychiatry is a dogmatic belief system, otherwise known as a religion. And Szasz makes the point (p. 28), "Herein lies one of the supreme ironies of modern psychotherapy; it is not merely a religion that pretends to be a science; it is actually a fake religion that seeks to destroy true religion." How he defines "true religion," he does not say.

Szasz goes on to explain how behavior previously viewed as "witchcraft," "possession," "malingering," and eye-of-the-beholder evil, evolved into an eye-of-the-beholder sickness called "mental illness."

In *The Myth of Psychotherapy* (p. 14), Szasz observes, "Aristotle casually remarks that private counselors do the same thing as men who address public assemblies...No one now believes that what politicians do constitutes a form of medical treatment, but many believe that what psychiatrists do does."

Szasz traces modern talk-therapy to Anton Mesmer. He writes (p. 48), "His therapeutic successes were phenomenal, since Mesmer's practice, like that of all psychotherapists before and since him, was devoted exclusively to persons who pretended to be ill." He adds (pp. 49-50), "At the beginning of the age of science, Mesmer failed to fool his colleagues in the hard sciences; today, psychiatrists and psychotherapists have succeeded where he failed." And he summarizes (p. 61), "Mesmer's scientific contemporaries succeeded in discovering that not only Mesmerism but modern psychiatry as well was fake science."

Szasz overlooks a valuable point when he reports (p. 201) on the bestowing of psychotherapy diplomas on bartenders who completed a night course, but offers no comment. Drawing attention to the lack of any significant difference between an ego-boosting conversation with a psychotherapist, and a similar conversation with a bartender, taxi driver or prostitute, would have added greatly to his thesis.

Szasz reiterates his previous conclusions in *The Therapeutic State* (p. 9): "Although mental illness is said to be an illness 'like any other,' and although psychiatry is said to be a medical specialty 'like any other,' it is obvious that neither assertion is true." He continues (p. 24), "If one

347

defines psychiatry conventionally, as the medical specialty concerned with the diagnosis and treatment of mental diseases, and if one believes, as I do, that there are no mental diseases, then one is, indeed, committed to opposing psychiatry as a specialty—not of medicine, but of mythology." And like all mythologies (p. 32): "Since psychiatry is a pseudoscience...psychiatrists are especially eager to be accepted as scientific experts...They have to do it by producing great quantities of gibberish."

As proof of that assertion, Szasz describes (pp. 32-33) a hoax lecture by a "Dr Fox" to "a group of psychiatrists, psychologists, and social-work educators," set up by experimenters "who coached a pretended doctor to teach charismatically and non-substantively on a topic about which he knew nothing." All 55 test subjects rated the eloquent gibberish "more favorable than unfavorable," using such terms as, "Excellent presentation...Good analysis of the subject...Knowledgeable." And ten years after the "Dr Fox" hoax, Szasz attended a lecture containing the following wisdom:

"Recall that clinical experience and science do incrementally define the selective use of innovations, while policy reflexly greets innovation with prophecies of fiscal doom. In retrospect, the actual gains for health might render such poor prophets a loss. Where policy seeks formulae for determining choice and guiding treatment, science understands the fundamental basis for variability in disease and response and the method for sequentially approximating precision in the clinical process."

The difference between the quoted psychobabble and the acknowledged hoax is that the author of the psychobabble was dead serious. Szasz writes, "When the *American Journal of Psychiatry* publishes such gibberish as if it were English and made sense...somebody ought to say...a psychiatric emperor is naked, especially when the emperor insists he is sporting the most splendid garments." And lest anyone think the doubletalker was a lightweight, Szasz identifies him as the president of the American Psychiatric Association, and the gobbledygook as his presidential address.

Szasz rejects the pretence that violent criminals can be "not guilty by reason of insanity." He writes (p. 82), "Just as illness is not a crime, so crime is not an illness." He elaborates (p. 31), "Criminals cannot be divided into two categories—that is, persons who break the law because they choose to and persons who break it because their 'mental illness' compels them to do so. All criminal behavior should be controlled by the criminal law, from the administration of which psychiatrists ought to be excluded."

On the subject of "the boundless hypocrisy of courtroom psychiatry," Szasz writes (p. 92), "The expert opinions of medical specialists are actually the semantic services of psychiatric prostitutes." He continues (p. 125), "In criminal trials such as the Hearst case, there are two types of psychiatrists: excusers and incriminators. The former, hired by the defense, are paid to offer psychiatric prevarications that tend to excuse the accused. The latter, hired by the prosecution, are paid to offer psychiatric prevarications tending to incriminate the accused. If a psychiatrist is unwilling to offer such testimony, he is not hired." Szasz states that (p. 121), "Trying to ascertain whether Patty Hearst has been brainwashed by having her examined by psychiatrists is like trying to ascertain whether holy water is holy by having it examined by priests." He points out that, for every psychiatrist who testifies in any trial, another psychiatrist can always be found who will contradict the first psychiatrist's testimony, thereby invalidating both opinions.

Szasz does not pull his punches. He asks (p. 214), "Were the persecutions of heretics an aberration of the Inquisition and a misuse of the inquisitors' clerical powers?...We now recognize each of these 'aberrations' and 'misuses' for what they were: integral parts of the Inquisition...The situation is the same with Institutional Psychiatry. It has no aberrations or misuses, because it is itself an aberration—of medicine." As an example of the absolute power psychiatrists wield, and the subjective nature of their diagnoses, Szasz cites the case (p. 81) of an immigrant psychiatrist with a limited knowledge of American idioms who diagnosed a woman as delusional, and sentenced her to involuntary incarceration in a madhouse, because she claimed to have butterflies in her stomach.

Despite his debunking of the pseudomedicine of psychiatry, Szasz does not criticize the pseudoscience from which it springs. In no legitimate discipline is it possible for two dissertations, so incompatible that for either one to be valid the other must be incompetent nonsense, both to receive Ph.D.s from the same department of the same university in the same year. That it can happen in psychology, Szasz apparently is unaware.

Psychiatrists are not born with a gullibility that would have P. T. Barnum, if he were still alive, beating a path to their doors. Their gullibility is a consequence of their quack-school conditioning, which programs them to believe they have an ability to detect when a patient is lying or fantasizing that the rest of the world lacks. If that were true, John Mack would not now be famous.

When a licensed, card-carrying psychiatrist such as Thomas Szasz informs the world that psychiatry is unmitigated humbuggery, that revelation should be taken as seriously as the confession by a Catholic priest such as Andrew Greeley that religion is similarly fraudulent. And so it might have been, if Szasz had stuck to his specialty. Unfortunately, he laces his books with irrelevant views I can only describe as extremist, anti-Democrat, anti-liberal fascism slightly to the right of Rush Limbaugh. For example, he gives higher priority to a redneck's right to own a device that has no function except facilitating homicide, than to law-abiding citizens' right not to be shot. And he rejects the right of the State to prevent Jehovah's Witnesses from killing their children by denying them lifesaving blood transfusions. Certainly that is the only way I can interpret his statement (p. 265) that Jehovah's Witnesses in the United States do not have freedom of religion. And it seems no coincidence that presidents castigated by Szasz are invariably Democrats. For while he devotes a whole chapter to spelling out what hypocritical, incompetent tyrants Jimmy and Rosalynn Carter were, in a book published in 1984 he gives no equal treatment to Ronald and Nancy Reagan. Anyone who can refer to "the failure and tragedy of evangelical liberalism as a psychological perspective and political ideology" (p. 168), while simultaneously calling himself a libertarian (p. 202), is oxymoronic to say the least.

Thomas Szasz is not a nice man (what fanatic right-winger is?). That is probably why, even though he has been drawing attention to the snake-oil status of psychiatry for forty years (his books continue to be reprinted), society is as shrink-infested as it has ever been. His attacks on psychiatry are perceived as stemming, not from the observable fact that it is a collection of self-contradictory inanities, but from its being incompatible with Szasz's version of fundamentalist puritanism. His sermons denouncing the power lust of the psychiatric priesthood might have achieved far wider distribution and acceptance if they had not been intertwined with sermons endorsing robber-baron capitalism. Had Szasz been able to put his political agenda on the back burner and concentrate on showing that psychiatry *does not work,* the abomination might have become an unpleasant memory like phrenology and phlebotomy long ago.

When I started reading Szasz, I expected that I would be able to compare him to Porphyry, author of the first definitive exposé of the absurdity of Christianity. It turns out that I agree with everything Szasz says about psychiatry, but I wish someone with more credibility had said it. Richard Nixon advocated a logical and long-overdue recognition of the reality of China—but he was still Nixon. Szasz advocates a logical and long-overdue recognition of psychiatry as pseudomedical humbuggery that differs in no significant way from theology or tarot reading—but he is still Szasz. As has been said of Madalyn Murray O'Hair, the god Yahweh, and Nixon: with friends like Szasz, who needs enemies?

## 77. *Child Sexual Abuse and False Memory Syndrome*
## Robert A. Baker, ed.

*Child Sexual Abuse* is as good as it possibly can be, considering that its thirty-four authors are all practitioners or administrators of psychological talk therapy who would face severe economic consequences if they acknowledged the reality that psychology is pseudoscience, psychiatry is pseudomedicine, multiple personality is playacting, hypnotism is delusion, psychogenic amnesia is a science fiction concept, sociology is to history what rap is to music, the MMPI is glorified tealeaf reading, and social workers are a cross between hospital orderlies and bartenders. But it more than achieves its objective of showing that, since some supposedly recovered memories are pure fantasy, every recovered memory should be viewed as probable fantasy until and unless it is independently corroborated.

Except for introductions and a conclusion by Baker, each of the 22 chapters is a reprint of an article first published between 1986 and 1996. More than one article draws a parallel between present-day convictions of sexual abuse based solely on recovered memories, and convictions of witchcraft based on similarly manipulated testimony in 1692 Salem. Several others demonstrate the similarities between recovered memories of sexual abuse and recovered memories of alien abductions, past lives and satanic cults.

The clearest statement of the problem Baker's contributors examine is by Richard Ofshe and Ithan Wetters (pp. 227-228): "Recently a new miracle 'cure' has been promoted by some mental health professionals—recovered memory therapy...In less than ten years' time this therapy, in its various forms, has devastated thousands of lives...Clients are essentially being tricked into believing that they are remembering events that never happened."

A particularly revealing chapter, by Michael Yapko, tells of a man who convinced his wife, his psychotherapist, and perhaps even himself, that he was suffering from post-traumatic stress as a consequence of having been a POW in Vietnam. When he committed suicide, his wife attempted to have his name added to the Vietnam War Memorial—only to learn that he had never been to Vietnam. While only one case, it establishes that a fantasizer's sincerity, observable suffering and sheer credibility do not constitute proof that traumatic memories necessarily have a factual basis. If John Mack had been aware of that case, he might have been less willing to accept his patients' apparent sincerity as all the proof he needed that they had been abducted by aliens.

The chapter by David Calof provides the strongest evidence that, on rare occasions, a recovered memory of sexual abuse may have a factual basis. (p. 88) He tells of a man who, *after* his doctor told him that his severe anal scarring was consistent with forced sex, "remembered" being consistently raped by an older cousin. Perhaps it is my reluctance to believe that *any* recovered memory is real that makes me wonder if a low-fiber diet and painful defecation can produce the same result. I also get the impression that Calof, in arguing for the legitimacy of some recovered memories, fails to distinguish between "repressed" memories, and events that remain out of conscious recollection for long periods but would immediately be recalled if the person was asked, "Did this ever happen...?" However, Calof's chapter is not valueless, and helps provide the book with necessary balance.

The chapter by Gail Goodman and others, in contrast, is unmitigated psychobabble from start to finish. As Robert Heinlein once observed, "If a person of normal intelligence, and a reasonably full education, cannot understand a piece of prose, then it *is* gibberish."

The chapter by Harrison Pope and James Hudson can be summarized by a line from its synopsis (p. 169): "Thus, present clinical evidence is insufficient to permit the conclusion that individuals can suppress memories of childhood sexual abuse." That conclusion is fully substantiated. In reporting on a study in which 267 out of 450 patients "with self-reported histories of sexual abuse who were currently in therapy" answered "yes" to the question, "Was there ever a time when you could not remember the forced sexual experience?" Pope and Hudson note that (p. 173), "A subject answering 'yes' might mean only that he or she gave no thought to the event during some period, or attempted to deny or minimize the event. No follow-up questions were asked to assess these possibilities." And in citing a book titled *Father-Daughter Incest*, they note that "all of the 40 women in the case series displayed clear and lasting memories of the abuse."

Several contributors refer to hypnosis as if it actually exists. However, when Barry Beyerstein and James Ogloff make the point (p. 22) that hypnotic memory enhancement procedures "are as likely to produce fantasy and confabulations that feel like valid recollections as they are to expose true hidden memories," they may be viewing the reality of hypnosis *per se* as irrelevant.

In contrast, Richard Ofshe and Ethan Walters, in an otherwise excellent article, say (p. 238), "Subjects can also be led to undergo somatic changes, such as causing warts to disappear at the direction of the hypnotist." They should have added, "or so we have been told." But they go on to make the valid point (p. 245) that "thousands of people have been hospitalized with the incorrect diagnosis of MPD—a condition that may not even exist...Compared to other fringe psychology and psychiatric excesses the ghastly ménage-a-trois—recovered memory therapy, multiple personality disorder, and satanic cult conspiracy—represent an exceptional problem...exceeding the relatively short life expectancy of other psychological quackeries."

By far the most incompetent chapter is by Donald Tayloe, who endorses the reality of repressed memory, hypnotic memory enhancement, psychogenic amnesia, and the guilt of George Franklin, convicted solely on the basis of his daughter's allegedly recovered memory, "for a murder he committed on September 22, 1969." Since the article was first published six months prior to Franklin's conviction being overturned, Tayloe's acceptance of the jury's verdict would be understandable—were it not that the fantasy nature of supposedly recovered memories was already sufficiently established to satisfy appeals courts, and should have been known to Tayloe. Later on, Tayloe describes how he convinced a jury to acquit a man named Bain who "accidentally" killed his wife, by authenticating the killer's *Twilight Zone* defense. Rather than let that stand unchallenged, Baker's introduction points out that (pp. 213-214), "Despite Tayloe's arguments, there is good reason to question: (1) whether or not Bain's amnesia was genuine; and (2) whether or not the hypnosis [used to recover Bain's memories] was real." I am guessing that, in including an article that presented a contrary viewpoint, Baker was counting on readers recognizing that Tayloe's obvious dogmatic gullibility concerning the Bain and Franklin cases would discredit his entire piece—as it does.

In chapters devoted to allegations of child sexual abuse not involving recovered memories, one set of contributors argues that the criterion for conviction should be raised from "some credible evidence" to "a preponderance of the evidence," while others argue that it should not because a few thousand lives destroyed by false allegations is a small price to pay for protecting children from genuine abuse. Tell that to the persons wrongly convicted.

Hollida Wakefield and Ralph Underwager confirm the findings of all of the book's contributors except the "truth is whatever I want it to be" dogmatists when they say (p. 430), "Claims of repressed memories of childhood sexual abuse recovered in the course of therapy are not supported by credible scientific data." Nonetheless, the refusal of the recovered-memory industry to recognize that sincerity does not imply truth, or that a patient's belief in a fantasy does not make it true, makes it unlikely that false memories will be universally recognized for what they are any time soon.

I was hoping that Baker's book would establish definitively that repressed memory does not exist and recovered memory does not exist. But none of the authors is willing to treat those conclusions as beyond dispute. Even so, it should be mandatory reading for every alleged therapist and prosecutor who still believes that fantasies put into a patient's mind by manipulative dogmatists are memories of events that really happened. I would make a similar recommendation concerning movie and television producers who foist "recovered memory" scripts on gullible viewers as nonfiction. But the rebroadcast, three years after George Franklin's conviction was overturned, of a movie maintaining the pretence that Franklin is a murderer, shows that the prostitute media are interested only in ratings and to hell with the facts. If Franklin has not already sued the network in question for repeating its libel without an updating disclaimer, he should certainly do so. While *Child Sexual Abuse* is perfect for its targeted audience of educated professionals, giving both sides of issues on which there really are two sides while avoiding the absurdity of giving equal treatment to what might be called geography and flat-earthism, I would like to see a book presenting the same information to John Q. Public. If that is not the kind of writing with which Baker feels comfortable, perhaps someone with more experience at presenting science to nonscientists might try? Martin Gardner and Joe Nickell come to mind.

## 78. *Return of the Furies: An Investigation Into Recovered Memory Therapy* by Hollida Wakefield and Ralph Underwager

Can all memory of a traumatic childhood event such as sexual abuse be repressed and then recovered twenty years later at the prompting of a psychotherapist whose other patients all tend to recover similar repressed memories? The authors of *Return of the Furies* show that, since no evidence has been found in sixty years of research that "repression" even exists, and that the techniques used to "recover" lost memories read like something out of *The Manchurian Candidate,* the answer is almost certainly No.

That recovered memory does not exist never has been and probably never will by proven, just as the nonexistence of spell-casting witches will never be proven. All that can happen is that the masses will eventually recognize nonsense for what it is. What has been established is that, (a) no allegedly recovered memory has ever been objectively validated, and (b) at least several dozen claims of recovered memories have been objectively falsified. Wakefield and Underwager quote (p. 163) and endorse an article in *Liberty*: "The quality of all evidence for recovered memories of childhood sexual molestation is precisely the same in every respect as the quality of the evidence for recovered memories of past lives and satanic rituals." They later report (p. 270), "The same factors are operating in people who report encounters with UFO aliens, in those who uncover memories of ritual satanic abuse, and in those who develop symptoms of multiple personality disorder."

While most of the sexual abuse allegations stemming from allegedly recovered memories are intrinsically plausible, in that they closely resemble verified cases of child abuse, there have also been hundreds of allegations of behavior so bizarre and preposterous, that the absence of any such stories from sources other than "recovered memories" should have been a clue that they were probably fantasies. What is most significant is that virtually all therapists who endorse recovered memory claims in principle show no ability to distinguish between a plausible allegation and one straight out of the imaginings of the Marquis de Sade. Of the four pages that list specific allegations of bizarre behavior by thirty-three recovered-memory patients (305-309), the winner for sheer imbecility must surely be, "Forced to have sex with a neighbor's dog and subsequently had a baby that was half dog."

Because they dare to question the literal truth of such allegations, the False Memory Syndrome Foundation has been identified by the recovered-memory cultists as "the real danger." The parallel with heretic-hating religious fundamentalists is hard to overlook. But then I strongly suspect that the recovered-memory cultists *are* religious fundamentalists. Underwager's failure to draw attention to the parallel can perhaps be attributed to his status as a Lutheran pastor.

The pushers of the recovered memory myth are as vicious and vindictive a cult as Scientology, targeting all opposition by whatever means are available. After Underwager started defending the falsely accused, the cultists launched a campaign of anonymous phone calls to prosecutors accusing him of "a variety of heinous acts, including sexual abuse. Since that time they have systematically tried to discredit us personally and professionally." (p. 5) While Underwager refused to be intimidated, many other potential defence witnesses refused to put themselves on the firing line by publicly contradicting the witch hunters. (p. 6)

The main weapon for propagating the recovered memory myth is *The Courage to Heal,* a piece of fiction by authors best described as the Charles Berlitzes of false memories. The book is the bible of the National Center for Prosecution of Child Abuse, an organization that believes that half of all women were sexually abused as children and that recovered memory is more valid than a ouija board. They assert (p. 42), "Children must always be believed at all costs, except when they deny that abuse has happened...People who don't believe the child must be protecting molesters and are probably molesters themselves." Can anyone doubt that the propagators of such paranoia are not sparking on all neurons?

In criticizing *The Courage to Heal* and its self-taught authors, Wakefield and Underwager quote several excerpts, including, "If you are unable to remember any specific instances...but still have the feeling that something abusive happened to you, it probably did," and, "If you think you were abused and your life shows the symptoms, then you were." (p. 134) By those criteria, a strong case could be made for the real existence of the Tooth Fairy, the Easter Bunny, the Great Pumpkin, and the serial killer known as God.

The widow of a man who died shortly after being falsely accused of sexually abusing his daughter wrote (p. 11), "There is no doubt in my mind that the stress he had suffered from the false accusations was at least partially responsible for his untimely death...In my opinion, the therapists who are promoting these false memories are guilty of murder."

But therapists are not the only criminals who can destroy innocent lives and rationalize that they are only doing their job. There is a national organization of prosecutors that is equally guilty. An American Bar Association task force in 1984 concluded (pp. 15-16), "The prosecutor is the single most powerful figure in the entire justice system, with absolute authority to determine who gets charged with what. Nobody, including judges, holds prosecutors accountable for their behavior; rather, they are immune from any attempt at redress or correction. It is a system with all the incentives awarded for making accusations and little interest in the accuracy of the decisions made. It pursues neither truth nor justice, but winning." To anyone who doubts the truth of those comments, and thinks that, even when prosecutors are wrong, their motives are admirable, I offer two words: Kenneth Starr.

Three hundred years ago, when Cotton Mather prosecuted alleged witches, he had the legitimate excuse that he lived in a culture in which the nonexistence of witchcraft was not widely recognized. Prosecutors who pursue child abuse allegations based on nothing but allegedly recovered memories have no such excuse. The fraudulence of recovered-memory "therapy" is now recognized by the courts, and even by most psychotherapists. The law permits "therapists" who manipulate patients into concocting false memories to be sued by victims of their patients' false allegations, and damages as high as $2.5 million have been awarded. But prosecutors who try to further their political careers by destroying lives in the absence of any valid evidence remain immune. It is time that immunity was withdrawn.

An interesting piece of information in *Return of The Furies* is that the assault on the Branch Davidian cult in Waco was ordered by the Attorney General when she was informed that seventeen children extracted from the cult's compound were telling stories of gross sexual abuse by the cult's leader. What Janet Reno was not told is that all seventeen had initially denied any kind of mistreatment, and had been coerced and manipulated for two months until they were finally pressured into saying what the mental health "experts" questioning them wanted to hear. In fact (p. 34), "The official Justice Department report on the event flatly concludes that the reports of child abuse are false." While David Koresh was primarily responsible for what happened at Waco, the fanatics who manipulate the vulnerable into concocting false memories are his unindicted coconspirators.

## 79. *Once Upon a Time: A True Story of Memory, Murder, and the Law* by Harry N. MacLean

"Mary Jane Larkin stood at the front of the classroom and looked out at the sea of small faces staring up at her. Her hands rested lightly on her wooden desk, the same one she had used for the past twenty-five years in teaching fourth and fifth grades at Foster City Elementary School. Next to her hands lay her open grade book, with the students' names neatly printed in alphabetical order, and pencils, bottles of glue and stacks of paper."

That fiction, for I can call it nothing else, is the first paragraph of what should have been the definitive debunking of the delusion put into the mind of Eileen Franklin by an alleged therapist, that she had "recovered" a suppressed memory of a murder she had witnessed twenty years earlier. Given MacLean's ignorance of how nonfiction should be written, it is hardly surprising that his combination of first class investigation and untalented hack writing, including whole chapters in diary-type present tense, did not receive the acclaim his analysis of the case that spawned the recovered memory hoax certainly deserved, and is already out of print.

Nonetheless, to anyone who can get past the author's apparent belief that he was writing a "nonfiction novel" (there is no such thing), *Once Upon a Time* is an extremely useful description of the case that gave the world a new superstition that in the decade it lasted destroyed thousands of lives. For the Franklin precedent predictably led copycat therapists to encourage their gullible patients to fantasize childhood sexual abuse, and the new masturbation fantasies only became unfashionable after several therapists whose patients accused innocent caregivers were successfully sued for damages as high as several million dollars.

The basics of the Franklin case can be summarized as follows. Eileen Franklin, a woman of whom the psychobabble description "disturbed" is perhaps not a misnomer, and whose own mother recognized her as a compulsive liar, was apparently having nightmares about witnessing a murder before ever consulting a therapist. And it was in a dream that she first "saw" her father kill her girlfriend. But it was her therapist who put her into the alleged state of hypnosis (there is no such condition), and by telling her to "think back," encouraged her to elaborate on her dream by incorporating all of the information about her childhood girlfriend's murder that had previously been published. (Eileen at no time provided accurate information not previously published.) At that point the therapist, no doubt visualizing her name in the Dictionary of Psychoquackery Biography, deliberately manipulated Eileen into believing that her fantasy was a true memory that she had "suppressed" (another psychobabble concept) for twenty years and had now "recovered."

Eileen reported her alleged recovered memory to the police. Her father was tried for murder, and he was convicted. Only after several other "recovered memory" allegations were definitively falsified did the courts recognize that the concept of recovered memory is not supported by a single verifiable case, and George Franklin's conviction was overturned. But because a fantasy that has been recalled to mind many times is indistinguishable from real memories, Eileen continues to believe (or so she maintains) in the reality of her alleged memories, as do many other victims of the mind manipulators, even though there have been sufficient instances where the tales patients were encouraged to mistake for recovered memories were proven to be pure fantasy, to conclude that recovered memory simply does not exist.

While I did not find the information in MacLean's book, I have to assume that the jury in George Franklin's trial was not allowed to learn of Eileen's habitual lying, her erratic behavior, her cocaine addiction, her drug overdose that either was or was not a suicide attempt, depending on who she was telling about it, or her prostitution conviction. Those special circumstances might have been sufficient to prevent them from accepting a new mythology (recovered memory) that should have presented a credibility problem even from a reliable witness. And the fact of George Franklin looking, to jurors who believed in the "honest face" myth, like someone capable of committing the crime with which he was charged (as Ted Bundy did not), may also have contributed to his conviction. As MacLean explains in his epilogue, "When the jury heard…that he had sex with his daughter, and all the rest [childbeating, etc], he became subhuman. Seeing him as a monster, the jurors were relieved of the normal anxiety one might have about making a mistake; if they incorrectly convicted him of the murder, he certainly had committed other sufficiently heinous crimes to warrant serious punishment." And Eileen's abusive treatment at her father's hands "immunized her from the effects of her lying and constantly shifting and changing stories." In other words, the jury saw that she was a liar, but accepted her lying as a consequence of her traumatic experiences.

The Franklin case became a movie starring Shelley Long, which unequivocally presented Eileen's fantasy and her father's guilt as fact. At the time, a jury verdict supported such a slander. The rerunning of the movie on a major network after Franklin's exoneration, with no disclaimer stating that the conviction had been overturned and that "recovered memory" was a questionable concept, is surely grounds for a truly punitive lawsuit against the prostitutes to whom truth never has and never will outweigh ratings. I certainly hope so.

MacLean's book was published before George Franklin's conviction was overturned. His line, "None of which means that she didn't see her father murder Susan Nason," was probably his attempt to appear objective even though the evidence he catalogued would have justified a conclusion that Franklin's conviction was indefensible. Had this book been written a few years later, when the fraudulence of the "recovered memory" psycho-drivel is an established fact, he would probably have been more willing to state categorically that, while the murder remains unsolved, there is no reason whatsoever to believe that Franklin did it.

## 80. *Demons of the Modern World*
## by Malcolm McGrath

"In his book McGrath explains in clear and careful detail how satanic ritual abuse and the widespread belief in Satanism was born, grew, developed, and proliferated from its birth in primitive times to the present day. The step from satanic conspiracies to false memories of childhood abuse was short and appallingly easy. McGrath shows the details of the birth, growth, and progress of such conspiracies in the minds of those who should have known better. From such a background of delusion, illusion, and the media's raucous reminders via subtle suggestions and disinformation, the average citizen was well prepared to accept the reality of demons from nether worlds and outer space haunting his mind and abducting his fellow citizens." (p. 10)

While I do not dispute the accuracy of the foregoing excerpt from the Foreword written by Robert Baker, I do not share Baker's opinion that this is a particularly good book. In fact *Demons of the Modern World* reads like a doctoral dissertation that was never rewritten into the non-boring format usually considered mandatory for published books. I see it as a dozen pages of cultural psychoanalysis of the kind of minds that can swallow such superstitious hogwash as Satanist cults, recovered memory, multiple personality and alien abductions, expanded into three hundred pages of psychobabble.

That does not mean that, on the rare occasions that McGrath actually says anything, what he says is not valid. Consider:

"By the mid-eighties Satan-hunting cops and journalists were claiming that as many as fifty thousand people a year were being secretly murdered by satanic cults that had millions of members. By the end of the decade, some 'experts' were describing conspiracies of extravagant size and complexity." (p. 93) "If the modern Satanism scare began with individual tales of molestation, it ended with tales of satanic conspiracy no less global in their proportions than the Red Scare of the McCarthy era or Cotton Mather's fear of the devil's invasion of New England." (p. 95) "Still, quack therapy aside, is there a danger to the demonic illusion? Only if we ignore the fact that it is an illusion...As long as the basic institutions of our society remain intact, the demonic illusion will always retain its official status as an illusion." (p. 291)

"Historians have suggested that one of the reasons for the collapse of the Salem witch panic was that eventually accusations began to fly too high up the social ladder. A similar process occurred with the recovered memory movement." (p. 133) "Also, people who had earlier attacked their parents for abusing them began to file lawsuits against their therapists for misleading them into believing false memories. By the mid-nineties, suits against some of the most prominent MPD therapists...resulted in million-dollar settlements." (p. 134) "The moment that recovered memory and MPD took on stories of satanic conspiracies, however, they sailed away from the safe harbor of unfalsifiability for good. Mass murders, animal mutilations, pornography rings; if they were true, all of these would have to leave hard evidence. When none was found, not only Satanism was called into question, but recovered memory therapy and MPD as well. By accepting or even suggesting stories of satanic ritual abuse, advocates of recovered memory therapy and MPD risked more than the possibility that the stories of Satanism might be disproven; they risked having the whole recovered memory edifice come crashing down on top of them." (pp. 157-158) "When the recovered memory and MPD therapy fell into disrepute, the last serious source of evidence of the existence of satanic cults disappeared." (p. 262)

"The role model for modern cases of MPD was the 1973 book *Sybil*, which became a TV movie three years later...Before deciding to publish the story in journalistic form, Cornelia Wilbur, Sibyl's (*sic*) therapist, had repeatedly tried to have her accounts of the case published in scholarly and medical journals. According to philosopher Ian Hacking, she was constantly rejected on the ground that multiple personality could not be taken seriously. Herbert Spiegel, one of Wilbur's colleagues…expressed his doubts that Sybil had multiple personalities to…the journalist who turned the story into a book, but she retorted, 'We have to call it a multiple. That's what the publisher wants. That's what will make it sell.'" (pp. 124-125) So the reason hogwash gets published is that the mass media are run by prostitutes interested only in profit and possessed of a depraved indifference to truth. So what else is new?

"Just as the institutional credibility of the Satanism scare began to collapse in the early 1990s, claims of another form of demons were on the upswing: aliens. Although the alien-abduction myth can be traced to different origins than the Satanism scare, by the mid-1990s it had assumed the structure of a classic demonic illusion." (p. 261) "Karl Marx once said, 'History always repeats itself, the first time as tragedy, the second time as farce.' If the Satanism scare can be seen as the tragic repetition of the great witch-hunt, there are good reasons to see the abduction craze as the farcical repetition of the Satanism scare." (p. 276)

Of John Mack, author of *Abduction*, McGrath observes (pp. 288-289), "Mack is an M.D. and a professor of psychiatry at Cambridge Hospital and Harvard Medical School…Yet his approach to alien abductions is not only unscientific, but antiscientific. According to Mack, letting go of science as we know it is both a necessary and desirable prerequisite for coming to terms with the aliens. This openly antiscientific approach is an innovation in abduction research and in UFO studies in general. Earlier attempts to prove the existence of UFOs and abductions had always cloaked themselves in an air of science. Hence the paradox: the most scientifically qualified of abduction writers presents one of the most antiscientific cases for the existence of abductions." Yet Harvard accepted this gullible incompetent's argument that he should not be fired. Go figure.

McGrath's thesis is valid. His presentation is significantly less readable than my own *The Disinformation Cycle* (Xlibris, 2002), Paul Kurtz's *The Transcendental Temptation* (Prometheus, 1986), or Michael Shermer's *Why People Believe Weird Things* (W.H. Freeman, 1997), on the same subjects. A marginal thumbs-up.

## 81. *The Hidden Book In the Bible*
## by Richard Elliott Friedman

In *The Hidden Book in the Bible,* Richard Friedman sets out to prove that the "J" portions of the Pentateuch, plus parts of Joshua, Judges, Samuel and Kings, ending with the accession of King Solomon, constitute a continuous narrative by a single author. Since the only person ever speculatively identified as J is the spokesman (*prophetes*) Nathan, who is known to have written biographies, long thought to be lost, of David and Solomon, Friedman's thesis is not as wild and speculative as it might seem.

There is evidence that J was written during the reign of Solomon's son (930-913 BCE), and scholarly consensus dates Judges and Samuel/Kings later than Deuteronomy (621 BCE). Nonetheless, Friedman does show a continuity of narrative, linguistic quirks and theme that add up to a powerful argument for his interpretation. The weakness is that Friedman also shows an inability to see evidence of interpolation in some stories that he attributes *in toto* to J, implying that his not detecting incompatibilities between J and the Judges/Kings history does not necessarily mean there are none.

For example, Friedman credits J with writing the portions of the Adam and Eve myth referring to the shame of nakedness. But the Jews did not acquire a nakedness taboo until Ezra invented one in post-Captivity times. J's original line was, "I heard the sound of you in the garden and was afraid, and I hid." The Redactor (R), whom Friedman agrees was Ezra, changed it to, "and was afraid *because I was naked,* and I hid." R wrote, "They sewed fig leaves together to make loincloths." J wrote, "Yahweh made skin garments for the human and his woman." Friedman attributes both versions of the invention of clothing to J.

Similarly, Friedman attributes the entire Shechem and Dinah tale to J, including references to circumcision, apparently failing to notice that, in J, circumcision began with Moses. It was the Priestly author (P) who backdated the custom to Abraham.

Friedman's competence is not in question. He was the first person to prove beyond a reasonable doubt that P and the final R, far from being the same person as was previously thought, were separated by the best part of two centuries.

Some of Friedman's translations are less than defensible. He consistently translates the dual-sex, generic plural *elohim* as the singular proper name, "God." Why? Since I have ruled out incompetence, I suggest that, like other god addicts, he is desperate to convince himself that a creature named God actually exists, and in order to maintain that delusion he is able to shut out the reality that his Bible makes no mention of any "God," only "gods."

Also, even though the Hebrew makes no such distinction, Friedman translates *melakhim* as "angels" when it refers to Levit's visitors at Sodom, but as "messengers" when the *melakhim* come from a human rather than from Yahweh. Again, I suggest that his need to believe in metaphysical creatures not mentioned in his Bible motivated him to put them there.

Friedman clings to the translation "coat of many colors" in defiance of the myriad of scholars who have long recognized that the Hebrew words really mean "coat with sleeves." Is Friedman right and they are all wrong? Perhaps. Or perhaps not.

In the story of Judah and Tamar, Friedman correctly distinguishes between *zownah* ("prostitute") and *kedeshah,* which he translates as "sacred prostitute." But *kedeshah* means literally "holy woman." Since a *kedeshah* copulated sacramentally with worshippers of her goddess, the non-connotative translation that does not impose modern religion's conditioned judgmentalism on an earlier culture would be "nun." (The A.V. translates both *zownah* and *kedeshah* in the story as "harlot.")

Friedman omits any mention of the plagues of Egypt or Balaam's ass from his J translation. But he explains in the textual notes that the doublets in the former are most likely by the eighth-century Elohist (E) and P, while in the latter any separation of J and E would have been a guess, and he preferred not to guess.

Friedman attributes the sentence, "And Isaac was sixty years old at their birth" (Jacob and Esau), to J, even though, in the context of a doting father (Abraham) sending a go-between to find a virgin bride for his nubile son, it is an absurdity. Admittedly Rebekah took several years to get pregnant. But, since Friedman recognized the statement that Isaac was married at 40 as not-J, surely it logically followed that the verse about his age when Rebekah gave birth was also not-J? Ages, particularly unrealistic ages, were a P fetish, not a Yahwistic one.

Friedman argues that, since the entire J chronicle calls the deity *Yahweh,* with the exception of chapters two and three of Genesis where it is called *Yahweh Elohim* the *Elohim* must have been added to those chapters by R at the point where J and P first came into conflict in order to accustom readers to the idea that Yahweh and Elohim were the same entity. While that makes sense, it raises the question why R did not change P passages in chapter one from *Elohim* to *Yahweh Elohim*.

**[I accept Friedman's reasoning on this point, and accordingly incorporated it into *The Judaeo-Christian Bible Fully Translated.* Thus Genesis 2:4b reads, "On the day that Yahweh [THE GODS] fashioned the land and the skies..." thereby indicating that [THE GODS] was R's interpolation. I similarly capitalized AVRUM and SARAY in all instances, to indicate that they represent R's amended spelling, not J's original.]**

Friedman also takes the position that only P made *Abram* and *Sarai* the original names of Abraham and Sarah, that J called them Abraham and Sarah from the start, and that R changed the names in the early part of J to make it consistent with P.

Friedman believes that J identified Noah's sons as Shem and Ham and Japheth, despite the fact that Noah cursed Canaan on account of "what his youngest son had done to him." The story becomes more logical and consistent with justice when the words italicized (by me) in Friedman's translation are viewed as editorial additions: "And Noah's sons who went out from the ark were Shem *and Ham* and Japheth and *Ham: he was the father of* Canaan." R changed the offender from Canaan to Ham because, in the P narrative and the R redaction, Noah's "youngest son" was not Canaan but Ham.

As for the nature of Ham's crime (actually Canaan's prior to R's alterations), Friedman accepts that it was looking at his father's nakedness—at a time when no nakedness taboo existed—since R also expurgated the description of "what his youngest son had done to him." Even the Talmud acknowledges that a verse has been deleted between Genesis 9:23 and 9:24, and identifies Ham/Canaan's crime as castrating his father, causing Shem and Japheth to turn their backs rather than look at Noah's "shame" (not his nakedness).

Friedman sticks to familiar English spellings, since changing them would serve little purpose. Thus he stays with *Gomorrah* and *Esau* even though both names begin with the semi-vowel *ayin,* and therefore should logically have the same initial. His transcribing *lvt* as "Lot" instead of correcting it to "Levit" is less defensible, since Levit is a common Jewish name and Lot (from Greek *Lōt*) is not.

Friedman attributes to J the lines, "Isaac went back and dug the water wells which they had dug in the days of his father Abraham and which the Philistines had stopped up after Abraham's death, and he called them by names, like the names that his father had called them." That passage harmonized the E narrative, in which Abraham dug and named some wells, with the J version in which the same wells were dug and named by Isaac. But in J's chronicle, in which Abraham did no such thing, there was no inconsistency that needed harmonizing. Is Friedman suggesting that J wrote the harmonizing passages out of foreknowledge that his story would one day be riffled together with a version by E that would not be composed until after J's death? If Friedman could miss such obvious inconsistencies in passages he attributes to a single author, it seriously undermines his claim that he can detect single authorship in writings not previously credited to the same author.

Friedman's determination to shut his eyes to all evidence that the biblical Yahweh is something less than a nice guy is the only explanation for his writing, "This conception of a God who is torn between justice and mercy informs the whole work. This is not the 'Old Testament God of wrath' that people frequently (and erroneously) imagine. It is more a picture of a God who can get angry but cannot stay angry." (p. 47) If Hitler had had a biographer who could rationalize like that, he would not have needed Goebbels.

It may yet happen that Friedman will eventually recognize the absurdity of continuing to believe in a religion that stands or falls on the veracity of a Bible he knows to be a creation of fallible human imaginations. But I withdraw the opinion expressed elsewhere that he is certain to do so.

## 82. *Jews Without Judaism: Conversations with an Unconventional Rabbi* by Rabbi Daniel Friedman

"Today more Jews are secular than religious. They may 'observe' a few of the rituals of Judaism, celebrating, albeit in the most minimal fashion, a Jewish holiday here and there, perhaps lighting Chanukah candles and participating in a seder at Passover. They may even belong to synagogues and temples, enroll their children in religious schools, celebrate a bar or bat mitzvah, engage rabbis to officiate at their weddings and funerals. But in their daily lives, the beliefs and requirements of Judaism have no bearing upon their decisions." (p. 12) In other words, by any legitimate definition, Jews are typical Americans.

And yet in an economic and social sense, Jews are not typical Americans. Despite constituting 2.3 percent of the American population, "Jews comprise over a third of the billionaires in this country, over a quarter of the multi-millionaires, and between a third and a half of the elite professionals in law, in journalism, in medicine, and in academia. More than one-third of America's Nobel Prize winners have been Jews. Jews occupy a disproportionate number of seats in Congress (37) and on the Supreme Court (2)." (p. 13)

So in case anyone thinks Jews are still an oppressed minority, even a rabbi agrees that they are not. Friedman states (p. 20), "American Jews know (even though they are hesitant to admit it) that their values and ideals are defined not by Judaism but by American liberalism; Judaism provides only an ethnic vocabulary for expressing the values they have already adopted. In the end, that renders Judaism irrelevant."

Friedman gives no indication of being a biblical scholar, and does not openly acknowledge that henotheists who had no belief in an afterlife wrote the Torah. But he is clearly aware of that reality, for, after describing rituals imposed on Jews by the Torah, he writes (p. 16), "*The rabbis* added bodily resurrection and life in the world to come as God's most precious gift to his loyal and obedient servants." (emphasis added)

*Jews Without Judaism*, and particularly the chapter on intermarriage, does illustrate one significant difference between humanistic Judaism and America's largest single religious sect. In his fictionalized interview with a couple planning a mixed marriage, Friedman nowhere implies that "My god can lick your god," or that one religion is more valid than another. I have yet to encounter a Catholic priest capable of such ecumenism.

In contrast, the religious Judaism that Friedman rejects and the redneck Christian Right follow identical practices in one significant element of observable behavior (p. 42): "This amounts to deciding what is true and then looking for evidence that God agrees. Whereas values that are actually demanded by the Bible are conveniently ignored."

Friedman's delineation of how he can be a Jew and a rabbi without believing in an imaginary playmate willing to grant him eternal life without passing GO and without collecting $200 is summarized in his answer to an addict's question, "Why do you call yourself a rabbi if you don't believe in God?" (p. 56): "As I understand Jewish experience, it is impossible to believe that an omnipotent and omnibenevolent God has been in charge of our destiny. Where was He during the Crusades? Where was He during the Inquisition? Where was He during the Holocaust?" (p. 57)

Any incurable godworshipper, Jewish, Christian or Muslim, who can rationalize a reason for an omnipotent, omnibenevolent Master of the Universe to countenance such atrocities, in order to retain belief in such a creature's existence, is one sick puppy.

## 83.  *No Man Knows My History:  The Life of Joseph Smith* by Fawn Brodie

"The Book of Mormon was a plagiarism of an old manuscript written by one Solomon Spaulding, which Sidney Rigdon had somehow secured from a printing house in Pittsburgh.  After adding much religious matter to the story, Rigdon determined to publish it as a newly discovered history of the American Indian.  Hearing of the young necromancer Joseph Smith...he visited him secretly and persuaded him to enact a fraudulent representation of its discovery." (p. 68)

After reporting that reasonably accurate account of the *Book of Mormon*'s true origin, Brodie then goes on to say, "Through the years the 'Spaulding theory' collected supporting affidavits as a ship does barnacles, until it became so laden with evidence that the casual reader was overwhelmed by the sheer magnitude of the accumulation.  The theory requires a careful analysis, because it has been so widely accepted." (p. 68)  She then concludes, "When heaped together without regard for chronology...and without any consideration of the character of either Joseph Smith or Sidney Rigdon, they seem impressive." (p. 442)  In other words, the character of the humbug who pretended to translate upside-down Egyptian funerary scrolls as the "Book of Abraham," and was conned into promulgating an alleged translation of the "Kinderhook plates," pseudo-hieroglyphs forged specifically to deceive him, was incompatible with his being a barefaced liar.  Sure.  And Santa Claus comes down the chimney on Mithra's birthday.

In fairness to Brodie, who updated her 1945 book in 1971, and died before the publication of two 1985 books that revealed twelve pages of the *Book of Mormon* to be in Solomon Spaulding's handwriting, her conclusion that Smith was not a plagiarist was less absurd in 1971 than it was in 1995, when her publisher decided to reprint her by then totally discredited interpretation in paperback.  The very fact that Brodie discussed the *Book of Mormon*'s Spaulding genesis and rejected it makes her biography particularly welcome to hardcore Mormons who think that truth is whatever the marks will swallow.  And even since the publication of *Joseph Smith and the Origins of the Book of Mormon* by D. Persuitte (Prometheus, 1985), and *Trouble Enough:  Joseph Smith and the Book of Mormon* by E. H. Taves (Prometheus, 1985), Brodie's credulous account continues to be cited as the definitive biography of the founder of a scam as blatant and consciously fraudulent as Scientology and televangelism.

Again in fairness to Brodie, she did not suppress any of the negative evidence, and her book is indeed a useful account of the *Book of Mormon*'s origins as a fictionalization of such evidence as came to the (true) author's attention in the early nineteenth century.  For example (pp. 34-37), she cites folklore and topography concerning Indian mounds and pre-existing Indians-as-Jews myths that in her view were Smith's inspiration for his fiction, as she acknowledges it to be.  And she quotes from seven affidavits declaring that the content and proper names found in the *Book of Mormon* were identical with excerpts from his novel that Solomon Spaulding had read in the signer's presence twenty years earlier.  But she then concludes that the affidavits were the result of false memories put into the signers' minds by an excommunicated Mormon, Philastus Hurlbut.  She writes (p. 446-447), "It can clearly be seen that the affidavits were written by Hurlbut, since the style is the same throughout."  In other words, the very fact that "Six recalled the names Nephi, Lamanite, etc.; six held that the manuscript described the Indians as descendants of the ten lost tribes; four mentioned that the great wars caused the erection of the Indian mounds; and four noted the ancient scriptural style," along with several recollections that Spaulding's archaic wording, "Now it came to pass," had grated on the witnesses long before they learned that Smith had repeated the wording unchanged, should be interpreted as evidence that the affidavits are valueless.  Apparently, when one person swears (p. 446), "I obtained the book [of Mormon], and on reading it, found much of it the same as Spaulding had written, more than twenty years before," such testimony can be taken seriously.  But when seven people say much the same thing, their testimony cannot be taken seriously. And despite the Mormon Church's open acknowledgement that Joseph Smith was semi-literate, Brodie insists that he was fully capable of concocting his entire pseudo-bible from scratch. Perhaps. But even if he could have done so, he did not.

Even someone who accepts Smith as the *Book of Mormon*'s author cannot read Brodie and continue to believe that Smith was anything but an imaginative fantasizer—unless of course the reader is a Mormon, in which case rationalizing away the evidence is no more difficult than rationalizing away the reality that the biblical god's official biography portrays him as the most sadistic, evil, megalomaniac serial killer in all fiction. To someone who can read a bible and see "God" as a good guy, reading *No Man Knows My History* and seeing Joseph Smith as a good guy is no big step.

## 84. *Great Quotations on Religious Freedom*
## Albert Menendez and Edd Doerr, eds

Fully evolved human beings, or "liberals" as they are more commonly called, do not need to be told that capital punishment is subhuman, or that the ability to mistake pre-human tadpoles with zero brainwave activity consistent with even minimal human thought for self-aware sentient beings is a form of intellectual retardation. If they did not already know that, they would be throwbacks to a stage of evolution roughly equivalent to Homo neanderthalensis. So *Great Quotations* will not tell them anything they did not already know about the difference between right and wrong. What it will tell them, very effectively, is which persons of state-of-the-species evolution said the same things they already know.

Theocracy, the enforcement of the religion of those with the biggest guns on persons who do not agree with it, has been supported historically by persons as rationally challenged as Jerry Falwell, Pat Robertson, Pat Buchanan, Ariel Sharon, Anthony Comstock, Cotton Mather, and Canada's western provinces redneck hate cult; as intellectually challenged as Ronald Reagan, George Bush Jr, Karol Wojtyla, Billy Graham, Danforth Quayle, Stockwell Day, and the perpetrators of *Touched By an Angel*; as intestinally challenged as William Jennings Bryan; as educationally challenged as creationists; and fanatics such as Torquemada, Khomeini, Hazballah, Hamaz, and bin Laden, of whom no comment is necessary.

In contrast, separation of church and state has been endorsed by Thomas Jefferson, James Madison, John Kennedy, and hundreds of others whose intelligence and rationality are not questioned even by the theofascists whose concept of morality is taken from the official biography of the most sadistic, evil, megalomaniac serial killer in all fiction, a fellow called "God." Instead of citing a long list of names, I will instead quote the words of some of the most prominent.

PRESIDENT JOHN ADAMS: The government of the United States is not, in any sense, founded on the Christian religion.

PRESIDENT JAMES MADISON: The establishment of the chaplainship in Congress is a palpable violation of equal rights as well as of Constitutional principles.

PRESIDENT THOMAS JEFFERSON: The clergy, by getting themselves established by law and ingrafted into the machine of government, have been a very formidable engine against the civil and religious rights of man.

PRESIDENT THEODORE ROOSEVELT: To discriminate against a thoroughly upright citizen because he belongs to some particular church, or because, like Abraham Lincoln, he has not avowed his allegiance to any church, is an outrage against that liberty of conscience which is one of the foundations of American life.

PRESIDENT BILL CLINTON: My administration has consistently opposed any action that seeks to provide public tax dollars in the form of vouchers to be used at private or religious schools.

PRESIDENT JIMMY CARTER: The government ought to stay out of the prayer business.

PRESIDENT JOHN KENNEDY: I believe in an America where the separation of church and state is absolute—where no Catholic prelate would tell the President (should he be Catholic) how to act and no Protestant minister would tell his parishioners for whom to vote.

WILLIAM BUTLER YEATS: Once you attempt legislation on religious grounds, you open the way for every kind of intolerance and religious persecution.

SENATOR EDWARD KENNEDY: I believe that religious witness should not mobilize public authority to impose a view where a decision is inherently private in nature or where people are deeply divided about whether it is.

GOVERNOR NELSON ROCKEFELLER: I do not believe it right for one group to impose its vision of morality on an entire society.

MARTHA KEGEL, ACLU activist: Creationism is a Bible story and teaching it as science in public schools violates the rights of religious minorities.

DANIEL DEFOE: And of all the plagues with which mankind are cursed ecclesiastic tyranny's the worst.

ALAN DERSHOWITZ: Let there be no mistake about the ultimate goal of the Christian right: to turn the United States into a theocracy ruled by Christian evangelicals.

CARDINAL JAMES GIBBONS: A civilian ruler dabbling in religion is as reprehensible as a clergyman dabbling in politics. Both render themselves odious as well as ridiculous.

JAMES MICHENER: Religious hatreds ought not to be propagated at all, but certainly not on a tax-exempt basis.

ELEANOR ROOSEVELT: I do not want church groups controlling the schools of our country. They must remain free.

SENATOR SAM ERVIN: Government is contemptuous of true religion when it confiscates the taxes of Caesar to finance the things of God.

POPE JOHN XXIII: Also among man's rights is the right to worship God in accordance with the right dictates of his own conscience.

HAZRAT MIRZA TAHIR AHMAD, author of *Murder in the Name of Allah*: People who persecute in the name of religion are totally ignorant of the essence of religion.

BALTHASAR HUEBMAIER, 16th century Anabaptist: The slayers of the heretics are the worst heretics of all.

GENERAL ROBERT E. LEE: Is it not strange that the descendants of those Pilgrim Fathers who crossed the Atlantic to preserve their own freedom of opinion have always proved themselves intolerant of the spiritual liberty of others?

BLAISE PASCAL: Men never do evil so completely as when they do it from religious conviction.

VOLTAIRE: The Inquisition is an admirable and wholly Christian invention to make the pope and the monks more powerful and turn a whole kingdom into hypocrites.

HASHEM AGHAJERI, Iranian reformer, July 2000: It is time for the institution of religion to become separated from the institution of government.

CARL SAGAN: The framers of the Bill of Rights had before them the example of England, where the ecclesiastical crime of heresy and the secular crime of treason had become nearly indistinguishable.

BENJAMIN FRANKLIN: When a religion is good, I conceive it will support itself; and when it does not support itself, and God does not take care to support it so that its professors are obliged to call for help of the civil power, 'tis a sign, I apprehend, of its being a bad one.

SUPREME COURT JUSTICE HARRY BLACKMAN for the majority, "Roe v. Wade": The states are not free, under the guise of protecting maternal health or potential life, to intimidate women into continuing pregnancies.

And on the side of theocracy:

JERRY FALWELL: The idea that religion and politics don't mix was invented by the Devil to keep Christians from running their own country.

JIMMY SWAGGART: It must never be forgotten that this is a Christian country based on the words of [his imaginary playmate].

PRESIDENT RONALD REAGAN, who believed in a geocentric universe and based foreign policy decisions on the dictates of an astrologer: God, the source of all knowledge, should never have been expelled from our children's classrooms.

*Great Quotations* contains 732 utterances similar to the foregoing. In summary, I can do no better than repeat the accurate description from the publisher's press release:

"This outstanding collection of memorable quotations on religious freedom—the most comprehensive ever assembled—covers many centuries of thought and a wide array of sources. On every page the reader will discover a wealth of thoughtful, wise, and sometimes impassioned statements by all manner of men and women on a subject that has moved the consciences of generations from the distant past to the present. Included are early church fathers, Enlightenment philosophers, popes, anticlerical European statesmen, journalists, famous writers, judges, twenty-six presidents of the United States, and many others. A special feature of this compilation is the inclusion of quotes from major judicial decisions, from 1872 to the present, that bear on religious liberty."

## 85. *Why People Believe Weird Things*
## by Michael Shermer

I sometimes wonder if I will ever find a book with which I am completely satisfied. Days ago I berated Philip Plait's excellent *Bad Astronomy* for being written at a comprehension level that made it possible to read and understand it while simultaneously watching *The Simpsons*. Now I find myself equally frustrated by Michael Sherman's excellent *Why People Believe Weird Things*, for aiming at a much more sophisticated audience, with the consequence that it must be read slowly and carefully and with full concentration. This book was not written to entertain. It was written to inform. Whether the evidence presented for Shermer's conclusions constitutes overkill, or simply thoroughness and attention to detail, is in the eye of the beholder. Certainly the book achieves its purpose, both in refuting nonsense beliefs, and in explaining, without resorting to psychobabble, why they are so prevalent.

If tabloid TV is any indication, the difference between science and pseudoscience is far from clear to most North Americans. Shermer defines the difference (pp. 33-34): "The search for extraterrestrial life is not pseudoscience because it is plausible, even though the evidence for it is thus far nonexistent…Alien abduction claims, however, are pseudoscience. Not only is physical evidence lacking but it is highly implausible that aliens are beaming thousands of people into spaceships hovering above the Earth without anyone detecting the spacecrafts or reporting the people missing."

Several chapters demolish the visiting-aliens delusion, and explain its origin. For example (p. 95), "The feedback loop was given a strong boost in late 1975 after millions watched NBC's *The UFO Incident*, a movie on Betty and Barney Hill's abduction dreams. The stereotypical alien with a large, bald head and big, elongated eyes, reported by so many abductees since 1975, was created by NBC artists for the program." As to why abduction tales have similarities that True Believers cite as corroboration (p. 97), "Yet I think we can expect consistencies in the stories since so many of the abductees go to the same hypnotist, read the same alien encounter books, watch the same science fiction movies, and in many cases even know one another and belong to 'encounter' groups."

On the pseudomedicine of "recovered memory" that in the decade it was fashionable destroyed thousands of innocent lives, Shermer notes (p. 110), "But what we appear to be experiencing…is not an epidemic of childhood sexual abuse but an epidemic of accusations. It's a witch craze, not a sex craze." That paragraph is accompanied by a chart showing that accusations against parents rose steadily from a few dozen in January 1992 to almost 12,000 in March 1994. "Fortunately, the tide seems to be turning in favor of the recovered memory movement being relegated to a bad chapter in the history of psychiatry." My one problem with that forecast is that I look forward to the day when *psychiatry* is relegated to a bad chapter in the history of medicine. "The parallels with Trevor-Roper's description of how a medieval witch craze worked can be eerie." (p. 112)

Fortunately, the steam was knocked out of the recovered memory delusion when (p. 111) "a six-member jury in Ramsey County, Minnesota, awarded $2.7 million" to a patient whose psychiatrist "planted false memories of childhood sexual abuse." The patient "was diagnosed with multiple personality disorder" [a psychobabble name for compulsive playacting]. The psychiatrist "'discovered' no less than 100 personalities" caused by the patient being "sexually abused by her mother, father, grandmother, uncles, neighbors, and many others." The patient's imaginary past, constructed by the psychoquack, "even included Satanic ritual abuse featuring dead babies being served as meals 'buffet style.' The jury didn't buy it."

On the ability of True Believers to see their own inadequacies in their opponents, Shermer writes (p. 114), "A subtle form of projection is at work when fundamentalists make the accusation that secular humanism and evolution are 'religions' or announce that skeptics are themselves a cult and that reason and science have cultic properties." I can confirm that. When I was trying to get *Mythology's Last Gods* published, the Canada Council referred it to a theologian (analogous to having James Randi's debunking of psychics evaluated by Uri Geller), and the theologian declared that it is the historians who start from predetermined conclusions and distort the evidence to make it fit.

Shermer classifies Ayn Rand's Objectivism as a cult. He writes (p. 123), "I have read *Atlas Shrugged*, as well as *The Fountainhead* and all of Rand's nonfiction works. I accept much of Rand's philosophy, but not all of it." That passage is something I might have written myself. What makes Objectivism a cult is that (p. 124), "But as soon as a group sets itself up as the final moral arbiter of other people's actions, especially when its members believe they have discovered absolute standards of right and wrong, it marks the beginning of the end of tolerance, and thus reason and rationality...Its absolutism was the biggest flaw in Ayn Rand's Objectivism, the unlikeliest cult in history."

One of Shermer's topics surprised me, since I was unaware that such a nonsense belief had sufficient adherents to warrant granting it the dignity of a rebuttal, but Shermer's statistics show that it is far more pervasive than most people realize. He writes (p. 131), "Of all the claims we have investigated at *Skeptic*, I have found only one that I could compare to creationism for the ease and certainty with which it asks us to ignore or dismiss so much existing knowledge. That is Holocaust denial. Further, the similarities between the two in their methods of reasoning are startling." And on pages 206-207, "The development of the Holocaust denial movement has striking parallels with the development of other fringe movements. Since deniers are not consciously modeling themselves after, for example, the creationists, we may be tracking an ideological pattern common to fringe groups trying to move into the mainstream." A particularly strong argument against the deniers is (p. 241) "During his trial, Eichmann never denied the Holocaust. His argument was that 'these crimes had been legalized by the state' and therefore the people that 'issued the orders' are responsible."

Shermer quotes 25 arguments for their position presented by creationists, and annihilates them one by one. For example (p. 147), the creationists claim, "Population statistics demonstrate that if we extrapolate backward from the present population using the current rate of population growth, there were only two people living approximately 6,300 years before the present." Shermer's rebuttal is, "Applying their model, we find that in 2600 B.C.E. the total population on Earth would have been around 600 people. We know with a high degree of certainty that in 2600 B.C.E. there were flourishing civilizations in Egypt, Mesopotamia, the Indus River Valley, and China. If we give Egypt an extremely generous one-sixth of the world's population, then 100 people built the pyramids."

Shermer's accounts of his appearances on the Phil Donohue and Oprah Winfrey shows prove only that everyone connected with such programs is as scientifically literate as a Canada goose. So what else is new? On psychic hotlines (p. 276), "The goal is to keep callers on the line long enough to turn a good profit but not so long that they refuse to pay the phone bill." And his invalidation of the concept of "race" includes the observable reality that (pp. 247-248), "Darwin noted that naturalists in his time cited anywhere from two to sixty-three different races of *Homo sapiens*. Today there are anywhere from three to sixty races, depending on the taxonomist...Europeans are an intermediate hybrid population of 65 percent Asian genes and 35 percent African genes...Recent research shows, in fact, that if a nuclear war exterminated all humans but a small band of Australian Aborigines, a full 85 percent of the variability of *Homo sapiens* would be preserved."

And in summary (p. 275), "More than any other, the reason people believe weird things is because they want to. It feels good. It is comforting. It is consoling. According to a 1996 Gallup poll, 96 percent of American adults believe in God." Since belief in "God" is a prerequisite for belief in an afterlife, without which one sixth of the human race would have to be institutionalized and diapered, that is hardly surprising. And Shermer states that, "Similarly, to the frequently asked question, 'What is your position on life after death?' my standard response is, 'I'm for it, of course.' The fact that I am *for* life after death does not mean I'm going to get it. But who wouldn't want it?" And whereas True Believers ask nontheists, "What have you got to lose?" Shermer notes that (p. 278), "by focusing on a life to come, we miss out what we have in this life." Pie in the sky when you die, anyone? All it will cost you is a lifetime of superstition, masochism, and autoreinforced braindeath.

## 86. *How We Believe: The Search for God in an Age of Science* by Michael Shermer

Michael Shermer writes (p. x), "If, in the process of learning how to think scientifically and critically, someone comes to the conclusion that there is no God, so be it—but it is not our goal to convert believers into nonbelievers...I would have thought that God's existence, from a scientific and rational perspective, remains an open question—it cannot be "proved" one way or the other."

Wrong! Religion does not stand or fall on the veracity of the claim that an entity exists with sufficiently advanced understanding of the laws of nature that it would seem to us not to be bound by those laws, a claim that indeed cannot be falsified. Religion stands or falls on the claim that a god has revealed its existence, and that it has specific qualities such as omniscience and omnipotence. All such claims have been traced to the same bible authors who also assured their readers (in seven places) that the earth is flat. The unsubstantiated testimony of the Judaeo-Christian Bible has as much credibility as the testimony of Baron Münchausen or Richard Nixon. Believing that God, the protagonist of a work of fiction, might exist, is no different from believing that Brobdingnagians, Lilliputians or Houyhnhnms might exist. And since omnipotence, the ability to create a number that is more than ten but less than nine, and omniscience, knowledge of future events that have not yet been caused *and are not predestined*, cannot exist, it follows that an entity possessed of such qualities cannot exist. And that which cannot exist does not exist.

The accusation that an agnostic is really a chicken atheist tends to come from dogmatic theists, persons who believe that anyone less brainwashed than themselves is lacking a positive quality that they equate with courage. In fact a better explanation for agnosticism, the view espoused by Michael Shermer, is political correctness, a desire to avoid appearing intolerant by admitting that he knows he is right and believers are wrong. Alternatively, he could be genuinely ignorant of the reality that religion has been as objectively falsified as Martian canals, even though the argument in the preceding paragraph proves that point by itself.

Shermer certainly recognizes the "god" hypothesis as indefensible. Yet he defends believers' right to hold dissenting opinions, as if knowledge and ignorance were equally meritorious. At least his unwillingness to offend the terminally ignorant does not go to the incredible lengths of Stephen Jay Gould, whose imbecilic *Rocks of Ages* categorized science and superstition as Non-Overlapping Magisteria, perspectives that could not be harmonized but were not really incompatible. Ignorance and knowledge *are* incompatible. The only true agnostic is someone who is unfamiliar with the falsifying evidence. Self-proclaimed agnosticism by anyone else is pure political correctness.

The inside cover of Shermer's book states that "Recent polls report that ninety-six percent of Americans believe in God." While I could conceivably have missed it, I certainly found no statement by Shermer supporting that statistic **[actually it was in his previous book. see review # 85]**, and it is more likely that the publisher was inserting the result of incompetent research. The actual number of believers is no higher than seventy percent,[1] and as many as one-third of those are merely reporting an alleged belief that has no relevance to their everyday lives. More significant is the finding of a study of one thousand scientists that only forty percent expressed a belief in a personal God, while only thirty percent believed in immortality. Among physicists and astronomers, belief was below twenty percent. And among National Academy of Science members, when "doubt" or "agnosticism" was factored in, belief dropped to seven percent. In short, the scientifically illiterate tend toward belief in a god, and the scientifically educated do not.

Nor is religion the only contrary-to-fact belief system that is widespread in a continent in which schools have been babysitting institutions in which teaching is illegal for more than fifty years. "A Gallup poll conducted in 1991 revealed that half of all Americans believe in astrology and almost as many believe in extrasensory perception, or ESP; a third believe in the lost continent of Atlantis and in ghosts; and a full two-thirds believe they have had a psychic experience." (p. 35)  Since ignorance and fuzzy thinking explains the prevalence of belief in parapsychology, it seems reasonable to conclude that the same qualities explain widespread belief in religion, even if Shermer is too politically correct to spell out such an implication.

In Shermer's appendix, he lists statistics on why people claim to believe in their particular god, or claim not to.  The only surprise is that a mere one percent of believers attribute the distinction between good and evil to a deity, meaning that without a Lawgiver to tell them which is which, evil would prevail because the masses would not know that there is a difference.  Unfortunately, that one percent includes religious nutcases like Falwell, Robertson and Bush Junior, who want to enslave the ninety-nine percent by giving the force of law to their brainwashing that only belief in their god can prevent the triumph of evil.

*How We Believe* is partly autobiography, partly a survey of religiosity in America, and partly an examination of human thinking.  While it is flawed, it is generally accurate and not always trivial.

1    A 2000 survey by the Graduate Center of the City University of New York found that 14.1 percent reported that they had "no religion."  It is a logical assumption that, for every person willing to go on record with such a politically incorrect admission, there was another who chose to protect himself from the foreseeable consequences of acknowledging a similar freedom from superstition.

## 87. *The Ghost in the Universe: God in Light of Modern Science* by Taner Edis

In his introduction to *The Ghost in the Universe*, Taner Edis observes (p. 14), "But today, conservative, magical, scripture-waving religion has become obviously false to the well-educated person...Now, I don't intend to spend time refuting such claims; they are too blatantly wrong." He continues (p. 17), "The complexities of life do not require intelligent design; accidents and blind mechanisms do the trick. Not only old-fashioned creationism but also more liberal attempts to find a progressive guiding hand in biology get nowhere." And as the icing on the cake he adds (p. 24), "It might even be said that proofs of God enjoy an immortality only truly bad ideas can aspire to."

In arguing for an evolutionary rather than creationist origin of life, Edis writes (pp. 51-54), "Evolution accounts for life as we see it, warts and all—without a designer...Now that creationism is relegated to the intellectual fringe, and that liberal religious people accept evolution, it would seem the Darwin wars are not relevant to today's questions about science and religion...When creationism collapsed, this was not because of philosophical difficulties or even because of a direct contradiction by data. There were always things which did not fit smoothly; parasites, for example, strained the picture of benign harmony, even if imperfections could be expected in a fallen world. In time, more uncomfortable facts like extinct species and an old earth accumulated...Populations adapt to their environment and exhibit good 'design' because genes promoting reproductive success have a better chance to make it to the next generation...Indeed, such uncompromising opposition to evolution as to concoct a bizarre 'creation science' as an alternative is largely an American, evangelical Protestant peculiarity."

On the failure of natural science to find any supporting evidence for a metaphysical First Cause or Intelligent Design, Edis observes (pp. 107-108), "The story is that Napoleon asked Laplace what part God played in his system, and Laplace answered, 'I have no need of that hypothesis.'...If we cannot fathom the divine reason behind why it rained last Sunday, or why the sky is blue, this is hardly a great challenge to religion. But when God vanishes from physics, indeed, from all natural science, it begins to look like there is no God after all. If there were a cosmic power, a divine purpose behind everything, we should see traces of God in our world. We do not." As for the reasoning behind the God hypothesis, Edis's comment is (p. 167), "It is like deciding Santa Claus exists because we cannot figure out who bought one of the presents under our Christmas tree."

On page 118: "Yahweh decides to cultivate a special nation from the seed of Abraham. He gathers the Jews out of Egypt, and leads them to a promised land which becomes theirs after some ethnic cleansing. He also promises that if the Jews try and become more godly, obey his laws, be nice to their neighbors, and slaughter whom he dislikes, they will become a privileged people." Reading that description of the authorized version of the origin of Western religion, I found myself wondering if it referred to ancient Jews under Joshua, or the modern Likud under prime ministers whose status as "freedom fighters" rather than "terrorists" is strictly a function of their being on "our" side.

As a physicist rather than a historian, Edis not surprisingly makes factual errors concerning historical documents. For example, he states (p. 121) that Deuteronomy was "written in or after the exile of the elites of Jerusalem to Babylon." But Deuteronomy was discovered behind a loose brick in Yahweh's temple in 621 BCE—a generation before the Babylonian Captivity. It is a minor, one might even say insignificant, error. But it draws attention to Edis's status as an amateur in at least one of the fields on which he comments. And, while he expresses disagreement with some of the conclusions of Burton L. Mack and John Dominic Crossan, he quotes or paraphrases them to a far greater extent than their indefensible speculations warrant.

373

On Muhammad (pp. 130-131): "Muhammad seems to have been especially concerned about poets who, in an age without television, could be influential in shaping public opinion. Fortunately, he was usually able to have them assassinated...Muhammad dealt with the last remaining Jewish tribe in Medina by a method straight out of the Bible...So the Muslims, led by Muhammad, killed all of the men, divided the property among themselves, and enslaved the women and children."

On the incompatibility of different religions (p. 133): "If we create God in our own image, it is no surprise that God should turn out to have so many different faces."

Edis asks (p. 153), "How does a Jewish apocalyptic prophet who was dead wrong about the coming Kingdom and who was crucified as a pest end up starting a Greek religion?" He suggests, (p. 167), "If Jesus was a failed apocalyptic prophet, a teacher, or a faith healer, it is hard to see why God would bother resurrecting him anyway." His explanation is (p. 169), "Believing in a peculiar but deeply meaningful creed is not unusual for a fringe religious group...Members of the early Christ cult were convinced Jesus *had to* be vindicated by their God, and converted the disaster of his death into a reason for missionary fervor." He summarizes (p. 172), "But it is still strange to watch Christianity dissolve into a vapid verbiage and contempt for truth which would be more at home in a California psychotherapy cult."

On the interchangeability of religion and parapsychology, Edis states (p. 179), "Miracles are the tabloid underside of religion. Tales of levitating saints and weeping statues belong with poltergeists, spirit-summoning mediums, and psychics predicting California will slide into the sea." He elaborates (p. 203), "For all the overheated rhetoric, the skeptics are correct: parapsychology, like homeopathy or astrology, survives not on the strength of its results but because of its appeal outside of the scientific community." And one of his subheadings is THIRTY-ONE FLAVORS OF ULTIMATE REALITY.

As for religion's pathetic attempts to harmonize an omnipotent, omnibenevolent god with the existence of AIDS, cancer, transportation accidents, earthquakes and religious wars, to say nothing of urine, excrement and menstruation, Edis's observations are as good as any I have seen before (p. 278): "After all, historically the biggest nuisance for theology has been the need to reconcile God and evil...Our Creator is supposed to be morally perfect and all-powerful, and *this* world was the best it could manage?...Most everyone can understand the pain of a parent who loses her child to disease, and hear God's maddening silence in response. We need not master arcane technical skills to see how theologians' excuses for the silence are absurd in the face of suffering."

Edis's book says little that is new. How well he says it, the reader can judge for himself on the basis on the quoted passages. I see history as the *only* discipline capable of ultimately freeing humankind from religion, no matter how valid philosophical arguments may be. Not everyone agrees.

## 88. *The Book Your Church Doesn't Want You To Read* edited by Tim C. Leedom.

"Knowing the reaction of established religion in the past to critique and examination, we anticipate a strong response from those who won't even read *The Book*. These leaders and followers continually take the attitude, 'don't bother me with facts; I've already made up my mind.'" (p. iii)

"Because of religion, more human beings have been murdered, tortured, maimed, denigrated, discriminated against, humiliated, hated and scorned than for any other reason in the totality of the history of man." (p. v)

After that promising opening, *The Book* moves steadily downhill. This well-meaning proof that "a little learning is a dangerous thing" is riddled with inaccuracies. For example, it credits Paul of Tarsus with adopting the Mithraic Sun-day to replace the Hebrew Sabbath. (p. 4) In fact Christianity had no sacred rest day until Constantine borrowed the Mithraic Sun-day three centuries after Paul.

It states that "Ancient man saw in his male offspring his own image and likeness, and his own existence as a father was proved by the person of his son." (p. 21) But men did not have any "father" concept until *c* 3500 BCE, millennia later than the author of that passage appears to believe. It lists the original concept of the zodiac, "circle of animals," as consisting of twelve houses. (p. 24) But the original zodiac contained thirteen houses. Ophiuchus was purged and its portion of the sky transferred to neighboring Scorpio sometime after the male revolution of *c* 3500 BCE.

While *The Book* correctly identifies the hexagram as "evolving later to become the Jewish Star of David," (p. 42) it gives its origin as a Hindu sun symbol. In fact it originated as a sex-glorification, with an up-pointing triangle representing the male genitalia superimposed on a down-pointing triangle representing the female genital orifice. And while it identifies the sixth-century Mandylion of Edessa as the source of all later portraits of Jesus, it does not mention that, prior to the sixth century, Jesus was acknowledged even by Christian apologists to have been ugly, deformed, and "not even of honest human shape." As for the statement that Jesus' Aramaic/Hebrew name was *Yehoshua*, (p. 147) the author of that item was either unaware or, as a practising pusher, deliberately concealed that the correct name was *Yahuwshua*, *Yahuw* being the Jewish god's proper name.

*The Book* also includes an excerpt from John Allegro's hypothesis that Jesus the Nazirite was "really" a deified mushroom, a theory so ridiculous that Robert Graves called it "an elaborate literary hoax." Graves was far too polite. Allegro's mushroom fantasy destroyed his reputation, and rightly so. The inclusion of such drivel says more about the editors' status as amateurs than perhaps anything else in the book.

On the good side, *The Book* includes a letter from Thomas Jefferson to his nephew (pp. 33-34), spelling out a methodology for examining the evidence on which religious claims are based, a methodology currently used by all legitimate scholars in all fields and rejected only by those disciplines (religion, parapsychology) that start from predetermined conclusions and distort the evidence until it fits. Articles by Bertrand Russell, Thomas Paine, Steve Allen and Joseph Campbell argue against the illogic of religious thinking, as do pieces by scholars whose names are less known to the masses, such as Gerald Larue, G. Vermes, A. J. Mattill Jr, Morton Smith, Robert Ingersoll and Joseph McCabe.

There is a lot of interesting reading in *The Book Your Church Doesn't Want You To Read*, but nothing to justify its pretentious title. Indeed, bible thumpers might be well advised to promote the book, so they can argue that, if this is the best case that can be made against religion, it has nothing to fear. Fortunately, there are books in print that do to religion what the first photographs of the Martian surface did to the "canals" delusion, including those of Martin Larson, Richard Friedman, McCabe and Ingersoll, not to mention *Mythology's Last Gods*.

375

## 89. *The Mythic Past:  Biblical Archaeology and the Myth of Israel*
## by Thomas L. Thompson

According to Thomas Thompson's *The Mythic Past* (p. 164), "We do not have evidence for the existence of kings named Saul, David or Solomon; nor do we have evidence for any temple in Jerusalem in this early period.  What we do know of Israel and Judah of the tenth century does not allow us to interpret this lack of evidence as a *gap* in our knowledge and information about the past, a result merely of the accidental nature of archaeology.  There is neither room nor context, no artifact or archive that points to such historical realities in Palestine's tenth century."  He even adds, "Jerusalem is not known to have been occupied in the tenth century."

That would certainly eliminate the empire of King David—except that Thompson's allegation is supported by no evidence whatsoever.

It is possible to prove anything if one can get people to believe his axioms.  Thompson speculates that *Hapiru* meant "bandit," even though most Egyptologists since James Breasted (*Ancient Records of Egypt*) and James Pritchard (*Ancient Near Eastern Texts Relating to the Old Testament*) translate it as "Easterner."  If he is right, then his claim that there was no invasion of the area that became Judah from across the Jordan becomes entirely plausible.  The desperate Amarna letters to Pharaoh Ikhenaton's viceroy from Jerusalem, Bethlehem and other cities demanding help to repel the invaders, can be dismissed as a request for more police to counter looters and vandals.  But while majority consensus has been wrong before, Thompson offers no evidence but his own conjecture that such is the case here.  However, the strongest argument against such a translation is Genesis 14:13, in which the Torah's earliest author, the Yahwist (J), refers to "Abr[ah]am the *habari*," surely a dialectal form of the same word.  That the Yahwist would describe Abraham, his alleged ancestor and hero, as a "bandit," carries speculation to the point of absurdity.

Thompson further speculates that Palestinian social ecology did not allow for central control of even small regions.  If that was so, then his conclusion that David could not have been more than a village chieftain whose hegemony did not extend beyond the walls of Jerusalem, whose population did not exceed two thousand, would appear to be proven.  Unfortunately, the assumptions leading to that conclusion appear to have originated in Thompson's own imagination.

That the Yahwist flattered his community by giving them mythical heroic ancestors and placing those ancestors in a cultural environment he himself invented is credible.  That the Elohist (E), writing independently 150 years later, borrowed from the same oral folk legends is credible.  That the Elohist also set his myths in a cultural environment that originated in the Yahwist's imagination is not credible.  Indeed, Thompson's reconstruction of the social and ecological conditions of the early second millennium BCE is entirely consistent with the one presented by J and E.  How Thompson can claim that the environment described in Genesis never existed, and then turn around and offer a substitute identical with the one he is repudiating, is impossible to guess.  And if no part of the (present) Torah was written until after the death of Alexander of Macedon, as Thompson maintains, it becomes impossible to understand why E would write a pro-Moshite, anti-Aaronid narrative at a time when the Moshite priesthood had been extinct since the sacking of Samaria by Assyria—or perhaps that did not really happen?  But to Thompson there is no problem, because to him J and E did not exist either.

Thompson's reconstruction of the composition of the Torah declares that it was written between 323 and 175 BCE, a long enough period for it to have been interpolated and redacted more than once, but not long enough for it to have originated as four separate documents that one or more redactors combined into a single narrative.  Thompson is therefore arguing for a single author, the same position taken by the most conservative religionists, with the exception that Thompson does not identify the author as Moses.  But the evidence of multiple authorship is overwhelming.

For example, J wrote that Noah took seven of each kind of "clean" (suitable for food) animal aboard the ark. P had him take two of each kind, making no distinction between clean and unclean because, in P's Torah, that distinction had not been invented in Noah's time. And that inconsistency can be found in consecutive verses. Clumsy editing by the Redactor is an acceptable explanation. A single author contradicting himself in successive lines is not. And where the Deuteronomist (D) wrote sympathetically of "the male lover who means as much to you as your own breath" (Deuteronomy 13:6), P described male homosexual coupling as "detestable" (Leviticus 18:22), and the final Redactor (R) inflicted a death penalty (Leviticus 20:13). That the same author wrote both Leviticus passages is a defensible interpretation. That he also wrote the Deuteronomy verse is not.

More than anything else, the ability to divide the story of Joseph's sale into slavery into two complete, self-contained accounts that contradict each other (P's Joseph was rescued from a well into which his brothers had tossed him and sold to an Egyptian by Midyanites. J's Joseph was sold by his brothers to Ishmaelites who then sold him to an Egyptian) is unique to a book compiled by editing conflicting narratives together. As Richard Friedman wrote (*Who Wrote the Bible?*, p. 59), "The very fact that it is possible to separate out two continuous stories like this is remarkable itself, and is strong evidence [for multiple authorship]. One need only try to do the same thing with any other book to see how impressive this phenomenon is."

Thompson's reference to "Mosaic monotheism that one might expect if one read the Bible as history" (p. 169), raises doubts that he has read the Torah himself. None of its four primary authors or its two redactors was a monotheist, although all were henotheists, partisans of a particular god while not denying the existence of others, as is clearly revealed in Exodus 9:14 by J; 12:12 by P (the Priestly author); 18:11 by E; Numbers 33:4 by R; and Deuteronomy 3:23 by D. He also seems unaware that the primary object of Israelite worship was not *El* but *elohim*, a dual-sex, generic plural meaning "the male and female gods."

Thompson is severely critical of Josephus for building vast theses out of half-vast evidence. What he does not explain is why that was reprehensible when Josephus did it but is not reprehensible when Thompson does it.

I do, however, find myself receptive to his theory that the Hyksos were not invaders who conquered Egypt from Asia, but were Delta pharaohs whose penetration of southern Egypt was miniscule, and whose alleged expulsion by Ahmose I was nothing more than a repudiation of Delta overlordship and reestablishment of Theban independence. But why Thompson thinks that proves there was never an Israelite state or a united monarchy is hard to comprehend.

One really annoying, irritating, grating point: On page after page after page, Thompson precedes words starting with a pronounced "h", such as "historical," with the article "an" rather than "a", as if *historical* started with a silent "h" like "honest." Whatever fatuous twit taught him to do that should take a course in remedial English. And while Thompson's book is not exactly riddled with run-on sentences, using a comma as if it were a conjunction, neither is it free of them. Typos, fortunately, are minimal. I detected only two in the first 200 pages.

According to theologian Don Cupitt, in a review in England's *The Observer*, "This book may be remembered as a landmark. It marks the end of 'Biblical archaeology.' The whole enterprise has been as big a mistake as it would be...to prove the historicity of King Arthur." So at least one scholar does not share my view that *The Mythic Past* is unmitigated hogwash. I can live with that.

While any falsification of the Bible merits close attention, and even unsubstantiated theorizing can have its merits, Occam's razor says that Thompson has simply replaced biblical fiction with a fiction of his own. His reconstruction of reality is a product of his imagination in one paragraph, and an affirmation of the Bible's social history in the next paragraph. One might even conclude, if such a theory was credible, that Thompson is only pretending to debunk the Bible in the hope of leading readers to an affirmation that it really is nonfiction.

377

I am now considering writing a history of early America, showing that there never was a revolutionary war, George Washington never existed, and the word "British" used to describe the overlords whom eighteenth century Americans repudiated really meant "tea drinkers." With the precedent of *The Mythic Past* to legitimize such undisciplined fantasizing, how could I go wrong?

## 90. *The Dawn of Human Culture*
## by Richard G. Klein with Blake Edgar

"Some might object that a neurological explanation for the explosion of culture after 50,000 years ago is simplistic biological determinism...Human brains had reached fully modern size many hundreds of thousands of years earlier, and skulls reveal little about the functioning of the brain underneath. There is nothing in the skulls of people from shortly before and after 50,000 years ago to show that a significant neurological change had occurred. The neurological hypothesis does, however, measure up to one important scientific standard: it is the simplest, most parsimonious explanation for the available archeological evidence...Other explanations for the origin of modern human behavior hypothesize that some radical social or demographic event sparked a behavioral revolution about 50,000 years ago. These explanations, however...offer no reason for why the momentous social or demographic change failed to occur tens of thousands of years earlier. Nominating a genetic mutation as the cause answers the 'why' question." (pp. 24-25)

In defining the difference between pre-human and human, Klein writes (p. 29) "The earliest representatives of the human line still looked and acted much like apes, and a casual observer might have mistaken them for a kind of chimpanzee. There was one essential difference, however: on the ground, they preferred to walk upright, on two legs. We know them today as the australopithecines, but in appearance and behavior, they could as well be called bipedal apes."

But while Klein names australopithecines as the first humans, he identifies *Homo ergaster* as the first "true human," the basis for the definition being that *ergaster* was the first hairless ape. Klein nowhere mentions the "aquatic ape" theory spelled out in the books of Elaine Morgan, but his reconstruction of hominid evolution is not incompatible with that theory. In order to harmonize the two, it is necessary to assume that *Homo* went into the water as a hairy *Homo habilis*, and came out as a hairless *Homo ergaster*, approximately 1.8 million years ago, with *habilis* disappearing as a separate species 1.6 million years ago. That would mean that such later dead ends as *erectus* and *neanderthalensis* had modern subcutaneous fat and hemispherical breasts, and mated face to face, but no evidence survives that they did not.

Mitochondrial DNA analysis has established that humans and neanderthals were separate species. Klein uses the analogy of dogs and wolves to make the point that interbreeding between the coexisting species may not have been impossible. But he effectively rebuts the alleged evidence that they ever did interbreed. (pp. 203-204)

While Klein makes no other reference to the origin of religion, he cites burial rituals practised by Cro-Magnons but never by neanderthals as evidence of belief in a "soul." And he makes clear that such a development was a consequence of a cultural "big bang" that he attributes to the above-mentioned unprovable genetic mutation within the brain. As retrograde a step as the invention of religion may have been, it indeed represented a capacity for reasoning (concocting explanations for observable reality) that did not previously exist. Indeed, it might even be argued that religion had survival value, since without religion-motivated mass murders (fifty million by the Christians alone) the human race might have overpopulated to the point where it could not feed itself and would already be extinct.

Since the size of the human brain has not increased in the last 130,000 years, the postulated genetic mutation presumably involved structural change. That being so, the difference between pre-50,000 and post-50,000-BCE brains must also be found in a comparison of human and bonobo brains. If the only changes such a comparison detected were in brain size and in factors that would have already been present in early *Homo sapiens*, would that invalidate Klein's thesis? I would be venturing far beyond my competence to suggest that it would. But it is a point I hope he will eventually clarify. And while Klein does not prove beyond a reasonable doubt that modern humans existed only in Africa 50,000 years ago, and subsequently colonized the world, eliminating all other hominid species, he does shoot down claims that modern skeletons found in Australia date from as much as 12,000 years earlier. (pp. 250-251)

Like many other specialists in a narrow field, Klein makes clumsy errors when he ventures outside of his expertise. In describing large axes as a status symbol used to attract a mate (p. 107), he speculates that, "When a female saw a large, well-made biface in the hands of its maker, she might have concluded that he possessed just the determination, coordination, and strength needed to father successful offspring"—1.5 million years before men first discovered their role in reproduction. And his tautological reference to a "male peacock" is equally clumsy. As for his use of the bigoted dating system, "A.D.," rather than the scientific equivalent "C.E.," no doubt he was not intentionally insulting this planet's five billion non-Christians, and will not use it again.

As a historian who has not taken even a single paleoanthropology course, I found *The Dawn of Human Culture* useful and informative. A reader with some undergraduate expertise might not. As a primer in the subject, whether for students or John Q Public, it is undoubtedly excellent. But I could not help concluding that the book's primary value will be as a text for paleoanthropology 101.

As an experiment, I would like to see Klein read his book to a colony of sea cabbages. If the cabbages show any sign of learning from it, he could then go even lower on the evolutionary scale and try teaching it to creationists. Just a suggestion.

## 91. *The Scars of Evolution: What Our Bodies Tell Us About Human Evolution* by Elaine Morgan

Elaine Morgan's earlier books, *The Descent of Woman* and *The Aquatic Ape,* spelled out Sir Alister Hardy's theory that the differences between humans and their proto-simian ancestors are best explained by postulating a period of human evolution that took place in an aquatic environment. In *Scars of Evolution,* Morgan catalogues the objections to that theory and rebuts them—very effectively. While acknowledging the unlikelihood of ever proving or disproving the Aquatic Ape Theory (AAT), she demonstrates that all alternative hypotheses are equally unverifiable and even more improbable. She reports that, at a 1987 international conference, "the pros and cons of AAT were publicly debated, and that event presented some kind of watershed. By now a number of scientists are prepared to endorse the theory as a tenable hypothesis. Others, while remaining neutral, have been willing to offer constructive criticism." (p. 179)

What makes AAT relevant to a rationalist publication is that it has been derided as pseudoscience comparable with Von Däniken and Velikovsky. Morgan shows that, while it may not be correct, it is demonstrably scientific.

On the issue of human hairlessness, a feature not found in any other former arboreal, savanna-dwelling primate, Morgan states (p. 77), "The basic question is Desmond Morris's: 'in what environment is nakedness at a premium?' There would seem to be a simple and logical answer: It has to be the environment in which nakedness is known most frequently to have evolved. That environment is water."

She elaborates (p. 94), "The strength of the aquatic case lies in the fact that those features of the human skin that are undeniably unique among primates—the nakedness, the underlying fat, greater elasticity, dearth of apocrines, proliferation of sebaceous glands—can all be paralleled in aquatic species, and none in grassland species."

In discussing the evolution and function of human palm sweating and its measurement by polygraphs, Morgan writes (p. 87), "But the machine does not detect lies—it only detects anxiety. You may feel a stab of apprehension before embarking on a perfectly true story if you know it sounds too improbable to be true. On the other hand, if you are a hard-boiled type with a hundred percent certainty of never being found out, you may lie your head off and still pass the test." While that comment is peripheral to Morgan's thesis, it provides a useful lesson to persons who continue to believe that polygraphs are "lie detectors."

Not being an anthropologist, I may conceivably be missing some profound weakness in AAT that would be apparent to an expert. But with that stipulation, I find the Aquatic Ape Theory plausible and consistent with the way evolution works, and the arguments presented in Morgan's books extremely convincing.

### AFTERWORD

Morgan's most recent books on AAT are *The Descent of the Child*, Oxford U.P., 1995, and *The Aquatic Ape Hypothesis*, Souvenir Press, 1997.

## 92. *Litany of Loons*
## by Jack Truett

"Frankly, I don't care one whit what people believe. That is their individual business and of no concern to me. But when a belief propels herds of them into damaging the rights and freedoms of others, when such belief continually and constantly results in hideous wars and slaughter, and when such belief is obviously destroying my country's future and its citizens' well being, then it damned well *is* my business." (p. 9) With those words, Jack Truett justifies writing a book designed to "help someone else, maybe even you, to avoid wasting your life-time stumbling fruitlessly on the same path." (p. 8) And he warns, "If you are a 'God' believer, I suggest you fasten your seat belt. Or, if you are determined to live in fear, ignorance, and slavery the rest of your life, you can toss this book in the trash can now." (p. 51) Sadly, the people most in need of liberation from self-inflicted mind-slavery are likely to do exactly that.

Truett explains the origin of tyranny as follows: "Perhaps one day the group encountered a beast of prey smack in the middle of their berry patch. The 'followers' would expect the King to do something...So he grabs a dry piece of brush...charges at the beast, yelling and raising hell...The predator growled threateningly even as he fled...but he *did* go...A few days later...there looms out of nowhere, a sudden violent storm...The bush shaking and yelling had scared off a vicious beast...what if...And, scared of the streaking, howling pandemonium around him, he grabbed the self same bush, shook it at the roaring storm...And the storm receded...And so was born the first 'incantation.'...The 'Adviser' had gained kudos galore, and it is very probable that he had also convinced himself of his 'magical' powers." (pp, 21-22)

Anyone who doubts the validity of that reconstruction need only look at a friend or acquaintance who thinks he/she is "psychic," to realize that as few as two or three fulfilled expectations in a lifetime will reinforce the believer's fantasy, while dozens of unfulfilled prophecies will not disillusion him. And I have met more than one megalomaniac priest incapable of doubting his power to send anybody who pissed him off to the Christian Hell.

On the origin of gods: "Awareness seeks answers, especially to avoid dangers. Explanations offered 'spirits.' Spirits demanded subservience. And thus Deity Religions were born." (p.38) To back up that assertion, Truett explains, "When one reads 'Holy' scripture from any of the world's religious writings, the 'God' that is described matches exactly the human Kings that ruled on Earth: Childish, self aggrandizing, temperamental, picayunish, demanding, conceited, and merciless...always ready with 'reward' offered for those who will literally kiss his ass." (p. 39)

After explaining that his disillusionment with religion began in World War II, when he found himself in a British dungeon still containing torture devices used by the Catholic Inquisition (*not* a peculiarly Spanish or splinter-sect phenomenon), Truett goes on to explain the evidence that falsifies religion far more effectively than any anecdotal experience. He reports the conclusions of a five-year examination of the King James Bible by a panel of various experts: "There are over 19,000 provable errors and self contradictions in God's Word, The Holy Bible." He adds his own comment, "Does that really sound like an Omnipotent, All Knowing God inspired the thing?" (p. 148)

Most of Truett's account of the older religions plagiarized by their bastard offspring, Christianity, is identical with what is to be found in *Mythology's Last Gods*. But in places his interpretation of the facts is very different. A truthseeker should read both and then decide who is right. Also, while I am not an expert on Muhammad and the origins of Islam, Ibn Warraq is, and Truett's reconstruction of the facts of history differs profoundly from Warraq's. Again, a seeker of truth should read both before reaching a tentative final conclusion. And Truett may be the first person with the intestinal fortitude to state openly that the position Muslims assume five times a day, presenting their rear ends to the sky, is an open invitation to their deity to shtup them up the brunzer.

While I have no quarrel with Truett's use of the words "hypnotized" and "mesmerized" to describe the cultural conditioning imposed on all of us on a regular basis by hidden persuaders, and *self-hypnotized* is surely the most accurate description one can come up with for incurable creationists, he reports a wartime anecdote that appears to indicate a belief that hypnotism as something more than heightened suggestibility actually exists. Apparently he has not read *They Call it Hypnosis*, Robert Baker's definitive debunking of the hypnotism delusion.

Truett's last chapter, in which he analyses the opening pages of Genesis, was clearly written before he encountered either *Mythology's Last Gods* or *The Judaeo-Christian Bible Fully Translated*, and was not rewritten to take the information in those publications into consideration. That is unfortunate, but it does not invalidate the rest of the book. But, assuming that he does not actually reject the multiple authorship of the Torah that explains the inconsistencies he found inexplicable, he should certainly give them a close reading before writing the second volume of his trilogy.

I sometimes got the impression that Truett received much of his information orally, and greater familiarity with the spoken than the written forms of such words as anything, bloodbath, brainwashing, countryside, everything, Hellfire, hindmost, humankind, leftover, lifetime, midnight, nevertheless, outcastes, outlays, reincarnation, selfsame, shamefaced, stranglehold, therein, thumbnail, uppermost, warlords, wedlock, whatever, wherein, widespread, without, wrongdoing, yourself, and a few others caused him to write the various parts or syllables as separate words, sometimes hyphenated. This is trivia, and only a nitpicker such as myself would even notice. But as a regular contributor to Truett's freethought journal, *Pagan Palaver* (P.O. Box 935, Somerville, TN 38068), I need to preempt any accusation of treating Jack's book less severely than that of an author with whom I am unacquainted, so that when I say this is a book worth reading, that evaluation can be accepted as objective, and not simple butt-kissing. And it *is* worth reading, despite technical errors that had reached triple digits by page 60—imperfect, but worth reading.

## 93. *They Call It Hypnosis*
## by Robert Baker

"In many ways the concept of hypnosis is analogous to some other mysteries that have confused and confounded scientists in the past—such as phlogiston, the ether wind, and 'N-rays.'" (p. 12)

Those are not hypnotism's only analogies. More than a century after historians proved the Judaeo-Christian bible to be a product of the human imagination, containing over 19,000 provable errors, the unlearned masses, encouraged by the prostitute media, continue to believe that the sadistic, megalomaniac serial killer known as "God" not only exists but is a nice guy. Decades after the Condon Report established that UFOs are misidentified mundane phenomena, the same media encourage the ignoranti to believe that humanoid aliens are abducting human beings. A decade after James Randi's Project Alpha proved that all claims for the reality of ESP come from gullible investigators incapable of detecting trickery, the masses continue to believe in paranormal phenomena, again urged on by media hacks to whom truth will never outweigh profit. And a decade after Robert Baker proved that "the phenomenon called 'hypnosis' does not exist, has never existed in the past, and will not exist in the future," (p. 17) the masses, whose only source of information is the vast wasteland, continue to believe that "hypnosis" is more real than demon possession or witchcraft.

I did not need Dr Baker to tell me that hypnotism does not exist. Forty years of touring with stage hypnotists who also practised hypnotherapy ultimately made such a conclusion inescapable. My book, *The Disinformation Cycle*, devotes a chapter to the subject. But the eight experiments described in that chapter cannot compare to the wealth and depth of information in Baker's definitive debunking of the hypnotism delusion. Anyone who continues to believe in hypnotism either has not read this book, or is unteachable.

Baker shows that (p. 24), "all of what we call 'hypnotic behavior' can be accounted for by a number of much simpler sorts of psychological processes that are well understood…When normal human beings close their eyes, go into a sleep-like trance state, and do strange and unusual things…the volunteers are merely complying with the hypnotist's requests, and…nothing other than suggestion and their own imagination is responsible for their behavior…As for the claimed therapeutic effectiveness of hypnosis and the many seemingly miraculous cures and events apparently due to the effects of hypnosis, in reality, these are due to a number of external social factors such as suggestion and conditioning interacting with internal psychological variables such as relaxation and imagination." And on the kind of hypnotic brainwashing popularized by the novel and movie, *The Manchurian Candidate*, Baker reports (p. 48), "Fortunately, the scenario described in the novel could never happen. Years of experimentation by the CIA has shown this sort of programming simply does not work and never will."

"Whenever it is argued that hypnosis does not exist or is nothing more than relaxation and suggestion, the other side responds, 'Well, what about those people that have surgery without anaesthetics, using only hypnosis?'" (p. 199) Baker devotes a whole chapter to answering that question, very effectively. In the case of Dr James Esdaile, one of the first practitioners of ostensibly painless operations under mesmerism, he writes (p. 200), "Esdaile reported how he mesmerized men in law court by making passes behind their backs…If you are skeptical of this, then you should be skeptical of his accounts of analgesia with mesmerism!"

On the fallacy that hypnotism can improve memory, Baker explains that a technique of relaxation and an instruction to "think back" can indeed enable an individual to remember unforgotten experiences in slightly greater detail. What it cannot do is guarantee that the added details are accurate: "At the moment we cannot tell whether a subject is telling the truth or is 'confabulating,' i.e., providing pseudomemories." He explains that "Confabulation is a tendency of ordinary, sane individuals to confuse fact with fiction and to report fantasized events as actual occurrences." (p. 194)  And in an experiment to test the validity of alleged past life memories, he encouraged one third of a group of sixty ·volunteers to believe that past life regression is valid, another third to be semi-skeptical, and the remaining third to view past life regression as nonsense.  The group given positive expectations produced the most past lives, while those to whom the concept was ridiculed produced the least.  "Regression subjects take cues as to how they are to respond from the person conducting the experiments and asking the questions...Therefore, it is hardly surprising that hypnotic age regression is a rich and inexhaustible source of confabulation." (p. 197)  But while Baker uses the word "hypnotic" in such passages, he had previously made clear that a person who is relaxed and treated as if he is hypnotized in all likelihood believes that he *is* hypnotized.

In the chapter, "The Uses and Misuses of Hypnosis," Baker debunks age regression by showing that equally convincing "memories" can be elicited by telling the subject he is traveling into the future.  He explains how the Betty and Barney Hill alien abduction hoax was created by a hypnotist prompting the subjects to concoct the kind of tale the hypnotist wanted to hear—and then persuading them that the confabulation was a genuine memory.  "This is, of course, one of the worst if not the worst misuse of so-called hypnosis." (p. 237)

Baker cites several novels and TV scripts about hypnotism, all of which credited hypnotists with powers that simply do not exist.  He makes the point that the dissemination of such disinformation, not always by persons who believe their scenarios could really happen, has much to do with hypnotism's reputation, not only as a force for potential evil, but as a state of being that exists outside of science fiction.  In 1990 as far as I am aware there was no novel about a hypnotist in print that showed the "author's viewpoint" character reaching the conclusion that hypnotism is a delusion.  Such a book is now available.  It is called *The Great Zubrick*.

I have to dispute Baker's contention (p. 15) that magician Kreskin's "powerful demonstrations of the power of suggestion and compliance in persons who are wide awake" are contributing to the delusion's demise.  In fact many in Kreskin's audience interpret his performance as evidence that hypnotism *does* exist, since what he demonstrates either *is* hypnotism, or is conscious playacting.  Kreskin is well aware that his volunteers are simulators, and his attempt to pass off fifth-rate playacting as something called "power of suggestion" is a hypocritical attempt to have his cake and eat it.  Baker is also too charitable to Dr Milton Erickson, a medical hypnotist whose claims of incredible successes went far beyond a gullible or naïve interpretation of observed phenomena, and are more reasonably viewed as imaginative fiction.  Baker questions Erickson's interpretations of his alleged successes, but not the claim that such things really happened.

Among the seventeen questions and answers that constitute Baker's summation of the book's content are the following (p. 289):

Q:  Then people under hypnosis don't become extraordinarily strong, are not able to remember accurately everything that ever happened to them, and are not able to have surgery without pain?

A:  They *do not* and *are not* able to do those things under hypnosis *any more than they can* do them when they are wide awake.

Q:  But why have there been all these claims about the mysterious powers and abilities people are supposed to acquire under hypnosis?

A:  Because some unscrupulous or naive people like to deceive and impress others and make them believe things that aren't true.  Salesmen do it all the time in order to sell us things.

So the next time someone tells you that hypnotism is a panacea for all ills, or conversely a diabolical power, ask yourself:  Would you buy a used car from that person?

## 94. *Franquin: Master Showman*
## by Jennie Rowley Lees

The preface of Jennie Lees's biography of stage entertainer Pat Quinn, "Franquin," describes how he became interested in hypnotism after his mother the psychic had a vision of a fatal accident thirty minutes before it happened. Lees describes the fable as if it were a true story and as if "psychic" were a legitimate concept. After a scientifically illiterate opening like that, I "foresaw" that Lees's book would be a gullible paean to a performer whose whole show depended on deluding the masses that his stooges were in a state of "hypnosis" that is now known to be nonexistent, and that is what it turned out to be. Whether Pat Quinn still believes, or ever believed, that hypnotism exists, the book does not make clear. He may well have recognized Lees as an uninformed ignoramus and fed her a line of bull, on the ground that one does that to any sucker. But his superstitious biographer certainly believes it.

In chapter two Lees credulously repeats Quinn's fantasy that his success as a cricketer was due to his ability to "wish" an opponent to trip over his feet and, by wishing, make it happen. Even if Quinn believes such drivel, and the endorsement of astrology in his first book makes that not improbable, Lees's willingness to report such nonsense as if it was true would have had P. T. Barnum beating a path to her door.

Lees's account of Quinn's childhood experiments with hypnotism might well be accurate. Most hypnotists start out as believers, and subjects willing to play along are not hard to find. Even at the height of his success, when his best laughs were produced by stooges (of whom this reviewer was one) reciting rehearsed scripts, Franquin may have still believed that at least some of his volunteers in some of his shows were genuinely hypnotized. He would have rationalized that, as long as the stooges were simulating what a genuinely hypnotized person would do, it was not really humbuggery. Certainly that was the rationalization that kept me working with stage hypnotists for many years, until remaining a believer became impossible. But the fantasy of Franquin hypnotizing a student who had snitched on him for smoking, into smoking in front of the housemaster, is unmitigated fiction. Even if Jennie Lees believes it happened, Quinn does not, for the obvious reason that anything that cannot happen did not happen.

In chapter 9, Lees quotes a newspaper review of one of Franquin's early performances in New Zealand (p. 104): "The only flaw in the evening's entertainment were the fairly lengthy and, at times involved explanations which preceded different sections of the performance." Since Lees made no comment on the passage, it probably did not cross her mind that detailed explanations of what was going to happen were necessary in order for Franquin to explain to the volunteers what they were to do in order to appear hypnotized.

*Franquin* is, despite the third-person narrator, very much an autobiography. Lees's gullible parroting of every fanciful absurdity Quinn spun for her, leaves little doubt that the rest of her book is also "his" version rather than an objectively researched account. Nonetheless, where paranormal superstitions are not involved, there is no reason to doubt its basic accuracy. Franquin's show business career was not unlike George M. Cohan's, or Harry Houdini's, or a thousand other performers who dragged themselves through rough times to success. And certainly he is important enough in Australian show business history to warrant a biography. It is unfortunate that his biographer's inability to separate sense from nonsense is comparable with Conan Doyle's.

Lees's bibliography lists books by such unquestioning believers as Ormond McGill and S. J. Van Pelt, as well as the nonsense novel, *Trilby*, and the long discredited *Search for Bridey Murphy*. Conspicuously absent are the names Druckman and Bjork, Spanos and Chaves, and Robert Baker's definitive 1990 book, *They Call It Hypnosis*, which proves to the satisfaction of all but incurable dogmatists that "hypnotism does not exist, has never existed in the past, and will not exist in the future."

Franquin was a first class entertainer, winning a reputation in Australia equal to that of Reveen in Canada. As a virtual emcee and ringmaster, he extracted performances from amateur actors that at least equaled the antics of the no-talent drones of *Whose Line Is It Anyway?* or professional wrestlers. It is therefore reasonable that a biographer would play down the reality that Quinn's performances were as much illusions as levitation or sawing a woman in half. No one would criticize David Copperfield for using trickery to make his audience think they were seeing the impossible. But a hagiographer who babbled that Copperfield really did make an assistant float in the air with nothing holding her up, rather than merely appearing to do so, would assuredly raise the question, "What color is the sky on your planet?" Lees's unbelievable ignorance raises the same question.

Nonetheless, to a reader who *can* separate the sense from the nonsense, this is a thoroughly detailed account of the life and career of an unquestioned giant in Australian show business.

## 95. *Hypnotism: A History*
## by Derek Forrest

There are two kinds of authors capable of writing a gullible, scientifically illiterate history of hypnotism theory and evolution, based on the delusion that a modified state of consciousness known as hypnosis actually exists. One is a stage hypnotist who needs to shut out the reality that his volunteer actors are simulators, since the alternative would be to recognize that he is a hypocrite and a humbug. The other is a psychoquack who practices cold reading in the belief that it is a branch of medicine, and shuts out the reality that his practice differs in no way from that of a tealeaf reader or a bartender. Derek Forrest is the latter kind.

Forrest not only endorses the legitimacy of Sigmund Freud's obscene and fully discredited "psychoanalysis" hoax. He even believes in the reality of multiple personality. Describing a particularly notorious case of multiple personality playacting, he writes (p. 259), "It may have been the case that these personalities were produced unwittingly by suggestions made by Prince [the hypnotist] during hypnosis...The other possibility, that Miss Beauchamp [the patient] deliberately deceived Prince for private reasons of her own, seems to be unlikely, as she was quite ignorant of the psychological literature and suffered much from the depredations caused by Sally [one of her personalities]." That rationalization, that a playactor could not have known how to deceive an alleged therapist, is typical of the lengths to which unteachables can go to reinforce their belief in the thing that is not. *Of course* the patient "deliberately deceived Prince." That is what compulsive playactors *do*.

Forrest's paragraph on Dr Milton Erickson is equally gullible. How anyone who has read Erickson's medical journal articles, in which he claimed results that simply could not happen, could fail to recognize that the man was an unmitigated liar, is incomprehensible.

In explaining why spectacular results in hypnotic experiments achieved in uncontrolled conditions are not replicable in a laboratory environment, Forrest states (p. 271), "The attitude of disciplined skepticism in the laboratory setting is not one calculated to allow the emergence of the dramatic and bizarre phenomena produced in response to the vivid urgings of the committed charismatic, such as a Du Potet. The really effective hypnotist has to exude confidence, enthusiasm, and certainty; in so doing, he will find it difficult to maintain the objective stance necessary for the scientist." In fairness, that is a more acceptable rationalization of why playactors do not perform well in the presence of persons willing to recognize acting when they see it, than the incredible claim of parapsychologists that ESP disappears in the presence of a skeptic because "psi is shy."

On the observable success of some applications of hypnotherapy, Forrest state (p. 272), "Much more seems to depend on non-hypnotic factors, such as the person's motivation, than to their susceptibility to hypnotic suggestion. In this context it is interesting to note that in the case of smoking, one or two sessions of hypnosis have been found to be sufficient to bring about cessation of the habit in 98 % of those who were successfully treated." What Forrest does not mention is the correlation between the effectiveness of the treatment and the size of the hypnotist's fee. The more money a smoker pays a hypnotherapist to cure his habit, the greater the smoker's incentive to see that the treatment works, not merely because he really wants to quit, but also because he does not wish to be perceived as a sucker.

On the effectiveness of hypnotic suggestion in suppressing pain, Forrest states (p. 273), "Research has shown that this relief is not simply the result of a placebo effect." WRONG. Hypnotherapy is by definition a placebo effect.

Forrest expresses the belief (p. 275) that, "Age regression may be upsetting if a 'return' is made to a traumatic past period; a bizarre suggestion to be carried out posthypnotically may cause embarrassment and even physical danger; sometimes subjects are not fully awakened after hypnosis;

and occasionally they return spontaneously to a trance-like state." Such beliefs might have been justifiable twenty years ago, before the publication of Robert A. Baker's *They Call It Hypnosis*, which established once and for all that "hypnotism does not exist, has never existed in the past, and will not exist in the future." That anyone could write such drivel today provides further proof that there are none so blind as those who will not see. On the good side, Forrest does recognize that "recovered memories" of childhood sexual abuse are fantasies put into the minds of their patients by therapists of whom the description "criminally incompetent" is too polite.

Forrest's history of the evolution of hypnotism theory from Mesmer to Charcot is more detailed than Peter Reveen's *The Superconscious World*, but also more credulous and less readable, for the good reason that Forrest did not have a skeptical editor (guess who?) looking over his shoulder and eliminating interpretations that are best described as naïve. Both books present a pro-hypnotism view of suggestion experimentation, but Reveen's 56 pages of history contain as much factual information as Forrest's 228 pages on the same subject, and a lot less superstitious hogwash. Both view Anton Mesmer favorably, even though he was, even by the standards of his time, as unwilling to face observable reality as Forrest still is.

This is not a book I can recommend. For a concise survey of hypnotism's origin and evolution, the place to go is the updated version of Reveen's 1987 book, now renamed *Hypnotism Then and Now*. It takes the position that hypnotism does exist (in a book designed to be sold at his show, Reveen could hardly do otherwise). But that does not invalidate most of the content. For the evidence that hypnotism does not exist, see *They Call It Hypnosis*.

## 96. *The Secret of Happiness*
## by Billy Graham

Nobody with a functioning human brain could expect a book by Billy Graham to be anything but superstitious drivel, and that is exactly what it is. And since Graham utilizes educated editors (ghostwriters?), it would be unrealistic to expect his own books to expose him as an ignorant hillbilly with all the sophistication of Gomer Pyle and Jed Clampett, even though that comparison is if anything unduly flattering.

Graham's second chapter is titled "Happiness through Poverty." Enough said.

In chapter three, "Happiness while Mourning," Graham states, "Nowhere has God promised anyone, even His children, immunity from sorrow, suffering, and pain. This world is a 'vale of tears,' and disappointment and heartache are as inevitable as clouds and shadows." To Graham the infliction of pain and suffering by an omnipotent god with the capacity to annul such evils is "his mysterious ways." And the Holocaust was Hitler's mysterious ways.

In chapter four, "Happiness through Meekness," Graham acknowledges that, "To most people today the word 'meek' brings to mind a picture of someone who is a weak personality, someone who allows everyone to walk over him. Meekness, in fact, in the popular mind is not seen as a desirable personality trait." But to Graham it is a desirable personality trait. Why? Because his imaginary playmate said so, therefore it must be true. Since the same bible that touts meekness as a virtue also states in seven places that the earth is flat, presumably that also must be true.

In "Happy though Hungry," Graham rationalizes, ""Well, to begin with, hunger is a sign of life. Dead men need no food, [sic] they crave no water." In other words, Graham's god inflicts hunger instead of death because he is a really nice guy. And if you believe that, I have a bridge for sale in Brooklyn that I think will interest you.

Graham touts "Happiness through Showing Mercy." I have no problem with that one. But in "Happiness in Purity," he equates *purity* with mindless, masochistic conformity to taboos on joyful, victimless behavior that god addicts view as *sinful* simply because "it's in the book."

"Happiness through Peacemaking" is likewise evidence that even a religious nutcase can be right sometimes. But "Happiness in Spite of Persecution" is another apology for a god that allows mass persecutions even though it has the power to prevent them.

Graham is a monster. He preaches a god that tortures taboo-breakers with flamethrowers for billions and billions of years in an underworld that can only be described as a sadist's dream, as well as executing a quarter-million men, women and children each and every day in reprisal for the alleged offence of their distant ancestors, but is nonetheless more admirable than Adolf Hitler. And he is on record as supporting a death penalty, an atrocity abolished by every sane government on earth, for persons who indulge in the victimless, nonprocreative, purely recreational activity that he mistakenly calls adultery. In fact, when the Torah author's "ten commandments" were composed, adultery was defined as the fraudulent impregnation of another man's wife, and was still so defined in the 1450s when Thomas Mallory had Lancelot copulating with Guinevere but not committing adultery with her, since he practised "courtly love," otherwise called *coitus interruptus*. One has to wonder if Graham's wishing for the permanent cancellation of such competitors as Jimmy Swaggert is purely self-serving? Or perhaps the target of his hatred is closer to home?

*The Secret of Happiness* was mind pablum for the braindead when it was first published in 1955, and the reprint is mind pablum for the braindead in 2002.

## 97. *Catholic Power vs. American Freedom*
## by George La Piana & John Swomley

"What's theirs is theirs. What's yours is negotiable." Those were Isaac Asimov's words (*The Roving Mind*, p. 13) concerning the theofascist right's demand that creationism be taught in schools, while refusing to teach evolution in their churches. And Robert Boston's unauthorized biography described the then-leader of the theofascist conspiracy to overthrow the Constitution and turn America into a Christian theocracy as *The Most Dangerous Man in America*. Both Asimov and Boston saw the threat to the separation of church and state as emanating predominantly from radical Protestantism. George La Piana shows that a far greater threat comes from the church with the longest history of totalitarianism, intolerance and fascist dictatorship in human history, the Roman Catholic.

Of all the evils being inflicted on the human race by the Catholic Church, the most pernicious and the most dangerous is its prohibition of sane population control, even including contraception. When Pope John XXIII ordered a feasibility study of continued prohibition, "the appointed lay commission voted 60 to 4 and the clergy 9 to 6 to change the papal teaching on birth control." A minority report drafted by Karol Wojtyla pointed out that, "If it should be declared that contraception is not evil in itself, then we should have to concede frankly that the Holy Spirit had been on the side of the Protestant churches." (p. 189) Paul VI, who had initially favored repealing prohibition, recognized that such a reversal would indeed diminish papal power, and wrote *Humanae Vitae*, endorsing the validity of the taboo. (p. 190) In other words, the Catholic Church continues to cause millions of child deaths a year, because the alternative would be to admit that right and wrong are something other than whatever the current pope says they are. And the decision to do so can be laid squarely at the feet of John Paul II, the most reactionary pope since Pius IX.

And the first act by the American religious right's puppet President after being treasonously appointed to the office he had failed to win by election, was to become the pope's accomplice in the mass murder of millions of third world children, by withholding financial aid from countries that did not enforce the Vatican Führer's ban on population control. At least Canada's religious right, despite holding Official Opposition status, will never be truly dangerous as long as eighty percent of the Canadian population continue to recognize it for the hate cult it clearly is.

And lest anyone think that the Republican Party's total surrender to the Theofascist Right is not a capitulation to the Catholic Church, co-author Swomley spells out the considerable evidence (pp. 169-171) that, "It seems clear that Jerry Falwell's Moral Majority and Pat Robertson's Christian Coalition would not have been organized if the Vatican through its agents, the American bishops, had not decided it was necessary to involve articulate right-wing Protestant leaders in their antiabortion campaign...The list of right-wing Catholics coopting Protestant fundamentalists is far from complete. However, it does reveal how clearly the Catholic right has led the Protestant fundamentalists...The Southern Baptist Convention came under right-wing control and accepted the Catholic position on abortion."

One could not legitimately oppose the enforcement of even a sectarian law code, if the code itself was objectively defensible. Swomley in particular recognizes this, and supports his opposition to turning the American government into a Vatican slave by showing that the laws the Catholic Church wants to impose on Catholics and non-Catholics alike are not only capricious, "whatever the pope wants," but also inconsistent. The Church's biggest Big Lie is that its teachings have never altered. But they have done a complete about-face on many issues, not the least of which is "right to life."

"If one examines Vatican dogma, it is only fetuses that have a 'right to life.' The pregnant woman whose life or health is endangered by the fetus has no right to life...Over the centuries the

Vatican has been involved in the direct or indirect slaughter of millions." (p.193) "Moreover, the Vatican's idea that human life begins at conception is an attempt to override the biblical position that human life begins with breathing. The Hebrew word that describes a human being is *nephesh*, the breathing one. It occurs 775 times in the Hebrew Bible. In Hebrew thought [and keep in mind that this is part of the *Catholic* Bible] an embryo or fetus is not a living human being because it does not breathe on its own." (p. 192) The about-face did not occur until 1869, when Pius IX declared abortion illegal from the moment of conception, even though previous popes had continued to maintain that abortion prior to quickening was acceptable after the American government banned it in 1830. And the Inquisition burned pregnant women to death on a regular basis. More than once, a woman gave birth on the stake, and the presiding priest tossed the baby into the flames.

Alleged "right to life" is simply part of the Vatican's power lust. "The abortion issue is also a cover for opposition to birth control…All family planning programs worldwide, other than [Vatican roulette] are opposed. The bishops were successful in dominating the Reagan and Bush administrations on this issue…Reagan's collaboration with the Vatican seriously impeded family planning activities in many countries, including curtailing the availability of contraceptives, thereby contributing to increasing the total world population…Nevertheless, it is important to note that the hierarchy and the Republican candidates misjudged the Catholic vote, which is far more progressive than the Vatican's." In other words, the threat to American freedom is not the several million American Catholics, but a foreign dictator whose control over those Catholics is less than he would wish. "This does not, however, minimize the Vatican threat to American democracy or its clear intent to establish its policies in the United States." (pp. 176-178)

La Piana compares the tactics of the Catholic Church with those of Soviet Russia: "This striking parallelism of claims and policies of a totalitarian state and a totalitarian church is not a casual coincidence." (p. 21) Like other fundamentalists (September 11 comes to mind), "The pope also insists that his authority to interpret what is moral must be placed ahead of the democratic judgment of people whose interpretation of the will of God differs from his." (p. 173) And like the Soviets, "One of the ways in which the Vatican exercises its totalitarianism in the United States is through censorship of the press so as to control public opinion. The Knights of Columbus, for example, has for many years tried to shut down any criticism of the Catholic church." (p. 179) And Catholic lobbyists have placed before many state legislatures a "Health Care Providers' Right of Conscience Act," "clearly designed to prevent normal medical service to women. For example, emergency contraception to a woman who has been raped would be denied, along with diagnosis and treatment if the rapist had AIDS." (pp. 243-244) "All over the country, bishops invite legislators and judges to hear their plea to put Catholic moral law ahead of civil law." (p. 264) "A 1992 newsletter of the Catholic Campaign declared that 'separation of church and state is a false premise that must finally be cast aside.'" (p. 166) This from the same tyranny that applauded church-state separation when the established religion was Protestantism, and separation increased the civil rights of Catholics. But that was then. This is now. Today, "American Catholics under the orders of their hierarchy have the religious and moral duty to undermine, by all means, the American system of religious freedom and equality of all religions before the law, and to bring about a union of the Catholic Church and the state in this country." (p. 71)

Sound familiar? Heil Wojtyla!

## 98. *Bible Prophecy: Failure or Fulfillment?*
## by Tim Callahan

*Bible Prophecy* is probably the best work of biblical analysis ever written by an author with no more qualifications for such a task than a baccalaureate in fine arts. Unfortunately, to a reviewer prematurely excited by the book's promising title, it is also a disappointment. I would have been satisfied with either a scholarly book that would be the definitive work on the subject and find a place in every university library, or a book written for the general public in language that a curable believer could understand without being swamped with obscure historical references that he could not begin to evaluate. *Bible Prophecy* tries to be both, and consequently succeeds in being neither.

Callahan's book is not, however, useless to readers of *The American Rationalist* [**in which this review was first published**], most of whom have less familiarity with the historical background in which biblical prophecies are set than, for example, Gerald Larue or Richard Friedman, but rather more than Jose McDoakes. And Callahan demonstrates clearly that prophecies about the political powers of the Middle East either failed, were written after the fact, or were logical consequences of events already in progress. While refusing to take the position that information cannot travel backward in time, as it would have to do for biblical prophecies to be valid, he nonetheless concludes that there is not a prophecy in the entire Bible that does not fit comfortably into one of those three categories. And he writes (p. 249), "That prophecies that were not fulfilled are assumed to be awaiting fulfillment—someday—highlights the impossibility of falsification built into the fundamentalist scenario."

I was initially uncomfortable that Callahan devoted whole chapters to debunking alleged prophecy fulfillments that only the most dogmatic fundamentalists tout as evidence of the Bible's veracity, thereby granting them an undeserved dignity. But then I remembered that for a long time competent scientists refused to treat the masturbation fantasies of Immanuel Velikovsky as worthy of rebuttal, and consequently the ignoranti drew the inference that Velikovsky could not be rebutted. Carl Sagan remedied that error, and what was commendable for Sagan must also be commendable for Callahan.

Nonetheless, for all practical purposes Callahan's book is aimed at the braindead incurables who wrote the books he rebuts, when he could have been more effective if he had aimed it at the curable believers who accept the fundamentalists' circular and spurious arguments at face value for lack of any falsifying evidence that they are able to comprehend. Thus instead of giving the highest priority to showing that prophecies of the destruction of Babylon were inaccurate in minor details, he should have concentrated on the major inaccuracies.

For example, (1) the division of Babylon between the Medes and the Persians, prophesied by the "writing on the wall," never happened. Babylon's Persian conqueror had defeated and subjugated the Medes more than a decade earlier. (2) The four empires prophesied to succeed Nebuchadnezzar included the nonexistent Median conquerors. (3) Babylon's conqueror was not a revived Assyria ("out of the north") as Jeremiah prophesied, but a Persian from the east. (4) Babylon's biblical conqueror, Darius the Mede, never existed. Callahan eventually mentions that, but only after his readers have been forced to struggle through so much less significant material that their attention span is likely to have been drained.

Callahan does point out at the appropriate point that Jeremiah's prophecy of a seventy-year captivity was falsified when the Jews were allowed to return home after forty-eight years. But he concludes that Jeremiah used "seventy years" as a metaphor for "a long time," downplaying the probability that Jeremyah (the book's author) really believed he had prophetic powers and fell into the common error of predicting the future instead of sticking to the much safer practice of "prophesying" the past.

Callahan's chapters on prophecies in the Old Testament that fundamentalists tout as being fulfilled, while trivial to the point of being boring, do contain enough meat to keep me reading. He proves himself well able to demolish the fundamentalists' rationalizations, including the fantasies of the imbecilic *The Late Great Planet Earth*, with the weapon of common sense, for the logical reason that his opponents are unarmed. It is when he ventures into the New Testament that the thin ice gives way and the realization that he is a well-meaning amateur hopelessly out of his depth becomes inescapable.

The inadequacy of Callahan's research is most apparent in his interpretation of the "little apocalypse" of Mark 13:5-27. Callahan treats it as a genuine preaching of Jesus, even though it was clearly a retroactive justification for the Nazirite sect's flight to the Decapolis that took place at the beginning of the war of 66-73 CE. Callahan not only imagines that Jesus' alleged instruction to his followers to flee into the hills referred to a hypothesized Last Days. He declares, "I do not see how that can be interpreted in any other way than Jesus' second coming at the end of the world." (p. 186) Yet several dozen books that Callahan could have and should have consulted would have told him that Jesus at no time ever predicted his own death and "second coming." Jesus believed that, as Messiah, he could not die until he had overthrown the Roman occupation and been crowned king of an independent Judea. His declaration that, "There are some standing here who will not experience death until they have seen Allah's theocracy established by force," was expected to be accomplished within weeks of the unilateral declaration of independence that in fact got him executed.

Among Callahan's inaccuracies are the following: He identifies the anonymous authors of Mark, Matthew and John as Jews, even though the account of Jesus' trial before the Sanhedrin is so incompatible with Jewish law that no Jew could have made such a blunder, and even though the educated language of John would have been beyond the capacity of a writer to whom Greek was a second language. He recognizes the problem of Matthew's dating Jesus' birth to the reign of Herod, and Luke's dating his birth ten years after Herod's death, but fails to grasp that, since the tales in each are totally dependent on the date being correct, no compromise is possible. Accepting fantasies of events that never happened as a clue to a historical birth date is surely naïve.

He accepts the early Christian belief that Micah prophesied a messiah born in Bethlehem, unaware that Micah was retroactively prophesying the birth of King David. He cites a passage in Mark as "the first sign we have of (Jesus') divinity," (p. 118) unaware that all NT authors except the fourth gospeller saw Jesus as nothing more than the successor and equal of David. And he completely misses the point (as did the gospel authors themselves) of an anecdote in which Jesus unequivocally conceded that he was not David's descendant. He does get right that, of 61 Old Testament passages cited by fundamentalists as prophecies of Jesus (p. 111), "There are none that are obviously about Jesus to the exclusion of all other explanations."

Callahan's analysis of Revelation is even more uninformed. He cites scholarly opinion that parts were written at about the time of the destruction of the temple, and other parts during the reign of Domitian, but rejects the earlier dating and expresses the view (p. 191) that, "There is no indication, however, of multiple authorship." How then does he explain the passages that show an awareness of the death of Nero in 68 CE and his rumored survival, coupled with a prophecy that the destruction of the temple that occurred in 70 CE could never happen? Only an author writing between those two dates could have made such a blunder. And only incompatible authorship can explain the inclusion of both the Essene dogma that, at the final judgment, only 144,000 celibate Essene monks will be saved, and the Nazirite doctrine that an uncountable number of Jews from the Dispersion (but no Christians) will make it into Jesus' theocracy.

It was the Essene author of July/August 70 CE who promised the Jews fighting for independence that the final battle of the war would take place at Armageddon, north of Jerusalem, and the Jews would win. In fact it took place at Masada, south of Jerusalem, and the Jews lost. By accepting the pretence that the "Armageddon" prophecy was written a generation after the fall of Masada, and therefore has not yet failed, Callahan gives aid and comfort to the fundamentalists who insist that it indeed refers to the distant future. With a little more research he could have shot that delusion down in flames. And in a book that many times strays from its central theme of prophecy, his comments on Revelation make no mention of the Nazirite redactor's denunciation of the Christians as "those who call themselves Jews and are not, but are a synagogue of the Satan."

Callahan mentions both the rumor that Nero survived his suicide, and Revelation's reference to the "beast" whose fatal wound had healed, but shows no awareness that the two are identical. And in equating the beast with the Christian "Antichrist" concept, he reveals an unawareness that neither of Revelation's authors was a Christian. (Nazirites, while accepting Jesus as Messiah, were Jews in every sense, not Christians). Given the Essene author's invective against all who do not wear Jewish *tefillim* on their foreheads, mistaking him for a Christian is quite an oversight. Since Callahan's entire analysis of Revelation is based on the false assumption that it is the work of a single, Christian author, writing later than the war that was the real subject of its prophecies, it is not surprising that his conclusions are as imaginative as those of the fundamentalists.

Even where Callahan is right, he is guilty of wasting opportunities. For example, he quotes a Revelation passage prophesying that the stars will fall out of the sky, but adds no comment. While he may have felt that the absurdity of objects millions of kilometers in diameter falling to earth is self-evident, if he was not going to draw attention to the absurdity, why did he write anything at all? He similarly quoted the prophecy that "The heavens are rolled up like a scroll," apparently failing to recognize that the author of the passages believed the sky to be a solid crystal dome to which the sun, moon and stars are attached.

Callahan's expressed belief that the first 39 chapters of Isaiah were written by the historical person of that name was widely held at the beginning of this century. But at a time when the publishers of both Catholic and Protestant bibles acknowledge that the additions to the original author's work begin with chapter 13, such a mistake is inexcusable.

While I have yet to read a book that is free of typos (including my own), the number of them in *Bible Prophecy,* some in quoted passages that should have been proofread even more thoroughly than the rest of the book, is simply unacceptable. And the typos are not limited to errors that only the author could have been expected to detect. Has Millennium Press never heard of line editors? Also, given Callahan's awareness that the Hebrew god had a proper name, Yahweh, and the inclusion in his bibliography of *The Jerusalem Bible,* which calls Yahweh Yahweh, his consistent quoting from translations that falsify Yahweh into "the LORD," along with references to "God" rather than "Yahweh" in his narrative, is irritating, to say the least.

I have probably overemphasized the weaknesses in a book that is far from worthless, in the hope of discouraging others from making the same mistakes. But his comment on fundamentalists and their messiah (p. 209) is a gem: "The Jesus they worship is part of a bait and switch scheme. Converts are lured in by the sanitized and rather Aryan Jesus with the soft-light halo frequently portrayed in Protestant churches. What they get is a harsh warrior-king...Jesus in Jack-boots." His description of "Mormons, Moonies, Scientologists, New-Agers and Hare Krishnas," as "junk food for the mind," (p. 215) makes me wish I had said it first. And he will get no argument from me when he says (p. 249), "Fundamentalists, when it comes right down to it, do not want their kids to think."

Given the overall validity of his book, Callahan's biggest mistake was rushing into print before he had completed even half of the research necessary for such a work. I made the same error of judgment, but was lucky enough to be unable to find a publisher for so long that very little of the first draft survived into the published version. Callahan was less fortunate. The definitive book on biblical prophecy has yet to be written. Perhaps Callahan will try again in ten or twelve years?

## 99. *Secret Origins of the Bible*
## by Tim Callahan

Tim Callahan's 1997 book, *Bible Prophecy: Failure or Fulfillment*, was riddled with technical errors, and not lacking factual errors. I suggested that he "try again in ten or twelve years." In fact he waited only five years, and I am happy to report that his new book is not merely better. It is good.

I was initially beset with trepidation on finding that Callahan's bibliography not only omits such names as Arnheim, Hoffman, Larson, and Loisy. It also does not list *Mythology's Last Gods* or *The Judaeo-Christian Bible Fully Translated*, two works that anyone writing about biblical origins should surely have consulted. I found myself wondering if perhaps he had learned nothing in the intervening years. But less than halfway through his introduction I became confident that this would be a valuable contribution to biblical scholarship, and I was not disappointed.

Callahan states in his introduction that his book will examine biblical narratives and seek answers to the following questions:

1)  Is the narrative literally true based on history, archaeology and science?
2)  Are there internal inconsistencies, anachronisms, or other internal clues that invalidate the narrative if it is to be considered historical or to be taken literally?
3)  Is the reasoning behind the narrative and the ethical beliefs derived from it based on a worldview that is foreign to our own sense of ethics?
4)  Is there a mythic meaning to the narrative that is quite different from what a literal interpretation of the narrative might imply?
5)  What social or political stance do believers derive from the biblical narrative, and how valid is their use of the Bible to back up their personal and political positions?

Those are essentially the same questions considered in *Mythology's Last Gods*, and I am not about to say that Callahan answers them more effectively than I did. But Callahan pays more attention to the sensibilities of the terminally ignorant, and it may be that persons who would not read *MLG* will be willing to read *Secret Origins*—and perhaps be cured of at least some of their contrary-to-fact beliefs. After all, I used to be a (moderate) god addict, and I was cured by an objective presentation of the evidence.

Callahan is particularly effective in detailing the evolution of what is still called the Documentary Hypothesis, even though the multiple authorship of the books once attributed to Moses has long ceased to be mere hypothesis and is acknowledged by all but incurable inerrantists to be proven fact. A better description would be Documentary Theory, since in science "theory" means a detailed reconstruction of cause and effect, not the speculation that the unlearned pretend it means, whereas "hypothesis" means a potential theory not yet adequately proven.

Callahan also justifies the Documentary Theory by the use of analogy. He points out that the line in Genesis, "These are the kings who reigned in the land of Edom before any king reigned over the Israelites," could only have been written after there was a king of Edom, as there was not until long after the death of Moses. He then makes the parallel, (p. 12), "Consider what is implied in the following statement: 'The Iroquois had a bicameral legislative council long before there was either a Senate or House of Representatives in the United States.' Such a statement could not have been written until there actually was a nation called the United States and until that nation had developed a bicameral legislative body called the Congress."

Precisely because *Secret Origins* is so detailed and thoroughly documented, it will have little appeal for nontheists who want evidence that religion is falsifiable, but want that evidence limited to "just the facts, ma'am, just the facts." But what is a turnoff for non-scholars is simultaneously an advantage for the liberal theologians Callahan hopes to influence.  His circumspection in criticizing religious beliefs makes the abundance of facts and reasoned arguments in his book a logical asset to any religious studies library.  This book may succeed in diminishing religious superstition, or at least the fundamentalist variety, where books that present the same evidence more bluntly do not.  It is imperfect, certainly on the minor issues (there are no major disagreements) where Callahan reaches conclusions that differ from my own.  But those imperfections are trivial.  *Secret Origins of the Bible* belongs in every university library.

## 100. Canada's Ayatollahs
### *Requiem For a Lightweight: Stockwell Day and Image Politics*, by Trevor W. Harrison
### *Think Big: My Adventures in Life and Democracy* by Preston Manning

"Who is Stockwell Day?" asks Trevor Harrison (p. 1) "Is he a living testament to the Peter Principle that every person rises to his or her level of incompetence?" That opening so succinctly sums up a would-be Ayatollah's rise and fall that it almost made it unnecessary to read the rest of the book—almost. If I had not continued reading, I would not have learned that Mark Steyn wrote in the *National Post*, "There are those who say Stockwell Day's just a crudely homophobic, Bible-thumping fundamentalist neanderthal. That's certainly why I supported him." (p. 65)

Stockwell Day is a religious fundamentalist. While there is no consensus on the point at which biblical literalism in defiance of the findings of a dozen sciences can legitimately be termed insanity, there is widespread agreement that someone who has brainwashed himself that the universe is six thousand years old and that dinosaurs coexisted with humans until they were drowned in a world-covering flood is not sparking on all neurons. As one commentator quipped, (p. 84), "Day believes the Flintstones is a documentary."

In justifying his employment as an uncertificated teacher, lacking even an undergraduate degree, in a fundamentalist school with an illegal curriculum "insensitive to blacks, Jews and natives," (pp. 4-5) Day stated, "God's law is clear. Standards of education are not set by government, but by God."

In Day's first campaign for the Alberta legislature, he declared that his life was "based on the supremacy of God and strong biblical principles." (p. 8) In the 1986 election, not being quite dumb enough to pretend that he was a teacher, he called himself an educational consultant. By stacking the nominating convention with supporters from his "Bentley Christian" cult and declaring that "the whole thing was birthed in prayer," he succeeded in stealing the nomination. "Declaring his campaign a 'moral crusade,' Day at first railed against homosexuality, pornography, the legal system's treatment of criminals, and the federal government." (pp. 8-9)

In 1990 he raised the spurious issue of condoms' unreliability in order to rail against the installation of dispensers in high school washrooms (p. 15), even though not even Day could be so stupid as to believe that adolescents will not copulate, with or without lifesaving, pregnancy-preventing condoms, in violation of fundamentalist taboos. He advocated banning Nobel laureate John Steinbeck's *Of Mice and Men* from school libraries. "Day supported a proposal to drop abortions from services insured by Medicare...Day also advocated work camps for young offenders and supported capital punishment, even for teenagers...He also argued for the banning of sex education in schools...Day also spoke out encouraging the government to invoke the 'notwithstanding clause' of Canada's Charter of Rights and Freedoms to overturn a Supreme Court of Canada ruling that homosexuals must be protected under Alberta's human-rights law." (p. 15-16)

In other words, anywhere except in the redneck anus of the universe, where his hatred of the human race made him a hero, Day would have been an embarrassment even to other bigots. What Harrison apparently failed to discover is that, when Day was Social Services Minister and presumably the architect of such an atrocity, the Alberta government put a contract on the old, the sick and the unemployed by reducing welfare payments to a level where recipients could either pay rent or buy food, but not both.

Harrison quotes one of Day's attempts to deny that he intended to make his fundamentalist religion the law of the land: (p. 56) "I am pro-life. But I would not seek to impose my views on the Canadian people. I would want issues such as these to be determined freely and democratically by the people, either through a referendum initiated by Canadians or a free vote of their representatives in the House of Commons." In other words, he would not turn Canada into a theocratic slave state unless he won majority control in Parliament by becoming Prime Minister. And he would not deny equal rights to minorities unless an intolerant majority permitted him to do so. Did Harrison not notice that Day's pretence that he would not make himself Ayatollah of Canada, while simultaneously using doubletalk to assure his fellow theocrats that he intended to do exactly that, is unmitigated LYING? Or did Harrison consider such an implication so self-evident that he did not need to add a personal comment, particularly since the chapter on the libel settlement that cost Alberta taxpayers almost $800,000 clearly identifies Day as a liar of the first magnitude?

Day became leader of the misnamed Canadian Alliance (allied with whom?), not because the Party's manipulators failed to recognize him as a no-talent hack, but because, with the whole country knowing and rejecting everything Preston Manning stood for, Day was unknown outside of Alberta, and there was at least a chance that he could win a federal election and turn Canada into a theocracy before the voters realized that he was a greater threat to the freedom of every Canadian than Manning, who at least was not a liar, had ever been.

Day lost his job when, like Clark, Turner and Campbell in the national parties, his election as Party Leader drew attention to his intellectual ineptitude that, as a subordinate, he had been able to conceal. He was replaced by another redneck (big surprise) who is on record as wanting to build a "fire wall" around Alberta to keep out such un-Albertan ideas as tolerance, human rights, equality of all Canadians, and plain humanity. If the new Head Bigot succeeds in taking his hate cult where Manning and Day could not, Day's defeat might actually turn out to be a bad thing.

Stockwell Day appears to have a dangerous future behind him. But the same thing was said of Ruholla Khomeini after his expulsion from Iran, and the unspeakable Nixon after his failure to win the California governorship. The price of liberty is eternal vigilance. Stockwell Day will be a jihad waiting to happen for a long time to come.

That Preston Manning agrees in principle with the foregoing evaluation should surprise no one. Manning created his hate cult for the avowed purpose of depriving women of sovereignty over their own bodies, denying homosexuals and others who did not grant blind, unquestioning obedience to the taboos of his imaginary playmate of basic human rights, and turning Canada into the slave state Alberta had been under theofascist Führers Aberhart and Manning senior. Not only did Day deprive Manning junior of the personal glory of achieving that ignoble purpose; he also so destroyed the new Taliban party that its chances of ever enslaving believers and nonbelievers alike is now vanishingly small.

But while Manning from the beginning of his political career made no secret of his plan to turn Canada into a theocracy in which anything deemed sinful by his fundamentalist religion would be criminalized, he did not direct his campaign toward fringe and reactionary religious organizations, as his successor, knowing that that was where his hate cult drew its support, made a point of doing. In *Think Big*, Manning writes (p. 315), that Stockwell Day "focused on two main groups: Progressive Conservatives who could be persuaded to join the Canadian Alliance and help pick its new leader; and a particular segment of the Christian community, namely, evangelical Protestants, conservative Catholics, and the pro-life organizations...Many of my religious convictions were similar to Stockwell's but I had always resisted campaigning directly for the 'Christian' vote." It was not that Manning did not want the 'Christian vote.' Rather, he recognized that making himself *the religious candidate* would "do more harm than good." (p. 316) And he was right. It was precisely because Canadians of moderate religious beliefs recognized Manning as a potential theocrat, and Day as nothing less than a would-be Ayatollah, that Canada remains free of a New Inquisition.

The first fourteen chapters of Manning's political autobiography say nothing on which any comment is necessary. He gives a detailed and presumably accurate description of the events in and relevant to his life that should be a useful reference to anyone interested in the rise and fall of extremism in Canada. His observations concerning current Alliance leader Stephen Harper are neither flattering nor derogatory. Presumably he still hopes that Harper can institute the enslavement of Canada to Manning's imaginary playmate in the sky that he and Day failed to achieve. He clearly has no ability to recognize that objective as neither achievable nor desirable. As Manning acknowledges (p. 148), one out of three Canadians told an Ipsos-Reid poll that their religious faith is not "very important" in their day-to-day lives—and if a third of the population were willing to go on record with such a politically incorrect position, the true figure must be considerably higher.

In commenting on the 2000 election in which the Liberal Party increased its majority, Manning's evaluation is (p. 363), "The Canadian people had been prepared to reject Chrétien and vote for a principled alternative; they just did not perceive that alternative in Stockwell Day and the Canadian Alliance."

Since Manning is not a liar, his admission that he saw a theocratic pseudo-political religion as a "principled alternative" shows just how dangerous his elevation to power would have been. When granting pre-human tadpoles with zero brainwave activity consistent with human thought a "right to life," while withholding the same status from lifeforms with at least minimal self-awareness, such as chickens, can be labeled a "principled alternative" to full human evolution, otherwise known as liberalism, one has to ask whether persons of such a mindset are really more evolved than the neanderthals. And just as religious fanatics everywhere are trying to swamp the socially responsible by outbreeding them, so, in a dangerously overpopulated world, Manning's depraved indifference to the human race's inability to feed itself is exemplified by his having five children.

After spelling out the disaster brought on his party by Stockwell Day's ineptitude, Manning asks (pp. 397-398), "Is there anything that can be learned from the root causes and events of this downward spiral that might assist in the recovery of the Alliance…especially by those of us in politics who also profess a Christian commitment?" My answer to that is: For Canada's sake, I sure hope not.

## 101. *Vulgarians at the Gate:  Trash TV and Raunch Radio*
## by Steve Allen

"Humans…cannot long survive, as a rational, emotionally healthy species, without a secure family structure…It is mostly the failed family, therefore, which has produced our present millions of prison inmates, rapists, drug addicts, burglars, muggers, sexual psychopaths, nonprofessional whores of both sexes, and general goofolas." (p. 37)

Come again?  Does Steve Allen actually endorse the "family values" propaganda of the lunatic right?  Sadly, the answer is yes.  Despite his claim to be a humanist, he apparently believed that a tolerant, pragmatic approach to living arrangements, and rejection of a religion-derived taboo code that defined morality as whatever a Sky Führer's dice-tossing scriptwriter said it was, are the cause of antisocial behavior ranging from nonconformist to criminally repugnant.

Allen's book is essentially a puritanical attack on human sexuality.  On page 35 he expresses hostility to "the witless spectacle known as *Married…with Children*—a deliberately vulgar situation comedy aired on the Fox network."  In my view, *Married…with Children* was one of television's few high points.  And he is as equally hostile to the delightfully funny *Sex and the City*. (pp. 45-51)

Allen also singles out *Just Shoot Me* as a target of his invective.  After describing an episode in which the characters, without resorting to Anglo-Saxon vulgarisms, discussed uncoerced, nonconsequential, victimless recreation, he comments (p. 94), "When company dollars send this kind of depravity into the home at the early hour of 9 P.M., something is dreadfully wrong with the equation."

On pages 141-142: "I doubt that anyone has ever seriously argued that the purpose of *Sports Illustrated* magazine's 'annual swimsuit issue' is to enlighten the public about the benefits of swimming or the latest developments in beach couture."  That would have been a valid criticism if Allen had been commenting on the presentation of visual art as sports-related.  But it appears in a paragraph about "the appeal of pornography."  Perhaps there is such a thing as pornography.  Some defining word is necessary to explain the difference between the artistically bankrupt images of *Hustler* and the tasteful eroticism of *Playboy*.  But *Sports Illustrated*?  Oh come now.

Allen denies taking the position "that sex itself is evil and that, therefore, almost all manifestations of it should be vigilantly discouraged." (p. 144)  But that disclaimer is inserted into a paragraph containing the expression, "sex, violence, and vulgarity," as if sadistic people-hurting and responsible joy-sharing are related acts that equally pollute the vast wasteland.  Perhaps if he had ever read *Stranger in a Strange Land*, it might have annulled his repressive cultural conditioning as it annulled mine.  Instead, Allen consistently harps that, even if sex is not evil, talking about it is.  Is this hypocrisy or what?

"In today's culture it sometimes seems that our entire society has become one massive occasion for sin." (p. 107)  In no dictionary that I consulted is *sin* defined as a synonym for unjust behavior, the unnecessary hurting of a nonconsenting victim.  Rather, it means disobedience to a decree from an imaginary lawgiver in the sky.  Yet Allen uses the word as if it refers to objective evil.  Clearly one does not need to swallow religion's fairy tales in order to be brainwashed by its pseudomorality.  Even persons who reject the alleged lawgiver have swallowed laws whose validity is totally dependent on the pretence, "God says so."

The last thing I expected when I started reading *Vulgarians at the Gate* was that I would be unable to praise it.  I expected an attack on the true obscenity that has made television the primary cause of the dumbing of North America:  the touting of superstitious hogwash for the sake of ratings, with depraved indifference to anything but profits.  I really wish I could say that, in between long harangues of anti-sex rhetoric, Allen also makes a lot of valid points.  But the closest I can come is to say that he makes a few valid points.  For example, his denunciations of Jerry Springer and Howard

Stern are totally justified. While I am in agreement that "the love of ratings is the root of all evil," (p. 36), the rest of the book is a vast disappointment.

This book has destroyed any respect I ever had for Steve Allen as a supposedly rational thinker. He was an ultra-conservative neanderthal who had no more ability to tell right from wrong on a rational basis (*e.g.*: Does an action unnecessarily hurt somebody?) than the religious nutcase currently polluting the White House.

## 102. *Ecohumanism*
### edited by Robert B. Tapp

Only a handful of *Ecohumanism*'s eighteen chapters make any useful or valid points. David Shafer, after spelling out the certainty of the extermination of the universe in a few billion years, either by heat loss through neverending expansion, or "inconceivably high temperatures not seen since the Big Bang" in a gravitation-induced Big Crunch, writes (p. 127), "So, for the Humanist, time is not on our side, and unless we are permanently lucky, sooner or later the likelihood is that we, and all life with us, will yield up forever the joys of existence in favor of total extinction. *But why should we rush it?*" In other words, Earth will be uninhabitable in billions of years, but that is no excuse for making it uninhabitable by air, water and population pollution within the foreseeable future.

Similarly, Gerald Larue echoes Thomas Malthus and Isaac Asimov in making the point (pp. 189-190) that, "Nevertheless it has now become clear that the irresponsible breeding of children, without plans or means for their support and education, is both immoral and unethical and contributes to the spread of hunger and malnutrition, particularly among women and children and the aged. Two hundred million children go to bed hungry every night. There is no estimate of how many elderly go to bed hungry each night...Without education and without the widespread use of birth control, world population will continue to expand and the number of impoverished elderly will grow proportionately."

Unfortunately, those are not the chapters with which this book begins, and there is a serious danger that many readers will give up before reaching them. I have no idea whether *Ecohumanism*'s first four chapters are written in psychobabble, sociobabble, theobabble, Faculty of Education doubletalk, or Etruscan. Perhaps they were intended as a primer for Gibberish 101?

Ninety percent of the book's value is to be found in the chapter, "The Impact of Population on Ecology," by John M. Swomley. He writes (pp. 167-168), "The World Health Organization estimates that 585,000 women die each year during pregnancy and childbirth. 'The death toll,' according to World Watch, 'underestimates the magnitude of the problem. For every maternal death as many as thirty women sustain oftentimes crippling and lifelong health problems related to pregnancy.' Moreover, many of these deaths and lifelong health problems could have been prevented by access to family planning services and safe, legal abortions.

"It is not only the deaths of thousands of women that make this a culture of death, but the projected deaths of some 23 million Africans because of the spread of HIV and the failure of governments to control it. The Vatican has also strongly opposed any funding of condoms to prevent the spread of these diseases."

After acknowledging that a big reason for continuing population growth is that more than forty percent of the present population are in or entering their prime breeding years, Swomley argues (pp. 165-166) that, "A second reason is the Vatican's persistent campaign for more births. This is evident in its worldwide opposition to contraceptive birth control and to abortion, but also in the direct appeals of the Pope...In the early 1980s, Pope John Paul II came to Nairobi and counseled Kenyans, whose population at that time was the fastest growing in Africa, probably in the world, to 'be fruitful and multiply.' The *New York Times* reported, 'In preparation for next month's Earth Summit in Rio de Janeiro, Vatican diplomats have begun a campaign to try to insure that the gathering's conclusions on the issue of runaway population growth are not in conflict with Roman Catholic teaching on birth control.' *Time* magazine reported: 'In response to the concerns of the Vatican, the Reagan administration agreed to alter its foreign aid program to comply with the church's teaching on birth control.' President Bush also blocked all U.S. funding for the United Nations Population program."

403

*Ecohumanism*, and particularly Swomley's chapter, makes clear that the greatest crime against humanity ever committed, the intentional furthering of the human race's inability to feed itself as a consequence of overpopulation, can be laid directly at the feet of a handful of identifiable criminals: the feebleminded tinpot Hitler in the Vatican, and the succession of Republican morons in the White House, who between them have already murdered more than twenty million human beings by starvation, malnutrition and disease, and if not stopped will double or triple that number.

Norman Spinrad wrote in *Greenhouse Summer*, "Greater than the courage to do right in the face of danger or adversity was the courage to commit a lesser evil to prevent a greater. And if the evil that needed preventing was the ultimate one, the death of all living things, then any means were justified to accomplish that end—anything at all." If there is no other way to save the human race from extermination, and certainly I can see no other, then to send the message that anthropocide will not be tolerated, it may be necessary to put Karol Wojtyla and George W. Bush on trial before the World Court for attempted anthropocide, and execute them, naked and without facemasks, on live television in South American prime time. Is there another solution? Perhaps. Or perhaps not.

# Praise for MYTHOLOGY'S LAST GODS
(Prometheus Books, 59 John Glenn Dr, Amherst NY, 1992)

"Here is a book you won't want to miss. As a scathing Biblical critique, certain to make a lasting impression, *Mythology's Last Gods* is truly a masterpiece."
Jon Nelson, *Atheists United*

"*Mythology's Last Gods* is provocative, intellectually stimulating, and a delight to read."
Norm R. Allen Jr, *Free Inquiry*

"Let us give hosanna—not to the Lord, but to William Harwood. Overall, *MLG* demonstrates that the Judeo-Christian Bible is full of errors of fact, bum guesses, prophecies *ex post facto*, excuses, and deliberate lies. The blurb says: 'It is the first book to critically analyze, and take issue with*, every section* of the Judeo-Christian Bible from a wholly skeptical, utterly scholarly perspective.' I agree."
Bernard Katz, *American Rationalist*

"This book is a solid work of scholarship that will enrich the erudition of anyone who reads it. Once you pick it up you won't be able to put it down."
Russ Roehm, *Humanist News and Views*

"This book will wipe out religion."
Steve Zarlenga, publisher, Books in Focus

"*Mythology's Last Gods* does an important job; it brings together a great deal of modern scholarship about comparative religion and mythology and applies it to the Bible to show the evolution of the idea of god."
Bill Cooke, *New Zealand Rationalist and Humanist*

"William Harwood has history, archaeology, textual criticism, science and common sense on his side…In *Mythology's Last Gods* William Harwood practices *history,* examining the evidence first and reaching conclusions based on it. His translations from the Bible may be the only translations where accuracy has not been sacrificed to religious sensibility. I guarantee that this book will erase any vestige of respect you may still have for the Judaeo-Christian tradition, and I recommend it highly."
Greg Erwin, *Humanist in Canada*

"There are four kinds of godworshippers: the stupid, the ignorant, the insane and the intestinally challenged. Nobody who has read *Mythology's Last Gods* can continue to plead ignorance."
Archimedes Fenton, *Here's Hughie*

## Praise for THE JUDAEO-CHRISTIAN BIBLE FULLY TRANSLATED
Imprintbooks, 5341 Dorchester Road, Suite 16, North Charleston, SC 29418

I shudder at the impossibility of doing justice to Dr Harwood's brilliant accomplishments in the two volumes he's published with the above title. In these he has presented exact translations of the oldest Biblical writings yet found. Therein lies the greatest debunking of Judaism and Christianity I have ever had the distinct pleasure of finding. No one can read the exact wording of these ancient writings without immediately knowing how drastically current "bibles" have been tampered with, deleted from, added to, and have thus purposely caused greatly evolved scriptures and "Gods" to become the hoaxes they now are. The best thing that could happen would be for these two volumes to become required reading for all pupils in our high schools and universities.

Right from the first line, it becomes more and more evident that early writers coining what would become the Bible firmly believed in and were describing *multiple gods, NOT just one.* But even more damaging to present day beliefs was their equal belief in, and referrals to "Goddesses." This stems from even older beliefs when gods had not yet been conceived of and *all gods were female, including what has come down to us as "Mother Earth."*

…there was NO virgin birth fable at all in the earlier versions of the "Bible"! The above examples are just the beginning of what should be required reading for all of our people.

Jack Truett, *Pagan Palaver*
Truett is the author of *Litany of Loons*, Xlibris 2002.

Harwood's biblical and extra-biblical translations and comments are a Herculean job—all are most interesting and informative. Interspersed among the texts are his bracketed inserts that are so illuminating I wish there were many more of them…Harwood shows quite clearly which authors were responsible for the various texts, *ones which led to the conflicting views we find in the Bible.* The conclusion is clear: such conflicts not only account for the many contradictions found in the Bible but also for the grievous error of the fundamentalists.

Now you can choose your biblical translations from two sources: traditional translations that are too often misleading and even false, and Harwood's. His "Translator's Prefaces" in both volumes are summaries that should be taught in every Bible class to show how the God's Word was compiled as well as to point out the intentional mistranslations of God's names, translations that give the illusion that there was only one God instead of many. As Harwood emphasizes, "The practice of bible-makers mistranslating a word that means 'the gods' as 'God' stems from the doublethink that, because the translators considered themselves monotheists, the bible authors must have done likewise. They did not." Thus Harwood explodes the Jewish and Christian myth of biblical monotheism.

Read and study—yes, study—these eye-opening books. You will realize that it is not the Bible but Dr. Harwood who is giving you proper revelations. Better still, there ought to be an air drop of the "Translator's Prefaces" from both volumes in all territories occupied by fundamentalists, much like what is done during war to convince an enemy to surrender.

Bernard Katz, *American Rationalist*
Katz is the author of *The Ways of an Atheist*, Prometheus 1999.

## Praise for THE DISINFORMATION CYCLE
Xlibris Corp., 436 Walnut Street, 11ᵗʰ Floor, Philadelphia, PA 19106

### NO WAY YOU CAN GO WRONG ON THIS ONE

Dr William Harwood's inimitable *The Disinformation Cycle* should be required reading for every high school and university in America. Unfortunately that will not happen, not in any of our lifetimes anyway. In fact anyone who has to move his or her lips and mumble every word they read won't understand most of it anyway. What a pity. Well researched and ideally presented, this book will acquaint you with such subjects as hoaxes, delusions, security beliefs, and North America's compulsory mediocrity.

I will not detract from the pleasure, as well as the enlightenment you'll get from reading this remarkable work, by quoting or attempting to explain his reasoning or hardcore evidence. But friends, it is a must-read for anyone capable of thinking. Perhaps the greatest compliment I can pay the author is this: "Darn! I wish I had written this book!"

Jack Truett, *Pagan Palaver*

This book does to the paranormal what *Mythology's Last Gods* did to religion.

Kaz Dziamka, *American Rationalist*

We need to liven it **[Positive Atheism]** up with stuff like this **[excerpt]**, and when I do, I prefer it to be tasteful—such as the above.

Cliff Walker, *Positive Atheism*

## Praise for HYPNOTISM THEN AND NOW
### by Peter Reveen, as told to William Harwood

The book is very well written, with a dry wit and a pleasant style. Make no mistake about it, however, Reveen is a skeptic and he definitely has a few bones to pick with the paranormal. He draws interesting connections between hypnosis, or a heightened state of suggestibility, and a long list of issues, most of which would be familiar to any reader of the *Skeptical Inquirer*. They include faith healing, acupuncture, homeopathy, phrenology, past-life regression, osteopathy, Uri Geller, Kreskin, Hitler, Jeane Dixon, N-Rays, channeling, and "New Age" religion.

He is extremely cutting in his criticism of past-life regression, especially as it relates to hypnosis. He also points out the numerous logical flaws in the reincarnation arguments (i.e.: Where do all the new souls come from?). He throws the cold light of reason on such common myths as the claimed ability of hypnosis to enhance memory, stating that hypnosis is just as likely to dredge up realistic fantasies as it is to recall "forgotten" facts.

The book is well researched and well worth reading. Some interesting insights are made that draw connections between hypnosis, suggestibility, and man's often monumental willingness to believe anything to which he is predisposed or attracted. Reveen's years of experience with suggestion and misdirection make him well qualified to argue the skeptic's perspective and his book is a worthwhile addition to any skeptic's bookshelf.

Lee Moller, *Rational Inquirer*

This review was written of *The Superconscious World*, of which *Hypnotism Then and Now* is an update. While the second edition contains added material, all of Mr Moller's comments are equally applicable to the current version.

# Books by William Harwood

## MYTHOLOGY'S LAST GODS: YAHWEH AND JESUS

Traces the origin of the "god" concept, the evolution of monotheism, the development of Judaism to the 1st century CE, and the evolution of Christianity from its monotheistic beginning to its promulgation as a three-god creed in 325 CE. (Prometheus, 1992)

## THE JUDAEO-CHRISTIAN BIBLE FULLY TRANSLATED, volume 1

The only uncensored translation of what the Hexateuch authors really wrote, showing which author wrote what, and placing their different versions of the same myth in parallel columns for easy comparison. The first useful translation for scholars. (Imprintbooks, 2002)

## THE JUDAEO-CHRISTIAN BIBLE FULLY TRANSLATED, volume 7

Gospels, Acts, Revelation, some Apocrypha, placing parallel gospel passages in adjacent columns to show borrowing from Mark and borrowing from a common source. (Imprintbooks, 2002)

## THE DISINFORMATION CYCLE: HOAXES, DELUSIONS, SECURITY BELIEFS, AND COMPULSORY MEDIOCRITY

The reality behind nonsense beliefs. Does to the paranormal what *Mythology's Last Gods* did to religion. (Xlibris, 2002)

## THE AUTOBIOGRAPHY OF GOD

Three gods, one of whom claims to be an extraterrestrial, relate conflicting accounts of human history. Not recommended for religion addicts. (Xlibris, 2002)

## UNCLE YESHU, MESSIAH

Historical novel describing the history of Jesus' family from his parents' marriage until the Bar Kokhba rebellion. Not recommended for persons who think Jesus was a nice guy. (Xlibris, 2001)

## THE BELOVED DISCIPLE AND THE HOUSE OF HIPPO

Historical fantasy; and futuristic science fiction. (Xlibris, 2001)

## A VISION OF MURDER AND HYPNAUGHTY BOY

A murder mystery set in the Jewish Mafia; and a back-stage showbiz adventure. (Xlibris, 2001)

## THE MYCENAEAN CHRONICLE

Was the Trojan War a religious crusade against heretics and infidels? Agamemnon says it was. And what happened when Moses asked for Agamemnon's help against their mutual enemies? This historical novel suggests some answers. (Xlibris, 2001)

## DICTIONARY OF CONTEMPORARY MYTHOLOGY

Religion, the paranormal, UFOs, urban legends, folk myths. Over 3500 entries (1stbooks, 2002)

# HYPNOTISM THEN AND NOW

(co-author with Peter Reveen)
The history of suggestion therapy, from pre-Mesmer to the present. (Imprintbooks, 2002)

# THERE BE NO SUN BUT YAHWEH, AND JESUS BE HIM'S PLANET

Science fiction short stories and novellas. (Imprintbooks, 2002)

# THE MAGICIANS: MOSES TO MERLIN

Historical and science fiction novellas, and more. (Imprintbooks, 2002)

# THE LAWLESS COURTENAYS

Devonshire and Yorkshire politics in the fifteenth and sixteenth centuries. (Imprintbooks, 2002)

# THREE SCREENPLAYS

A Vision of Murder; Hypnaughty Boy; And Now, Heee-eres Jesus! (Imprintbooks, 2002)

# THE GREAT ZUBRICK

Novel about the principal and crew of a touring concert hypnotist show. (Xlibris, 2002)

# THE GREAT ZUBRICK TELEPLAY

Five part miniseries based on the novel. (Imprintbooks, 2002)

# A HUMANIST IN THE BIBLE BELT

Collected papers, 1974-2002. (1stbooks, 2003)

The above books can be ordered from
www.Prometheusbooks.com
www.Xlibris.com
www.1stbooks.com
www.Booksurge.com
or from the special orders desk at any book store.

## Publications by William Harwood

1. "Education: A Faculty of Incompetence," University of Calgary *Gauntlet*, Sept 17, 1974.
2. "Alice in Faculty of Education Land," *Mensa Canada Communications*, 1.2. 1982, pp. 3-4.
3. "Gods, Goddesses and Bibles: The Canonization of Misogyny," *The Humanist*, May 1985, pp. 23-25.
4. "Priestly Power and the Role of Sin," *Humanist in Canada*, Autumn 1985, pp. 20-21.
5. "The Uzziah Syndrome," *Humanist in Canada*, Autumn 1986, pp. 24-25.
6. "Yahweh: A Morally Retarded God," *Free Inquiry*, Fall 1986, p. 59.
7. "The Vanishing Gods," book review, *Humanist in Canada*, Autumn 1986, pp. 31-32.
8. "The Search For Psychic Power," book review, *Humanist in Canada*, Autumn 1990, p. 36.
9. "Jesus the Nazirite: Fact or Fiction?" *American Rationalist*, Sep/Oct 1990, pp. 39-40.
10. "Can a Theist Be Moral?" *Humanist in Canada*, Autumn 1992, pp. 22 & 36.
11. *Mythology's Last God: Yahweh and Jesus*, Prometheus Books, Amherst NY, 1992.
    reviewed in *Shofar*, Summer 1992.
    reviewed in *Reasonings*, Sep. 1992
    reviewed in *New Zealand Rationalist and Humanist*, 1992
    reviewed in *Humanist News and Views*, Nov. 1992.
    reviewed in *Mensa Bulletin*, Dec. 1992.
    reviewed in *Humanist in Canada*, Spring 1993.
    reviewed in *Atheists United News Letter*, March 1993.
    reviewed in *Biblical Theology*, 1993.
    reviewed in *Free Inquiry*, Fall 1993.
    reviewed in *American Rationalist*, Nov/Dec 1993.
12. "On the Wild Side," book review, *Humanist in Canada* Spring 1993, p. 38.
13. "Secrets of the Amazing Kreskin," book review, *Humanist in Canada*, Winter 1993/1994, pp. 38-39.
14. "The Ten Commandments: What They Originally Said," *Humanist in Canada*, Spring 1997, pp. 10-11.
15. "Deception and Self-Deception," book review, *American Rationalist*, July/Aug 1997, p. 38.
16. "Once Upon a Time in the Sky," book review, *American Rationalist*, Nov/Dec 1997, pp. 78-79.
17. "Why I Am Not a Terrorist," *American Rationalist*, Jan/Feb 1998, pp. 94-95.
18. "Memes: The New Pseudo-Science," book review, *American Rationalist*, Jan/Feb 1998, p. 96.
19. "The Koran," book review, *American Rationalist*, Mar/Apr 1998, pp. 114-115.
    reprinted in *Midwest Book Review*, Nov. 2002.
20. "Who Wrote the Gospels?" book review, *American Rationalist*, July/August 1998, pp. 11-12.
21. "My Dog Can Lick Your Dog," satire, *American Rationalist*, July/August 1998, pp. 13-14.
22. "Virus of the Mind," book review, *Humanist in Canada*, Autumn 1998, pp. 35-36.
23. "The Bible: Divine or Human Origin," book review, *American Rationalist*, Nov/Dec 1998, pp. 13-14.
24. "Enemies: The Rationalist View," book review, *American Rationalist*, Nov/Dec 1998, pp. 15-17.
25. "Sin is Whatever My Führer Says It Is," *American Rationalist*, Jan/Feb 1999, pp. 14-15.
26. "Bible Prophecy: Failure or Fulfillment?" book review, *American Rationalist*, Mar/April 1999, p 10-12.
27. "The Roving Mind," book review, *American Rationalist*, May/June 1999, pp. 14-15.
    reprinted in *Midwest Book Review*, February 2002.

28    "Eight Little Piggies," book review, *American Rationalist,* July/August 1999, pp. 13-15.

29    "The Wreck of the Titanic Foretold?" book review, *American Rationalist,* Sep/Oct 1999, pp. 14-15.

30    "Biological Exuberance," book review, *American Rationalist,* Nov/Dec 1999, pp. 20-21.

31    "The Scars of Evolution," book review, *American Rationalist,* Nov/Dec 1999, p. 21.

32    "Autobiography of God," excerpt 1, satire, *American Rationalist,* Mar/April 2000, pp. 9-10.

33    "Where Was Jesus Born?" *Soar,* June 2000, pp. 1-3.

34    "Is Religiosity a Form of Unsanity?" *Freethinker* (UK), June 2000, p. 10.

35    "Autobiography of God," excerpt 2, satire, *American Rationalist,* May/June 2000, pp. 13-15.

36    "Who Killed Jesus?" book review, *New Zealand Rationalist & Humanist,* Winter 2000, pp. 22-23.

37    "Confirmation:  The Hard Evidence of Aliens Among Us," book review, *Rational Inquirer,* July 2000, pp. 9-10.
       reprinted in *Midwest Book Review,* October 2002.

38    "The Inconsistency of Round-Earth Religionists," *The Skeptical Review,* July/August 2000, pp. 5 & 9.

39    "Sodom and Khomorah: When, Where and How?" *Freethought Perspective,* Aug/Sep 2000, p. 11.

40    "Luke:  Christianity's First Revisionist," *Ulster Humanist* (UK), Sep/Oct 2000, p. 26.

41    "Freedom of Conscience:  A Baptist/Humanist Dialogue," book review, *Pagan Palaver,* Fall 2000, pp. 71-76.

42    "Conflicting Attitudes Toward Homosexuality in the Judaeo-Christian Bible," *Pagan Palaver,* Fall 2000, pp. 77-81.

43    "If Bible Religion is True, Then the Earth is Flat," *The Freethought Exchange,* Oct 2000, pp. 23-24.

44    "The Multiple Authorship of the Books Attributed to Moses," *The Skeptical Review,* Sep/Oct 2000, pp. 2-3.

45    "Revelation:  Invective Against the Christians," *Freethought Perspective,* Oct/Nov 2000, pp. 13-15.

46    "The Sunken Kingdom:  The Atlantis Mystery Solved," book review, *Indian Skeptic* (Madras), Oct 15, 2000, pp. 18-19.

47    "Moses versus Aaron:  Their Comparative Status to Different Pentateuch Authors," *Bibles Review,* Oct 2000, pp. 17-18.

48    "Restoring the Goddess," book review, *American Rationalist,* July/August 2000, pp. 10-11.

49    "Autobiography of God," excerpt 3, satire, *American Rationalist,* July/August 2000, pp. 9-10.

50    "Hitler's Pope," book review, *Free Inquiry,* Fall 2000, pp. 66-67.

51    "The Dark Side of Human History," book review, *Humanist in Canada,* Autumn 2000, p. 35.

52    "Galactic Rapture," book review, *The Ulster Humanist,* Nov/Dec 2000, p. 24.
       reprinted in *Midwest Book Review,* August 2002.

53    "Do American Sex Laws Violate the First Amendment?" *Freethought Perspective,* Dec/Jan 2000/2001, pp. 7-8.

54    "Close Encounters with the Religious Right," book review, *Pagan Palaver,* Winter 2000/2001, pp. 48-51.
       partially reprinted in *Humanist News and Views,* Nov. 2000, p. 7.

55    "The Great Deception," book review, *Pagan Palaver,* Winter 2000/2001, pp. 52-55.

56    "Biblical Nonsense," *Pagan Palaver,* Winter 2000/2001, pp. 64-66.

57    "Entities," book review, *Humanist in Canada,* Winter 2000/2001, pp. 34-35.

58    "Oldies But Baddies," review of Dale Carnegie, Napoleon Hill, Norman Vincent Peale, *The Match,* Winter 2000/2001, pp. 25-27.

reprinted in *Midwest Book Review*, Nov. 2002.

59    "Modern Religion and Bible Religion: the Differences," *American Rationalist*, Jan/Feb 2001, pp. 9-10.

60    "The Happy Heretic," book review, *Ulster Humanist*, Jan/Feb 2001, p. 23.

61    "Jesus' Miracles: Anything Eliyah did, Jesus did better," *Ulster Humanist*, Jan/Feb 2001, p. 11.

62    "The Death of Hananias and Sapphire: An Explanation," *Freethought Perspective,* Feb/Mar 2001, pp. 20-21.

63    "The Origin and Evolution of the Doctrine of Predestination," *American Rationalist*, Mar/Apr 2001, pp. 3-4.

64    "Autobiography of God," excerpt 4, satire, *American Rationalist*, Mar/Apr 2001, pp. 4-5.

65    "Messiah: What the title meant to Jesus the Jew," *Ulster Humanist*, March/April 2001, pp. 19-20.

66    "Social Incest: Taboo in the Kibbutz," *Freethought Perspective*, Apr/May 2001, pp. 11-12.

67    "Rocks of Ages," book review, *Midwest Book Review*, May 2001.

68    "Merely Mortal," book review, *Freethinker*, May 2001, p. 13
reprinted in *Midwest Book Review*, June 2002.

69    "From the Wandering Jew to William Buckley," review of religious chapters, *Ulster Humanist*, May/June 2001, p. 25.

70    "From the Wandering Jew to William Buckley," review of paranormal chapters, *Pagan Palaver*, Summer 2001, pp. 26-28.

71    "No Sex Please: We're Masochists," *Pagan Palaver*, Spring 2001, pp. 91-104.

72    "God and Santa Claus: A Letter to a Little Girl," *New Zealand Rationalist and Humanist*, Autumn 2001, p. 24.
reprinted in *Ulster Humanist*, November-December 2001, p. 18.

73    "Papal Sin," book review, *Australian Humanist*, Winter 2001, pp. 18-19.
reprinted in *Midwest Book Review*, October 2002.

74    "Autobiography of God," excerpt 5, satire, *American Rationalist*, May/June 2001, pp. 12-13.

75    "The Quest For the Historical Muhammad," book review, *American Rationalist*, May/June 2001, pp. 11-12.
partially reprinted in *The Match*, Winter 2000-2001, p. 38.

76    "The Mythic Past," book review, *Journal of Higher Criticism*, Spring 2001, pp. 144-147.
reprinted in *Humanistic Judaism*, Spring/Summer 2001, pp. 42-43.

77    "The Second Coming and the Wandering Jew: Salvaging a Failed Prophecy," *Freethought Perspective*, June 2001, p. 6.

78    "The Torah and its God," book review, *Midwest Book Review*, June 2001.

79    "The Gospels & Acts," book review, *Midwest Book Review*, June 2001.

80    "Has Religion Been Disproven?" *Free Inquiry*, Summer 2001, pp. 62-63.

81    "Why Yahweh's Name Became Taboo," *Ulster Humanist*, July/August 2001, pp. 18-19.

82    "Judgment Day For The Shroud of Turin," book review, *Midwest Book Review*, July 2001.

83    "Deconstructing Jesus," book review, *Midwest Book Review*, July 2001.

84    *The Beloved Disciple and The House of Hippo*, Xlibris, Philadelphia, 2001.

85    *A Vision of Murder and Hypnaughty Boy*, Xlibris, Philadelphia, 2001.

86    "The Demon-Haunted World, *and* Billions and Billions," book review, *American Rationalist*, July/August 2001, pp. 11-13.

87    "Queen Jane's Version," book review, *American Rationalist*, July/August 2001, pp. 13-14.
reprinted in *Midwest Book Review*, August 2001.

88    "God Unmasked," book review, *Midwest Book Review*, August 2001.

89   "The Lost Gospel and Who Wrote the New testament?" book review, *Midwest Book Review*, August 2001.

90   "Psychics and Other Time Travelers," *Indian Skeptic*, August 2001, pp. 17-22.

91   "The Most Dangerous Man in Canada," *Pagan Palaver*, Summer 2001, pp. 83-91.

92   "Greetings, Carbon-Based Bipeds," book review, *Freethought Perspective*, Sept. 2001, pp. 14-15.

93   "The Vanquished Gods," book review, *Midwest Book Review*, Sept 2001.

94   "Skeptical Odysseys," book review, *Midwest Book Review*, Sept 2001.

95   "Would You Believe?" book review, *Humanist in Canada*, Autumn 2001, pp. 36-38.

96   "The Third Reich," book review, *Midwest Book Review*, Oct. 2001.

97   "Atheism: A Reader," book review, *Secular Nation*, Oct-Dec 2001, p. 29.

98   "Bible Belt or Loony Bin: Is There a Difference?" *Secular Nation*, Oct-Dec 2001, pp. 13-14.

99   "Autobiography of God," excerpt 6, satire, *American Rationalist*, Sep/Oct 2001, pp. 13-14.

100  "Naturalism and religion," book review, *Midwest Book Review*, Nov. 2001.

101  "The Ways of an Atheist," book review, *Midwest Book Review*, Nov. 2001.

102  *Uncle Yeshu, Messiah*, Xlibris, Philadelphia, 2001.

103  *The Mycenaean Chronicle*, Xlibris, Philadelphia, 2001.

104  "Autobiography of God," Yahweh/Pan excerpt, *Pagan Palaver*, Fall 2001, pp. 61-70.

105  "The Jewish Parochialism of the *Our Father*," *Pagan Palaver*, Fall 2001, pp. 59-61.

106  "The Test of Love," book review, *Freethought Perspective*, Dec. 2001, pp. 6-7.

107  "The Betrayal of America," book review, *Midwest Book Review*, Dec. 2001.

108  "Why Atheism?" book review, *Midwest Book Review*, Dec. 2001.

109  "Autobiography of God," excerpt 7, satire, *American Rationalist*, Nov/Dec 2001, pp. 12-14.

110  *The Judaeo-Christian Bible Fully Translated*, volume 1, Imprintbooks, North Charleston SC, 2002.

111  *The Judaeo-Christian Bible Fully Translated*, volume 7, Imprintbooks, North Charleston SC, 2002.
     reviewed in *American Rationalist*, Mar/Apr 2002.
     reviewed in *Pagan Palaver*, #19, Summer 2002.
     reviewed in *Open Society*, Summer 2002.
     reviewed in *Ulster Humanist* Dec/Jan 2002/03.

112  *Dictionary of Contemporary Mythology*, 1st Books Library, Bloomington IN, 2002.

113  *The Disinformation Cycle*, Xlibris, Philadelphia, 2002.
     reviewed in *Pagan Palaver*, #20, 2002.

114  "The Jesus Mysteries," book review, *Midwest Book Review*, Jan. 2002.

115  "The Jesus Puzzle," book review, *Midwest Book Review*, Jan. 2002.

116  "Three books by Thomas Szasz," book review, *Midwest Book Review*, Feb. 2002.

117  "Demons of the Modern World," book review, *Midwest Book Review*, March 2002.

118  "Jesus 100 Years Before Christ," book review, *Midwest Book Review*, March 2002.

119  "Final Séance," book review, *Skeptical Inquirer*, March/April 2002, pp. 52-53.

120  "Sweet Jesus," book review, *Midwest Book Review*, April 2002.

121  "Jews Without Judaism," book review, *Midwest Book Review*, April 2002.

122  "No Man Knows My History," book review, *Midwest Book Review*, April 2002.

123  *Hypnotism Then and Now*, co-author with Peter Reveen, Imprintbooks, North Charleston SC, 2002.

124  *There Be No Sun But Yahweh, And Jesus Be Him's Planet*, Imprintbooks, North Charleston SC, 2002.

125  "Judas's Betrayal: Why the Author of Mark Invented It," *Ulster Humanist*, Mar/Apr 2002, pp. 18-19.

126    *The Autobiography of God*, Xlibris, Philadelphia, 2002.
      reviewed in *Pagan Palaver*, Autumn 2002.
      reviewed in *Open Society*, 2003.

127    *The Magicians*, Imprintbooks, North Charleston SC, 2002.

128    "A Rebel to his Last Breath," book review, *Freethought Perspective*, April 2002, pp. 8-9.

129    "Bandits, Prophets and Messiahs, *and* Whoever Hears You Hears Me," book review, *Journal of Higher Criticism*, Spring 2001, pp. 156-158.

130    "The Autobiography of God," excerpt 8, satire, *American Rationalist*, Mar/Apr 2002, pp. 13-14.

131    "The in the Sky," book review, *Midwest Book Review*, May 2002.

132    "Great Quotations on Religious Freedom," book review, *Midwest Book Review*, May 2002.

133    "The Hidden Jesus," book review, *Midwest Book Review*, May 2002.

134    "The Yahwist's Eden," *Ulster Humanist*, May/June 2002, p. 20.

135    "The Hidden Book in the Bible," book review, *The Open Society* (NZ), Autumn 2002, pp. 22-23.

136    "Hyperspace: It's Not a Free lunch," *Truth Seeker E-Zine*, May 2002.

137    "Jesus' Deification: When and Why?" *Freethought Perspective*, July 2002, pp. 8-9.

138    "Lying About Hitler," book review, *Humanist in Canada*, Summer 2002, pp. 36-38.
      reprinted in *Midwest Book Review*, July 2002.

139    *The Lawless Courtenays*, Imprintbooks, North Charleston SC, 2002.

140    *Three Screenplays*, Imprintbooks, North Charleston SC, 2002.

141    "Monism versus Dualism: The *Job* Compromise," *Ulster Humanist*, July/Aug 2002, p. 20.

142    "Three UFO books," book review, *American Rationalist*, July/Aug 2002, pp. 12-14.

143    "Once Upon a Time: A True Story," book review, *Midwest Book Review*, August 2002.

144    "Science: Good, Bad and Bogus," book review, *Midwest Book Review*, August 2002.

145    "Why People Believe Weird Things," book review, *Midwest Book Review*, August 2002.

146    "The Book Your Church Doesn't Want You To Read," book review, *Midwest Book Review*, September 2002.

147    "Bad Astronomy," book review, *Midwest Book Review*, September 2002.

148    "The Ghost in the Universe," book review, *Midwest Book Review*, September 2002.

149    "Litany of Loons," book review, *Midwest Book Review*, September 2002.

150    "Alien Visitors? I Don't Think So," *Pagan Palaver*, Summer 2002, pp. 114-128.

151    *The Great Zubrick Teleplay*, Imprintbooks, North Charleston SC, 2002.

152    "Child Sexual Abuse and Recovered Memory," book review, *American Rationalist*, Sep/Oct 2002, pp. 9-11

153    "Autobiography of God," excerpt 9, satire, *American Rationalist*, Sep/Oct 2002, p. 13.

154    "They Call It Hypnosis," book review, *Midwest Book Review*, October 2002, pp. 9-11.

155    "Was Nathanael the Beloved Disciple?" *Freethought Perspective*, Oct. 2002, pp. 7-8.

156    "The Our Playmate," satire, *Freethought Perspective*, p. 12.

157    "Moses and Joshua: The Problem of Chronology," *Open Society*, Spring 2002, pp. 16-17.

158    "The Dawn of Human Culture," book review, *Ulster Humanist*, Oct/Nov 2002, p. 23.

159    "Franquin: Master Showman," book review, *Midwest Book Review*, Nov 2002.

160    "The Secret of Happiness," book review, *Midwest Book Review*, Nov 2002.

161    "How We Believe," book review, *Midwest Book Review*, Nov 2002.

162    "Return of the Furies," book review, *American Rationalist*, Nov/Dec 2002, pp. 12-13.

163    "Literature Quiz," *American Rationalist*, Nov/Dec 2002, p. 13.

164    "Hypnotism: A History," book review, *Midwest Book Review*, Dec 2002.

165    "The Supernatural, the Occult and the Bible," *Midwest Book Review*, Dec 2002.

166 "Encyclopedia of Claims, Frauds, and Hoaxes of the Occult and Supernatural," book review, *Midwest Book Review*, Dec 2002.

167 "Vulgarians at the Gates," book review, *Midwest Book Review*, Dec 2002.

168 "Six Novels About Jesus," book review, *Midwest Book Review*, Dec 2002.

169 "Is God a Petulant Little Boy?" *American Atheist*, Winter 2002/2003.

170 "Is Agnosticism Defensible?" *Australian Humanist*, Summer 2002/2003.

171 "Nineteenth Century Religions," *Positive Atheism*, 2002.

172 *The Great Zubrick*, Xlibris, Philadelphia, 2002.

173 "George W. Bush's Ultimate Hero," *Freethought Perspective*, Jan 2003.

174 "Understanding the Hadith," book review, *Freethought Perspective*, Jan 2003.

175 "H. L. Mencken on Religion," *Freethinker*, Jan 2003.

176 "The Rise and Fall of Jesus," book review, *Midwest Book Review*, Jan 2002.

177 "Joshua: The Man They Called Jesus," book review, *Midwest Book Review*, Jan 2003.

178 "The UFO Mystery Solved," book review, *Midwest Book Review*, Jan 2003.

179 "The Loch Ness Monster Mystery," book review, *Midwest Book Review*, Jan 2003.

180 "Ecohumanism," book review, *Midwest Book Review*, Jan 2003.

181 "The Real Jesus," book review, *Midwest Book Review*, Jan 2003.

182 *A Humanist in the Bible Belt*, 1stbooks, 2003.

## IN PRODUCTION

183 "Joshua: A Jewish Hitler," *Pagan Palaver*.

184 "Is Pithecanthropus Troglodytus Alive in Alberta?" *Pagan Palaver*.

185 "Is This 1984—Or What?" *Sea of Faith* (UK).

186 "666 and Armageddon: Who and When?" *Ulster Humanist*.

187 "Moses and DeMille: Parting the Red Sea," *Open Society*.

188 "Secret Origins of the Bible," book review, *American Rationalist*.

189 "Jesus the Nazirite: Real Person or Literary Creation?" *American Rationalist*.

190 "Catholic Power vs. American Freedom," book review, *Humanist in Canada*.

191 "Canada's Ayatollahs: Manning and Day," book review, *Humanist in Canada*.

192 "Do Laws Permitting Religions to Kill Children Violate First Amendment?" *Free Inquiry*.

193 "Alien Abductions," book review, *Skeptic*.

194-196 "Autobiography of God," Pan excerpts 1-3, *Pagan Palaver*.

197-201 "Autobiography of God," excerpts 10-14, *American Rationalist*.

# ABOUT THE AUTHOR

William Harwood, Ph.D., M.Litt. (Cambridge), is a member of the editorial board of *Free Inquiry*, a contributing editor of *American Rationalist*, and author of eighteen books. He has written over 150 articles for *F.I., A.R., The Humanist, Humanist in Canada,* and a dozen other periodicals in seven countries. On completing a Ph.D. in Religious Studies, he applied for several advertised lectureships in universities in Canada and elsewhere, in the naïve belief that the author of a book disproving religion had any more hope of being hired by religion professors whose economic security depended on their maintaining the pretence that religion cannot be disproven, than Ralph Nader would have had of becoming publicity director of Volkswagen after writing *Unsafe at Any Speed.* After giving up the uneven struggle, he settled for teaching high school, before the reality became inescapable that North American schools are babysitting institutions in which any teacher who attempted to teach anything would be purged before he could raise the question, "How come nobody else is doing that?

Dr. Harwood was born in the world's largest culturally deprived environment, Australia, among people who lived in constant terror that some day, somewhere, someone might actually *do* something. He stayed there just long enough to recognize that there is something terribly wrong with a country that rejects moderate, pragmatic, middle-of-the road politics, and instead alternates between governments of the far right and the far left. Later, as a research student at Cambridge University, he discovered that the world's largest insane asylum, England, has a similar deficiency.

Before settling in Canada, Dr Harwood toured Australia, New Guinea, Fiji, Bermuda and the USA as advertising manager for three hypnotic stage shows, including one that became a household name in Canada. Between shows, he obtained graduate degrees from universities in Canada, England and the USA, and spent eight years as a teacher. He joined Mensa for intelligent conversation, and left when he failed to find any. His most satisfying bread-and-butter job was general factotum at a now defunct private gambling establishment in Calgary, not unlike the one depicted in *A Vision of Murder* and mentioned briefly in *The Great Zubrick.* His proudest moment came when he learned that he was listed in the biographical dictionary, *Who's Who in Hell?,* alongside Steve Allen, Woody Allen, Isaac Asimov, Stephen Jay Gould, Stephen Hawking, Thomas Jefferson, Abraham Lincoln, Thomas Paine, Carl Sagan, and many of his other heroes. Unfortunately, the pride somewhat dissipated when he learned that the dictionary's compiler was not a braindead fundamentalist (tautology) but a humanist. He is currently a resident of what is politely called Canada's Bible Belt, but is more accurately described as the redneck anus of the universe, where a theofascist majority believe that a referendum can give them the right to deprive minorities of basic human rights, strip women of sovereignty over their own bodies, and restore such barbarisms as capital punishment and probably heretic burning.

He started life as a Protestant (Methodist father, Anglican mother), and turned Catholic when he discovered that Protestantism is repudiated by its own Bible. He remained Catholic until he took his first ancient history course at the University of Calgary, and learned that fifty other virgin-born savior gods had risen from the dead on the third day centuries and even millennia before Jesus. (His desperate search for rebuttal evidence initially led him to discover that all claims of a god revealing its existence have been traced to the same authors who also assured their readers that the earth is flat, and eventually gave him the material to write *Mythology 's Last Gods.*) But recognizing the falseness of religion at an intellectual level was not immediately sufficient. More than three years later, on a

Sunday morning at Cambridge, England, he ate breakfast, got dressed, and opened the front door to go to mass, when it suddenly hit him:

"If I participate in this 5,000-year-old Egyptian god-eating ritual even once more, I will throw up." At that point he was cured—totally, permanently, irreversibly. For a while, however, he did maintain the neo-agnostic "Hughieist" philosophy attributed to a character in *The Great Zubrick* and later spelled out in The *Autobiography of God* for others who were not ready to abandon all metaphysics cold turkey. Today he is an unequivocal nontheist: "Gods do not exist, have never existed in the past, and will not exist in the future."

His only encounter with religion since then occurred when he attended what he was promised would be a purely secular memorial, only to find himself subjected to fundamentalist triple-god worship, conducted by a megalomaniac tinpot Hitler priest, that even a Unitarian would have recognized as an abominable and detestable crime against sanity. He may forgive the individual whose lie that there would be no religion put him in such an intolerable position—some day.

But while he accepts ignorance as a curable perversion, he has little tolerance for self-inflicted braindeath. The intestinally challenged bully whose reaction to his book disproving religion was an arrogant demand that he "pray," is now someone he used to know. He is, as Isaac Asimov used to say, in his late youth, and is between marriages (isn't everybody?). He has one daughter and two grandchildren.

Dr Harwood's first book, *Mythology's Last Gods,* based on his doctoral dissertation, was published in 1992, his first four novels in 2001, and his bible translation and *Dictionary of Contemporary Mythology* in 2002. *A Humanist in the Bible Belt* is his eighteenth book.

He considers North America the most near-perfect society on earth and, despite the treasonous coup d'etat that made a morally, rationally, intellectually and educationally challenged puppet of the Theofascist Right President even though the American people rejected him by almost a million votes, and despite the election of a hate cult manipulated by the most dangerous mad dogs Canadian politics has ever produced as Canada's Official Opposition, he would not wish to live anywhere else.

Printed in the United States
1168400001B/57-58